Polymer Processing and Surfaces

Polymer Processing and Surfaces

Editor

Michal Sedlačík

MDPI • Basel • Beijing • Wuhan • Barcelona • Belgrade • Manchester • Tokyo • Cluj • Tianjin

Editor
Michal Sedlačík
Centre of Polymer Systems
Tomas Bata University in Zlín
Zlín
Czech Republic

Editorial Office
MDPI
St. Alban-Anlage 66
4052 Basel, Switzerland

This is a reprint of articles from the Special Issue published online in the open access journal *Polymers* (ISSN 2073-4360) (available at: www.mdpi.com/journal/polymers/special_issues/polym_process_surf).

For citation purposes, cite each article independently as indicated on the article page online and as indicated below:

LastName, A.A.; LastName, B.B.; LastName, C.C. Article Title. *Journal Name* **Year**, *Volume Number*, Page Range.

ISBN 978-3-0365-3626-2 (Hbk)
ISBN 978-3-0365-3625-5 (PDF)

© 2022 by the authors. Articles in this book are Open Access and distributed under the Creative Commons Attribution (CC BY) license, which allows users to download, copy and build upon published articles, as long as the author and publisher are properly credited, which ensures maximum dissemination and a wider impact of our publications.

The book as a whole is distributed by MDPI under the terms and conditions of the Creative Commons license CC BY-NC-ND.

Contents

About the Editor vii

Preface to "Polymer Processing and Surfaces" ix

Yuhui Zhou, Li He and Wei Gong
Effect of Organic Cage Nucleating Agent Structure on Nucleating Efficiency and the Structure-Property Relationship
Reprinted from: *Polymers* 2020, 12, 1975, doi:10.3390/polym12091975 1

Yanjun Lu, Wang Luo, Xiaoyu Wu, Bin Xu, Chunjin Wang and Jiajun Li et al.
Fabrication of Micro-Structured LED Diffusion Plate Using Efficient Micro Injection Molding and Micro-Ground Mold Core
Reprinted from: *Polymers* 2020, 12, 1307, doi:10.3390/polym12061307 15

Nur Sharmila Sharip, Hidayah Ariffin, Tengku Arisyah Tengku Yasim-Anuar, Yoshito Andou, Yuki Shirosaki and Mohammad Jawaid et al.
Melt- vs. Non-Melt Blending of Complexly Processable Ultra-High Molecular Weight Polyethylene/Cellulose Nanofiber Bionanocomposite
Reprinted from: *Polymers* 2021, 13, 404, doi:10.3390/polym13030404 31

Wei Li, Jie Wu, Zhengqiao Zhang, Lanjuan Wu and Yuhao Lu
Investigation on the Synthesis Process of Bromoisobutyryl Esterified Starch and Its Sizing Properties: Viscosity Stability, Adhesion and Film Properties
Reprinted from: *Polymers* 2019, 11, 1936, doi:10.3390/polym11121936 47

Hongxia Li, Jianqun Yang, Shangli Dong, Feng Tian and Xingji Li
Low Dielectric Constant Polyimide Obtained by Four Kinds of Irradiation Sources
Reprinted from: *Polymers* 2020, 12, 879, doi:10.3390/polym12040879 65

Yan Wu, Xinyu Wu, Feng Yang and Jiaoyou Ye
Preparation and Characterization of Waterborne UV Lacquer Product Modified by Zinc Oxide with Flower Shape
Reprinted from: *Polymers* 2020, 12, 668, doi:10.3390/polym12030668 79

Konstantinos Karvanis, Soňa Rusnáková, Ondřej Krejčí and Milan Žaludek
Preparation, Thermal Analysis, and Mechanical Properties of Basalt Fiber/Epoxy Composites
Reprinted from: *Polymers* 2020, 12, 1785, doi:10.3390/polym12081785 91

Guo Feng, Feng Jiang, Zi Hu, Weihui Jiang, Jianmin Liu and Quan Zhang et al.
Pressure Field Assisted Polycondensation Nonaqueous Precipitation Synthesis of Mullite Whiskers and Their Application as Epoxy Resin Reinforcement
Reprinted from: *Polymers* 2019, 11, 2007, doi:10.3390/polym11122007 109

Mirjana Rodošek, Mohor Mihelčič, Marija Čolović, Ervin Šest, Matic Šobak and Ivan Jerman et al.
Tailored Crosslinking Process and Protective Efficiency of Epoxy Coatings Containing Glycidyl-POSS
Reprinted from: *Polymers* 2020, 12, 591, doi:10.3390/polym12030591 121

Itziar Otaegi, Nora Aranburu, Maider Iturrondobeitia, Julen Ibarretxe and Gonzalo Guerrica-Echevarría
The Effect of the Preparation Method and the Dispersion and Aspect Ratio of CNTs on the Mechanical and Electrical Properties of Bio-Based Polyamide-4,10/CNT Nanocomposites
Reprinted from: *Polymers* **2019**, *11*, 2059, doi:10.3390/polym11122059 139

Sumita Swar, Veronika Máková and Ivan Stibor
The Covalent Tethering of Poly(ethylene glycol) to Nylon 6 Surface via *N,N*-Disuccinimidyl Carbonate Conjugation: A New Approach in the Fight against Pathogenic Bacteria
Reprinted from: *Polymers* **2020**, *12*, 2181, doi:10.3390/polym12102181 155

Haifeng Cai, Yang Wang, Kai Wu and Weihong Guo
Enhanced Hydrophilic and Electrophilic Properties of Polyvinyl Chloride (PVC) Biofilm Carrier
Reprinted from: *Polymers* **2020**, *12*, 1240, doi:10.3390/polym12061240 171

Jinsu Gim, Eunsu Han, Byungohk Rhee, Walter Friesenbichler and Dieter P. Gruber
Causes of the Gloss Transition Defect on High-Gloss Injection-Molded Surfaces
Reprinted from: *Polymers* **2020**, *12*, 2100, doi:10.3390/polym12092100 189

Iva Rezić and Ana Kiš
Design of Experiment Approach to Optimize Hydrophobic Fabric Treatments
Reprinted from: *Polymers* **2020**, *12*, 2131, doi:10.3390/polym12092131 205

Yanqing Li, Jian Hao, Jinfeng Zhang, Wei Hou, Cong Liu and Ruijin Liao
Improvement of the Space Charge Suppression and Hydrophobicity Property of Cellulose Insulation Pressboard by Surface Sputtering a ZnO/PTFE Functional Film
Reprinted from: *Polymers* **2019**, *11*, 1610, doi:10.3390/polym11101610 221

Gangqiang Zhang, Jiewen Hu, Tianhui Ren and Ping Zhu
Microstructural and Tribological Properties of a Dopamine Hydrochloride and Graphene Oxide Coating Applied to Multifilament Surgical Sutures
Reprinted from: *Polymers* **2020**, *12*, 1630, doi:10.3390/polym12081630 237

Alaa Mohammed, Gamal A. El-Hiti, Emad Yousif, Ahmed A. Ahmed, Dina S. Ahmed and Mohammad Hayal Alotaibi
Protection of Poly(Vinyl Chloride) Films against Photodegradation Using Various Valsartan Tin Complexes
Reprinted from: *Polymers* **2020**, *12*, 969, doi:10.3390/polym12040969 251

Thalita M. C. Nishime, Robert Wagner and Konstantin G. Kostov
Study of Modified Area of Polymer Samples Exposed to a He Atmospheric Pressure Plasma Jet Using Different Treatment Conditions
Reprinted from: *Polymers* **2020**, *12*, 1028, doi:10.3390/polym12051028 269

Vijay Kakani, Hakil Kim, Praveen Kumar Basivi and Visweswara Rao Pasupuleti
Surface Thermo-Dynamic Characterization of Poly (Vinylidene Chloride-Co-Acrylonitrile) (P(VDC-co-AN)) Using Inverse-Gas Chromatography and Investigation of Visual Traits Using Computer Vision Image Processing Algorithms
Reprinted from: *Polymers* **2020**, *12*, 1631, doi:10.3390/polym12081631 289

Petr Slobodian, Pavel Riha, Robert Olejnik and Jiri Matyas
Accelerated Shape Forming and Recovering, Induction, and Release of Adhesiveness of Conductive Carbon Nanotube/Epoxy Composites by Joule Heating
Reprinted from: *Polymers* **2020**, *12*, 1030, doi:10.3390/polym12051030 313

Mónica Fuensanta, María Agostina Vallino-Moyano and José Miguel Martín-Martínez
Balanced Viscoelastic Properties of Pressure Sensitive Adhesives Made with Thermoplastic Polyurethanes Blends
Reprinted from: *Polymers* **2019**, *11*, 1608, doi:10.3390/polym11101608 327

Valentina Sabatini, Tommaso Taroni, Riccardo Rampazzo, Marco Bompieri, Daniela Maggioni and Daniela Meroni et al.
PA6 and Halloysite Nanotubes Composites with Improved Hydrothermal Ageing Resistance: Role of Filler Physicochemical Properties, Functionalization and Dispersion Technique
Reprinted from: *Polymers* **2020**, *12*, 211, doi:10.3390/polym12010211 351

Janusz Musiał, Serhiy Horiashchenko, Robert Polasik, Jakub Musiał, Tomasz Kałaczyński and Maciej Matuszewski et al.
Abrasion Wear Resistance of Polymer Constructional Materials for Rapid Prototyping and Tool-Making Industry
Reprinted from: *Polymers* **2020**, *12*, 873, doi:10.3390/polym12040873 371

Martin Vasina, Katarina Monkova, Peter Pavol Monka, Drazan Kozak and Jozef Tkac
Study of the Sound Absorption Properties of 3D-Printed Open-Porous ABS Material Structures
Reprinted from: *Polymers* **2020**, *12*, 1062, doi:10.3390/polym12051062 383

Romana Daňová, Robert Olejnik, Petr Slobodian and Jiri Matyas
The Piezoresistive Highly Elastic Sensor Based on Carbon Nanotubes for the Detection of Breath
Reprinted from: *Polymers* **2020**, *12*, 713, doi:10.3390/polym12030713 397

Petr Slobodian, Pavel Riha, Robert Olejnik and Michal Sedlacik
Ethylene-Octene-Copolymer with Embedded Carbon and Organic Conductive Nanostructures for Thermoelectric Applications
Reprinted from: *Polymers* **2020**, *12*, 1316, doi:10.3390/polym12061316 409

About the Editor

Michal Sedlačík

Michal Sedlačík received his Ph.D. degree from Tomas Bata University in Zlín (TBU) in 2012. From that same year, he has since been Lecturer at the Faculty of Technology at TBU. He successfully defended his habilitation work entitled "Novel Approaches to Design of Intelligent Fluids" in 2016. Currently, he is a senior researcher in the Nanomaterials and Advanced Technologies Group at Centre of Polymer Systems, TBU. He is a member of numerous international scientific societies such as American Chemical Society, The Society of Rheology, and The Society of Plastics Engineers. He shares authorship of 75 papers with h-index = 23. His current research interests include synthesis and properties of intelligent systems, elastomers, and electromagnetic shielding composites.

Preface to "Polymer Processing and Surfaces"

Polymer processing and surfaces are considered as key parameters for developing unique materials for various applications. At the same time as developing the next generation of polymeric and composite materials, the material resources, utility properties of the product, its economy aspects, as well as environmental issues can be optimized. The surface of the filler or final product is of particular interest with respect to the interaction with their surroundings; hence its modification can positively affect the stimuli-responsive character, adhesion, absorption, tribological properties, or biomedical applications among others.

This book, which consists of 26 articles written by research experts in their topic of interest, reports the most recent research on polymer processing with an emphasis on the surface properties. Several novel and fascinating methods related to the polymer fabrication, biointerfaces, surface modification, characterization, sensitivity to the surroundings, and applications are introduced.

Michal Sedlačík
Editor

Article

Effect of Organic Cage Nucleating Agent Structure on Nucleating Efficiency and the Structure-Property Relationship

Yuhui Zhou [1], Li He [2,3,*] and Wei Gong [4,*]

1. School of Chemistry and Chemical Industry of Guizhou University, Guiyang 550025, China; huihuisabrina@aliyun.com
2. Department of Polymer Material and Engineering, College of Materials and Metallurgy, Guizhou University, Guiyang 550025, China
3. National Engineering Research Center for Compounding and Modification of Polymer Materials, Guiyang 550025, China
4. School of Materials and Architectural Engineering of Guizhou Normal University, Guiyang 550025, China
* Correspondence: lihe@gzu.edu.cn (L.H.); gw20030501@163.com (W.G.)

Received: 28 July 2020; Accepted: 18 August 2020; Published: 31 August 2020

Abstract: Three types of organic cage compounds, namely, cucurbit[6]uril (Q[6]), hemicucurbit[6]uril (HQ[6]), and β-cyclodextrin (BC), with different cavity structures as heterogeneous nucleation agents were selected for a polypropylene (PP) foaming injection molding process. The experimental results showed that Q[6] with a "natural" cavity structure possessed the best nucleation efficiency of these three cage compounds. The nucleation mechanism of organic cage compounds was explored through classical nucleation theory, molecular structure, and in situ visual injection molding analysis.

Keywords: heterogeneous nucleation; cell morphology; injection molding foaming; composite materials; visualization

1. Introduction

The specific strength [1–5] and dimensional stability of foaming materials [6] are directly related to cell density, which is determined by the number of cells present during the foaming process. Therefore, the quality of foaming materials can be effectively improved by regulating nucleation in the foaming process [7–12].

According to classical theory [13–16], the nucleation process of gas in melt proceeds in two modes, namely, homogeneous nucleation and heterogeneous nucleation. In homogeneous nucleation, the second-phase component in a single homogeneous phase forms a stable second phase through accumulation and dispersion fluctuation over a critical size. Homogeneous nucleation occurs when no nucleating agent is contained in the polymer melt, or when the content of the nucleating agent is lower than its solubility in the melt. However, in actual material preparation, all types of additives are used, impurities are common (except for gas and polymer), and various internal tissue defects emerge. These external particles or rough surfaces contribute to heterogeneous nucleation in the foaming process. With regard to the phase change process during material processing and preparation, almost all nucleation behaviors are non-uniform nucleation processes. Thus, studying heterogeneous nucleation is crucial for the preparation of plastic foaming materials.

Research has shown that adding nano-inorganic particles to a polymer as a heterogeneous nucleating agent can effectively improve the quality of the cell morphology, as well as the impact toughness and heat insulation performance of microfoaming materials [17–20]. However, the nucleation efficiency of nano-inorganic powders is reduced because of their high surface density and easy agglomeration [21,22]. In addition, the compatibility between an inorganic nucleating agent and a matrix resin is poor, and excessive addition degrades the performance of the foaming materials. Therefore, the design and development of highly effective organic nucleating agents with a special structure are essential for improving the foaming behavior and properties of polymers.

Thus far, no study has investigated the application of organic caged compounds as heterogeneous nucleating agents in polymer foaming. In the current study, polypropylene (PP) was used as the polymer substrate. PP and cucurbit[6]uril (Q[6]), hemicucurbit[6]uril (HQ[6]), and β-cyclodextrin (BC) composite foaming materials were prepared through microcellular injection. The PP foaming behavior of organic caged and traditional nucleating agents was observed in situ using a visual injection molding device under the condition of adding the same amount of nucleating agent particles. The aim was to analyze and compare the nucleating performance of each organic cage nucleating agent. The formation and growth of cells during the foaming process of each composite material were summarized, and the heterogeneous nucleation mechanism of the caged compound was verified.

2. Experimental Setup

2.1. Materials

Nano-sized Q[6] and HQ[6] were prepared according to the procedures described in previous literature [23,24]. β-cyclodextrin (BC) was of reagent grade and was used without further purification, and silicon dioxide (SiO_2) was obtained from Sinopharm Group (Shanghai, China). PP T30S with a melt flow index of 3.2 g/10 min at 230 °C/2.16 kg and a density of 0.906 g/cm^3 was obtained from SINOPEC, Beijing, China. The blowing agent, azodicarbonamide (AC), was obtained from Hanhong Co., Wuhan, China, with a gas production of 220 mL/g.

2.2. Nanocomposite Preparation

Q[6], HQ[6], BC, and PP were vacuum dried at 80 °C for 8 h before use. Then, PP with different Q[6], HQ[6], and BC contents was melt-extruded using a twin-screw extruder with an increasing extrusion temperature profile (175~195 °C). The Q[6], HQ[6], and BC contents prepared for the master batch nanocomposites were 0.25, 0.5, 1.0, 2.0, 3.0, 5.0, and 7.0 wt%, and were thereafter coded as PQ0.25, PQ0.5, PQ1.0, PQ2.0, PQ3.0, PQ5.0, and PQ7.0; PH0.25, PH0.5, PH1.0, PH2.0, PH3.0, PH5.0, and PH7.0; and PB0.25, PB0.5, PB1.0, PB2.0, PB3.0, PB5.0, and PB7.0, respectively.

2.3. Injection Molding Foaming

The foaming samples were prepared using a microcellular injection foaming molding machine equipped with a volume-adjustable cavity. The extrusion temperature profile from hopper to nozzle was 165~175 °C, and the expansion ratio remained constant, controlled by the sample thickness expanding from 3.5 to 4.0 mm. The gas (N_2) content was 1%, which was determined by the percentage of AC gas production.

2.4. In Situ Foaming Visualization

The setup of the batch foaming visualization system, as illustrated in Figure 1, was used to observe the in situ foaming behavior of the aforementioned polymer blowing agent system. The system consisted of a double screw system, a data acquisition system for pressure measurement (i.e., a data acquisition board and a computer), and an optical system (i.e., an objective lens, a light source, and a high-speed camera).

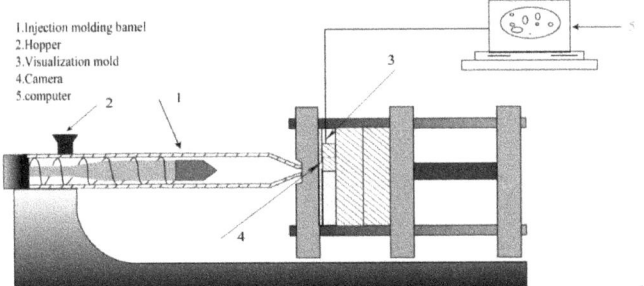

Figure 1. Schematic diagram of the visual injection molding device.

2.5. Morphological Analysis

The morphologies of the foamed samples were observed with a scanning electron microscope (SEM) at an accelerating voltage of 25.0 KV (KYKY-2800B, Zhongke Co., Beijing, China). The samples were first frozen in liquid nitrogen for 3 h, and then transversely fractured in a liquid nitrogen atmosphere. Before SEM testing, the samples were sprayed gold for 60 s. Both the cell size and density were determined from the SEM micrographs. The cell size and cell density were calculated via image analysis. The cell density was calculated from the following equation:

$$N = \left[\frac{nM^2}{A}\right]^{\frac{3}{2}} \left[\frac{\rho}{\rho_f}\right] \quad (1)$$

where N_0 is the cell density (cells/cm^3), n is the number of cells on the SEM micrograph, M is the magnification factor, and A is the area of the micrograph (cm^2). The densities of unfoamed (ρ) and foamed (ρ_f) samples were measured via the water displacement method, in accordance with ASTM D792.

2.6. Transmission Electron Microscope (TEM) Analysis

The morphology of Q[6] was determined by a transmission electron microscope (Tecnai G2 F20, FEI Co., Portland, OR, USA) at 200 KV. Q[6] was prepared using acetone to form a relatively dilute suspension solution, and the solution was then dropped onto the copper grids before the TEM analysis.

3. Results and Discussion

3.1. Cell Morphology

SEM images of the fracture surface are shown in Figures 2–4. The corresponding cell morphology parameters, including the average cell sizes and cell densities, are presented in Figure 5. As expected, the PP foam presented a poor cell structure, characterized by a large cell size and non-uniform cell distribution. The average cell size decreased with the addition of Q[6], HQ[6], and BC nanoparticles to the PP. Specifically, the cell size of the pure PP decreased from 55 μm to 26 μm when the PQ2 nanocomposite foam was used. The cell density increased from 1.9×10^6 cells/cm^3 for the pure PP foam to 1.3×10^7 cells/cm^3 for the composite foams. When the Q[6] content increased to 1.0 wt%, an obvious improvement in the cell structure was observed. However, further increments in Q[6] content did little to change the cell morphology.

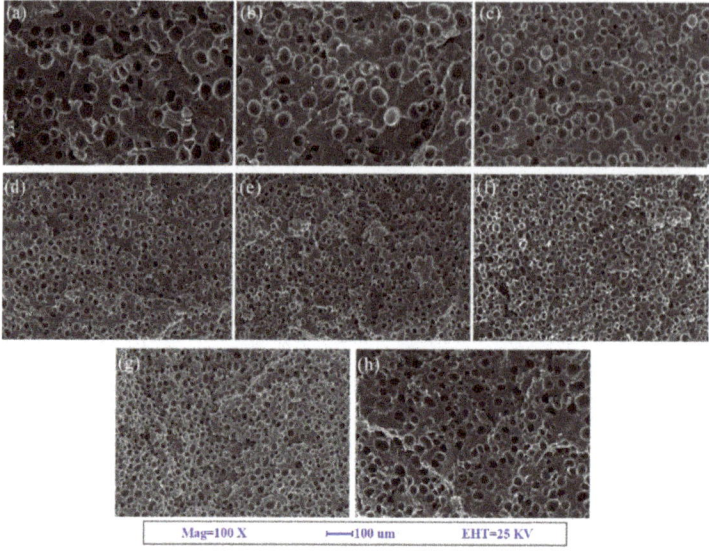

Figure 2. SEM images of the polypropylene (PP) and PP + cucurbit[6]uril (Q[6]) (PQ) composite foams: (**a**) PP, (**b**) PQ0.25, (**c**) PQ0.5, (**d**) PQ1, (**e**) PQ2, (**f**) PQ3, (**g**) PQ5, and (**h**) PQ7.

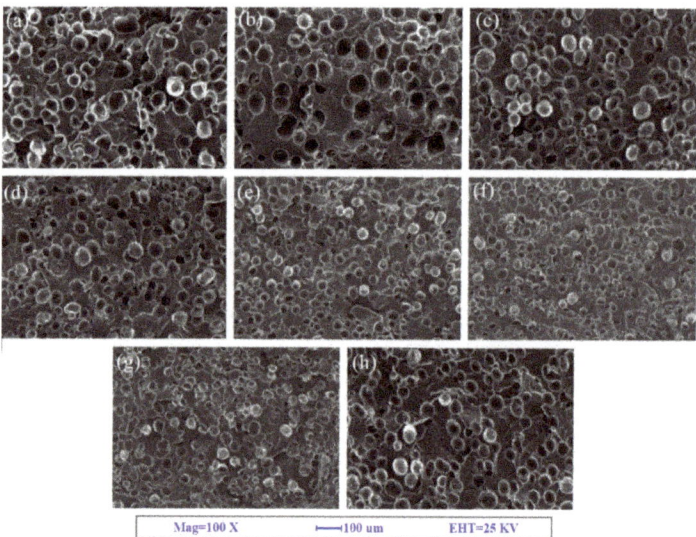

Figure 3. SEM images of the PP and PP + hemicucurbit[6]uril (HQ[6]) (PH) composite foams: (**a**) PP, (**b**) PH0.25, (**c**) PH0.5, (**d**) PH1, (**e**) PH2, (**f**) PH3, (**g**) PH5, and (**h**) PH7.

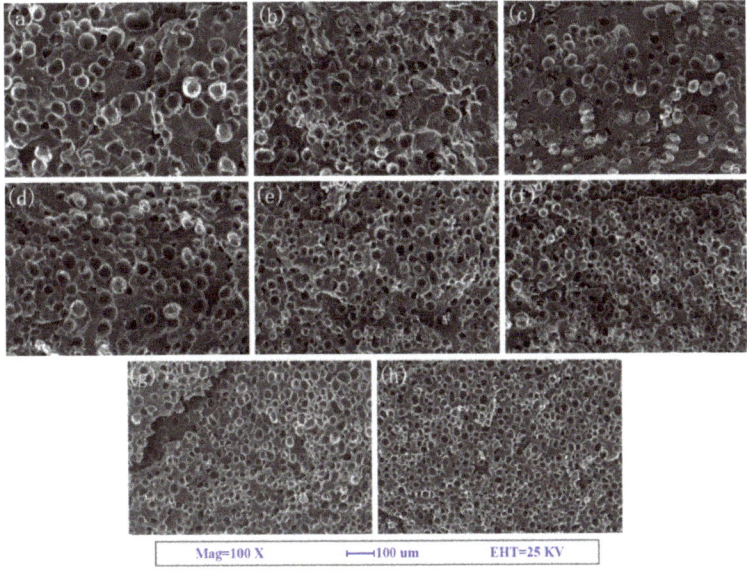

Figure 4. SEM images of the PP and PP + β-cyclodextrin (BC) (PB) composite foams: (**a**) PP, (**b**) PB0.25, (**c**) PB0.5, (**d**) PB1, (**e**) PB2, (**f**) PB3, (**g**) PB5, and (**h**) PB7.

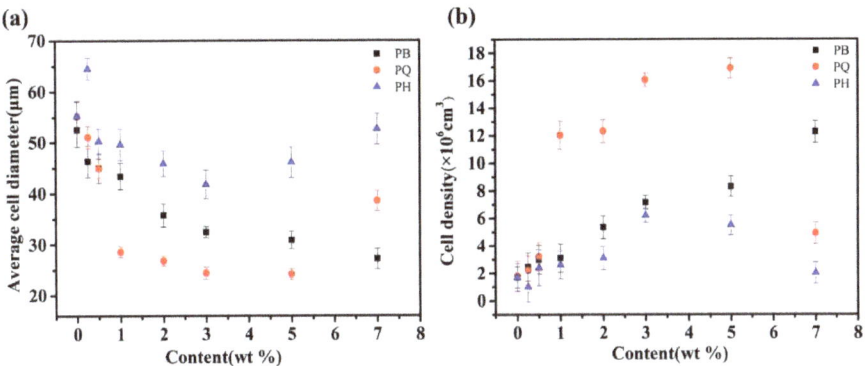

Figure 5. Average cell diameters (**a**) and cell densities (**b**) of the foamed PQ, PH, and PB.

When HQ[6] was added to PH at 3.0 wt%, the morphology of the cells improved. The cell diameter was 41.3 μm, and the cell density was $6.2 \times 10^6/cm^3$. However, when the added HQ[6] exceeded 3.0 wt%, the PH cells began to deteriorate, and the cell diameter increased sharply. For example, when the content was 7.0 wt%, the cell parameter was close to that of the pure PP.

With the increase in BC content, the average size of the PB bubbles decreased and the cell density increased. When the maximum amount of BC was 7.0 wt%, the minimum average cell size of the PB was 27.1 μm and the maximum cell density was $1.22 \times 10^7/cm^3$.

3.2. Discussion on Nucleating Efficiency and Structure-Property Relationship

Q[6] (Figure 6) is a macrocyclic compound synthesized with glycoluril and formaldehyde. It has a highly polarizable carbonyl-rich portal and a hydrophobic interior cavity. The structure of HQ[6] is similar to the structure of Q[6], i.e., it cuts halfway along the equator of the cage structure of Q[6]. The glycoside urea unit bridged by the methylene unit makes HQ[6] lose its structural rigidity and become a flexible macro-ring structure. Under the strong external force of melt shear, the material easily twists and deforms into a continuous phase interface with a radian. BC [25,26] is a cyclic oligosaccharide, consisting of glucopyranose units in a torus-like structure, with hydrophilism on the outer surface and relative hydrophobicity in their internal cavity. The molecular structures of the three cage compounds and TEM diagram of Q[6] are shown in Figure 7.

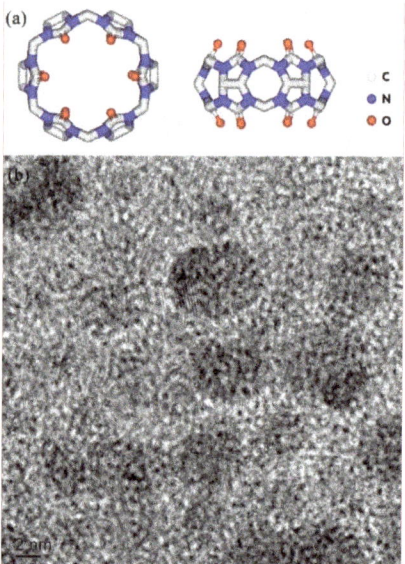

Figure 6. (a) Molecular structure and (b) TEM image of Q[6].

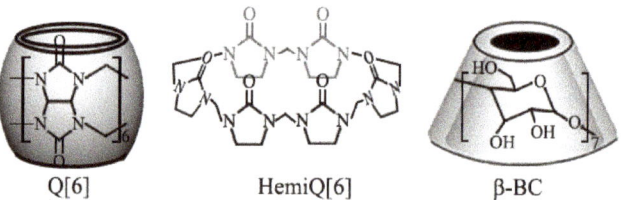

Figure 7. Molecular structures of Q[6], HQ[6], and BC.

According to classical nucleation theory, the Gibbs free energy (Gibbs) barrier of gas conuclei can be expressed as

$$\Delta G_{hom} = -(P_{bub} - P_{sys})V_g + \sigma_{lg}A_{lg} \tag{2}$$

The Gibbs free energy barrier of heterogeneous nucleation can be expressed as follows:

$$\Delta G_{het} = -(P_{bub} - P_{sys})V_{bub} + \sigma_{lg}A_{lg} + \sigma_{sg}A_{sg} - \sigma_{sl}A_{sl} \tag{3}$$

where P_{bub} is the gas pressure in the bubble; P_{sys} is the ambient system pressure; A is the surface area; σ is the interface energy; and l, g, and s are the melt, gas, and solid phases, respectively.

When a gas is nucleated on a solid nucleating agent, the term $\sigma_{sl}A_{sl}$ can reduce the energy barrier of the system, because no new generation is required at the solid–melt interface. When the boundary area, A_{sl}, is fixed, the larger the interface energies of σ_{lg} and σ_{sg}, the larger the heterogeneous nucleation barrier, ΔG_{het}. A traditional nucleating agent relies on $\sigma_{sl}A_{sl}$ to reduce the nuclear barrier. The gas and melt interface energy, $\sigma_{lg}A_{lg}$, is the key factor that restricts the nuclear barrier. If $\sigma_{lg}A_{lg}$ can be reduced, then the nucleation efficiency will improve.

Among the three cage compounds, Q[6] exerted a greater improvement effect on the PP at a low amount of addition, which means that Q[6] had the highest nucleation efficiency. On the one hand, BC contains many hydroxyl functional groups with a relatively strong molecular polarity. The polarity difference between the PP, which is a non-polar polymer, and BC was the largest in this study. Under the same experimental conditions, the interface energy value of BC as the nucleating agent system was the largest. Q[6] and HQ[6] are similar in chemical composition and have the same polarity. Their molecular polarity is weaker than that of BC. On the other hand, the hydroxyl groups in BC generate hydrogen bonds with one another, resulting in the "sealing end" of BC molecules and forming a special 3D structure. In this structure, one end of BC is open and the other end is nearly closed. Compared with Q[6], the BC with an approximate cone structure had a larger A_{lg} value, i.e., $\Delta G_{het}(BC) > \Delta G_{het}(Q[6])$. Under the strong external force of melt shear, HQ[6] was easily twisted and deformed into a continuous phase interface, and $\Delta G_{het}(HQ[6]) > \Delta G_{het}(Q[6])$.

In summary, the structural composition of Q[6] endows the material with a low surface density and reduces the gas nuclear energy barrier. However, HQ[6] and BC have high nuclear barriers and a relatively poor nucleation performance because of their high surface density and polarity (Figure 8).

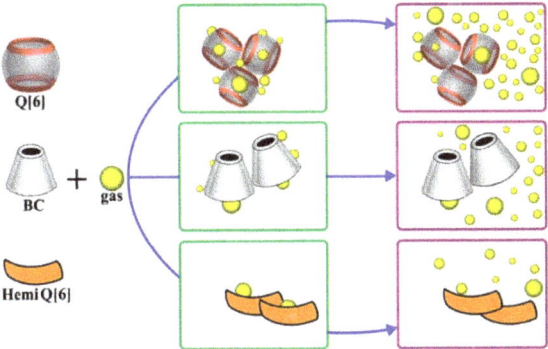

Figure 8. Schematic diagram of heterogeneous nucleation of Q[6], HQ[6], and BC.

Silicon dioxide (SiO_2), a traditional nucleation agent, was used to compare the nucleation performance of the organic caged and traditional nucleation agents (Figures 9 and 10). SiO_2 is an inorganic nucleating agent with a solid spherical structure, and it is commonly used to prepare polymer foaming materials. The composite of SiO_2 and PP was labeled PS in this study. The weight of the nucleating agent added with PP in the experiment was converted into the number of particles for the scientific characterization of the nucleation performance of the various nucleating agents. Trends of the changes in average cell diameter and density with the number of nucleating agent particles were obtained and are shown in Figure 10.

The average cell diameter of PB decreased gradually with the increase in the number of nucleation sites. PQ, PH, and PS presented the same variation trends with the number of nucleation sites. Specifically, with the increase in the number of nucleation sites, the average cell diameter decreased and then increased, whereas the average cell density increased and then decreased. This result was obtained because the addition of the nucleating agent promoted the heterogeneous nucleation of cells. When the added content was increased, the agglomeration effect of the nucleating agent particles reduced the number of effective nucleation sites in the PP, thus diminishing cell improvement.

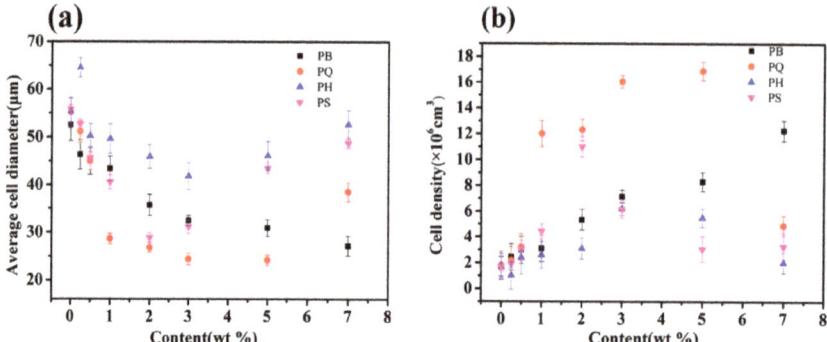

Figure 9. The average cell diameters (**a**) and cell densities (**b**) of the foamed PP composites with different contents.

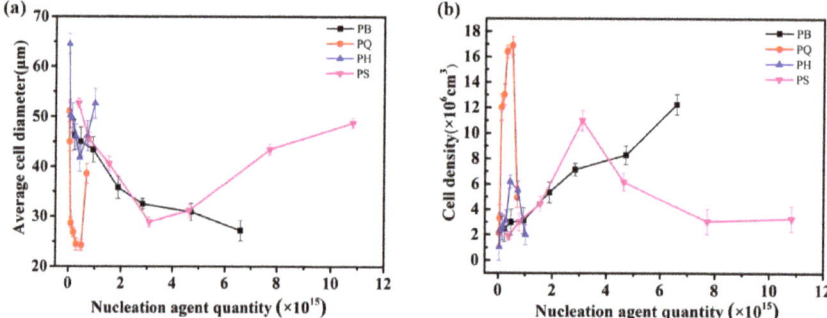

Figure 10. Average cell diameters (**a**) and cell densities (**b**) of the foamed PP composites with different nucleation agent quantities.

The improvement effect of HQ[6] on the PP cell morphology was limited. Compared with HQ[6], the PQ, PB, and PS series achieved a better improvement in PP cell morphology and obtained a smaller cell size and a higher cell density. Moreover, the number of nucleating particles required to achieve the same improvement differed. For example, when the PP cell size was reduced by 50%, the number of particles required for each nucleating agent differed by several times. The number of nucleation sites for PQ, PB, and PS was 0.19×10^{15}, 6.5×10^{15}, and 3.0×10^{15}, respectively. Among the three series, PQ required the least nucleating particles to obtain the same bubble density, whereas PS and PB required the most. This experimental result proves that the nucleation efficiency of Q[6] was higher than that of the traditional nucleating agent, SiO_2. According to the preceding theoretical analysis, SiO_2 is a solid sphere, and its contact surface with the gas and melt is a continuous interface with a high density. It relies on the term $\sigma_{sl}A_{sl}$ to reduce the nuclear barrier. However, the gas–melt interface energy term is the key factor that restricts the nuclear barrier of the bubble shape. Unlike in Q[6], which is composed of a single atomic layer, $Alg(SiO_2) > Alg(Q[6])$. Therefore, the nucleation barrier $\Delta G_{het}(SiO_2) > \Delta G_{het}(Q[6])$ is required for gas nucleation on the surface. A many of nucleating agent particles are needed for SiO_2 to achieve the same improvement effect.

3.3. In Situ Foaming Visualisation

The visual injection molding equipment developed by our research group can perform real-time monitoring of the growth process of cells in actual injection molding. The formation and growth of cells in the actual dynamic injection molding process of polymer foam materials can be determined by directly observing and comparing the results of actual injection molding.

In the current experiment, PP and the nucleating agent composites containing the same amount of nucleating agent particles were prepared using a mixer. The specific added content is shown in Table 1 (the particles are treated approximately as spheres).

Table 1. Parameters of nucleating agents.

	BC	Q[6]	HQ[6]	SiO_2
density (g/cm^3)	1.37	1.44	1.23	1.12
particle size (nm)	114	204	326	104
single particle volume ($\times 10^{-15}$ cm^3)	0.775	4.443	5.774	0.589
single particle weight ($\times 10^{-15}$ g)	1.062	6.220	6.929	0.647
add weight per 100 g PP (g)	0.085	0.5	0.55	0.05

Figure 11 shows images of the bubble growth of the pure PP and the PQ composites in the visual injection molding equipment. The growth of bubbles was relatively slow, and the number of cells was small in the injection molding process of pure PP. After the addition of nucleating agent Q[6], the bubble growth rate of PQ significantly increased. The number of PQ cells increased by nearly two orders of magnitude compared with the number of pure PP cells. This result indicates that the addition of a heterogeneous nucleating agent significantly reduced the nuclear energy barrier of gas, made the growth of cell nucleation easy, and increased the cell density significantly.

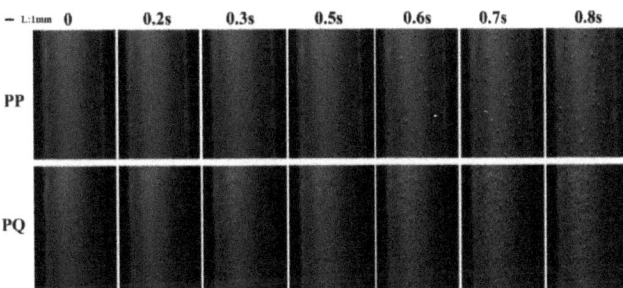

Figure 11. Pictures of the cells of the pure PP and the PQ composites changing over time.

Figure 12 shows the variation trends in the number and diameter of cells of the PP and PP composite systems over time. The number of cells in all of the experimental systems increased initially and then stabilized, which was a manifestation of gas nucleation from continuous growth to complete nucleation. The growth rate of the number of cells reflected the rate of nucleation and was related to the nuclear barrier of the system. The faster the growth, the easier the nucleation and the lower the nuclear energy barrier of the system. Hence, the slope of the bubble growth phase could be compared to that of the nuclear barrier. The higher the slope, the faster the nucleation and the lower the nuclear barrier. The slope of the bubble growth stage of the pure PP was the lowest, and the degree of bubble equilibrium was significantly lower than that of the other experimental systems, indicating that the nuclear energy barrier of the pure PP was the largest. After the addition of the same amount of nucleating agent particles, the growth rate of the number of balanced bubbles and the number of bubbles in each composite system became significantly higher than that of the PP. The PQ system had the highest slope and the fastest nucleation during the growth stage of the number of cells, followed by the PS, PB, and PH systems. The experimental results showed that Q[6] had the lowest nuclear energy barrier among all of the nucleating agents. A comparison of the number of equilibrium cells in each composite system showed that PQ had the largest number of cells, which was about nine times that of the pure PP. The number of cells increased and decreased successively in the order of PS, PB, and PH.

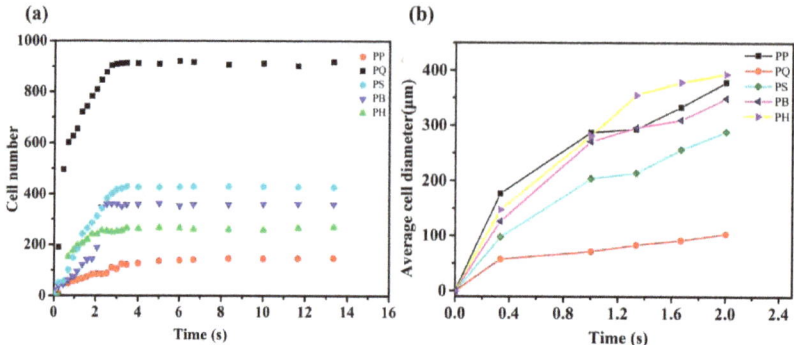

Figure 12. The average cell diameter (**a**) and cell density (**b**) of the foamed PP composites with different times.

As shown in Figure 13, new cells were generated near the previously nucleated and growing cells, despite the rapid gas consumption from cell growth in these regions. This phenomenon was less pronounced when the pure PP sample was foamed. The cavity of Q[6] can also become a natural "air pocket". As nucleated bubbles expand, their growth causes tangential stretching on the surface [27], resulting in a reduction in nearby pressure. R_{cr} and ΔG_{het} are reduced [28,29] accordingly to promote the nucleation of new bubbles and the growth of the original gas cavity, thereby increasing the final material bubble density. The nucleation efficiency of PQ with cavitation was significantly higher than that of the PP. The results of the experiment once again confirmed the conclusion that the cavity not only reduced the nuclear barrier, but also stimulated the growth of secondary bubbles. The in situ observation of the growth of the secondary bubbles revealed that the potential mechanism of nucleation was enhanced by the nucleating agents, and comprehensively explained the mechanism of heterogeneous nucleation promoted by fillers.

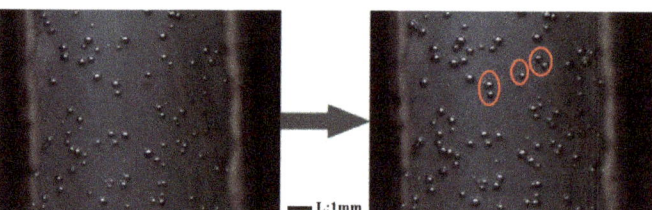

Figure 13. Photos of the secondary bubble growth of the PQ composites.

In addition, the visualized foaming behavior was observed under the same injection molding conditions, but with an extended mold opening time to verify the experimental conclusions further. The relation curve between the change in cell number and the delay time of each composite system is shown in Figure 14. At the end of the injection, when the mold opening time was prolonged, the temperature in the mold cavity decreased, and the melt viscosity increased. In other words, prolongation of the mold opening time increased the nuclear energy barrier and made bubble nucleation difficult. The higher the bubble reduction rate, the more difficult bubble nucleation became and the higher the nuclear energy barrier. As indicated in Figures 14 and 15, the PQ system had the lowest rate of bubble number reduction, followed by the PS, PB, and PH systems. The pure PP system had the highest rate of bubble number reduction, which also reflected the order of the heterogeneous nuclear energy barrier of the systems. The nucleating agent of the PQ system had the lowest nuclear energy barrier among all of the nucleating agents, and this finding is consistent with the results of the microporous injection foam.

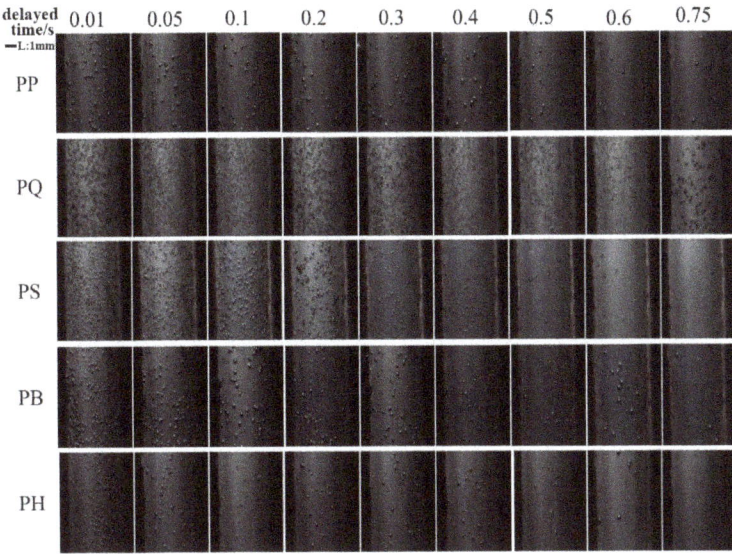

Figure 14. Pictures of the bubbles of the PP composite with different mold opening times.

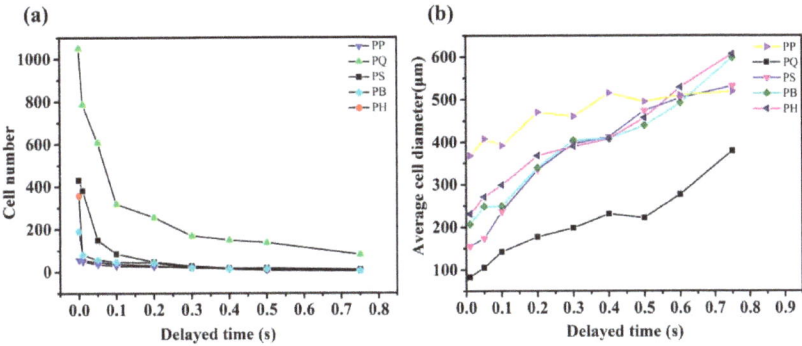

Figure 15. The cell number (**a**) and cell density (**b**) of the foamed PP composite with different mold opening times.

4. Conclusions

Analysis of the heterogeneous nucleation mechanism showed that $\sigma_{lg}A_{lg}$ was the key factor that restricted bubble nucleation. When the value of $\sigma_{lg}A_{lg}$ was reduced, the efficiency of nucleation improved. Organic cage compounds Q[6], HQ[6], and BC with different cavity structures were synthesized. A natural "cavity" was observed. This cavity was a large area of the "natural" gas/solid interface of Q[6], and it effectively reduced the value of $\sigma_{lg}A_{lg}$, decreased the heterogeneous nuclear barrier, and had the highest nucleation efficiency. The growth rule of the bubble obtained by visual injection molding was consistent with the conclusion from the actual injection molding. The conclusion that the low surface density of the nucleating agent could reduce the nuclear energy barrier and improve nucleation efficiency was verified again. This work provides a reference for the application of organic cage compounds in polymer foaming materials.

Author Contributions: Methodology, L.H.; Writing-Review & Editing, W.G. and Investigation, Writing-Original Draft Preparation, Y.Z. All authors have read and agreed to the published version of the manuscript.

Funding: This research was funded by the Science and Technology Foundation of Guizhou Province, grant number [20191083].

Conflicts of Interest: The authors declare no conflict of interest.

References

1. Nam, P.H.; Maiti, P.; Okamoto, M. Foam processing and cellular structure of polypropylene/clay nanocomposites. *Polym. Eng. Sci.* **2002**, *9*, 1907–1918. [CrossRef]
2. Suh, K.W.; Park, C.P.; Maurer, M.; Tusim, M.H.; Genova, R.D.; Broos, R.; Sophiea, D.P. Lightweight cellular plastics. *Adv. Mater.* **2000**, *12*, 1779–1789. [CrossRef]
3. Kumar, V. Microcellular polymers: Novel materials for the 21st century. *Cell. Polym.* **1993**, *12*, 207–223.
4. Ma, X.; Tu, R.; Cheng, X.; Zhu, S.; Ma, J.; Fang, T. Experimental study of thermal behavior of insulation material rigid polyurethane in parallel, symmetric, and adjacent building facade constructions. *Polymers* **2018**, *10*, 1104. [CrossRef] [PubMed]
5. Mao, H.J.; He, B.; Guo, W.; Hua, L.; Yang, Q. Effects of nano-$CaCO_3$ content on the crystallization, mechanical properties, and cell structure of PP nanocomposites in microcellular injection molding. *Polymers* **2018**, *10*, 1160. [CrossRef] [PubMed]
6. Spina, R. Technological characterization of PE/EVA blends for foam injection molding. *Mater. Design* **2015**, *84*, 64–71. [CrossRef]
7. Li, J.; Zhang, A.; Zhang, S.; Gao, Q.; Zhang, W.; Li, J. Larch tannin-based rigid phenolic foam with high compressive strength, low friability, and low thermal conductivity reinforced by cork powder. *Compos. Part B* **2019**, *156*, 368–377. [CrossRef]
8. Wang, G.; Liu, X.; Zhang, J.; Sui, W.; Jang, J.; Si, C. One-pot lignin depolymerization and activation by solid acid catalytic phenolation for lightweight phenolic foam preparation. *Ind. Crops. Prod.* **2018**, *124*, 216–225. [CrossRef]
9. Ma, Y.; Gong, X.; Liao, C.; Geng, X.; Wang, C.; Chu, F. Preparation and characterization of DOPO-ITA modified ethyl cellulose and its application in phenolic foams. *Polymers* **2018**, *10*, 1049. [CrossRef]
10. Yang, J.; Ye, Y.; Li, X.; Lu, X.; Chen, R. Flexible, conductive, and highly pressure-sensitive graphene-polyimide foam for pressure sensor application. *Compos. Sci. Technol.* **2018**, *164*, 187–194. [CrossRef]
11. Sun, G.; Wang, W.; Zhang, C.; Liu, L.; Wei, H.; Han, S. Fabrication of isocyanate-based polyimide foam by a post grafting method. *J. Appl. Polym. Sci.* **2017**, *134*, 44240. [CrossRef]
12. Sun, G.; Wang, W.; Wang, L.; Yang, Z.; Liu, L.; Wang, J.; Ma, N.; Wei, H.; Han, S. Effects of aramid honeycomb core on the flame retardance and mechanical property for isocyanate-based polyimide foams. *J. Appl. Polym. Sci.* **2017**, *134*, 45041. [CrossRef]
13. Tucker, A.S.; Ward, C.A. Critical state of bubbles in liquid-gas solutions. *J. Appl. Phys.* **1975**, *46*, 4801–4806. [CrossRef]
14. Forest, T.W.; Ward, C.A. Effect of a dissolved gas on the homogeneous nucleation pressure of a liquid. *J. Chem. Phys.* **1977**, *66*, 2322–2330. [CrossRef]
15. Colton, J.S.; Suh, N.P. Nucleation of microcellular thermoplastic foam with additives: Part I: Theoretical considerations. *Polym. Eng. Sci.* **1987**, *27*, 485–492. [CrossRef]
16. Colton, J.S.; Suh, N.P. The nucleation of microcellular thermoplastic foam with additives: Part II: Experimental results and discussion. *Polym. Eng. Sci.* **1987**, *27*, 493–499. [CrossRef]
17. Nofar, M.; Majithiya, K.; Kuboki, T.; Park, C.B. The foamability of low-melt-strength linear polypropylene with nanoclay and coupling agent. *J. Cell. Plast.* **2012**, *48*, 271–287. [CrossRef]
18. Li, J.; Zhang, G.; Li, J.; Zhou, L.; Jing, Z.; Ma, Z. Preparation and properties of polyimide/chopped carbon fiber composite foams. *Polym. Adv. Technol.* **2017**, *28*, 28–34. [CrossRef]
19. Abbasi, H.; Antunes, M.; Velasco, J.I. Effects of carbon nanotubes/graphene nanoplatelets hybrid systems on the structure and properties of polyetherimide-based foams. *Polymers* **2018**, *10*, 348. [CrossRef]
20. Collais, D.I.; Baird, D.G. Tensile toughness of microcellular foams of polystyrene, styrene-acrylonitrile copolymer and polycarbonate, and the effect of dissolved gas on the tensile toughness of the same polymer matrices and microcellular foams. *Polym. Eng. Sci.* **2010**, *35*, 1167–1177. [CrossRef]

21. Yang, J.T.; Huang, L.Q.; Zhang, Y.F.; Chen, F.; Fan, P.; Zhong, M.Q.; Yeh, S. A new promising nucleating agent for polymer foaming: Applications of ordered Mesoporous silica particles in polymethyl methacrylate supercritical carbon dioxide microcellular foaming. *Ind. Eng. Chem. Res.* **2013**, *52*, 14169–14178. [CrossRef]
22. Zhai, W.T.; Wang, J.; Chen, N.; Naguib, H.E.; Park, C.B. The orientation of carbon nanotubes in poly(ethylene-co-octene) microcellular foaming and its suppression effect on cell coalescence. *Polym. Eng. Sci.* **2012**, *52*, 2078–2089. [CrossRef]
23. Miyahara, Y.; Goto, K.; Oka, M.; Inazu, T. Remarkably Facile Ring-Size Control in Macrocyclization: Synthesis of Hemicucurbit[6]uril and Hemicucurbit[12]uril. *Angew. Chem. Int. Ed.* **2004**, *43*, 5019–5022. [CrossRef] [PubMed]
24. Wu, F.; Wu, L.H.; Xiao, X.; Zhang, Y.Q.; Xue, S.F.; Tao, Z.; Day, A. Locating the Cyclopentano Cousins of the Cucurbit[n]uril Family. *J. Org. Chem.* **2012**, *77*, 606–611. [CrossRef] [PubMed]
25. Mhlanga, S.D.; Mamba, B.B.; Krause, R.W.; Malefetse, T.J. Removal of organic contaminants from water using nanosponge cyclodextrin polyurethanes. *J. Chem. Technol. Biotechnol.* **2007**, *82*, 382–388. [CrossRef]
26. Alongi, J.; Poskovic, M.P.; Visakh, M.; Frache, A.; Malucelli, G. Cyclodextrin nanosponges as novel green flame retardants for PP, LLDPE and PA6. *Carbohydr. Polym.* **2012**, *88*, 1387–1394. [CrossRef]
27. Leung, S.N.; Wong, A.; Wang, L.C.; Park, C.B. Mechanism of extensional stress-induced cell formation in polymeric foaming processes with the presence of nucleating agents. *J. Supercrit. Fluids* **2012**, *63*, 187–198. [CrossRef]
28. Leung, S.N.; Leung, S.N.; Li, H. Numerical simulation of polymeric foaming processes using modified nucleation theory. *Plast., Rubber Compos.* **2006**, *35*, 93–100. [CrossRef]
29. Leung, S.N.; Wong, A.; Park, C.B.; Zong, J.H. Ideal surface geometry of nucleating agents to enhance cell nucleation in polymer foaming. *J. Appl. Polym. Sci.* **2008**, *108*, 3997–4003. [CrossRef]

© 2020 by the authors. Licensee MDPI, Basel, Switzerland. This article is an open access article distributed under the terms and conditions of the Creative Commons Attribution (CC BY) license (http://creativecommons.org/licenses/by/4.0/).

Article

Fabrication of Micro-Structured LED Diffusion Plate Using Efficient Micro Injection Molding and Micro-Ground Mold Core

Yanjun Lu [1,*], Wang Luo [1], Xiaoyu Wu [1], Bin Xu [1,*], Chunjin Wang [2], Jiajun Li [1] and Liejun Li [3,*]

1. Guangdong Provincial Key Laboratory of Micro/Nano Optomechatronics Engineering, College of Mechatronics and Control Engineering, Shenzhen University, Shenzhen 518060, China; 1810293039@email.szu.edu.cn (W.L.); wuxy@szu.edu.cn (X.W.); lijiajunszu@163.com (J.L.)
2. Partner State Key Laboratory of Ultra-precision Machining Technology, Department of Industrial and Systems Engineering, The Hong Kong Polytechnic University, Hong Kong; chunjin.wang@polyu.edu.hk
3. Guangdong Key Laboratory for Processing and Forming of Advanced Metallic Materials, School of Mechanical and Automotive Engineering, South China University of Technology, Guangzhou 510640, China
* Correspondence: luyanjun@szu.edu.cn (Y.L.); binxu@szu.edu.cn (B.X.); liliejun@scut.edu.cn (L.L.); Tel.: +86-755-86950053

Received: 10 April 2020; Accepted: 4 June 2020; Published: 8 June 2020

Abstract: In this paper, a new style of micro-structured LED (light-emitting diode) diffusion plate was developed using a highly efficient and precise hybrid processing method combined with micro injection molding and micro-grinding technology to realize mass production and low-cost manufacturing of LED lamps with excellent lighting performance. Firstly, the micro-structured mold core with controllable shape accuracy and surface quality was machined by the precision trued V-tip grinding wheel. Then, the micro-structured LED diffusion plate was rapidly fabricated by the micro injection molding technology. Finally, the influences of micro injection molding process parameters on the illumination of the micro-structured diffusion plate were investigated. The simulated optical results show that the illumination of the micro-structured diffusion plate can achieve a maximum value when the V-groove depth and V-groove angle are designed to be 300 μm and 60°, respectively. The experimental results indicate that the developed micro-structured diffusion plate may improve the illumination by about 40.82% compared with the traditional diffusion plate. The prediction accuracy of the designed light efficiency simulation method was about 90.33%.

Keywords: micro-structure; diffusion plate; micro injection molding; grinding

1. Introduction

LED lamps are widely used in the lighting field due to their advantages of being energy efficient and environmentally friendly and producing a uniform and soft light. [1,2]. To further improve the luminous efficiency of LED lamps and reduce their production and manufacturing costs, many researchers have focused on the fabrication of the micro-array structure on the diffusion lampshade of LED lamps [3,4], which can reduce the diffusion and scattering of light to improve the utilization rate of light energy.

At present, the fabrication process of the micro-structured diffusion plate mainly includes silk-screen printing, hot pressing, and micro injection molding. Silk-screen printing can be used to directly fabricate a diffusion plate and has the advantages of low manufacturing cost and high production efficiency. It can print a uniform macroscopic dot structure, resulting in uniform light. However, the silk-screen printing process was complex, and the generated ink volatilization polluted the environment [5]. Previous studies have shown that the micro-structure was first machined on the surface of a mold and then was replicated to the polymer workpiece surface to rapidly fabricate the micro-structured polymer diffusion plate by hot pressing or micro injection molding technologies [6,7] to realize the mass production and manufacture of micro-structured diffusion plates. Due to its simple manufacturing process and low production cost, micro injection molding is suitable for the mass production and manufacture of micro-structured polymers. Besides, it is almost free from the limitation of plastic parts' geometry. Therefore, the micro injection molding technology has become the main forming process for micro-structured polymer products [8,9]. For example, the relationship between mold micro-manufacturing and the micro injection molding process was studied to optimize the demolding phase [10]. The hybrid method combining optical micro metrology and injection molding process monitoring was proposed to verify the quality of 3D micro-molded components [11]. The effects of milling strategies and cutting parameters on the mold core were investigated to reveal the relationship between mold topography and the ejection force in micro injection molding [12]. The relationship between different surface topography parameters and the ejection force in micro injection molding was analyzed to describe the friction behavior at the polymer–tool interface [13].

The forming quality of the micro-structured polymer products mainly depends on the machining quality and shape accuracy of the micro-structured mold core. The common micro-machining technologies of micro-structured mold core include electrochemical etching [14], laser processing [15], electrical discharge machining (EDM) [16], and mechanical micro-cutting [17,18], ect. Although chemical etching can ensure the shape accuracy of a nanoscale structure [14], it was impossible to efficiently and precisely machine a 3D micro-structure at the micron scale. In order to greatly improve the mold etching precision, the production cost would be greatly increased. Besides, the chemical etching solution would also pollute the environment. High-efficiency laser processing is environmentally friendly [15], but the equipment cost is considerably expensive and it is very difficult to control the shape accuracy of the 3D micro-structure. Although the micro-EDM technology can produce micro-structures of less than 100 microns in size [16], it was limited by conductive mold materials. Unfortunately, the surface quality of the 3D micro-structure was not satisfactory, and the EDM oil polluted the environment. A micro-structure with high surface quality can be processed by the mechanical diamond cutting [17,18], but this method was limited to soft metal materials with low hardness. Besides, the depth-to-width ratio of the machined micro-structure was also relatively low. Therefore, under the premise of environmental protection and high efficiency, it is very difficult to process a micro-structured mold with high shape accuracy and surface quality at the micron scale. The previously developed micro-grinding technology may be used to machine high-quality microarray structures with controllable shape accuracy on the surface of hard and brittle mold materials [19,20]. Moreover, the environmentally friendly and pollution-free processing method has a low production cost and was a simple operation. At present, there are no relevant research reports on the fabrication of micro-structured LED diffusion plates using micro-grinding and micro injection molding. Therefore, the precision trued grinding wheel tool is first proposed to machine a micro-array structure with high surface quality and shape accuracy on the surface of mold core, and then the 3D micro-structure derived from mold core surface will be duplicated to the acrylic polymer lampshade by micro injection molding to rapidly fabricate a high-quality micro-structured LED diffusion plate, leading to an improvement in light efficiency of LED lamp.

In this paper, the hybrid processing method combining with micro-grinding and micro injection molding is proposed to efficiently fabricate a micro-structured LED diffusion plate with high surface quality and forming accuracy. Firstly, the optical simulation method was designed to simulate the distribution of light intensity on the micro-structured diffusion plate to obtain optimal micro-structure size parameters. Then, according to the micro-structure parameters obtained by simulation, the V-shaped groove array structure with high shape accuracy was fabricated by precision micro-grinding technology on the surface of the mold steel. Finally, the V-grooved array structure on the surface of the mold core was rapidly replicated to the surface of the acrylic polymer diffusion plate by the micro injection molding process to efficiently produce a new type of micro-structured LED diffusion plate. The influences of micro injection molding process parameters on the light intensity of the LED diffusion plate were investigated to obtain the optimal injection process parameters.

2. The Optical Design and Light Efficiency Simulation of the Micro-structured LED Diffusion Plate

LED diffusion plate material is usually polymethyl methacrylate (PMMA), which has excellent optical properties such as high white light penetration, high refractive index, and high light transmittance. Its Mohs hardness and refractive indexes are 95 kg/cm^2 and 1.49, respectively. According to previous investigations [21,22], the optical performance can be effectively improved by fabricating the micro-structures on the surface of the LED diffusion plate. In order to efficiently design the optical structure for the LED diffusion plate, the TracePro software was employed to simulate the luminous flux and illuminance of traditional and micro-structured LED diffusion plate surfaces to evaluate their light efficiencies.

According to the light-emitting principle of the diffusion plate, it is very feasible for the micro-structure to be designed as a V-shaped groove. A common LED lamp mainly contains an LED light source, a lampshade, and a diffusion plate (see Figure 1a). The ray of light emitted by the light source is refracted from the surface as it travels through the diffusion plate. The outside diameter and thickness of the diffusion plate sample were 50 and 1.3 mm, respectively. The LED light source was a lamp bead with a rated power of 3 W and package size of 10 mm in diameter and 1 mm in thickness. The distance between the light source and the diffusion plate was 18 mm, and the simulated light ray was visible light with a wavelength of 400–700 nm. The wavelength of the simulated light ray was set as 450 nm, and the number of experimental light rays was 10,000. As seen from Figure 1a, the side of the diffusion plate with regular micro-array structure was regarded as the incident surface, and the backside was set as the light-emitting surface. The V-groove angle, V-groove depth, and V-groove width of the micro-array structure on the surface of diffusion plate sample were labeled as θ, h, and L, respectively (Figure 1b). Refraction and reflection would be produced as the incident ray travels through the diffusion plate. The incident ray, reflected ray, and refracted ray were expressed as i, r, and γ, respectively. The optical travel principles of the micro-structured and traditional diffusion plates are shown in Figure 1b,c. As observed from Figure 1b,c, in qualitative terms, compared with the traditional surface, the micro-structured surface may produce a more refracted and reflected ray to improve the light efficiency of the LED lamp. The simulated luminous flux and illuminance derived from the diffusion plate were recorded to evaluate the light efficiency.

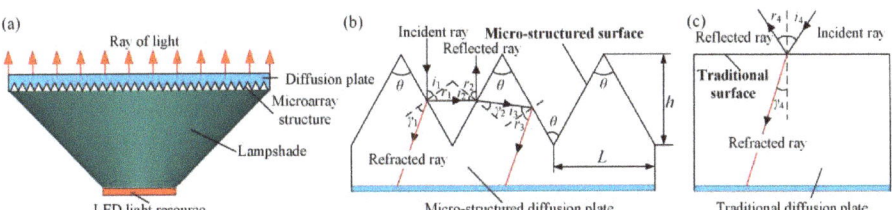

Figure 1. (**a**) Schematic diagram of the LED lamp. (**b**) Optical travel principle of the microstructured diffusion plate. (**c**) Optical travel principle of the traditional diffusion plate.

In order to obtain the optimal microgroove structure sizes, the V-groove depth h of 100–400 µm and V-groove angle θ of 60°–150° were designed and changed to calculate the theoretical illuminance E (see Table 1). Table 1 shows the simulated illuminances of the LED diffusion plate under different micro-structure parameters.

Figure 2 shows the influence of the designed V-groove angle θ on the irradiance and illuminance of the micro-structured LED diffusion plate. A brighter color indicates a higher illumination intensity. The diameter of the diffusion plate was 50 mm, and the light source was located directly below the diffusion plate. As seen from the irradiance and illuminance image of the diffusion plate, areas of lower light intensity of the area were further from the center of light source and resulted in darker images. It can also be seen that the highest light intensity was located in the center area. When the V-groove depth h remained consistent, the light intensity decreased with the increase of V-groove angle θ. When the V-groove angle was 60°, the light intensity was the highest, at this time, the highest illuminance E reached 138.07 Klux.

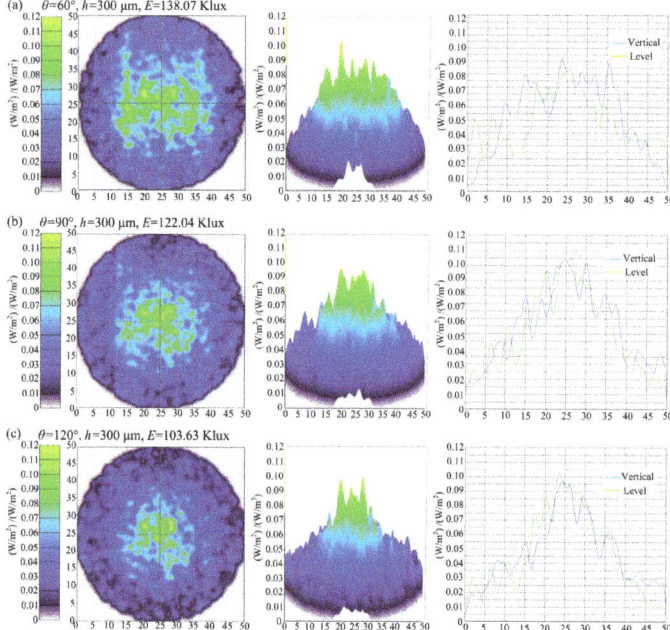

Figure 2. The simulated irradiances and illuminances of micro-structured LED diffusion plates depending on the designed V-groove angle θ: (**a**) $\theta = 60°$; (**b**) $\theta = 90°$; (**c**) $\theta = 120°$.

Figure 3 shows the influence of the designed V-groove structure parameters on simulated illuminances. It is shown that the influence of the V-groove angle on luminous flux was greater than the influence of V-groove depth. Within the V-groove angle range of 60 to 150°, a smaller V-groove angle resulted in a greater illumination intensity (see Figure 3a). It can be observed that when the V-groove depth was between 100 and 400 µm, there was little change in illumination intensity as the V-groove depth increased (see Figure 3b). The illumination intensity was significantly improved by designing a specific V-groove structure on the surface of the diffusion plate. Compared with the simulated illumination $E_{traditional}$ of 94.42 Klux for the traditional diffusion plate, the maximum illumination E_{max} of 138.07 Klux for the micro-structured diffusion plate may be theoretically enhanced by about 46.23%. Greater groove depths cause the grinding wheel to wear faster, necessitating frequent dressing and truing for the grinding wheel. Therefore, comprehensively considering the actual processing conditions and simulation results, the optimal micro-groove parameters of a V-groove depth of 300 µm and V-groove angle of 60° were chosen in the experiments.

Table 1. Simulated illuminances of LED diffusion plates under different V-groove structure parameters.

No.	V-Groove Depth h (μm)	V-Groove Angle θ (°)	Illuminance E (Klux)
1	0	0	94.42
2	100	60	134.29
3	100	90	121.64
4	100	120	103.56
5	100	150	95.49
6	200	60	132.82
7	200	90	123.42
8	200	120	102.55
9	200	150	95.22
10	300	60	138.07
11	300	90	122.04
12	300	120	103.63
13	300	150	94.49
14	400	60	132.79
15	400	90	122.02
16	400	120	102.07
17	400	150	94.45

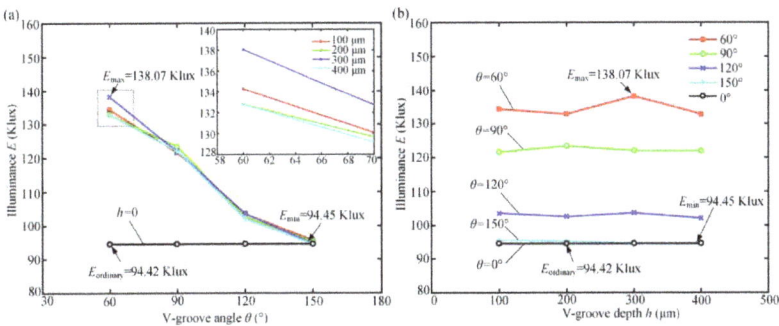

Figure 3. The simulated illuminance E as influenced by the designed V-groove structure parameters: (a) V-groove angle θ; (b) V-groove depth h.

3. Experimental Details

The surface quality and shape accuracy of V-groove structure machining of the mold core surface mainly depend on the dressing and truing accuracy of CBN (Cubic boron nitride) grinding wheel V-tip. Therefore, before the grinding experiment, the previously developed precision dressing and truing technology [23,24] was first implemented to produce a V-tip grinding wheel with a V-tip angle of 60°. Then, the trued #600 resin-bonded CBN grinding wheel was employed to machine regular V-grooved array structures with the V-groove depth of 300 μm on the surface of the mold core. Next, the micro-structured LED diffusion plate was fabricated by micro injection molding technology based on a micro-ground mold core. Finally, the light efficiencies of traditional and micro-structured LED diffusion plates were comparatively tested.

3.1. Precision Truing of the V-Tip CBN Grinding Wheel

Figure 4 shows the V-tip truing principle and photo of the CBN (Cubic boron nitride) grinding wheel. The machine tool is a three-axis precision CNC plane grinder (SMART-B818 III, Chevalier, Taiwan, China) with a minimum feed of 1 μm. Firstly, a macroscopic V-shaped tip grinding wheel was rapidly formed through rough truing along a V-shape interpolation truing path using the #80 coarse oilstone truer (green SiC) (see Figure 4a). Then, the #600 oilstone was used for fine truing to obtain a

V-tip with high shape accuracy and sharp micro-grain cutting edges. The theoretical V-tip truing angle was set as 60°. The V-tip angle of the trued grinding wheel was labeled as β.

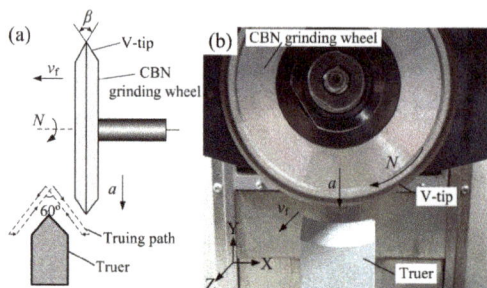

Figure 4. The precision truing of the V-tip CBN grinding wheel: (**a**) V-tip truing principle; (**b**) truing photo.

3.2. Precision Grinding of the Micro-structured Mold Core

In this study, S136H mold steel with the hardness of 30–35 HRC was chosen as the mold core material due to its excellent wear resistance and machinability, and it was polished to a mirror-like surface finish before micro-structure grinding. Figure 5 shows the V-groove array structure grinding principle and an experimental photo of the mold core. The precision trued V-tip CBN grinding wheel was driven by the CNC system to grind the mold steel installed on the horizontal worktable by the fixture (see Figure 5a). The V-groove was machined on the surface of mold steel by replicating the V-tip shape of the CBN grinding wheel along the set horizontal reciprocating cutting path. When a V-groove was machined, the grinding wheel was moved at a set pace along the Z-axis direction to conduct the machining of the second V-groove. Finally, the micro-structured mold core with regular and controllable V-grooved array structures was fabricated (see Figure 5b). According to previously determined fundamental experimental parameters, to ensure the grinding quality and production efficiency, the precision grinding conditions of the micro-structured mold core, including wheel speed N, depth of cut a, and feed speed v_f, were chosen and are shown in Table 2. The V-groove angle θ and V-groove depth h were set as 60° and 300 μm, respectively. The V-groove angle and V-groove depth of the ground micro-structured mold core surface are expressed as α and H, respectively (see Figure 5c).

Figure 5. The precision grinding of the micro-structured mold core: (**a**) grinding principle and path; (**b**) experimental photo; (**c**) schematic diagram of the V-groove structure.

Table 2. The precision grinding conditions of the micro-structured mold core.

CNC Grinding Machine	SMART B818 III
Grinding wheel	#600 CBN resin-bonded grinding wheel Wheel speed N = 3000 r/min
Workpiece	Super-mirror mold steel (S136H)
Rough grinding	v_f = 1000 mm/min; a = 5 μm, Σa = 290 μm
Fine grinding	v_f = 1000 mm/min; a = 1 μm, Σa = 10 μm
Cooling fluid	Water-soluble coolant
V-groove machining parameters	V-groove angle α = 60°, V-groove depth H = 300 μm

3.3. Micro injection Molding of the Micro-structured LED Diffusion Plate

The efficient micro injection molding technology was developed to replicate the V-grooved array structure of the mold core surface to the PMMA surface to precisely form the micro-structured diffusion plate sample through a micro injection molding machine (6/10P, Babyplast, Hospitalet Llobregat, Spain). Figure 6 shows a micro injection molding photo and the molding principle of a micro-structured LED diffusion plate. Firstly, the ground V-groove array structured mold core was installed on the rear mold, and the PMMA (8817, Degussa, Frankfurt, Germany) polymer particles were put into the hopper (see Figure 6a). Then the polymer particles were driven by a pneumatic device into a plasticizing chamber. Next, the plasticized and melted polymer flowed into the front mold through the pouring gate. Finally, the close-fitting rear mold and front mold were separated after cooling and holding pressure (see Figure 6b). Thus, the V-groove structure of the mold core surface was copied on the polymer workpiece surface to fabricate the V-shaped injection molded sample (see Figure 6c).

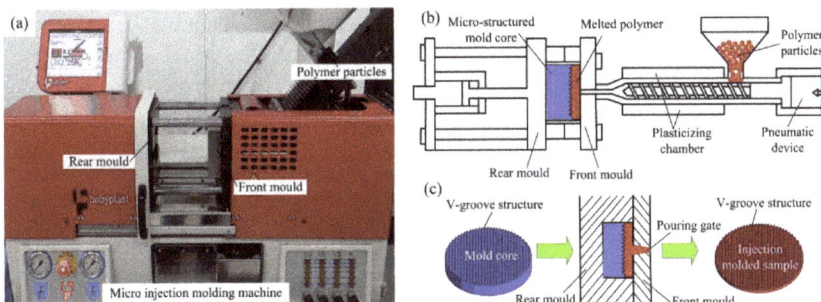

Figure 6. The micro injection molding of the micro-structured diffusion plate: (**a**) experimental photo; (**b**) schematic diagram of micro injection molding; (**c**) molding principle of the injection molded sample.

The micro-forming quality of the injection molded diffusion plate sample can affect the light efficiency of the LED lamp. Therefore, the optimal micro injection molding process parameters needed to be determined to achieve the maximum light efficiency of the micro-structured diffusion plate. To do this, the influences of the micro injection molding process parameters on the forming accuracy and surface quality of the micro-structured polymer sample needed to be studied. Under general injection molding conditions, the ideal mold temperature for PMMA injection is about 40–70 °C. According to previous experimental results [19], the molding quality of a micro-structured polymer was satisfactory at room temperature in this experiment. Therefore, the mold temperature was not considered as an injection molding process parameter in this work. The common micro injection molding process parameters, including melt temperature, injection speed, injection pressure, holding pressure, and holding time, were chosen and experimental designed in this work as listed in Table 3. Under each process parameter, ten micro injection molded samples were captured for testing to reduce the experimental error.

Table 3. Micro injection molding process parameters.

Process Parameters	Level
Melt temperature T (°C)	240, 245, 250, 255, 260
Injection speed V (mm/s)	40, 44, 48, 52, 56
Injection pressure P_i (MPa)	8.0, 8.5, 9.0, 9.5, 10.0
Holding pressure P_h (MPa)	5.0, 5.5, 6.0, 6.5, 7.0
Holding time T (s)	1.5, 2.5, 3.5, 4.5, 5.5

3.4. Light Efficiency Testing of the Micro-structured LED Diffusion Plate

In order to obtain maximum light efficiency, the digital illuminometer (1335, TES, Taiwan, China) with the measurement resolution of 0.01–0.1 Klux was employed to test the illuminance of micro-structured LED diffusion plates under different micro injection molding process parameters. The average value of five measured illumination values was regarded as the average illumination E. The light source was a commercial LED embedded lamp (LEH0103010, OPPLE, Shanghai, China) with the rated power of 3 W, rated voltage of 220 V, and an rated current of 0.025 A. Figure 7 shows the light efficiency testing photos of the traditional and micro-structured LED diffusion plates. Here, the injection molding process parameters of melt temperature, injection speed, injection pressure, holding pressure, and holding time were 245 °C, 40 mm/s, 8 MPa, 5 MPa, and 2.5 s, respectively. It is shown that the illuminance values for the traditional and micro-structured LED diffusion plates were 89.4 and 115.2 Klux, respectively. This indicates that the illuminance of the micro-structured diffusion plate may be improved by about 28.8% compared with the traditional one.

To evaluate the machining quality of the micro-ground mold core and the micro-forming quality of the injection molded LED diffusion plate sample, a probe stepper (D-300, KLA-Tencor, California, USA) was adopted to capture the section profiles of the micro-structured mold core and polymer workpieces. A 3D laser scanning microscope (VK-X250K, Keyence, Osaka, Japan) was used to test the 3D topographies of the micro-structured mold core and micro-formed polymer samples. High-resolution scanning electron microscopy (SEM, FEI Quanta 450 FEG and Apreo S, FEI Company, Hillsboro, USA) was employed to detect the surface topographies of the micro-structured mold core and injection-molded polymer samples. Through the measured section profile curves, the V-groove structure parameters of the micro-ground mold core and micro-formed diffusion plate sample can be obtained. The V-groove structure parameters of the micro-structured mold core are described as width V-groove angle α, V-groove depth H, and V-tip radius R. The V-groove structure parameters of the micro-structured diffusion plate sample are described as V-groove angle θ, V-groove depth h, and V-tip radius r. The presented result was the average value of three measurements.

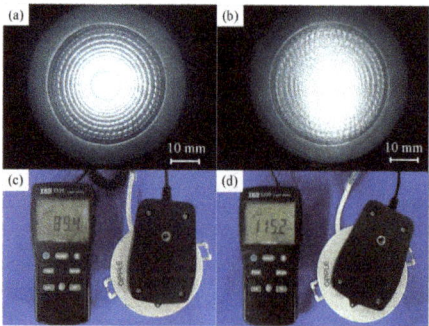

Figure 7. The light efficiency testing photos of the LED diffusion plates: (**a**) traditional diffusion plate; (**b**) micro-structured diffusion plate; (**c**) Light intensity of traditional diffusion plate; (**d**) Light intensity of micro-structured diffusion plate.

4. Results and Discussion

4.1. Truing Accuracy of the Grinding Wheel V-Tip

Generally, the grinding wheel tip will slightly wear in the grinding process, but the tip angle of the grinding wheel will not change at the macro-level. To reduce wear and improve shape accuracy of the V-groove machining, the grinding wheel was trued at regular intervals to ensure the processing quality and efficiency. Figure 8 shows the V-tip section shape of #600 CBN grinding wheel after precision V-tip truing. The V-tip section shape of the trued grinding wheel can be obtained by cutting and replicating a V-groove on the surface of the carbon graphite plate [25]. It is shown that the trued V-tip angle β was about 60.5°. The V-tip truing angle error was only 0.5° compared with the theoretical V-tip angle of 60°. Therefore, the previously developed grinding wheel V-tip truing technology [19] can ensure the shape accuracy of micro-structure grinding machining, facilitating micro-forming accuracy of the micro-structured polymer diffusion plate in the micro injection molding process.

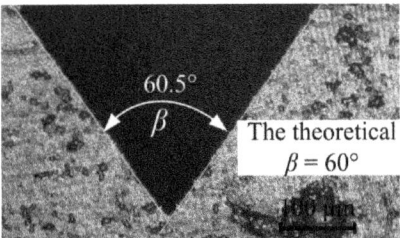

Figure 8. The V-tip section shape of trued grinding wheel.

4.2. Photos and Surface Topographies of the Micro-ground Mold Core and the Injection Molded Micro-structured Diffusion Plate

Figure 9 shows the photographs of the micro-ground mold core and the injection molded micro-structured LED diffusion plate. It is shown that the regular and smooth V-groove array structures without spacing were machined on the surface of mold steel mold core by micro-grinding technology (see Figure 9a). After micro injection molding, the inverted V-shaped groove array structures were fabricated on the surface of PMMA polymer to form the micro-structured diffusion plate by replicating the micro-structure characteristic derived from the mold core. The measured surface roughness R_a values of the V-grooved surface for the mold core and diffusion plate were 0.42 μm and 0.148 μm, respectively. It can be concluded that the surface quality of the injection molded micro-structured diffusion plate was significantly higher than that of the mold core. Under the action of injection pressure, this is because the molten polymer produces contraction after a period of compression and cooling in the process of micro injection molding, leading to a dense and smooth surface. Therefore, the proposed precision grinding and micro injection molding technologies have the potential for the mass production and manufacturing of micro-structured LED diffusion plates.

Figure 10 shows the SEM photos of the micro-structured mold core under vertical and inclined orientations. By observing the surface topographies of the mold core along vertical and inclined orientations, it is found that the microscopic V-groove structures were considerably regular and integral. Specifically, the V-groove top had hardly any damage, and the contour of the V-groove bottom was clearly identifiable. There was an existing arc radius at the V-groove bottom due to the inevitable V-tip truing error and micro-grinding machining error. Therefore, the V-groove array structure with regular and controllable shape accuracy can be machined on the surface of mold steel mold core through precision micro-grinding technology.

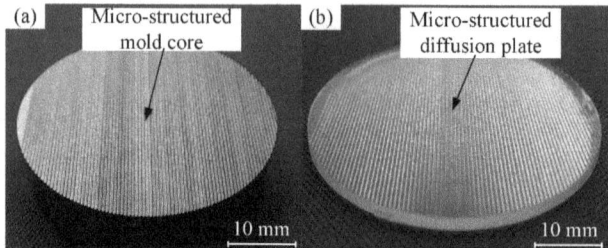

Figure 9. Photographs of the micro-structured mold core and diffusion plate: (**a**) micro-ground mold core; (**b**) injection molded diffusion plate sample.

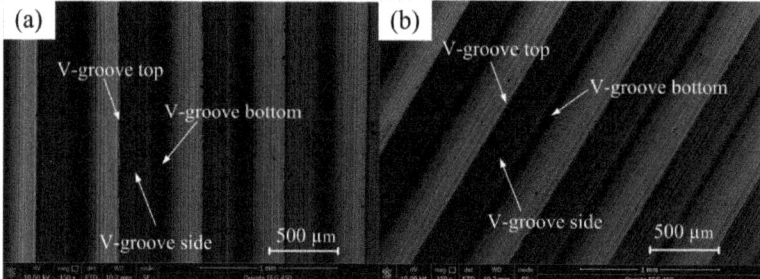

Figure 10. SEM photos of the micro-structured mold core under different orientations: (**a**) vertical; (**b**) inclined.

Due to a large number of diffusion plate samples, it is not necessary to show all the SEM photos. Therefore, the diffusion plate sample prepared under one group of process parameters was selected for SEM image characterization. Here, the injection molding process parameters of melt temperature, injection speed, injection pressure, holding pressure, and holding time were 250 °C, 40 mm/s, 8.0 MPa, 5.0 MPa, and 2.5 s, respectively. Figure 11 shows the SEM photos of the injection molded micro-structured diffusion plate sample under vertical and inclined orientations. It can be seen that the injection molded polymer sample surface was rather smooth, and the micro-formed V-groove array structure was also highly regular and integral. Moreover, there was no burr or damage at the top and bottom of the V-groove. This indicates that a micro-structured polymer sample with high injection molding quality can be obtained by using the proposed micro injection molding technology along with a micro-structured mold core with high shape accuracy. Compared with the SEM morphology of the mold core shown in Figure 10, it is found that the V-groove side surface of the micro-structured polymer diffusion plate was smoother. The melt viscosity of PMMA is mainly governed by pressure, so the injection pressure was increased to reduce the viscosity and increase the fluidity, giving the melted polymer good fluidity and contractility and facilitating the fabrication of a smooth micro-structured diffusion plate.

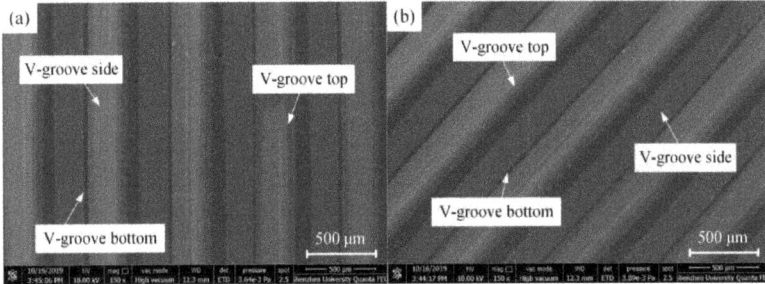

Figure 11. SEM photos of the injection molded micro-structured diffusion plate sample under different orientations: (**a**) vertical; (**b**) inclined.

4.3. The Section Profiles of the Micro-structured Mold Core and Diffusion Plate

The probe stepper was employed to capture the section profile curves of the micro-structured mold core and diffusion plate sample, as shown in Figure 12. As seen from Figure 12a, the measured average V-groove depth H and V-groove angle α of micro-ground mold core were 293.8 μm and 60.11°, respectively. Compared with the theoretical designed V-groove depth of 300 μm and V-groove angle of 60°, the micro-machining errors of V-groove depth and V-groove angle were only 6.2 μm and 0.57°, respectively. The average V-tip radius R of the ground V-groove structure was 3.22 μm. By comparing the theoretical with the actual V-groove profile curves of the mold cores, the shape error of micro-structure machining could be calculated. The shape error of the micro-structured mold core is mainly concentrated at the V-groove top due to the ineradicable V-tip arc radius. The absolute difference between the peak value and valley value of the V-groove profile curve was defined as the shape accuracy [19,25]. Thus, the shape accuracy of the micro-ground mold core was calculated to be 9.2 μm. The results indicate that a micro-structured mold core with high shape accuracy can be manufactured by micro-grinding technology, which was beneficial to produce an injection molded diffusion plate with high forming quality. The depth, angle, and tip radius of the three V-groove profiles on the front of the mold core and injection molded part was measured to calculate the average value to evaluate the shape accuracy. Figure 12b shows the profile curve of the micro-structured injection-molded diffusion plate sample. At this point, the micro injection molding process parameters, namely, the melt temperature, injection speed, injection pressure, holding pressure, and holding time were 250 °C, 40 mm/s, 8.0 MPa, 5.0 MPa, and 2.5 s, respectively. As shown in Figure 12b, the measured average V-groove depth h and V-groove angle θ of the injection-molded micro-structured diffusion plate sample were 265.2 μm and 60.03°, respectively. Compared with the V-groove depth and V-groove angle of the micro-ground mold core shown in Figure 12a, the micro-forming errors of V-groove depth and V-groove angle of the micro-structured diffusion plate were 28.6 μm and 0.08°, respectively. The average V-tip radius r of the micro-formed diffusion plate was 3.34 μm. As a result, compared with the V-tip radius of the mold core, the micro-forming error of the V-tip radius for the micro-structured diffusion plate was calculated to be 0.12 μm. Therefore, the developed micro injection molding using a micro-ground mold core can precisely fabricate a micro-structured diffusion plate with high molding accuracy. The replication rate was defined as the ratio of the micro-structural depth of the micro injection molded part to the micro-structural depth of the mold core [26,27]. Therefore, according to the average V-groove depth of the micro-structured mold core and diffusion plate sample, the replication rate of micro injection molded polymer sample was computed to be about 90.26%. The average standard deviations of the V-groove depth, V-tip angle, and V-tip radius of mold core were 5.89 μm, 0.83 μm, and 0.09 μm, respectively. The average standard deviations of the V-groove depth, V-tip angle, and V-tip radius of the diffusion plate were 2.82 μm, 0.57 μm, and 0.17 μm, respectively. It can be seen that the standard deviations for the micro-structured diffusion plate were relatively less than those of mold core.

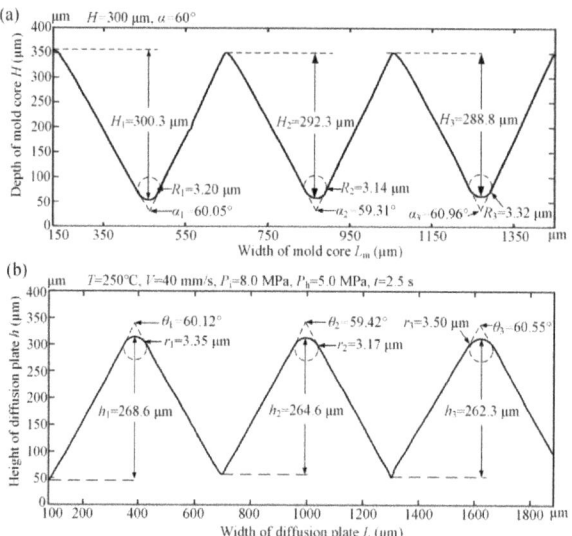

Figure 12. The section profile curves of the micro-structured mold core and diffusion plate sample: (**a**) mold core; (**b**) diffusion plate sample.

4.4. Light Efficiency Analysis of the Micro-structured LED Diffusion Plate

In order to study the influences of melt temperature, injection speed, injection pressure, holding pressure, and holding time on the light efficiency of the micro-structured diffusion plate to determine the optimal micro injection molding process conditions, the relationships between average illumination E of the micro-structured LED diffusion plate and different micro injection molding process parameters were determined and are described in Figure 13. It is shown that the measured average illumination $E_{traditional}$ of the traditional diffusion plate was 89.4 Klux. As observed in Figure 13, the average illumination E values ranged from 115.3 to 119.2 Klux, 116.4 to 119.2 Klux, 116.9 to 120.6 Klux, and 118.0 to 122.7 Klux with the change of melt temperature, injection speed, injection pressure, and holding time, respectively (see Figure 13a–c,e). There was a variation of about 13% in optical performance. This was mainly attributed to the replication of the V-groove geometry controlled by micro injection molding process parameters. Because the V-groove has a tip arc, the differences in the replication of the groove geometry eventually lead to differences in optical performance. There were no significant changes in the average illumination when the melt temperature, injection speed, injection pressure, or holding time increased. The average illumination E ranged from 110.8 to 125.9 Klux with the change of holding pressure (see Figure 13d). This indicates that the average illumination had an obvious variation with the variation of holding pressure. The average illumination E of the injection molded micro-structured diffusion plate reached a maximum value of 125.9 Klux where the optimal micro injection molding process parameters, namely the melt temperature, injection speed, injection pressure, holding pressure, and holding time, were 240 °C, 40 mm/s, 8.0 MPa, 6.5 MPa, and 2.5 s, respectively. The experimental results show that the holding pressure had the greatest effect on the light efficiency of the micro-structured diffusion plate, while the other micro injection molding process parameters had little effect. Because the illumination of the traditional diffusion plate was 89.4 Klux, the maximum illumination of the micro-structured diffusion plate may be improved by about 40.82%. For the micro-structured diffusion plate, the experimentally measured illumination ranged from 110.8 to 125.9 Klux, which was very close to the simulated theoretical illumination of 138.07 Klux. Therefore, it was feasible to design the micro-structured LED diffusion plate through the light efficiency simulation method.

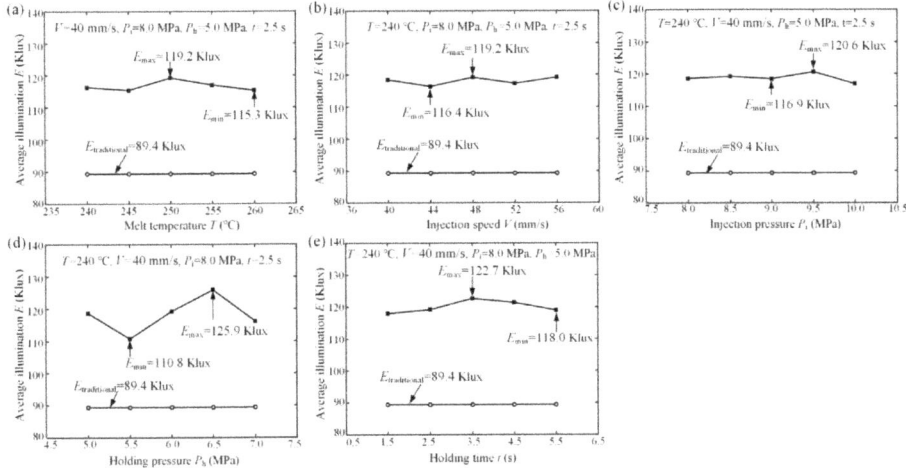

Figure 13. The average illuminations E of LED diffusion plates depending on micro injection molding process parameters: (**a**) melt temperature T; (**b**) injection speed V; (**c**) injection pressure P_i; (**d**) holding pressure P_h; (**e**) holding time t.

Figure 14 shows the prediction accuracies of traditional and micro-structured diffusion plates. When the V-groove depth h and V-groove angle θ were designed as 300 μm and 60°, respectively, the illumination value of the micro-structured diffusion plate determined by the designed optical simulation method was 138.07 Klux. The experimental maximum illumination value of the injection molded micro-structured diffusion plate was 125.9 Klux. Therefore, the prediction accuracy for the micro-structured diffusion plate was about 90.33%. The simulated and measured experimental illumination values of the traditional diffusion plate were 94.42 and 89.4 Klux, respectively. Therefore, the prediction accuracy for the traditional diffusion plate was about 94.38%. As a result, the designed light efficiency simulation method with high prediction accuracy may be used to conduct the structural design of a micro-structured LED diffusion plate. The proposed efficient micro injection molding technology based on a micro-ground mold core can realize the mass production and manufacturing of micro-structured LED diffusion plate products with high forming accuracy.

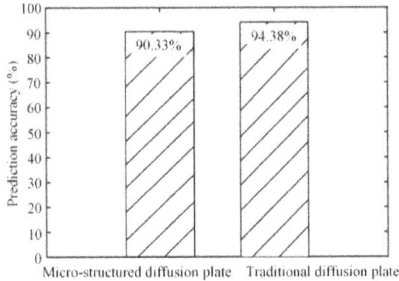

Figure 14. The prediction accuracies of traditional and micro-structured diffusion plates.

5. Conclusions

In this paper, the light efficiency simulation method was designed to obtain the optimal micro-structure parameters of a micro-structured LED diffusion plate. High-precision micro-grinding technology was proposed for the fabrication of regular and controllable micro-grooved array structures on the surface of a mold core. Efficient micro injection molding was employed to rapidly fabricate a micro-structured diffusion plate with high forming accuracy. The proposed hybrid fabrication method can realize mass production and manufacturing of micro-structured LED diffusion plates due to its low manufacture cost. The main results are summarized as follows:

1. Through the light efficiency simulation method, the optimal micro-groove parameters, namely V-groove depth and V-groove angle, were designed to be 300 μm and 60°, respectively.
2. The machining errors of the V-groove depth and V-groove angle for the micro-ground mold core were 6.2 μm and 0.57°, respectively. The micro-forming errors of the V-groove angle and V-tip radius for the micro-structured diffusion plate in micro injection molding were 0.08° and 0.12 μm, respectively. The shape accuracy of the micro-ground mold core and the replication rate of the micro injection molded diffusion plate were 9.2 μm and 90.26%, respectively.
3. The holding pressure has the greatest effect on the light efficiency of the micro-structured diffusion plate. When the melt temperature, injection speed, injection pressure, holding pressure, and holding time were 240 °C, 40 mm/s, 8.0 MPa, 6.5 MPa, and 2.5 s, respectively, the measured illumination of the micro-structured diffusion plate reached a maximum value of 125.9 Klux, improving the illumination by about 40.82% against the traditional diffusion plate.
4. The prediction accuracies of the designed light efficiency simulation method for the traditional and micro-structured diffusion plates are 94.38% and 90.33%, respectively.

Author Contributions: Conceptualization, Y.L.; data curation, Y.L. and W.L.; formal analysis, W.L.; funding acquisition, Y.L., X.W., and L.L.; investigation, W.L. and J.L.; Methodology, Y.L., B.X., and C.W.; supervision, X.W., B.X., and L.L.; validation, C.W.; writing—original draft, Y.L.; writing—review & editing, Y.L. and W.L. All authors have read and agreed to the published version of the manuscript.

Funding: This project was funded by the National Natural Science Foundation of China (grant No. 51805334), the Science and Technology Planning Project of Guangdong Province (grant Nos. 2016A040403043 and 2017A010102003), the major project for industrial technology of Guangzhou City (grant No. 201902010018), and the Fundamental Research Funds for the Shenzhen University (grant No. 2018031). The authors also gratefully acknowledge the instrumental analysis center of Shenzhen University (Xili Campus).

Conflicts of Interest: The authors declare no conflict of interest.

References

1. Ramírez, M.G.; Sirvent, D.; Morales-Vidal, M.; Ortuño, M.; Martínez-Guardiola, F.J.; Francés, J.; Pascual, I. LED-Cured Reflection Gratings Stored in an Acrylate-Based Photopolymer. *Polymers* **2019**, *11*, 632.
2. Lu, L.B.; Zhang, Z.; Guan, Y.C.; Zheng, H.Y. Enhancement of Heat Dissipation by Laser Micro Structuring for LED Module. *Polymers*. **2018**, *10*, 886. [CrossRef] [PubMed]
3. Chien, C.H.; Chen, C.C.; Chen, T.; Lin, Y.M.; Liu, Y.C. Thermal deformation of microstructure diffuser plate in LED backlight unit. *J. Soc. Inf. Display.* **2016**, *24*, 99–109. [CrossRef]
4. Chien, C.H.; Chen, Z.P. The study of integrated LED-backlight plate fabricated by micromachining technique. *Microsyst. Technol.* **2009**, *15*, 383–389. [CrossRef]
5. Pardo, D.A.; Jabbour, G.E.; Peyghambarian, N. Application of Screen Printing in the Fabrication of Organic Light-Emitting Devices. *Adv. Mater.* **2020**, *12*, 1249–1252. [CrossRef]
6. Weng, C.; Wang, F.; Zhou, M.; Yang, D.; and Jiang, B. Fabrication of hierarchical polymer surfaces with superhydrophobicity by injection molding from nature and function-oriented design. *Appl. Surf. Sci.* **2018**, *436*, 224–233. [CrossRef]
7. Liao, Q.H.; Zhou, C.L.; Lu, Y.J.; Wu, X.Y.; Chen, F.M.; Lou, Y. Efficient and Precise Micro-Injection Molding of Micro-Structured Polymer Parts Using Micro-Machined Mold Core by WEDM. *Polymers* **2019**, *11*, 1591. [CrossRef]
8. Maghsoudi, K.; Jafari, R.; Momen, G.; Farzaneh, M. Micro-nanostructured polymer surfaces using injection molding: A review. *Mater. Today. Commun.* **2017**, *13*, 126–143. [CrossRef]
9. Masato, D.; Sorgato, M.; Lucchetta, G. Analysis of the influence of part thickness on the replication of micro-structured surfaces by injection molding. *Mater Design.* **2016**, *95*, 219–224. [CrossRef]

10. Masato, D.; Sorgato, M.; Parenti, P.; Annoni, M.; Lucchetta, G. Impact of deep cores surface topography generated by micro milling on the demolding force in micro injection molding. *J. Mater. Process. Tech.* **2017**, *246*, 211–223. [CrossRef]
11. Baruffi, F.; Calaon, M.; Tosello, G. Micro-injection moulding in-line quality assurance based on product and process fingerprints. *Micromachines* **2018**, *9*, 293. [CrossRef] [PubMed]
12. Parenti, P.; Masato, D.; Sorgato, M.; Lucchetta, G.; Annoni, M. Surface footprint in molds micromilling and effect on part demoldability in micro injection molding. *J. Manuf. Process.* **2017**, *29*, 160–174. [CrossRef]
13. Sorgato, M.; Masato, D.; Lucchetta, G. Effects of machined cavity texture on ejection force in micro injection molding. *Precis. Eng.* **2017**, *50*, 440–448. [CrossRef]
14. Kim, J.H.; Mirzaei, A.; Kim, H.W.; Kim, S.S. Facile fabrication of superhydrophobic surfaces from austenitic stainless steel (AISI 304) by chemical etching. *Appl. Surf. Sci.* **2018**, *439*, 598–604. [CrossRef]
15. Zhou, C.L.; Ngai, T.W.L.; Li, L.J. Wetting behaviour of laser textured Ti_3SiC_2 surface with micro-grooved structures. *J. Mater. Sci. Technol.* **2016**, *32*, 805–812.
16. Zhou, C.L.; Wu, X.Y.; Lu, Y.J.; Wu, W.; Zhao, H.; Li, L.J. Fabrication of hydrophobic Ti_3SiC_2 surface with micro-grooved structures by wire electrical discharge machining. *Ceram. Int.* **2018**, *44*, 18227–18234. [CrossRef]
17. Xu, S.L.; Shimada, K.; Mizutani, M.; Kuriyagawa, T. Fabrication of Hybrid micro/nano-textured surfaces using rotary ultrasonic machining with one-point diamond tool. *Int. J. Mach. Tools Manuf.* **2014**, *86*, 12–17. [CrossRef]
18. Wang, J.S.; Zhang, X.D.; Fang, F.Z.; Chen, R.T. Diamond cutting of micro-structure array on brittle material assisted by multi-ion implantation. *Int. J. Mach. Tools Manuf.* **2019**, *137*, 58–66. [CrossRef]
19. Lu, Y.J.; Chen, F.M.; Wu, X.Y.; Zhou, C.L.; Lou, Y.; Li, L.J. Fabrication of micro-structured polymer by micro injection molding based on precise micro-ground mold core. *Micromachines.* **2019**, *10*, 253. [CrossRef]
20. Lu, Y.J.; Li, L.J.; Xie, J.; Zhou, C.L.; Guo, R.B. Dry electrical discharge dressing and truing of diamond grinding wheel V-tip for micro-grinding. In Proceedings of the The 5th International Conference on Mechatronics, Materials, Chemistry and Computer Engineering (ICMMCCE 2017), Chongqing, China, 24–25 July 2017; Atlantis Press: Paris, France, 2017.
21. Li, P.; Xie, J.; Cheng, J.; Jiang, Y.N. Study on weak-light photovoltaic characteristics of solar cell with a microgroove lens array on glass substrate. *Opt. Express.* **2015**, *23*, 192–203. [CrossRef]
22. Kim, D.S.; Kim, D.H.; Jang, J.H. A nanoscale conical polymethyl methacrylate (PMMA) sub-wavelength structure with a high aspect ratio realized by a stamping method. *Opt. Express.* **2013**, *21*, 8450–8459. [CrossRef] [PubMed]
23. Lu, Y.J.; Xie, J.; Si, X.H. Study on micro-topographical removals of diamond grain and metal bond in dry electro-contact discharge dressing of coarse diamond grinding wheel. *Int. J. Mach. Tools Manuf.* **2015**, *88*, 118–130. [CrossRef]
24. Xie, J.; Xie, H.F.; Luo, M.J. Dry electro-contact discharge mutual-wear truing of micro diamond wheel V-tip for precision micro-grinding. *Int. J. Mach. Tools Manuf.* **2012**, *60*, 44–51. [CrossRef]
25. Xie, J.; Zhuo, Y.W.; Tan, T.W. Experimental study on fabrication and evaluation of micro pyramid-structured silicon surface using a V-tip of diamond grinding wheel. *Precis. Eng.* **2011**, *35*, 173–182. [CrossRef]
26. Seo, Y.S.; Park, K. Direct patterning of micro-features on a polymer substrate using ultrasonic vibration. *Microsyst. Technol.* **2012**, *18*, 2053–2061.
27. Lu, Y.J.; Chen, F.M.; Wu, X.Y.; Zhou, C.L.; Zhao, H.; Li, L.J.; Tang, Y. Precise WEDM of micro-textured mould for micro-injection molding of hydrophobic polymer surface. *Mater. Manuf. Process.* **2019**, *34*, 1342–1351. [CrossRef]

© 2020 by the authors. Licensee MDPI, Basel, Switzerland. This article is an open access article distributed under the terms and conditions of the Creative Commons Attribution (CC BY) license (http://creativecommons.org/licenses/by/4.0/).

Article

Melt- vs. Non-Melt Blending of Complexly Processable Ultra-High Molecular Weight Polyethylene/Cellulose Nanofiber Bionanocomposite

Nur Sharmila Sharip [1], Hidayah Ariffin [1,2,*], Tengku Arisyah Tengku Yasim-Anuar [2], Yoshito Andou [3], Yuki Shirosaki [4], Mohammad Jawaid [1], Paridah Md Tahir [1] and Nor Azowa Ibrahim [5]

1. Institute of Tropical Forestry and Forest Products (INTROP), Universiti Putra Malaysia, UPM Serdang, Selangor 43400, Malaysia; nursharmilasharip@gmail.com (N.S.S.); jawaid@upm.edu.my (M.J.); parida.introp@gmail.com (P.M.T.)
2. Department of Bioprocess Technology, Faculty of Biotechnology and Biomolecular Sciences, Universiti Putra Malaysia, UPM Serdang, Selangor 43400, Malaysia; tengkuarisyah@gmail.com
3. Department of Biological Functions and Engineering, Graduate School of Life Science and Systems Engineering, Kyushu Institute of Technology, 2-4 Hibikino, Wakamatsu-ku, Kitakyushu, Fukuoka 808-0196, Japan; yando@life.kyutech.ac.jp
4. Department of Applied Chemistry, Faculty of Engineering, Kyushu Institute of Technology, 1-1 Sensui-cho, Tobata-ku, Kitakyushu, Fukuoka 804-8550, Japan; yukis@che.kyutech.ac.jp
5. Department of Chemistry, Faculty of Science, Universiti Putra Malaysia, UPM Serdang, Selangor 43400, Malaysia; norazowa@upm.edu.my
* Correspondence: hidayah@upm.edu.my; Tel.: +603-9769-7515

Abstract: The major hurdle in melt-processing of ultra-high molecular weight polyethylene (UHMWPE) nanocomposite lies on the high melt viscosity of the UHMWPE, which may contribute to poor dispersion and distribution of the nanofiller. In this study, UHMWPE/cellulose nanofiber (UHMWPE/CNF) bionanocomposites were prepared by two different blending methods: (i) melt blending at 150 °C in a triple screw kneading extruder, and (ii) non-melt blending by ethanol mixing at room temperature. Results showed that melt-processing of UHMWPE without CNF (MB-UHMWPE/0) exhibited an increment in yield strength and Young's modulus by 15% and 25%, respectively, compared to the Neat-UHMWPE. Tensile strength was however reduced by almost half. Ethanol mixed sample without CNF (EM-UHMWPE/0) on the other hand showed slight decrement in all mechanical properties tested. At 0.5% CNF inclusion, the mechanical properties of melt-blended bionanocomposites (MB-UHMWPE/0.5) were improved as compared to Neat-UHMWPE. It was also found that the yield strength, elongation at break, Young's modulus, toughness and crystallinity of MB-UHMWPE/0.5 were higher by 28%, 61%, 47%, 45% and 11%, respectively, as compared to the ethanol mixing sample (EM-UHMWPE/0.5). Despite the reduction in tensile strength of MB-UHMWPE/0.5, the value i.e., 28.4 ± 1.0 MPa surpassed the minimum requirement of standard specification for fabricated UHMWPE in surgical implant application. Overall, melt-blending processing is more suitable for the preparation of UHMWPE/CNF bionanocomposites as exhibited by their characteristics presented herein. A better mechanical interlocking between UHMWPE and CNF at high temperature mixing with kneading was evident through FE-SEM observation, explains the higher mechanical properties of MB-UHMWPE/0.5 as compared to EM-UHMWPE/0.5.

Keywords: ultra-high molecular weight polyethylene; cellulose nanofiber; bionanocomposite; melt-blending; ethanol mixing

Citation: Sharip, N.S.; Ariffin, H.; Yasim-Anuar, T.A.T.; Andou, Y.; Shirosaki, Y.; Jawaid, M.; Tahir, P.M.; Ibrahim, N.A. Melt- vs. Non-Melt Blending of Complexly Processable Ultra-High Molecular Weight Polyethylene/Cellulose Nanofiber Bionanocomposite. *Polymers* **2021**, *13*, 404. https://dx.doi.org/10.3390/polym13030404

Received: 28 October 2020
Accepted: 3 December 2020
Published: 27 January 2021

Publisher's Note: MDPI stays neutral with regard to jurisdictional claims in published maps and institutional affiliations.

Copyright: © 2021 by the authors. Licensee MDPI, Basel, Switzerland. This article is an open access article distributed under the terms and conditions of the Creative Commons Attribution (CC BY) license (https://creativecommons.org/licenses/by/4.0/).

1. Introduction

Ultra-high molecular weight polyethylene (UHMWPE) is a long linear engineered thermoplastic with extremely high molecular weight of approximately 3×10^6 g/mol [1]. It possesses high resistance against impact, fatigue, chemical corrosion and abrasion, which stemmed from effective load transfer to its long linear backbone. This polymer also has a remarkable self-lubricating, low friction coefficient and good biocompatibility [2–4] that enable its application in various fields including aerospace and industrial machineries (i.e., pipes, panels, bars, gears), microelectronics and joint replacement or also known as arthroplasty (i.e., hip liner, tibial inserts) [5–7]. However, relatively low Young's modulus and surface hardness of UHMWPE could limit the sustainability of this polymer against wear as a result of contact and slip with harder counterpart such as metal under repeated motion [8]. This results in abrasion where generated debris in turns may accelerate cracks leading to component loosening and failure [4,5].

Various studies have been conducted involving fillers incorporation in UHMWPE matrix with the aim to improve its abrasion and wear through Young's modulus enhancement. The fillers used ranged from inorganic to organic and natural fibers such as carbon nanofibers, hydroxyapatite as well as nanocellulose [9,10]. Besides improving the stiffness, the presence of fillers in polymer matrix could play a role in mitigating wear through its act as solid lubricant by rolling or sliding at interface between the contacted surfaces [11,12]. Nevertheless, this mechanism of solid lubrication is greatly dependent on the filler properties and size, where it could also become a third body abrasive that further abrade the UHMWPE surface, or further trigger the inflammation due to fillers cytotoxicity [10,13]. For instance, nanocellulose filler has been proven beneficial in enhancing wear resistance of UHMWPE and exhibit good biocompatibility against osteoblast cells MC3T3-E1 [5,11]. The nanocellulose debris was reported to serve as solid lubricant between metal and polymer surface, thus prevented further abrasion of UHMWPE, with relatively low wear volume as compared to neat UHMPWE. Additionally, nanocelluloses are biocompatible and non-toxic by which it can be used in many biomaterials application such as for wound dressings materials [14–16], scaffold for bone or tissue regeneration [17–19], carrier for drug deliveries [20–24] and many more [25–28]. These properties of nanocellulose make it an excellent material as UHMWPE fillers, particularly for artificial joint application.

Common method for nanocellulose composites fabrication in various matrices is through solution processing and melt blending, by which the latter is comparably easy, as well as industrially and economically viable [29–31]. In melt blending, nanocellulose is introduced and mixed with polymer in molten state [32–34]. Nonetheless, unlike most thermoplastic polymers, fabricating UHMWPE composites via conventional melt processing methods is extremely difficult. Viscous flow state of melt UHMWPE is not attainable even with increases in temperature, and it maintains in non-uniform or non-continuous rubberlike state. This is attributed by its higher theoretical viscous flow temperature as compared to its decomposition temperature [35], as a result of numerous chain entanglements contributed by its extremely high molecular weight. In fact, its melt viscosity could be up to 1×10^8 Pa.s which is about 2500 times higher than high density polyethylene (HDPE) [36]. Similarly, UHMWPE composite fabrication by solution mixing is not convenient either, attributed to inertness of UHMWPE that is resilient to any reaction with acids, alkalis and organic solvent as well as biological reaction [37].

Wang et al. (2016) produced UHMWPE nanocellulose composites by mixing UHMWPE and nanocellulose in ethanol. The solution was continuously mixed until the ethanol was completely evaporated. This process aids in nanocellulose drying without the occurrence of aggregation up to 0.5 wt.% cellulose nanocrystals (CNC) loading. Nevertheless, results showed that better nanocellulose dispersion with higher micro-hardness was achieved through melt processing as compared to the ethanol mixing process. Yet, no information was given on the mechanical properties such as tensile strength, modulus and elongation of UHMWPE/CNC produced by the two different processes [11]. In our previous study, we fabricated UHMWPE/cellulose nanofiber (CNF) through melt blending process in triple screw kneading extruder. Despite homogenous filler dispersion and optimized parameters, the resulted tensile strength with 3 wt.% CNF loading was found decreased. In consideration that there is lack information on the mechanical properties of the UHMWPE/nanocellulose composites fabricated through different processing techniques, hence this study was conducted to investigate the effect of UHMWPE/CNF bionanocomposites blending process (melt and non-melt blending) on its mechanical and crystallinity properties.

2. Materials and Methods

2.1. Materials

Ultra-high molecular weight polyethylene (UHMWPE) were purchased from Sigma-Aldrich (ST. Louis, MO, USA) in the form of fine powder with particle size of 96 ± 20 μm. The molecular weight, melting point and density of the polymer was 3×10^6–6×10^6 g/mol, 138 °C and 0.94 g/mL, respectively. Meanwhile, 2 wt.% cellulose nanofiber (CNF) of 53.4 ± 9 nm diameter sizes was purchased from ZoepNano Sdn. Bhd. (Serdang, Malaysia) in slurry form. Absolute ethanol 99.8% AR grade was purchased from John Kollin Corporation (Midlothian, UK).

2.2. Bionanocomposite Fabrication and Moulding

Non melt-blending (ethanol mixing) process was conducted according to Wang et al. (2016) with some modification (Figure 1) [11]. About 10 wt. % UHMWPE-CNF (0.5 wt.% CNF in UHMWPE) was added into ethanol and mechanically stirred by using JLT Series Flocculators (Velp Scientifica, Usmate, Italy) at 120 rpm speed. The experiment was conducted at room temperature until the solvent was completely evaporated before being dried at temperature 50 °C overnight.

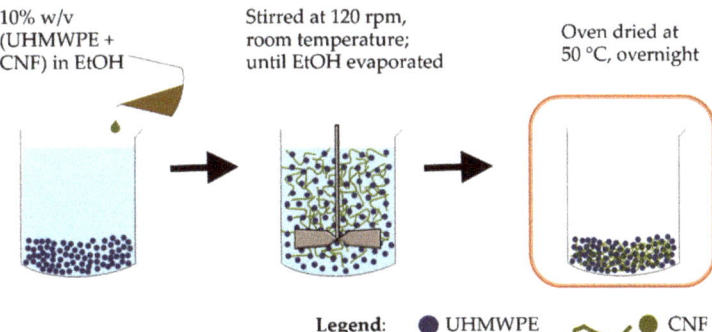

Figure 1. UHMWPE/CNF bionanocomposites by ethanol mixing.

For comparison, UHMWPE/CNF bionanocomposite of same composition was melt blended by using triple screw kneading extruder at Kyushu Institute of Technology, Fukuoka, Japan with optimized condition of 150 °C and 60 rpm rotational speed [38]. Meanwhile, UHMWPE without filler was subjected to both blending process and denoted as MB-UHMWPE/0 and EM-UHMWPE/0. The summary of produced bionanocomposites is as presented in Table 1.

Table 1. UHMWPE/CNF bionanocomposite samples and control.

Sample	Blending Process	CNF Content (wt.%)	Remarks
Neat-UHMWPE	none	0	control for blending effect
MB-UHMWPE/0	Melt blending	0	control for CNF addition effect by melt blending
MB-UHMWPE/0.5	Melt blending	0.5	-
EM-UHMWPE/0	Ethanol mixing	0	control for CNF addition effect by ethanol mixing
EM-UHMWPE/0.5	Ethanol mixing	0.5	-

CNF: cellulose nanofiber, UHMWPE: ultra-high molecular weight polyethylene, MB: melt blend, EM: ethanol mixed.

All samples were molded into 10 cm × 10 cm × 1 mm sheet by direct compression at 175 °C and 15 MPa for 45 min [39].

2.3. Characterization of Bionanocomposites

2.3.1. Mechanical Analysis

The tensile properties was conducted by using a compact tensile and compression tester IMC-18E0 (Imoto Machinery Co., Ltd., Kyoto, Japan). Eight specimens of samples were subjected to a tensile tester with crosshead speed of 50 mm/min (ASTM D638).

2.3.2. Morphological Analysis

The morpholocal analysis was carried out by using a high-resolution field-emission scanning electron microscopy (FESEM) (FEI Nova NanoSEM 230, FEI Company, Hillsboro, OR, USA) with accelerating voltage of 10 kV. Sample specimens subjected to tensile testing were analyzed for surface fracture and fiber matrix inter-relations. The tensile fractured samples were coated with platinum using a vacuum sputter coater prior to FESEM observation.

2.3.3. X-Ray Diffraction Analysis

The crystallinity was measured by using a MiniFlex 600 X-ray diffractometer (XRD) (Rigaku Co., Tokyo, Japan) at 40 kV and 10 mA at room temperature. Cu Kα radiation (λ = 1.54 Å) was used as the X-ray source while the diffraction angle was scanned at 2θ from 3° to 50° at a rate of 20°/min. The crystallinity index (CrI) was calculated and determined based on this equation:

$$CrI = (I_{total} - I_{am})/I_{total} \times 100 \tag{1}$$

which I_{total} and I_{am} are the intensity of highest peak in crystalline region and amorphous region, respectively [40,41]. For example, in this study the I_{total} for highest crystalline peak was approximately at 2θ = 22 while I_{am} was at about 2θ = 21, representing the peak of the amorphous point.

2.4. Statistical Analysis

Statistical analysis was conducted by using Statistical Analysis Software (SAS®) University Edition through one-way ANOVA and Duncan's Multiple range test at $p < 0.05$.

3. Results and Discussion

3.1. Mechanical Properties

The mechanical properties of the polymers and the bionanocomposites samples are as presented in Table 2. The tensile strength of MB-UHMWPE/0 reduced by almost half of the value exhibited by Neat-UHMWPE while no significant difference was observed in EM-UHMWPE/0 sample. Yet, an opposite trend was observed when incorporating 0.5 wt.% CNF into the polymer matrix. Addition of CNF through ethanol mixing reduced the tensile strength by 34% from 55.4 MPa (EM-UHMWPE/0) to 36.6 MPa (EM-UHMWPE/0.5) whereas by melt blending, the reduction was only 11% which was from 31.8 MPa (MB-UHMWPE/0) to 28.4 MPa (MB-UHMWPE/0.5). Significant improvements in yield strength and Young's modulus of MB-UHMWPE/0 sample by 15% and 25% were observed as compared to Neat-UHMWPE sample. These two mechanical parameters were also found increased in MB-UHMWPE/0.5 with 26% and 52% higher than Neat-UHMWPE. In other hand, yield strength, elongation, Young's modulus and toughness of ethanol mixed samples were almost similar to Neat-UHMWPE except for elongation and toughness of EM-UHMWPE/0.5 that was 33% and 53% lower, respectively. On contrary, the toughness of melt-blended samples were largely affected by the process in which 48% and 31% lower value were obtained as compared to Neat-UHMWPE. Nevertheless, the toughness of MB-UHMWPE/0.5 (168.4 ± 3.2 J/m^3) was 31% higher EM-UHMWPE/0.5 sample (116.5 ± 5.8 J/m^3).

Table 2. Effect of blending process and CNF addition on the mechanical properties of UHMWPE.

	Neat UHMWPE Properties	Mechanical Properties (% Difference Compared to Neat-UHMWPE)			
		MB-UHMWPE/0	MB-UHMWPE/0.5	EM-UHMWPE/0	EM-UHMWPE/0.5
Tensile strength (MPa)	62.1 ± 5.0	31.8 ± 3.1 * (−49)	28.4 ± 1.2 * (−54)	55.4 ± 3.3 (−11)	36.6 ± 1.1 * (−41)
Yield strength (MPa)	20.6 ± 0.6	23.8 ± 0.8 * (+15)	25.9 ± 0.5 * (+26)	20.0 ± 0.2 (−3)	20.3 ± 0.1 (−2)
Elongation (%)	691.1 ± 37.4	726.0 ± 45.2 (+5)	749.6 ± 2.0 * (+8)	694.2 ± 40.6 (0)	465.1 ± 4.2 * (−33)
Young's modulus (MPa)	267.9 ± 22.4	334.3 ± 14.4 * (+25)	407.9 ± 24.3 * (+52)	254.1 ± 17.4 (−5)	276.8 ± 0.2 (+3)
Toughness (J/m^3)	245.7 ± 21.3	128.8 ± 40.7 * (−48)	168.4 ± 3.2 * (−31)	224.3 ± 17.3 (−9)	116.5 ± 5.8 * (−53)

MB: melt blend, EM: ethanol mixed, UHMWPE: ultra-high molecular weight polyethylene. Asterisk (*) indicates significant difference of samples with neat UHMWPE ($p < 0.05$).

The mechanical properties of the samples can be explained based on the representative stress-strain curve in Figure 2. Neat-UHMWPE exhibited a tough and ductile behavior, which was in agreement with published reports on UHMWPE characteristics [42–45]. An identical profile was observed on EM-UHMWPE/0 indicated that subjecting UHMWPE to ethanol at room temperature could less likely affect the UHMWPE polymer structure evident from minimal changes in mechanical properties and similar hardening or cold drawing behavior of the samples in uniaxial tension. This can be supported by published reports stated that UHMWPE is inert and resilient to any reaction with acids, alkalis and organic solvent as well as biological reaction [37,46,47]. The curves in Figure 2 also showed that the incorporation of filler through ethanol mixing was ineffective that the common filler stiffening effect in polymer matrix was not observed. On the contrary, processing through melt blending enabled significant improvement of yield strength and Young's modulus which was very notable from the lower strain regime of stress-strain curves shown in the figure. It can be suggested that better UHMWPE-CNF adhesion was achieved through melt blending as compared to ethanol mixing. This is supported by published report stating that infiltration of melt polymer with addition of shear force during melt blending could results in smaller filler agglomerates and better interaction [48].

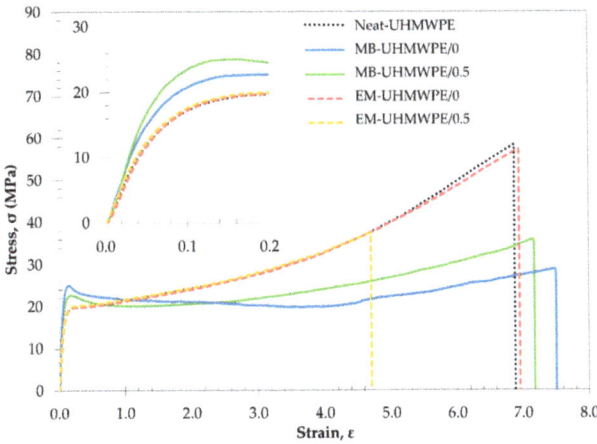

Figure 2. Representative engineering stress-strain curve of the neat UHMWPE and UHMWPE/CNF bionanocomposites showing differences in Young's modulus (inlet with lower strain regime), yield strength and fracture strain.

The information on toughness could also be obtained from the area under the stress-strain curve in Figure 2. As mentioned before, even though melt blending process reduced the toughness of the polymer (MB-UHMWPE/0 as compared to Neat-UHMWPE), the value when incorporating CNF (MB-UHMWPE/0.5) was apparently higher than the one produced through ethanol mixing (EM-UHMWPE/0.5). In respect to their respective polymer subjected to the same blending process without addition of CNF (MB-UHMWPE/0 and EM-UHMWPE/0), melt blending process enabled improvement in toughness of MB-UHMWPE/0.5 sample by 25%, whereas reduction by 48% was observed on ethanol mixed bionanocomposite sample (EM-UHMWPE/0.5). This proved the efficiency of melt blending for producing UHMWPE/CNF bionanocomposites as penetration of molten polymers into the fillers during high temperature processing could results in better mechanical interlock between the matrix and filler even without any chemical bonding presence [49,50].

In term of tensile strength, the reduced value in melt blending samples could be attributed to some molecular weight reduction and chain scission during melt processing. Although UHMWPE is thermally stable up to 400 °C, chain scission of the polymer may occur at lower temperature due to mechanically-initiated breaks such as shear forces [51–53], which is very likely to occur in melt blending. However, recrystallization of newly formed shorter chain could contribute to increased crystallinity and toughness besides enhancing diffusion of polymer for improved chain entanglement [54–56], providing better intrinsic properties of the composites. This was confirmed by the hardening behavior of samples as presented in Figure 3. Both melt blended samples (MB-UHMWPE/0 and MB-UHMWPE/0.5) exhibited lower hardening profile as compared to Neat-UHMPWE and ethanol mixed (EM-UHMPW/0 and EM-UHMWPE/0.5). According to Kurtz (2016), the hardening behavior of UHMWPE is sensitive to its molecular weight in which lower molecular weight exhibited lower hardening profile and vice versa [57,58]. Albeit this, the tensile strength of the CNF-incorporated melt blended sample (MB-UHMWPE/0.5) differed by only 8 MPa as compared to the one produced through ethanol mixing (EM-UHMWPE/0.5). It is also important to note that the tensile strength of MB-UHMWPE/0.5 sample which was 28.4 ± 1.0 MPa surpassed the minimum requirement of standard specification for fabricated UHMWPE for surgical implant, ASTM F648-14 which is 27 MPa [59].

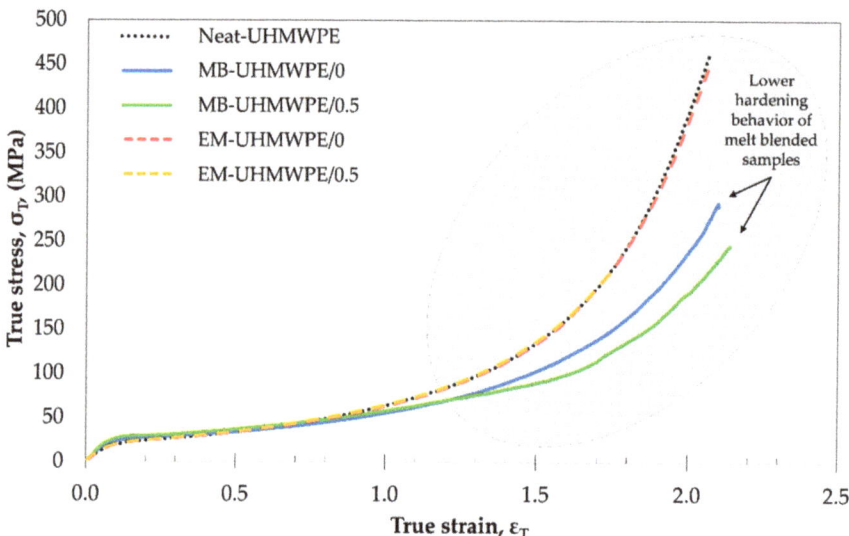

Figure 3. The true stress-strain curve of the neat UHMWPE and UHMWPE/CNF bionanocomposites showing hardening or cold drawing portion behavior in uniaxial tension.

3.2. Morphological Properties

The morphology of CNF and UHMWPE was observed under 1×10^5 and 200 times magnification using scanning electron microscope, accordingly (Figure 4). The average diameter for single nanocellulose fiber and UHMWPE resin was about 53 ± 9 nm and 96 ± 20 µm, respectively.

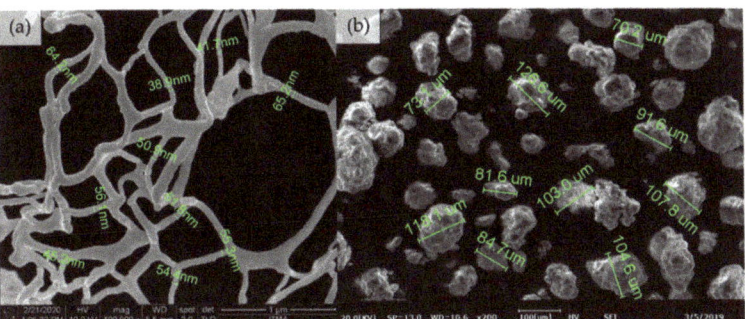

Figure 4. Scanning electron microscope images of (**a**) CNF and (**b**) UHMWPE.

Meanwhile, the Neat-UHMWPE film appeared white and subjecting the polymer to different blending process did not affect its color appearance (Figure 5). The absence of color changes in melt blended polymer without CNF (MB-UHMWPE/0) proved that processing the polymer (with combination of heat and mechanical stress) at temperature 150 °C did not lead to UHMWPE thermal decomposition. Absence of notable changes on the color appearance was also observed with incorporation of 0.5% CNF via ethanol mixing (EM-UHMWPE/0.5). In the meantime, processing the bionanocomposites with 0.5 wt.% CNF through melt blending resulted in yellowish MB-UHMWPE/0.5 sample, suggesting some effect of heat degradation on CNF. A combination of heat and mechanical force exerted on CNF during melt blending might be a contributing factor thus explained the different in MB-UHMWPE/0.5 appearance regardless of same amount of filler loading with EM-UHMWPE/0.5. According to Heggset et al. (2017) [60], one of indicators for nanocellulose decomposition at high temperature was color changes in which the percentage of color changes increased with increased of temperature (110 °C to 150 °C). The appearance of yellow/brownish/black from colorless/white was deemed associated to thermal oxidation in the presence of oxygen. In oxidation and hydrolysis reactions, aldehyde and carboxyl groups were formed and the resulted carbonyl group generated in the cellulose chains influenced its color appearance [61,62]. Formation of furan type compounds in thermal degradation of carbohydrates was also a responsible factor for the color changes in cellulose due to high temperature [63,64].

The fractured section of the MB-UHMWPE/0.5 and EM-UHMWPE/0.5 samples were observed to investigate the configuration of CNF filler in UHMWPE matrix by both processes. It is important to note that the diameter of CNF in MB-UHMWPE/0.5 increased to 71 ± 14 nm (Figure 6e) from 53.4 ± 9 nm of its initial size (Figure 4a), whereas the diameter of CNF in EM-UHMWPE/0.5 sample was about the same (52 ± 5 nm) (Figure 6f). Increased in CNF diameter size by melt blending was attributed to the CNF fast drying in triple screw kneading extruder aided by high temperature processing. Meanwhile, a notable appearance of mesh-like CNF can be observed covering the fractured polymer of EM-UHMWPE/0.5 sample (Figure 6b,d). In comparison, CNF in MB-UHMWPE/0.5 were embedded and fractured along with the polymer (Figure 6a,c).

Figure 5. The visual appearance of polymer and UHMWPE/CNF bionanocomposites samples produced through melt blending and ethanol mixing.

The explanation to the increased diameter of CNF during drying can be shown by the schematic representation in Figure 7, whereby it is shown that the removal of water molecules during drying leads to the formation of capillary forces exerted on the hydrophilic cellulose. This capillary effect causes the adjacent fibers to be drawn together and formed strong hydrogen bonding and hence, caused increase in diameter size [65–67]. Capillary tension increases with the increase in vapor pressure, which is affected by the temperature increment [68–70]. Processing bionanocomposites through melt blending at high temperature (150 °C) caused the increment in capillary tension resulting in bigger CNF diameter size as compared to the one processed through ethanol mixing at room temperature. Even though CNF drying was also occurred in ethanol mixing, the presence of alcohol reduced the interfacial tension of the liquid-water interface. This was due to the disruption of hydrogen bond network corresponded to the decrease of water-water hydrogen bond [71]. Ethanol also possesses lower surface tension at 25 °C (22 × 10^{-3} J/m^2) which is much lower than water at higher temperature of 100 °C (58.9 × 10^{-3} J/m^2) [72]. The use of alcohols such as ethanol, methanol and butanol in nanocellulose drying provides more interfibrillar distance than water only due to their higher molecular size as compared to water molecules. This is beneficial in reducing interfibrillar contacts and adhesion between nanocellulose fibers [73,74].

Mesh-like observations of the CNF in UHMWPE matrix bionanocomposites can be schematically viewed in Figure 8. The melt blending process enables mixing of CNF in molten state of UHMWPE thus allowing penetration of filler into matrix particle (Figure 8a). The penetration of filler into matrix and the shrinkage of the polymer during cooling developed mechanical interlocking between filler and matrix when molded, thus resulted in better mechanical properties [49,75]. Through ethanol mixing method, the CNF could not penetrate into the non-molten UHMWPE matrix. Instead, the UHMWPE particles resided in between the mesh-like CNF (Figure 8b) and combined through continuous mixing whilst solvent evaporated.

Figure 6. Fracture surface of UHMWPE/CNF prepared by (**a**,**c**,**e**) melt blending and (**b**,**d**,**f**) ethanol mixing at 10,000×, 50,000× and 100,000× magnification, respectively.

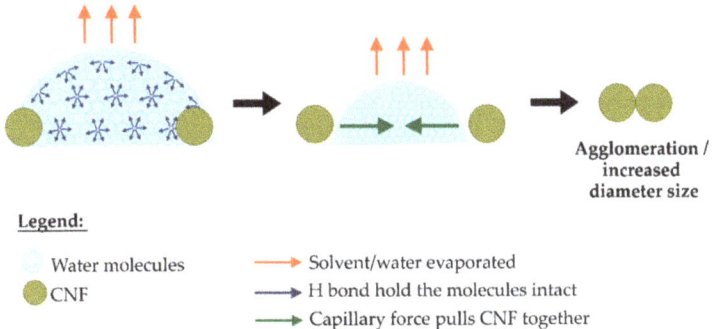

Figure 7. Schematic representation of CNF drying in bionanocomposites blending process.

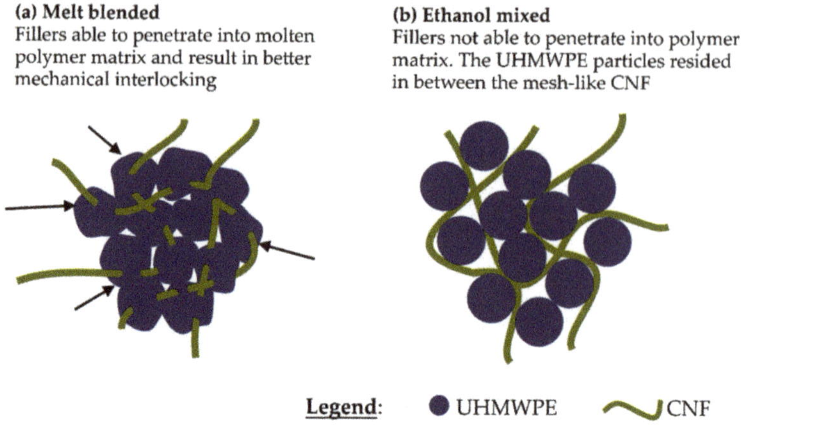

(a) Melt blended
Fillers able to penetrate into molten polymer matrix and result in better mechanical interlocking

(b) Ethanol mixed
Fillers not able to penetrate into polymer matrix. The UHMWPE particles resided in between the mesh-like CNF

Legend: ● UHMWPE ∼ CNF

Figure 8. Illustration of CNF interaction with UHMWPE matrix in (**a**) melt blended (MB-UHMWPE/0.5) and (**b**) ethanol mixed (EM-UHMWPE/0.5) samples.

3.3. Crystallinity Evaluation

The x-ray diffraction pattern of UHMWPE and CNF are presented in Figure 9. The calculated crystallinity index of the two materials in respect to their respective total peak height were 78% and 50%, in the same order. The UHMWPE peak was seen to exhibit sharp increase of crystalline peak befitted long and unbranched polymer crystalline structure and arrangement [76,77], while CNF peak shows semi crystalline pattern indicating existence of crystalline and amorphous region of the cellulose chains [78,79]. Additionally, all polymer and bionanocomposites samples exhibited similar pattern with two prominent diffraction peaks centered at around 22.0° and 24.4° of 2θ, which correspond to (110) and (200) reflection of polyethylene in orthorhombic phase [80–82]. The diffraction peak of the filler could not be observed due to low percent loading [81,83] and overlapped peak of UHMWPE with CNF at around 22° in 2θ, which was in agreement with reported studies involving nanocellulose filler in polyethylene matrix [40,84]. Reduction in amorphous region was observed between 18° to 21° for samples fabricated through melt blending suggesting an improvement in chain entanglement and improved crystallinity stemmed from chain scission occurrence [54,55].

MB-UHMWPE/0 and EM-UHMWPE/0 had crystallinity of 83% and 78, respectively. When incorporated with 0.5% CNF, the melt-blended MB-UHMWPE/0.5 had a slight increment in crystallinity to 86%, while the crystallinity of ethanol-mixed EM-UHMWPE/0.5 remained. The increment of crystallinity for melt-blended samples with and without CNF is in correlation with the possible occurrence of polymer chain scission. Even in temperature lower than its decomposition temperature, chain scissioning is possible due to mechanically initiated breaks caused by other factors including shear forces [51–53]. According to Fu and co-workers, chain scissioning of UHMWPE lead to recrystallization of newly formed shorter chain hence contributed to increased [54–56]. The formation of shorter chain also enhanced the diffusion of polymer and improved the chain entanglement resulting in higher crystallinity. On another note, the addition of cellulose nanomaterials also may act as nucleating agent, which has been previously [30,40,85–87] As a nucleating agent, CNF as natural fiber could induce more formation of crystallites in the polymer matrix [29].

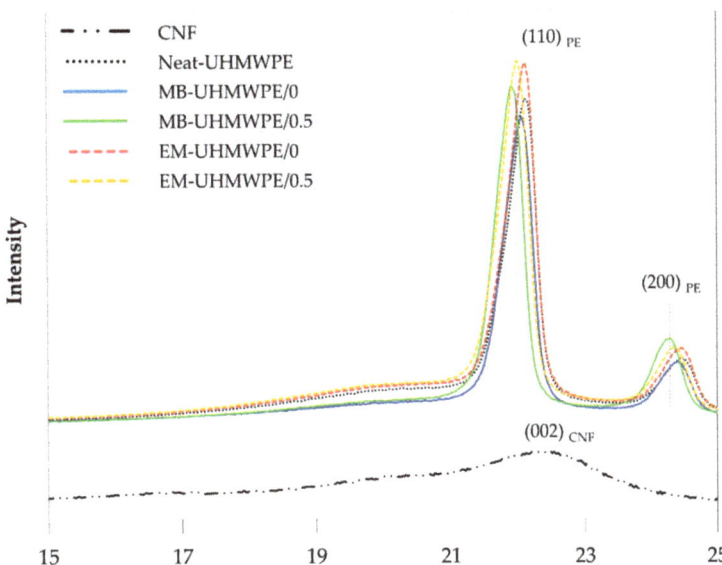

Figure 9. X-ray diffractogram of UHMWPE/CNF bionanocomposites.

4. Conclusions

Comparison in mechanical properties of UHMWPE/CNF bionanocomposites fabricated through non-melt blending (ethanol mixing) and melt blending process revealed that bionanocomposites from the latter method had better properties. Significantly higher yield strength, elongation at break, Young's modulus, toughness and crystallinity by 28%, 61%, 47%, 45% and 11%, respectively, were achieved through melt blending as compared to ethanol mixing. A better mechanical interlocking between UHMWPE and CNF was seen through FE-SEM micrographs indicating a good blending of CNF with the polymer matrix, assisted by the use of elevated temperature and kneading. Lower tensile strength (22%) of melt-blended UHMWPE/CNF was recorded, indicating the occurrence of chain-scission during melt blending as evidenced by the reduction in strain hardening. Nevertheless, the tensile strength value surpassed the minimum requirement of standard specification for fabricated UHMWPE in surgical implant application. The results demonstrated melt blending as a better fabrication process for producing UHMWPE/CNF compared to ethanol mixing, with the advantage of being easily scalable for larger scale processing.

Author Contributions: Conceptualization, H.A. and N.S.S.; data curation, H.A. and Y.S.; formal analysis, N.S.S.; funding acquisition, H.A. and Y.A.; investigation, N.S.S and T.A.T.Y.-A.; methodology, H.A., N.S.S. and T.A.T.Y.-A.; project administration, H.A.; resources, H.A., N.A.I., Y.A. and Y.S.; software, N.S.S.; supervision, H.A., Y.S., M.J., P.M.T. and N.A.I.; validation, H.A.; writing—original draft, N.S.S.; writing—review and editing, H.A. and M.J. All authors have read and agreed to the published version of the manuscript.

Funding: This research was funded by Ministry of Higher Education (MOHE), Malaysia through HICOE research grant (Vote No.:6369111).

Acknowledgments: The authors would like to thank Ministry of Higher Education (MOHE, MALAYSIA) for providing fund for this project through Higher Institution Centre of Excellence (HICoE) research grant as well as Universiti Putra Malaysia (UPM, MALAYSIA) and Japan Student Services Organization (JASSO, JAPAN) for provision of scholarship to the first author.

Conflicts of Interest: The authors declare no conflict of interest.

References

1. Manley, M.T. Highly Cross-Linked and Annealed UHMWPE. In *UHMWPE Biomaterials Handbook: Ultra High Molecular Weight Polyethylene in Total Joint Replacement and Medical Devices*, 3rd ed.; Elsevier Inc.: Waltham, MA, USA, 2016; pp. 274–292.
2. Khalil, Y.; Kowalski, A.; Hopkinson, N. Influence of energy density on flexural properties of laser-sintered UHMWPE. *Addit. Manuf.* **2016**, *10*, 67–75. [CrossRef]
3. Paxton, N.C.; Allenby, M.C.; Lewis, P.M.; Woodruff, M.A. Biomedical applications of polyethylene. *Eur. Polym. J.* **2019**, *118*, 412–428. [CrossRef]
4. Chukov, D.I.; Stepashkin, A.A.; Gorshenkov, M.V.; Tcherdyntsev, V.V.; Kaloshkin, S.D. Surface modification of carbon fibers and its effect on the fiber-matrix interaction of UHMWPE based composites. *J. Alloys Compd.* **2014**, *586*, S459–S463. [CrossRef]
5. Li, Y.; He, H.; Huang, B.; Zhou, L.; Yu, P.; Lv, Z. In situ fabrication of cellulose nanocrystal-silica hybrids and its application in UHMWPE: Rheological, thermal, and wear resistance properties. *Polym. Compos.* **2017**, 1–13. [CrossRef]
6. Wang, J.; Cao, C.; Yu, D.; Chen, X. Deformation and Stress Response of Carbon Nanotubes/UHMWPE Composites under Extensional-Shear Coupling Flow. *Appl. Compos. Mater.* **2018**, *25*, 35–43. [CrossRef]
7. Raghuvanshi, S.K.; Ahmad, B.; Siddhartha; Srivastava, A.K.; Krishna, J.B.M.; Wahab, M.A. Effect of gamma irradiation on the optical properties of UHMWPE (Ultra-high-molecular-weight-polyethylene) polymer. *Nucl. Instrum. Methods Phys. Res. B* **2012**, *271*, 44–47. [CrossRef]
8. Laux, K.A.; Sue, H.J.; Montoya, A.; Bremner, T. Wear behavior of polyaryletherketones under multi-directional sliding and fretting conditions. *Tribol. Lett.* **2015**, *58*, 1–13. [CrossRef]
9. Sharip, N.S.; Ariffin, H. Polymeric Composites for Joint Replacement. In *Nanostructured Polymer Composites for Biomedical Applications*; Elsevier: Amsterdam, The Netherlands, 2019; pp. 385–404.
10. Baena, J.; Wu, J.; Peng, Z. Wear performance of UHMWPE and reinforced UHMWPE composites in arthroplasty applications: A review. *Lubricants* **2015**, *3*, 413–436. [CrossRef]
11. Wang, S.; Feng, Q.; Sun, J.; Gao, F.; Fan, W.; Zhang, Z.; Li, X.; Jiang, X. Nanocrystalline cellulose improves the biocompatibility and reduces the wear debris of ultrahigh molecular weight polyethylene via weak binding. *ACS Nano* **2016**, *10*, 298–306. [CrossRef]
12. Choudhury, N.R.; Kannan, A.G.; Dutta, N.K. Novel nanocomposites and hybrids for lubricating coating applications. In *Tribology of Polymeric Nanocomposites: Friction and Wear of Bulk Materials and Coatings*; Elsevier: Oxford, UK, 2008; Volume 55, pp. 501–542.
13. Liu, Y.; Sinha, S.K. Wear performances and wear mechanism study of bulk UHMWPE composites with nacre and CNT fillers and PFPE overcoat. *Wear* **2013**, *300*, 44–54. [CrossRef]
14. Poonguzhali, R.; Khaleel Basha, S.; Sugantha Kumari, V. Novel asymmetric chitosan/PVP/nanocellulose wound dressing: In vitro and in vivo evaluation. *Int. J. Biol. Macromol.* **2018**, *112*, 1300–1309. [CrossRef]
15. Lu, T.; Li, Q.; Chen, W.; Yu, H. Composite aerogels based on dialdehyde nanocellulose and collagen for potential applications as wound dressing and tissue engineering scaffold. *Compos. Sci. Technol.* **2014**, *94*, 132–138. [CrossRef]
16. Poonguzhali, R.; Basha, S.K.; Kumari, V.S. Synthesis and characterization of chitosan-PVP-nanocellulose composites for in-vitro wound dressing application. *Int. J. Biol. Macromol.* **2017**, *105*, 111–120. [CrossRef]
17. Singh, B.N.; Panda, N.N.; Mund, R.; Pramanik, K. Carboxymethyl cellulose enables silk fibroin nanofibrous scaffold with enhanced biomimetic potential for bone tissue engineering application. *Carbohydr. Polym.* **2016**, *151*, 335–347. [CrossRef]
18. Bhattacharya, M.; Malinen, M.M.; Lauren, P.; Lou, Y.R.; Kuisma, S.W.; Kanninen, L.; Lille, M.; Corlu, A.; GuGuen-Guillouzo, C.; Ikkala, O.; et al. Nanofibrillar cellulose hydrogel promotes three-dimensional liver cell culture. *J. Control. Release* **2012**, *164*, 291–298. [CrossRef]
19. Afrin, S.; Karim, Z. Nanocellulose as Novel Supportive Functional Material for Growth and Development of Cells. *Cell Dev. Biol.* **2015**, *4*, 154.
20. Bhandari, J.; Mishra, H.; Mishra, P.K.; Wimmer, R.; Ahmad, F.J.; Talegaonkar, S. Cellulose nanofiber aerogel as a promising biomaterial for customized oral drug delivery. *Int. J. Nanomed.* **2017**, *12*, 2021–2031. [CrossRef]
21. Kolakovic, R.; Peltonen, L.; Laukkanen, A.; Hirvonen, J.; Laaksonen, T. Nanofibrillar cellulose films for controlled drug delivery. *Eur. J. Pharm. Biopharm.* **2012**, *82*, 308–315. [CrossRef]
22. Kolakovic, R.; Peltonen, L.; Laukkanen, A.; Hellman, M.; Laaksonen, P.; Linder, M.B.; Hirvonen, J.; Laaksonen, T. Evaluation of drug interactions with nanofibrillar cellulose. *Eur. J. Pharm. Biopharm.* **2013**, *85*, 1238–1244. [CrossRef]
23. Masruchin, N.; Park, B.D.; Causin, V. Dual-responsive composite hydrogels based on TEMPO-oxidized cellulose nanofibril and poly(N-isopropylacrylamide) for model drug release. *Cellulose* **2018**, *25*, 485–502. [CrossRef]
24. O'Donnell, K.L.; Oporto-Velásquez, G.S.; Comolli, N. Evaluation of Acetaminophen Release from Biodegradable Poly (Vinyl Alcohol) (PVA) and Nanocellulose Films Using a Multiphase Release Mechanism. *Nanomaterials* **2020**, *10*, 301. [CrossRef] [PubMed]
25. Rojas, J.; Bedoya, M.; Ciro, Y. Current trends in the production of cellulose nanoparticles and nanocomposites for biomedical applications. In *Cellulose-Fundamental Aspects and Current Trends*; IntechOpen: London, UK, 2015; pp. 193–228.
26. Sharip, N.S.; Ariffin, H. Cellulose nanofibrils for biomaterial applications. *Mater. Today Proc.* **2019**, *16*, 1959–1968. [CrossRef]
27. Abdul Khalil, H.P.; Bhat, A.H.; Abu Bakar, A.; Tahir, P.M.; Zaidul, I.S.; Jawaid, M. Cellulosic nanocomposites from natural fibers for medical applications: A review. In *Handbook of Polymer Nanocomposites. Processing, Performance and Application: Volume C: Polymer Nanocomposites of Cellulose Nanoparticles*; Springer: Berlin/Heidelberg, Germany, 2015; pp. 475–511.
28. Lin, N.; Dufresne, A. Nanocellulose in biomedicine: Current status and future prospect. *Eur. Polym. J.* **2014**, *59*, 302–325. [CrossRef]
29. Dufresne, A. Cellulose nanomaterial reinforced polymer nanocomposites. *Curr. Opin. Colloid Interface Sci.* **2017**, *29*, 1–8. [CrossRef]

30. Herrera, N.; Mathew, A.P.; Oksman, K. Plasticized polylactic acid/cellulose nanocomposites prepared using melt-extrusion and liquid feeding: Mechanical, thermal and optical properties. *Compos. Sci. Technol.* **2015**, *106*, 149–155. [CrossRef]
31. Mondal, S. Review on Nanocellulose Polymer Nanocomposites. *Polym. Plast. Technol. Eng.* **2018**, *57*, 1377–1391. [CrossRef]
32. Jonoobi, M.; Harun, J.; Mathew, A.P.; Oksman, K. Mechanical properties of cellulose nanofiber (CNF) reinforced polylactic acid (PLA) prepared by twin screw extrusion. *Compos. Sci. Technol.* **2010**, *70*, 1742–1747. [CrossRef]
33. Oksman, K.; Mathew, A.P.A.P.A.P.; Bismarck, A.; Rojas, O.; Sain, M. Melt compounding process of cellulose nanocomposites. In *Handbook of Green Materials*; World Scientific Publishing: Singapore, 2014; pp. 53–68.
34. Oksman, K.; Aitomäki, Y.; Mathew, A.P.; Siqueira, G.; Zhou, Q.; Butylina, S.; Tanpichai, S.; Zhou, X.; Hooshmand, S. Review of the recent developments in cellulose nanocomposite processing. *Compos. Part. A Appl. Sci. Manuf.* **2016**, *83*, 2–18. [CrossRef]
35. Zhang, H.; Liang, Y. Extrusion Processing of Ultra-High Molecular Weight Polyethylene. In *Extrusion of Metals, Polymers and Food Products*; IntechOpen: London, UK, 2018.
36. Hikosaka, M.; Tsukijima, K.; Rastogi, S.; Keller, A. Equilibrium triple point pressure and pressure-temperature phase diagram of polyethylene. *Polymer* **1992**, *33*, 2502–2507. [CrossRef]
37. Macuvele, D.L.P.; Nones, J.; Matsinhe, J.V.; Lima, M.M.; Soares, C.; Fiori, M.A.; Riella, H.G. Advances in ultra high molecular weight polyethylene/hydroxyapatite composites for biomedical applications: A brief review. *Mater. Sci. Eng. C* **2017**, *76*, 1248–1262. [CrossRef]
38. Sharip, N.S.; Ariffin, H.; Andou, Y.; Shirosaki, Y.; Bahrin, E.K.; Jawaid, M.; Tahir, P.M.; Ibrahim, N.A. Process Optimization of Ultra-High Molecular Weight Polyethylene/Cellulose Nanofiber Bionanocomposites in Triple Screw Kneading Extruder by Response Surface Methodology. *Molecules* **2020**, *25*, 4498. [CrossRef]
39. Kurtz, S.M.; Muratoglu, O.K.; Evans, M.; Edidin, A.A. Advances in the processing, sterilization, and crosslinking of ultra-high molecular weight polyethylene for total joint arthroplasty. *Biomaterials* **1999**, *20*, 1659–1688. [CrossRef]
40. Yasim-Anuar, T.A.T.; Ariffin, H.; Norrrahim, M.N.F.; Hassan, M.A.; Tsukegi, T.; Nishida, H. Sustainable one-pot process for the production of cellulose nanofiber and polyethylene/cellulose nanofiber composites. *J. Clean. Prod.* **2019**, *207*, 590–599. [CrossRef]
41. Segal, L.; Creely, J.J.J.; Martin, A.E.E.; Conrad, C.M.M. An Empirical Method for Estimating the Degree of Crystallinity of Native Cellulose Using the X-Ray Diffractometer. *Text. Res. J.* **1959**, *29*, 786–794. [CrossRef]
42. Mohseni Taromsari, S.; Salari, M.; Bagheri, R.; Faghihi Sani, M.A. Optimizing tribological, tensile & in-vitro biofunctional properties of UHMWPE based nanocomposites with simultaneous incorporation of graphene nanoplatelets (GNP) & hydroxyapatite (HAp) via a facile approach for biomedical applications. *Compos. Part. B Eng.* **2019**, *175*, 107181.
43. Mohagheghian, I.; McShane, G.J.; Stronge, W.J. Impact perforation of monolithic polyethylene plates: Projectile nose shape dependence. *Int. J. Impact Eng.* **2015**, *80*, 162–176. [CrossRef]
44. Spiegelberg, S.; Kozak, A.; Braithwaite, G. Characterization of Physical, Chemical, and Mechanical Properties of UHMWPE. In *UHMWPE Biomaterials Handbook: Ultra High Molecular Weight Polyethylene in Total Joint Replacement and Medical Devices*, 3rd ed.; Elsevier Inc.: Waltham, MA, USA, 2016; pp. 531–552.
45. Lozano-Sánchez, L.M.; Bagudanch, I.; Sustaita, A.O.; Iturbe-Ek, J.; Elizalde, L.E.; Garcia-Romeu, M.L.; Elías-Zúñiga, A. Single-point incremental forming of two biocompatible polymers: An insight into their thermal and structural properties. *Polymers* **2018**, *10*, 391. [CrossRef]
46. Salari, M.; Taromsari, S.M.; Bagheri, R.; Ali, M.; Sani, F.; Mohseni Taromsari, S.; Bagheri, R.; Faghihi Sani, M.A.; Taromsari, S.M.; Bagheri, R.; et al. Improved wear, mechanical, and biological behavior of UHMWPE-HAp-zirconia hybrid nanocomposites with a prospective application in total hip joint replacement. *J. Mater. Sci.* **2019**, *54*, 4259–4276. [CrossRef]
47. Kurtz, S.M. A Primer on UHMWPE. In *UHMWPE Biomaterials Handbook: Ultra High Molecular Weight Polyethylene in Total Joint Replacement and Medical Devices*, 3rd ed.; Elsevier Inc.: Waltham, MA, USA, 2016; pp. 1–6.
48. Dantas de Oliveira, A.; Augusto Gonçalves Beatrice, C. Polymer nanocomposites with different types of nanofiller. In *Nanocomposites—Recent Evolutions*; IntechOpen: London, UK, 2019; pp. 103–128.
49. Ahmad, M.; Wahit, M.U.; Kadir, M.R.A.; Dahlan, K.Z.M.; Uzir Wahit, M.; Abdul Kadir, M.R.; Mohd Dahlan, K.Z.; Wahit, M.U.; Kadir, M.R.A.; Dahlan, K.Z.M. Mechanical, rheological, and bioactivity properties of ultra high-molecular-weight polyethylene bioactive composites containing polyethylene glycol and hydroxyapatite. *Sci. World J.* **2012**, *2012*, 1–13. [CrossRef]
50. Wang, M. Developing bioactive composite materials for tissue replacement. *Biomaterials* **2003**, *24*, 2133–2151. [CrossRef]
51. Gol'dberg, V.M.; Zaikov, G.E. Kinetics of thermooxidation of polymers in processing. In *Polymer Yearbook*; Pethrick, R.A., Ed.; Harwood Academic Publishers GmbH: Glasgow, UK, 1989; pp. 87–124.
52. El'darov, E.G.; Mamedov, F.V.; Gol'dberg, V.M.; Zaikov, G.E. A kinetic model of polymer degradation during extrusion. *Polym. Degrad. Stab.* **1996**, *51*, 271–279. [CrossRef]
53. Gol'dberg, V.M.; Zaikov, G.E. Kinetics of mechanical degradation in melts under model conditions and during processing of polymers-A review. *Polym. Degrad. Stab.* **1987**, *19*, 221–250. [CrossRef]
54. Fu, J.; Doshi, B.N.; Oral, E.; Muratoglu, O.K. High temperature melted, radiation cross-linked, vitamin e stabilized oxidation resistant UHMWPE with low wear and high impact strength. *Polymer* **2013**, *54*, 199–209. [CrossRef]
55. Fu, J.; Ghali, B.W.; Lozynsky, A.J.; Oral, E.; Muratoglu, O.K. Ultra high molecular weight polyethylene with improved plasticity and toughness by high temperature melting. *Polymer* **2010**, *51*, 2721–2731. [CrossRef]
56. Fu, J.; Ghali, B.W.; Lozynsky, A.J.; Oral, E.; Muratoglu, O.K. Wear resistant UHMWPE with high toughness by high temperature melting and subsequent radiation cross-linking. *Polymer* **2011**, *52*, 1155–1162. [CrossRef]

57. Kurtz, S.M. From Ethylene Gas to UHMWPE Component: The Process of Producing Orthopedic Implants. In *UHMWPE Biomaterials Handbook: Ultra High Molecular Weight Polyethylene in Total Joint Replacement and Medical Devices: Third Edition*; Elsevier Inc.: Waltham, MA, USA, 2016; pp. 7–20.
58. Haward, R.N. Strain hardening of high density polyethylene. *J. Polym. Sci. Part. B Polym. Phys.* **2007**, *45*, 1090–1099. [CrossRef]
59. *Standard Specification for Ultra-High-Molecular-Weight Polyethylene Powder and Fabricated Form for Surgical Implants*; ASTM F648-14; ASTM International: West Conshohocken, PA, USA, 2014.
60. Heggset, E.B.; Chinga-Carrasco, G.; Syverud, K. Temperature stability of nanocellulose dispersions. *Carbohydr. Polym.* **2017**, *157*, 114–121. [CrossRef]
61. Yatagai, M.; Zeronian, S.H. Effect of ultraviolet light and heat on the properties of cotton cellulose. *Cellulose* **1994**, *1*, 205–214. [CrossRef]
62. Łojewska, J.; Missori, M.; Lubańska, A.; Grimaldi, P.; Zięba, K.; Proniewicz, L.M.; Congiu Castellano, A. Carbonyl groups development on degraded cellulose. Correlation between spectroscopic and chemical results. *Appl. Phys. A* **2007**, *89*, 883–887. [CrossRef]
63. Nie, S.P.; Huang, J.G.; Hu, J.L.; Zhang, Y.N.; Wang, S.; Li, C.; Marcone, M.F.; Xie, M.Y. Effect of pH, temperature and heating time on the formation of furan from typical carbohydrates and ascorbic acid. *J. Food Agric. Environ.* **2013**, *11*, 121–125.
64. Sharip, N.S.; Ariffin, H.; Hassan, M.A.; Nishida, H.; Shirai, Y. Characterization and application of bioactive compounds in oil palm mesocarp fiber superheated steam condensate as an antifungal agent. *RSC Adv.* **2016**, *6*, 84672–84683. [CrossRef]
65. Peng, Y.; Gardner, D.J.; Han, Y. Drying cellulose nanofibrils: In search of a suitable method. *Cellulose* **2012**, *19*, 91–102. [CrossRef]
66. Zimmermann, M.V.G.; Borsoi, C.; Lavoratti, A.; Zanini, M.; Zattera, A.J.; Santana, R.M.C. Drying techniques applied to cellulose nanofibers. *J. Reinf. Plast. Compos.* **2016**, *35*, 682–697. [CrossRef]
67. Ras, R.H.A.; Tian, X.; Bayer, I.S. Superhydrophobic and Superoleophobic Nanostructured Cellulose and Cellulose Composites. In *Handbook of Nanocellulose and Cellulose Nanocomposites*; Kargarzadeh, H., Ahmad, I., Thomas, S., Dufresne, A., Eds.; Wiley-VCH Verlag GmbH & Co. KGaA: Weinheim, Germany, 2017; pp. 731–760.
68. Yaws, C.L.; Satyro, M.A. Vapor pressure—Organic compounds. In *The Yaws Handbook of Vapor Pressure*; Elsevier: Oxford, UK, 2015; pp. 1–314.
69. Wiener, H. Vapor pressure-temperature relationships among the branched paraffin hydrocarbons. *J. Phys. Colloid Chem.* **1948**, *52*, 425–430. [CrossRef]
70. Petrucci, R.H.; Herring, F.G.; Bissonnette, C.; Madura, J.D. *General Chemistry: Principles and Modern Applications*; Pearson: London, UK, 2017.
71. Biscay, F.; Ghoufi, A.; Malfreyt, P. Surface tension of water-alcohol mixtures from Monte Carlo simulations. *J. Chem. Phys.* **2011**, *134*, 044709. [CrossRef]
72. Petrucci, R.H.; Herring, F.G.; Bissonnette, C.; Madura, J.D. Intermolecular forces: Liquids and solids. In *General Chemistry: Principles and Modern Applications*; Pearson: London, UK, 2017.
73. Hanif, Z.; Jeon, H.; Tran, T.H.; Jegal, J.; Park, S.-A.; Kim, S.-M.; Park, J.; Hwang, S.Y.; Oh, D.X. Butanol-mediated oven-drying of nanocellulose with enhanced dehydration rate and aqueous re-dispersion. *J. Polym. Res.* **2018**, *25*, 191. [CrossRef]
74. Sehaqui, H.; Zhou, Q.; Berglund, L.A. High-porosity aerogels of high specific surface area prepared from nanofibrillated cellulose (NFC). *Compos. Sci. Technol.* **2011**, *71*, 1593–1599. [CrossRef]
75. Baliga, B.R.; Reddy, P.; Pandey, P. Synthesis and Wear Characterization of CNF-UHMWPE Nanocomposites for Orthopaedic Applications. *Mater. Today Proc.* **2018**, *5*, 20842–20848. [CrossRef]
76. Fu, J.; Jin, Z.-M.; Wang, J.-W. Highly Crosslinked UHMWPE for Joint Implants. In *UHMWPE Biomaterials for Joint Implants: Structures, Properties and Clinical Performance*; Springer: Singapore, 2019; pp. 21–68.
77. Véronique Migonney Materials Used in Biomaterial Applications. In *Biomaterials*; John Wiley & Sons: London, UK, 2014.
78. Sepahvand, S.; Jonoobi, M.; Ashori, A.; Gauvin, F.; Brouwers, H.J.H.; Yu, Q. Surface modification of cellulose nanofiber aerogels using phthalimide. *Polym. Compos.* **2020**, *41*, 219–226. [CrossRef]
79. Pennells, J.; Godwin, I.D.; Amiralian, N.; Martin, D.J. Trends in the production of cellulose nanofibers from non-wood sources. *Cellulose* **2020**, *27*, 575–593. [CrossRef]
80. Myasnikova, L.; Baidakova, M.; Drobot'ko, V.; Ivanchev, S.; Ivan'kova, E.; Radovanova, E.; Yagovkina, M.; Marikhin, V.; Zubavichus, Y.; Dorovatovskii, P. The Crystalline Structure of Nascent Ultra High Molecular Weight Single Particles and Its Change on Heating, as Revealed by in-situ Synchrotron Studies. *J. Macromol. Sci. Part B* **2019**, *58*, 847–859. [CrossRef]
81. Ortiz-Hernández, R.; Ulloa-Castillo, N.A.; Diabb-Zavala, J.M.; la Vega, A.E.-D.; Islas-Urbano, J.; Villela-Castrejón, J.; Elías-Zúñiga, A. Advances in the Processing of UHMWPE-TiO2 to Manufacture Medical Prostheses via SPIF. *Polymers* **2019**, *11*, 2022. [CrossRef] [PubMed]
82. Xu, H.-J.; An, M.-F.; Lv, Y.; Wang, Z.-B.; Gu, Q. Characterization of Structural Knot Distributions in UHMWPE Fibers. *Chin. J. Polym. Sci.* **2016**, *34*, 606–615. [CrossRef]
83. Sui, G.; Zhong, W.H.H.; Ren, X.; Wang, X.Q.Q.; Yang, X.P.P. Structure, mechanical properties and friction behavior of UHMWPE/HDPE/carbon nanofibers. *Mater. Chem. Phys.* **2009**, *115*, 404–412. [CrossRef]
84. Mannan, T.M.; Soares, J.B.P.; Berry, R.M.; Hamad, W.Y. In-situ production of polyethylene/cellulose nanocrystal composites. *Can. J. Chem. Eng.* **2016**, *94*, 2107–2113. [CrossRef]
85. Farahbakhsh, N.; Roodposhti, P.S.; Ayoub, A.; Venditti, R.A.; Jur, J.S. Melt extrusion of polyethylene nanocomposites reinforced with nanofibrillated cellulose from cotton and wood sources. *J. Appl. Polym. Sci.* **2015**, *132*, 41857. [CrossRef]

86. Kargarzadeh, H.; Huang, J.; Lin, N.; Ahmad, I.; Mariano, M.; Dufresne, A.; Thomas, S.; Gałęski, A. Recent developments in nanocellulose-based biodegradable polymers, thermoplastic polymers, and porous nanocomposites. *Prog. Polym. Sci.* **2018**, *87*, 197–227. [CrossRef]
87. Yasim-Anuar, T.A.T.; Ariffin, H.; Norrrahim, M.N.F.; Hassan, M.A.; Andou, Y.; Tsukegi, T.; Nishida, H. Well-Dispersed Cellulose Nanofiber in Low Density Polyethylene Nanocomposite by Liquid-Assisted Extrusion. *Polymers* **2020**, *12*, 927. [CrossRef]

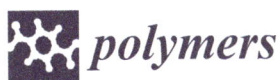

Article

Investigation on the Synthesis Process of Bromoisobutyryl Esterified Starch and Its Sizing Properties: Viscosity Stability, Adhesion and Film Properties

Wei Li [1,*], Jie Wu [1], Zhengqiao Zhang [1], Lanjuan Wu [1] and Yuhao Lu [2]

1. College of Textiles and Garments, Anhui Polytechnic University, Wuhu 241000, Anhui, China; w364363324@sina.com (J.W.); 1914530057zzq@sina.com (Z.Z.); christinewu77@sina.com (L.W.)
2. Hefei Safood Starch Co. Ltd., Hefei 230000, China; yhaoluu@163.com
* Correspondence: liw@ahpu.edu.cn or fangzhiliweiwu@sina.com; Tel.: +86-183-5531-6328

Received: 28 October 2019; Accepted: 19 November 2019; Published: 25 November 2019

Abstract: To confirm the suitable synthesis process parameters of preparing bromoisobutyryl esterified starch (BBES), the influences of the synthesis process parameters—amount of 2-bromoisobutyryl bromide (BIBB), amount of catalyst (DMAP), reaction temperature and reaction time—upon the degree of substitution (DS) were investigated. Then, to produce a positive effect on the properties of graft copolymers of BBES prepared in the near future, a series of BBES samples were successfully prepared, and their sizing properties, such as apparent viscosity and viscosity stability, adhesion, and film properties, were examined. The BBES granules were characterized by Fourier transform infra-red spectroscopy and scanning electron microscopy. The adhesion was examined by determining the bonding forces of the sized polylactic acid (PLA) and polyester roving. The film properties were investigated in terms of tensile strength, breaking elongation, degree of crystallinity, and cross-section analysis. The results showed that a suitable synthesis process of BBES was: reaction time of 24 h, reaction temperature of 40 °C, and 0.23 in the molar ratio of 4-dimethylaminopyridine to 2-bromoisobutyryl bromide. The bromoisobutyryl esterification played the important roles in the properties of the starch, such as paste stabilities of above 85% for satisfying the requirement in the stability for sizing, improvement of the adhesion to polylactic acid and polyester fibers, and reduction of film brittleness. With rising DS, bonding forces of BBES to the fibers increased and then decreased. BBES (DS = 0.016) had the highest force and breaking elongation of the film. Considering the experimental results, BBES (DS = 0.016) showed potential in the PLA and polyester sizing, and will not lead to a negative influence on the properties of graft copolymers of BBES.

Keywords: bromoisobutyryl esterification; cornstarch; synthesis process; past stability; adhesion; film properties

1. Introduction

Starch as one of the most abundantly occurring organic polymers in nature [1], and also as a renewable [2], biodegradable [3], and economical [4–6] raw material, has widespread industrial applications [7,8]. Starch has a general formula of $(C_6H_{10}O_5)_n$, including two types of D-glucan macromolecules, i.e., linear amylose and branched amylopectin [9,10], and consists of interconnected anhydroglucose units, each of which includes three hydroxyls and are linked together by a-D-glucosidic bonds [11]. It can be modified enzymatically, physically or chemically to meet various requirements. Granule forms of starch can remain unaffected during the chemical modification where hydroxyls are converted to other functionalities [12]. Generally, chemical modification of starch can be performed by esterification reaction [13] or grafting reaction with various monomers [14,15]. The most common method for synthesizing grafted starch copolymers is radical polymerization, but this method commonly suffers from lack of control of the graft density and length, and easily forms unattached homopolymer. Recently, controlled radical polymerization such as electron transfer atom transfer radical polymerization (ARGET ATRP) has received increasing attention [16–18]. The significant advantages of this graft copolymerization technique include controlled graft density and length, narrow molecular weight distribution, and the formation of no homopolymer impurities. However, there is little research on this polymerization method being applied for the grafting of starch to modify the starch. Prior to the graft copolymerization by ARGET ATRP, the starch is modified to form starch macroinitiator for further conducting the ARGET ATRP technique. Currently, one of the most commonly used reagents for modifying the starch to prepare the starch macroinitiator is 2-bromoisobutyryl bromide [2,19,20].

As is well known, starch has widespread applications in the textile field as a sizing agent and n paper making as a surface-sizing agent [21,22]. However, the use of native starches in these applications is limited due to low resistance to shear and high temperatures, as well as high tendency towards retrogradation [23]. This low resistance can result in the degradation of starch macromolecules, producing the instability of paste viscosity [24,25]. The instable viscosity can lead to the instability of size pick-up for warp sizing or surface sizing, thereby exerting an adverse effect on sizing and subsequently reducing the quality of fibrous goods. The large tendency towards retrogradation not only makes the paste microheterogeneous, thereby inducing an incomplete wetting and spreading out of the paste onto the fiber surfaces, leading to a low adhesion of starch to fibers, but also results in brittle starch film. Strong adhesion has been perceived to be a valuable behavior for starch adopted for sizing in textiles and paper-making [26]. It has become a highly important factor of analyzing the quality of starch-based sizes. Furthermore, the film covering the yarn surfaces plays an important role in protecting the warps from mechanical abrasions, thereby ensuring the smooth running of the weaving [27]. In paper-making, the film can enhance the paper flexibility. As a result, the toughness of starch film occupies a very stringent place for the quality of fibrous products. Fortunately, chemical modification of starch such as traditional graft copolymerization [28,29] can involve the alteration of the physical and chemical characteristics of native starch, thereby improving the adhesion and film properties of starch. Nevertheless, there are no applications of grafted starch prepared by controlled ARGET ATRP technique in warp sizing or paper-making. In addition, if the starch macroinitiator prepared can display a positive effect on viscosity stability, adhesion and film properties, it would be quite meaningful for ensuring the positive effect on the properties generated by the grafted starch macroinitiator. Therefore, whether starch macroinitiators show an adverse influence or a positive one should be ascertained. If the starch macroinitiator shows a positive influence, a suitable modification level also needs to be confirmed.

For example, due to the similarity with polyethylene terephthalate, easy processing as well as environment-friendliness, polylactic acid (PLA) acquired from renewable resources has been extensively explored for fiber applications [30]. Nowadays, if the PLA warps or PLA filaments are woven directly, it will produce some serious problems due to poor cohesion, loose tows, and easily occurrence of entanglement and bonding, and thus the PLA warps or PLA filaments must be sized before weaving. In addition, the sizing process of polyester warps also needs to be conducted before the weaving. Obviously, hydrophobic 2-bromoisobutyryl substituents can be introduced into the starch molecules to prepare the starch macroinitiator by starch esterification. The 2-bromoisobutyryl substituents that contain ester groups are conducive to enhancing the van der Waals force at the interfaces of starch adhesive layers with PLA fibers or polyester ones, owing to chemical similarity with the esters in PLA or polyester molecular chains; thereby, they can be expected to improve the adhesion. However, it is indubitable that excessive introduction of the hydrophobic substituents will reduce the water-dispersibility of starch and adversely affect the adhesion since the paste used in sizing warps is only a water-based adhesive. Accordingly, due to the indeterminacy of the influence on the adhesion generated by the 2-bromoisobutyryl substituents, we ought to evaluate the influence on the adhesion of starch to PLA and polyester fibers, and confirm whether it is an adverse influence or a positive one on the adhesion.

Currently, there is little research about the synthesis process of starch macroinitiator, i.e., bromoisobutyryl esterified cornstarch (BBES). Therefore, an important aim of this work is to obtain suitable synthesis process parameters for preparing the BBES, as shown in Scheme 1. Moreover, there is no study that has been conducted on the properties of BBES, such as viscosity stability, adhesion and film properties. Accordingly, another objective is to reveal whether bromoisobutyryl esterification is able to enhance viscosity stability, adhesion of cornstarch to PLA and polyester fibers and film properties compared with acid-converted starch (ACS) as a control. Furthermore, the suitable level of starch modification for the BBES shall also be determined since the functional ability of a modified starch is correlated with the degree of substitution (DS) [31]. The obtained starch samples in this work were characterized by the uses of Fourier transform infra-red (FTIR) spectroscopy and Scanning electron microscopy (SEM). In addition, the measurements on DS values of BBES samples, apparent viscosity and viscosity stability of cooked starch paste, adhesion to PLA and polyester fibers, tensile strength and breaking elongation of the films, as well as film characterization with SEM and X-ray diffraction (XRD) were also performed.

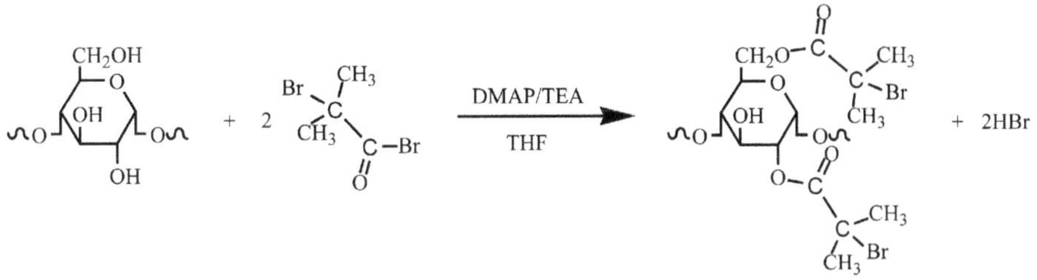

Scheme 1. Bromoisobutyryl esterification of cornstarch for the synthesis of BBES.

2. Experimental

2.1. Materials

Natural cornstarch with a moisture content of 12.8 wt. % and an apparent viscosity of 46 mPa·s, was purchased from Shandong Hengren Industry and Trade Co., Ltd. (Shandong, China). Before use, the starch was refined to remove the protein substance [32] and subsequently acid-converted with hydrochloric acid to lower its excessive viscosity [33] to 19.5 mPa·s. Tetrahydrofuran (THF) and triethylamine (TEA) were dried with molecular sieve. 2-bromoisobutyryl bromide (BIBB) and 4-dimethylaminopyridine (DMAP) (used directly), sodium hydroxide and methanol, were purchased from Aladdin Industrial Corporation (Shanghai, China). Anhydrous ethanol was obtained from (Shanghai Maclean Biochemical Technology Co., Ltd., Shanghai, China) The pure PLA roving (520 tex) and polyester one (365 tex), adopted for adhesion measurement, were kindly supplied by BBCA Group Co. Ltd. (Bengbu, China) and Anhui Huamao Group Co., Ltd. (Anqing, China), respectively.

2.2. Preparation Method and Characterization of BBES for Studying Its Synthesis Process

2.2.1. Preparation

The BBES was synthesized by the reaction between ACS and BIBB in a THF medium.

Briefly, dried ACS (2 g) and a certain amount of DMAP were added into a 250 mL four-necked flask connected to a mechanical stirrer, and dispersed with 100 mL of THF under mechanical agitation. Then, the TEA was added into the dispersion, and the reaction system was mechanically stirred for 30 min under N_2 atmosphere. Subsequently, a certain amount of BIBB was added dropwise in an ice-cold bath under mechanically stirred. The reaction mixture was stirred for 1 h in the bath and heated up to a given temperature. Afterwards, the mixture was reacted at the given temperature for a certain time. Then, the final product was filtered, washed thoroughly with anhydrous ethanol, followed by freeze-dried in vacuum, powdered, and sieved with a 100-mesh sieve.

2.2.2. Measurement of DS

DS indicated the number of the hydroxyls per anhydroglucose unit substituted by the 2-bromoisobutyryl substituents. Back titration with HCl preceded by alkali saponification [34,35] was employed to determine the DS. The DS value and reaction efficiency were calculated with the following Equations (1) and (2):

$$DS = \frac{162B}{150 - 150B} \quad (1)$$

$$RE = \frac{DS}{M/n} \quad (2)$$

where B and 150 denote the percentage content (%) and molecular mass of 2-bromoisobutyryl substituents, respectively, and M and n are the moles of BIBB and anhydroglucose residues of starch, respectively.

2.2.3. Infrared Spectral Analysis

The Fourier transform infra-red (FTIR) spectra of ACS and BBES samples were acquired on an IRPrestige-21 FTIR Spectrometer (Shimadzu Co. Ltd., Kyoto, Japan) by using KBr disk technique to prove the successful introduction of new functional 2-bromoisobutyryl substituents in the starch molecules after bromoisobutyryl esterification. The spectra were collected in a wavenumber range of 500–4000 cm^{-1} with the running condition of 4 cm^{-1} spectral resolution.

2.3. Preparation Method of BBES under a Suitable Synthesis Process

Dried ACS (120 g) and anhydrous sodium sulfate (30 g) were added into a 500 mL four-necked flask connected to a mechanical stirrer and dispersed with THF to form a 30% dispersion. Then, the TEA (molar ratio of TEA to BIBB was 2:1) was added dropwise into the dispersion, and the dispersion was mechanically stirred for 60 min under N$_2$ atmosphere. After the temperature of the reaction system had been cooled to 0–8 °C, certain amounts of BIBB and DMAP (dissolved with THF, respectively) were slowly added dropwise. The reaction was kept for 1 h at the above temperature and then heated up to 40 °C for 24 h. Finally, the final product was filtered, washed thoroughly with anhydrous ethanol, followed by freeze-dried in vacuum, powdered and sieved with a 100-mesh sieve.

2.4. Scanning Electron Microscopy (SEM) Analysis

The surface morphologies of the granular ACS and BBES samples as well as the cross-sections of their film samples were observed by means of a scanning electron microscope (Hitachi S-4800, Tokyo, Japan). Prior to SEM imaging, the samples were sprayed gold to avoid charge accumulation.

2.5. Apparent Viscosity and Viscosity Stability

The apparent viscosity of gelatinized starch paste (6% (w/w)) was recorded using an NDJ-79 Rotary Viscometer (Tongji Electrical Machinery Plant, Shanghai, China) at 95 °C by the method described in the literature [36]. For each paste, two individual determinations of the viscosity were conducted, and the average value was reported. Viscosity stability was determined and calculated using the method described in the work [37].

2.6. Adhesion Test

Pure PLA and polyester roving were applied as adherents to examine the adhesion using the legal standard method (FZ/T 15001-2008) in China to determine the adhesion of a sizing material to fibers. The determination of the adhesion contains three steps (as shown in Scheme 2): (1) the formation of the 1% starch aqueous paste by thoroughly stirring the dispersion and heated it to 95 °C for 1 h; (2) PLA or polyester roving carefully wound onto a rectangular steel frame, was immersed in the paste for 5 min, and subsequently the immersed wet roving was atmospherically dried; (3) After storing at 65% relative humidity (RH) and 20 °C for 24 h, bonding forces of dried roving were measured on YG026D Electronic Strength Tester (Ningbo Textile Instrument Factory, Zhejiang, China) with an initial chuck-distance of 100 mm and a drawing speed of 50 mm/min at the above ambient condition [38]. For each case, the mean values of 20 successful tests were reported.

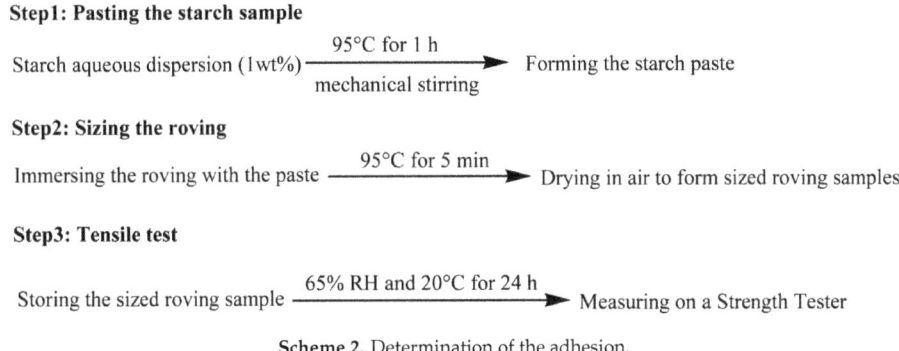

Scheme 2. Determination of the adhesion.

2.7. Surface Tension

A DCAT 21 Automatic Tensiometer (Dataphysics Co., Ltd., Filderstadt, Germany) was used to determine the surface tension of cooked starch paste. A 1 wt. % starch aqueous paste was prepared by heating the 1 wt. % starch aqueous dispersion to 95 °C and maintaining at this temperature for 1 h. After cooling down, the tension was determined in duplicate at room temperature.

2.8. Preparation and Measurement of Starch Films

2.8.1. Preparation

The film was formed by drying cooked starch paste that was cast onto a polyester film at 65% RH and 20 °C according to the method described in our previous work [39]. Briefly, a 6% (w/w) starch aqueous paste (400 mL) was formed by preparing the starch aqueous suspension and heating the suspension to 95 °C for 1 h under mechanical agitation. Afterwards, the paste was cast onto a smooth polyester film (650 mm in length and 400 mm in width) spread on a glass plate and dried at the above ambient condition for forming the starch film.

2.8.2. Measurement

Then, the films prepared were cut into 200 mm × 10 mm strips. After storing at 20 °C and 65% RH for 24 h, tensile strength and elongation at break of the films were measured on a YG026D Electronic Strength Tester in an initial chuck-distance of 100 mm and a stretching speed of 50 mm/min in accordance with the ASTM D 882-02 method. For each set of data, the averages of test results of twenty strips were reported.

2.9. X-Ray Diffraction (XRD) Analysis of Starch Film

The XRD patterns of ACS and BBES films were recorded on an XRD-6000 X-ray Diffractometer (Shimadzu Co., Japan) equipped with a wavelength of 0.154 nm CuKa radiation at an X-ray generator setting of 40 kV and 30 mA. The scanning region of diffraction was registered with a speed of 4°/min and a 2θ angular range of 5° to 45° with an angular step of 0.02°.

3. Results and Discussion

3.1. Process Research of Synthesizing BBES

To ascertain the suitable synthesis process of BBES, this work will mainly carry out research on the influences of the synthesis process parameters—amount of BIBB, amount of catalyst (DMAP), reaction temperature, and reaction time—upon the DS.

To confirm that the synthesis process of BBES in Section 2.2 could be adopted to successfully prepare the BBES samples, we performed the esterification of ACS with BIBB in a THF medium, and subsequently tested the chemical compositions of the granular product prepared by scanning electron microscopy with an energy-dispersive X-ray spectrometer (SEM-EDS) [40], as shown in Figure 1. It can be seen that the elemental composition of the product mainly comprised O, C and Br, while the ACS mainly comprised O and C. The elemental composition of Br demonstrated the successful preparation of BBES, thereby laying a foundation for subsequent research on the synthesis process of BBES.

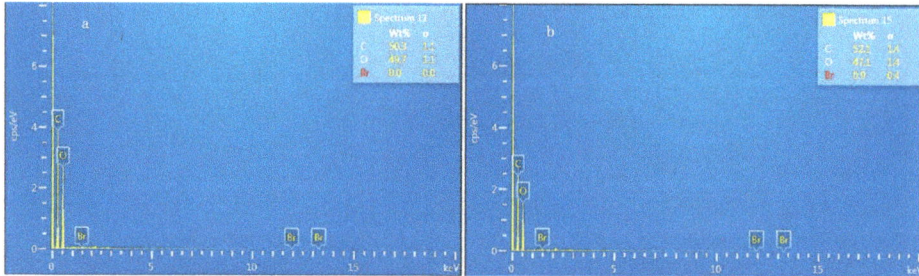

Figure 1. SEM-EDS images of (**a**) ACS and (**b**) BBES.

The successful introduction of 2-bromoisobutyryl substituents into starch molecules after esterification was demonstrated by FTIR spectra of BBES and ACS, as shown in Figure 2a,b, respectively. It can be seen that there were differences in the wavenumber range of 3600–3100 cm^{-1}, which were mainly attributed to the decreases of the OH groups after the esterification. In addition, the FTIR spectrum of the BBES (Figure 2a) showed an absorption bond at the wavenumber of 1737 cm^{-1}, which indicated the characteristic absorption band of carbonyl groups [20,41,42] in 2-bromoisobutyryl substituents. There was no sign of this bond in the spectrum of the ACS (Figure 2b). This observation confirmed the successful introduction of 2-bromoisobutyryl substituents into the starch molecules.

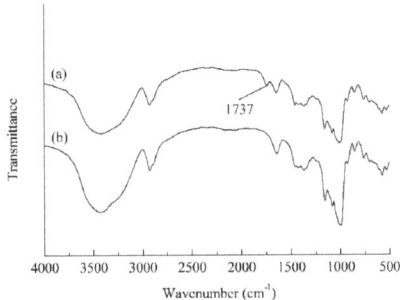

Figure 2. FTIR spectra of BBES with a DS value of 0.107 (**a**) and ACS (**b**).

3.1.1. Influence of the Amount of BIBB on the DS

When the other conditions were fixed, it can be seen from Figure 3 that the DS of BBES increased as the molar ratios of BIBB to anhydroglucose residues were raised from 0.35 to 2.8. This is mainly attributed to the fact that as the amount of BIBB increases, the concentration of BIBB in the reaction system gradually increases, and the active hydroxyl groups on the macromolecular chains of ACS will have more chances to react with the BIBB, so that more hydroxyl groups in the starch molecules are esterified, thereby causing a gradual increase in the extent of esterification.

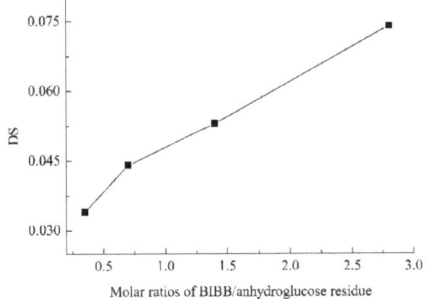

Figure 3. Influence of the amount of BIBB on the DS of BBES.

3.1.2. Influence of the Amount of DMAP on the DS

Under the following fixed conditions: reaction temperature of 30 °C, reaction time of 24 h, and molar ratio (BIBB/ACS) of 2.8; the influence of the amount of DMAP (denoted by the molar ratio of DMAP to BIBB) on the DS is depicted in Figure 4. As observed, when the other conditions were fixed, as the molar ratio of DMAP/BIBB was increased from 0.11 to 1.38, the DS showed a trend that first increased and then decreased. When the molar ratio was 0.23, the DS of BBES reached its maximum value of 0.074, which indicated that the most appropriate molar ratio of DMAP/BIBB was 0.23. The DS value gradually decreased once the molar ratio of DMAP/BIBB was above 0.23, which might be attributed to the fact that this might occur as a side reaction when the catalyst concentration was increased, thereby producing an adverse effect on the DS.

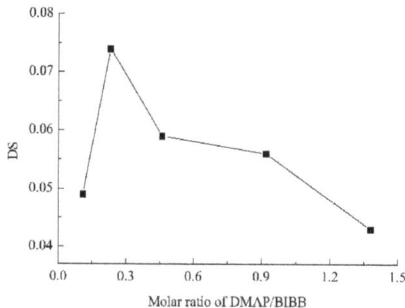

Figure 4. Influence of the amount of DMAP on the DS of BBES.

3.1.3. Influence of Reaction Temperature on the DS

To avoid the gelatinization of starch granules in the aqueous dispersion, a reaction temperature range of 20–60 °C was selected for the study. The influence of reaction temperature on the DS is represented in Figure 5 under the following conditions: 24 h reaction time, 2.8 molar ratio of BIBB/anhydroglucose residues, and 0.23 molar ratio of DMAP/BIBB. It can be found that the DS was dependent on the reaction temperature, and with the rise in the temperature range from 20 to 60 °C, DS showed a gradually increased tendency. With the increase in the temperature, the thermal motion of the starch macromolecules and the BIBB molecules increases, and the probability of collision between them increases, so that the DS of BBES gradually increases. However, when the temperature exceeds 40 °C, the increment in the DS lowers. This may be attributed to the fact that, as the temperature is higher than 40 °C, the excessive temperature may result in decreased activity of the DMAP. Accordingly, we concluded that the appropriate reaction temperature under the given process conditions was 40 °C.

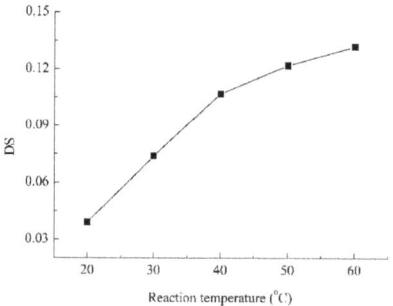

Figure 5. Influence of reaction temperature on the DS of BBES.

3.1.4. Influence of Reaction Time on the DS

Figure 6 depicts the influence of reaction time on the DS under the following conditions: 30 °C reaction temperature, 2.8 molar ratio of BIBB to anhydroglucose residues, and 0.23 molar ratio of DMAP to BIBB. Under the above conditions, it can be seen that the DS of BBES exhibited a proportional relationship with the reaction time, i.e., the DS increased as raising the reaction time. In the initial stage of the reaction, BIBB reacts with DMAP to form the intermediate. After a certain reaction time, the intermediates reach a certain concentration and subsequently reacting with the hydroxyl groups on the starch chains. Therefore, with the extension of the reaction time, BIBB reacts with DMAP to form new intermediates for continuous reaction with hydroxyls, so that the DS gradually increases with the extension of the reaction time from 0 to 24 h.

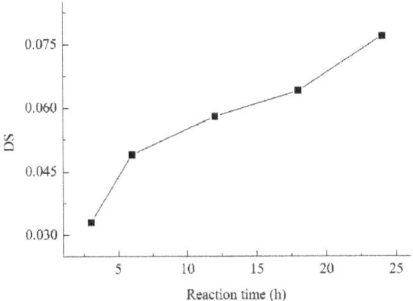

Figure 6. Influence of reaction time on the DS of BBES.

3.2. SEM and DS Analyses of the BBES Samples with the Suitable Synthesis Process

After investigating the influences of the synthetic process parameters upon the DS, we obtain the suitable synthesis process parameters of BBES: reaction time of 24 h, reaction temperature of 40 °C, and 0.23 in molar ratio of catalyst (DMAP) to BIBB. Accordingly, to study the properties of BBES, such as adhesion-to-fibers and film properties, etc., a series of BBES samples with different DS values were prepared.

SEM technique has become an important means of clearly recording the granule morphology of a modified starch [43]. Figure 7 shows the SEM images of ACS and BBES granules. As can be seen, the ACS granules were round or polygonal shapes with various sizes (Figure 7a) [44]. Nevertheless, the morphology of BBES granules was changed in comparison with the ACS. As shown in Figure 7b, the surfaces of the BBES granules underwent some visible damage, indicating the successful bromoisobutyryl esterification of granular ACS, and the modification mainly occurs on the hydroxyls of granule surfaces, as the reaction temperature is 40 °C.

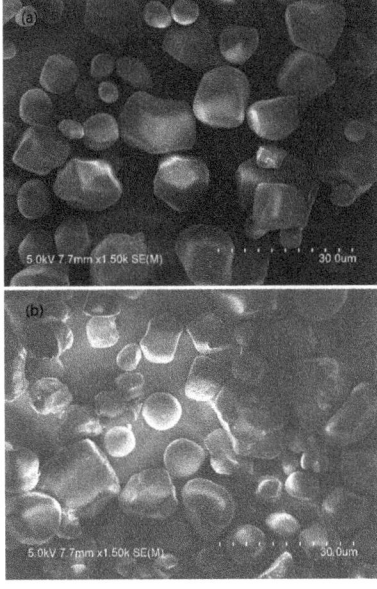

Figure 7. SEM images of ACS granules (**a**) and BBES ones (**b**).

Using the suitable synthetic process parameters, the DS values and reaction efficiencies of granular BBES samples prepared were measured and denoted by plotting the DS values and reaction efficiencies versus moles of BIBB, as illustrated in Figure 8. It can be found that the DS values and reaction efficiencies were dependent on the moles of the BIBB used in the esterification, i.e., with the increases in the moles of BIBB from 0 to 0.065, the DS values gradually increased from 0 to 0.036, which indicated that BIBB with a esterification level range from 0 to 0.036 could be obtained by the reaction of ACS with BIBB in a THF medium. In addition, as the moles of BIBB increased, the efficiencies gradually decreased. As the esterification goes on, active sites are gradually occupied by the substituents introduced and there are not adequate sites for introducing new 2-bromoisobutyryl substituents into starch molecules. Consequently, the efficiencies decrease with the increase in the moles of BIBB.

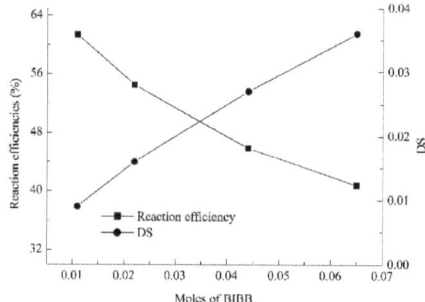

Figure 8. Degrees of substitution and reaction efficiencies of BBES samples prepared.

3.3. Influence of Bromoisobutyryl Esterification

3.3.1. Influence on Apparent Viscosity and Its Stability

Figure 9 depicts the influence of bromoisobutyryl esterification on apparent viscosity and its stability of cooked ACS paste. As observed, the esterification was able to raise the paste viscosity compared with that of the ACS (DS = 0) paste. In addition, with the rises in DS values, the viscosities of cooked BBES pastes gradually increased. In addition, the BBES samples were superior to their counterpart ACS in stability (82.1% paste stability for ACS). As the total DS raised, the stabilities of the BBES pastes exhibited an increased tendency and were all above 85%, which fulfilled the requirement of ≥85% in paste stability for achieving a stable size pick-up during warp sizing [45]. This suggested the bromoisobutyryl esterification of ACS did not produce a negative effect on the paste stability, and the BBES paste could satisfy the requirement in the paste stability for confirming the stability of size pick-up during the sizing process.

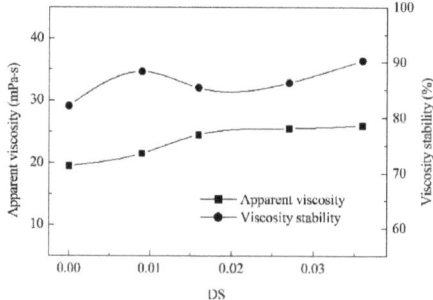

Figure 9. Apparent viscosity and its stability of gelatinized starch paste.

3.3.2. Influence on Adhesion

The influence of bromoisobutyryl esterification on adhesion of ACS to PLA and polyester fibers was estimated by plotting the bonding forces versus DS values, as shown in Figures 10 and 11, respectively. It can be seen that the bonding forces of ACS (DS = 0) to PLA and polyester fibers were 47.4 N and 107.8 N, respectively, and BBES was superior to ACS in the bonding forces to PLA and polyester fibers. This meant that the esterification was capable of ameliorating the adhesion of ACS to PLA and polyester fibers. The forces of BBES were correlated with the DS. With the rise in the DS, the forces of BBES to PLA and polyester fibers increased, and when the DS was 0.016, the forces reached their maximum values, at 50.6 N for PLA fibers and 116.8 N for polyester ones. When the DS was above 0.016 in the range of 0.016 to 0.036, the forces gradually decreased to 48.2 N for PLA fibers and 110.5 N for polyester ones, still showing a slight increase compared with those of ACS. This implies that bromoisobutyryl esterification will not produce an adverse effect on the adhesion of grafted BBES prepared in future based on the BBES. Based these results, the suitable DS of BBES for improving the adhesion of ACS to both fibers was 0.016.

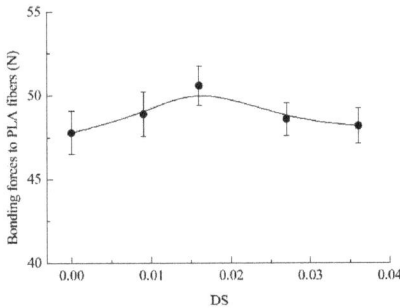

Figure 10. Influence of bromoisobutyryl esterification on adhesion of starch to PLA fibers.

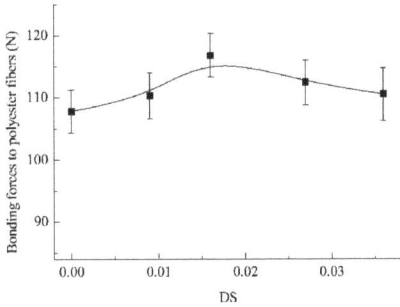

Figure 11. Influence of bromoisobutyryl esterification on adhesion of starch to polyester fibers.

To ascertain the influence of bromoisobutyryl esterification on adhesion, the characteristics of the gelatinized starch paste must be given attention. It is generally known that starch granules in the aqueous dispersion heated at a high temperature will swell due to adsorbing the water, and the hydrogen bonds of starch will be disrupted; sequentially the crystalline structure of the granules will be destroyed to form an amorphous structure, subsequently forming a gelatinized starch paste [46]. The paste formed can be regarded as a biphasic system, simultaneously containing disperse and continuous phases [47]. The disperse phase is swollen remnants of the granules that mainly consist of the amylopectins, and the continuous one is the solution of the soluble starch coins that is mainly comprised of linear amyloses [48]. Linear amyloses in low-temperature aqueous paste tend to orient themselves in a parallel fashion and approach each other closely enough to form aggregates [49]. Furthermore, amylose molecules are also able to produce co-crystallization with the linear branches of amylopectins [50] in the aqueous paste by the hydrogen bonding between starch hydroxyls [51]. These phenomena can lead to paste retrogradation of starch [11]. At high concentrations, the retrogradation can convert the paste into a gel that is composed of a three-dimensional network held together by hydrogen bonding [11]. It is generally accepted that a gelled paste may lose the fluidity and produce incomplete wetting and spreading of the paste onto the fiber surfaces [52]. According to the failure position, the failure of an adhesive joint commonly contains cohesive failure and interfacial one [53], which denote failures that occur wholly within the matrix of an adhesive layer formed by adhesives such as starch, and exactly at the interfaces between the layers and fibers, respectively. The interface failure and high internal stresses within the matrix of the adhesive layer [38] are prone to occur around unwetted or outspread areas, thereby generating a damage to adhesion [49].

Undoubtedly, steric hindrance generated by the 2-bromoisobutyryl substituents derivatized into starch molecules can obstruct the parallel fashion and closely approach between starch amyloses in the paste, and lower the co-crystallization between the amyloses and linear branches of amylopectin, thereby favoring the diminishing of paste retrogradation. Consequently, the 2-bromoisobutyryl substituents derivatized may be expected to improve the wetting and spreading of the paste on the fiber surfaces. In addition, as broadly accepted, wetting and spreading of an adhesive liquid onto a given solid surface is closely related to surface tension [38]. A decreased tension favors wetting and spreading, and commonly induces an improved adhesion [50]. For this reason, the influence of 2-bromoisobutyryl substituents on the surface tension of cooked starch paste was investigated, as depicted by indicating the tension versus DS in Figure 12. As observed, the tension of cooked ACS (DS = 0) paste was 66.7 mN/m. The substituents had much impact on the tension, and the tension was lowered after the introduction of the substituents. The tension of the BBES paste depended on the DS, and it gradually decreased in the DS range of 0.009–0.036. This meant that 2-bromoisobutyryl substituents were necessary to provide the derivatives with stronger surface activity. A decreased tension also favors the amelioration of the wetting and spreading of the pastes onto the fiber surfaces. The amelioration will lower the probability of interfacial failure mentioned above, inducing an improvement in the adhesion. Furthermore, the hindrance is also able to raise intermolecular distance between starch chains, and reduce hydrogen bonding between starch hydroxyls, thereby exerting an internal plasticization on the matrix of starch adhesive layers. Therefore, the internal stresses within the matrix of the adhesive will be reduced after the derivatization of the 2-bromoisobutyryl substituents. Additionally, 2-bromoisobutyryl substituents contain ester groups, and display a hydrophobic characteristic, which will increase the van der Waals force at the interfaces between the layers and PLA and polyester fibers, due to the chemical similarity with the esters in PLA and polyester chains. This also favors improving the adhesion. These reasons indicate that bromoisobutyryl esterification provides a positive effect on the adhesion. Nevertheless, it is doubtless that the hydrophobic 2-bromoisobutyryl substituents derivatized can lower the water-dispersibility of starch. It has been verified that the worse water-dispersibility can cause incomplete wetting and spreading [54], thereby leading to a negative influence on the adhesion. Consequently, the combination of the positive and negative effects may be expected to reduce the probability of interfacial failure and internal stresses, favoring the adhesion. In addition, due to the combination effects, the adhesion of BBES to PLA and polyester fibers presents a trend that first increases and then decreases with increasing esterification level.

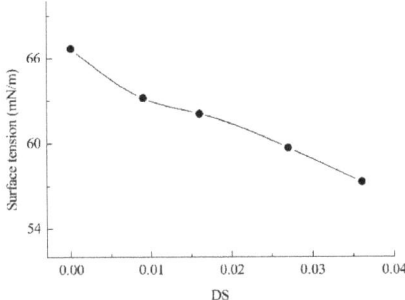

Figure 12. Influence of the 2-bromoisobutyryl substituents on the surface tension of cooked starch paste.

3.3.3. Influence on Film Properties

During the sizing process, a portion of starch paste forms a starch film around the surfaces of sized yarns [55]. The film can provide a protective effect for the warps from mechanical abrasions, and thus enhancing their weavability [27]. Therefore, desirable film must be flexible and stretchable [27] for withstanding repeated and extensive drawing, impacting, friction, and bending actions in weaving [56]. As a result, the tensile properties of starch film were determined for evaluating the protection against the actions.

The influence of bromoisobutyryl esterification on tensile strength and breaking elongation of ACS film is shown in Figure 13. Compared with the ACS film, the films cast from cooked BBES pastes exhibited higher breaking elongation and lower tensile strength. This implied that derivatizing 2-bromoisobutyryl substituents onto the backbones of starch can decrease the brittleness of starch film. The maximal elongation (3.63%) of the BBES film was observed at the DS value of 0.016 compared with the elongation of 2.64% for the ACS film. In addition, breaking elongation and tensile strength of the BBES films were related to the DS. With increasing DS, the elongation gradually increased, reaching its highest value when the DS was 0.016, and subsequently gradually decreasing. In addition, the strength gradually reduced as the DS increased. Therefore, the BBES (DS = 0.016) film with a higher elongation is preferable to the brittle ACS film for the applications in warp sizing and surface coating, and it will not produce an adverse effect on the film properties of grafted BBES prepared in future based on the BBES.

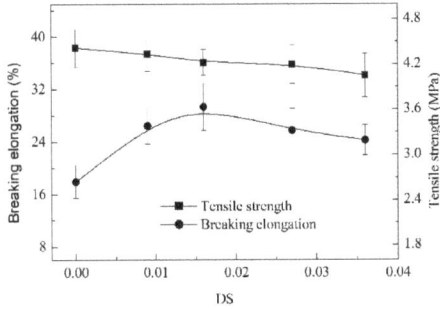

Figure 13. Influence of bromoisobutyryl esterification on tensile strength and breaking elongation of ACS films.

During formation of the film, the amyloses in aqueous paste parallelly arrange and approach closely with each other, and subsequently form a three-dimensional network structure held together by hydrogen bonding between the hydroxyls [57]. This is the main reason starch film shows the characteristics of strong brittleness and small deformation [58]. Undoubtedly, the steric hindrance of

2-bromoisobutyryl substituents favors blocking the arrangement in an orderly way of the amyloses during film formation and preventing the growth of crystal on nucleus. As a consequence, it may be expected to lower the degree of crystallinity ascertained by X-ray diffraction of the starch film, as depicted Figure 14. By comparing the X-ray diffraction patterns of the ACS film (Figure 14a) and BBES film (Figure 14b), we found that the intensity of the peaks for BBES film was lower than the intensity for the ACS one, although the location of the peaks was almost the same. The reduction of the intensity of the starch crystal peak showed that the substituents introduced lowered the formation of the crystalline structure of starch film, indicating that BBES film had a lower crystallinity (14.6%) than ACS film (21.7%). As a result, a high elongation and a low strength are exhibited for the BBES film by diminishing intermolecular hydrogen bonding compared with ACS film. This meant that the bromoisobutyryl esterification was able to lower the brittleness of starch film, providing a toughening effect for the film. The reduced brittleness provided by the esterification can be confirmed by observing the SEM images of the cross-section of ACS and BBES film samples, as shown in Figure 15. It can be observed from Figure 15a that ACS film exhibits a strong brittleness, while BBES film represents a lower brittleness, as can be seen from Figure 15b, confirming that the bromoisobutyryl esterification produced the effect of lowering the brittleness of the film.

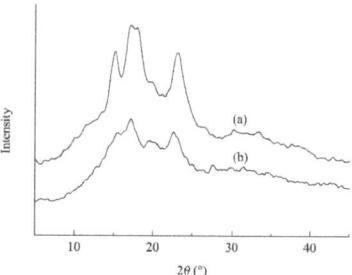

Figure 14. X-ray diffraction of (**a**) ACS film and (**b**) BBES (with a DS of 0.016) one.

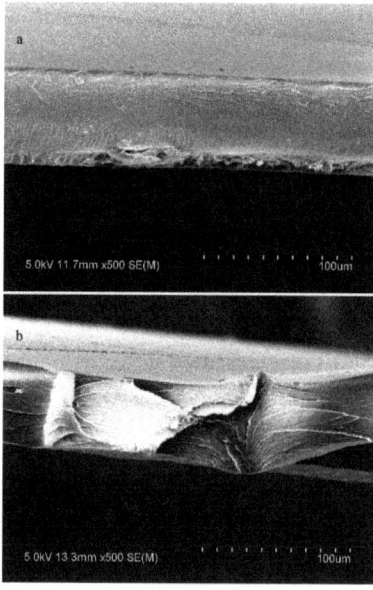

Figure 15. SEM images of the cross-section of ACS film (**a**) and BBES (with a DS of 0.016) one (**b**).

4. Conclusions

Through the investigation to the influences of the synthesis process parameters—amount of BIBB, amount of catalyst (DMAP), reaction temperature and reaction time—upon the DS, we obtain suitable synthesis process parameters for BBES: reaction time of 24 h, reaction temperature of 40 °C, and 0.23 in molar ratio of DMAP to BIBB. Then, to study the properties of BBES such as adhesion-to-fibers and film properties, etc., for producing a positive effect on the properties of grafted BBES prepared in the near future, a series of BBES samples with the DS values of 0.009, 0.016, 0.027 and 0.036 and reaction efficiencies of 61.3%, 54.5%, 46.0% and 40.9%, were prepared according to the suitable synthesis process parameters. The successful introduction of the 2-bromoisobutyryl substituents was demonstrated by FTIR technique. It could be concluded that introducing 2-bromoisobutyryl substituents into starch molecules via bromoisobutyryl esterification of ACS with BIBB generating the positive effects on the properties of ACS, such as paste stabilities of above 85%, improvement in the adhesion of ACS to PLA and polyester fibers, and reduction of film brittleness. The stabilities of above 85% indicated that the introduction of the substituents could meet the requirement in the stability for warp sizing. The adhesion of BBES to both fibers was correlated with the DS, and the forces presented a trend that first increased and then decreased as increasing DS. When the DS was 0.016, the highest force was obtained. The breaking elongation of the BBES film was higher than the ACS film, and its tensile strength and degree of crystallinity were inferior to those of ACS film, indicating that 2-bromoisobutyryl substituents derivatized could lower the film brittleness (was verified by the SEM analysis of the cross-sections of ACS and BBES film samples). Considering the results of BBES with respect to paste stability, adhesion, and film properties, BBES with a DS of 0.016 showed potential in the applications of PLA and polyester sizing, and will not produce an adverse effect on the properties of grafted BBES prepared in the near future based on the BBES.

Author Contributions: W.L. conceived and designed the experiments, analyzed the data and wrote the manuscript; J.W. performed the determination of the DS and property measurements; Z.Z. performed the property measurements; L.W. performed the characterization; Y.L. contributed reagents and materials.

Funding: This work was financially supported by the Natural Science Foundation of Anhui Province (No. 1908085ME124), Key Research and Development Program of Anhui Province (No. 201904a06020001), Pre-research Project of National Natural Science Foundation of China (No. 2018yyzr08), Science and Technology Planning Project of Wuhu City (No. 2018pt04), China.

Conflicts of Interest: The authors declare no conflict of interest.

Abbreviations

degree of substitution	DS
bromoisobutyryl esterified cornstarch	BBES
Scanning electron microscopy	SEM
polylactic acid	PLA
4-dimethylaminopyridine	DMAP
2-bromoisobutyryl bromide	BIBB
acid-converted starch	ACS
electron transfer atom transfer radical polymerization	ARGET ATRP
Fourier transform infra-red	FTIR
X-ray diffraction	XRD
tetrahydrofuran	THF
triethylamine	TEA
relative humidity	RH
scanning electron microscopy with an energy-dispersive X-ray spectrometer	SEM-EDS

References

1. Ramaraj, B. Crosslinked poly (vinyl alcohol) and starch composite films. II. Physicomechanical, thermal properties and swelling studies. *J. Appl. Polym. Sci.* **2007**, *103*, 909–916. [CrossRef]

2. Nurmi, L.; Holappa, S.; Mikkonen, H.; Seppälä, J. Controlled grafting of acetylated starch by atom transfer radical polymerization of MMA. *Eur. Polym. J.* **2007**, *43*, 1372–1382. [CrossRef]
3. Cano, A.; Fortunati, E.; Cháfer, M.; Kenny, J.M.; Chiralt, A.; González-Martínez, C. Properties and ageing behaviour of pea starch films as affected by blend with poly(vinyl alcohol). *Food Hydrocoll.* **2015**, *48*, 84–93. [CrossRef]
4. Chen, J.; Liu, C.; Chen, Y.; Chen, Y.; Chang, P.R. Comparative study on the films of poly(vinyl alcohol)/pea starch nanocrystals and poly(vinyl alcohol)/native pea starch. *Carbohydr. Polym.* **2008**, *73*, 8–17. [CrossRef]
5. Xu, J.Y.; Krietemeyer, E.F.; Finkenstadt, V.L.; Solaiman, D.; Ashby, R.D.; Garcia, R.A. Preparation of starch-poly-glutamic acid graft copolymers by microwave irradiation and the characterization of their properties. *Carbohydr. Polym.* **2016**, *140*, 233–237. [CrossRef]
6. Yao, K.H.; Cai, J.; Liu, M.; Yu, Y.; Xiong, H.G.; Tang, S.W.; Ding, S.Y. Structure and properties of starch/PVA/nano-SiO2, hybrid films. *Carbohydr. Polym.* **2011**, *86*, 1784–1789. [CrossRef]
7. Daniel, J.R.; Whistler, R.L.; Roper, H. Starch. In *Ullmann's Encyclopedia of Industrial Chemistry*; Wiley–VCH Verlag GmbH and Co.: Weinheim, Germany, 1997; pp. 721–726.
8. Maurer, H.W.; Kearney, R.L. Opportunities and challenges for starch in the paper industry. *Starch-Stärke* **1998**, *50*, 396–402. [CrossRef]
9. Shamai, K.; Bianco-Peled, H.; Shimoni, E. Polymorphism of resistant starch type III. *Carbohydr. Polym.* **2003**, *54*, 363–369. [CrossRef]
10. Bansal, A.; Kumar, A.; Latha, P.P.; Ray, S.S. Expanded corn starch as a versatile material in atom transfer radical polymerization (ATRP) of styrene and methyl methacrylate. *Carbohydr. Polym.* **2015**, *130*, 290–298. [CrossRef]
11. Wurzburg, O.B. *Modified Starches: Properties and Uses*; CRC Press: Boca Raton, FL, USA, 1986; pp. 3–16.
12. Han, T.L.; Kumar, R.N.; Rozman, H.D.; Noor, M.A.M. GMA grafted sago starch as a reactive component in ultra violet radiation curable coatings. *Carbohydr. Polym.* **2003**, *54*, 509–516. [CrossRef]
13. Li, W.; Wu, J.; Cheng, X.G.; Wu, L.J.; Liu, Z.; Ni, Q.Q.; Lu, Y.H. Hydroxypropylsulfonation/caproylation of corn starch to enhance its adhesion to PLA fibers for PLA sizing. *Polymers* **2019**, *11*, 1197. [CrossRef]
14. Fanta, G.F. Starch Graft Copolymers. In *Polymeric Materials Encyclopedia*; Salamone, J.C., Ed.; CRC Press: Boca Raton, FL, USA, 1996; pp. 7901–7910.
15. Jenkins, D.W.; Hudson, S.M. Review of vinyl graft copolymerization featuring recent advances toward controlled radical-based reactions and illustrated with chitin/chitosan trunk polymers. *Chem. Rev.* **2001**, *101*, 3245–3273. [CrossRef]
16. Król, P.; Chmielarz, P. Synthesis of PMMA-b-PU-b-PMMA tri-block copolymers through ARGET-ATRP in the presence of air. *Express Polym. Lett.* **2013**, *7*, 249–260. [CrossRef]
17. Yamamoto, S.I.; Matyjaszewski, K. ARGET-ATRP synthesis of thermally responsive polymers with oligo (ethylene oxide) units. *Polym. J.* **2008**, *40*, 496–497. [CrossRef]
18. Hu, L.; Hu, Y.Y.; Chang, J.; Zhang, C.K. Synthesis and self-aggregation of PTRIS-co-MMA polymer films via ARGET-ATRP. *Mater. Lett.* **2014**, *120*, 79–81. [CrossRef]
19. Chalid, M.; Handayani, A.S.; Budianto, E. Functionalization of starch for macro-initiator of atomic transfer radical polymerization (ATRP). *Adv. Mater. Res.* **2014**, *1051*, 90–94. [CrossRef]
20. Wang, L.L.; Shen, J.N.; Men, Y.J.; Wu, Y.; Peng, Q.H.; Wang, X.L.; Yang, R.; Mahmood, K.; Liu, Z.P. Corn starch-based graft copolymers prepared via ATRP at the molecular level. *Polym. Chem.* **2015**, *6*, 3480–3488. [CrossRef]
21. Maran, J.P.; Sivakumar, V.; Sridhar, R.; Immanuel, V.P. Development of model for mechanical properties of tapioca starch based edible films. *Ind. Crop. Prod.* **2013**, *42*, 159–168. [CrossRef]
22. Mali, S.; Grossmann, M.V.E.; García, M.A.; Martino, M.N.; Zaritzky, N.E. Mechanical and thermal properties of yam starch films. *Food Hydrocoll.* **2005**, *19*, 157–164. [CrossRef]
23. Waterschoot, J.; Gomand, S.V.; Fierens, E.; Delcour, J.A. Starch blends and their physicochemical properties. *Starch-Stärke* **2015**, *67*, 1–13. [CrossRef]
24. Zhou, Y.Y. *Theory of Textile Warp Sizes*; China Textile& Apparel Press: Beijing, China, 2004; pp. 115–260.
25. Davidson, V.J.; Paton, D.; Diosady, L.L.; Larocque, G.J. Degradation of wheat starch in a single-screw extruder: Characteristics of extruded starch polymers. *J. Food Sci.* **1984**, *49*, 453–458. [CrossRef]
26. Li, W.; Zhang, Z.Q.; Wu, J.; Xu, Z.Z.; Liu, Z. Phosphorylation/caproylation of cornstarch to improve its adhesion to PLA and cotton fibers. *RSC Adv.* **2019**, *9*, 34880–34887. [CrossRef]

27. Behera, B.K.; Gupta, R.; Mishra, R. Comparative analysis of mechanical properties of size film. I. Performance of individual size materials. *Fibers Polym.* **2008**, *9*, 481–488. [CrossRef]
28. Beliakova, M.K.; Aly, A.A.; Abdel-Mohdy, F.A. Grafting of poly (methacrylic acid) on starch and poly (vinyl alcohol). *Starch-Stärke* **2004**, *56*, 407–412. [CrossRef]
29. Meshram, M.W.; Patil, V.V.; Mhaske, S.T.; Thorat, B.N. Graft copolymers of starch and its application in textiles. *Carbohydr. Polym.* **2009**, *75*, 71–78. [CrossRef]
30. John, M.J.; Anandjiwala, R.; Oksman, K.; Mathew, A.P. Melt-spun polylactic acid fibers: Effect of cellulose nanowhiskers on processing and properties. *J. Appl. Polym. Sci.* **2012**, *127*, 274–281. [CrossRef]
31. Bello-Pérez, L.A.; Bello-Flores, C.A.; Nunez-Santiago, M.D.C.; Coronel-Aguilera, C.P.; Alvarez-Ramirez, J. Effect of the degree of substitution of octenyl succinic anhydride-banana starch on emulsion stability. *Carbohydr. Polym.* **2015**, *132*, 17–24. [CrossRef]
32. Zhu, Z.F.; Zhuo, R.X. Controlled release of carboxylic-containing herbicides by starch-g-poly (butyl acrylate). *J. Appl. Polym. Sci.* **2001**, *81*, 1535–1543. [CrossRef]
33. Zhu, Z.F.; Cheng, Z.Q. Effect of inorganic phosphates on the adhesion of mono-phosphorylated cornstarch to fibers. *Starch-Stärke* **2008**, *60*, 315–320. [CrossRef]
34. Bismark, S.; Zhu, Z.F. Amphipathic starch with phosphate and octenylsuccinate substituents for strong adhesion to cotton in warp sizing. *Fibers Polym.* **2018**, *19*, 1850–1860. [CrossRef]
35. Varavinit, S.; Chaokasem, N.; Shobsngob, S. Studies of flavor encapsulation by agents produced from modified sago and tapioca starches. *Starch-Stärke* **2001**, *53*, 281–287. [CrossRef]
36. Li, W.; Xu, Z.Z.; Wang, Z.Q.; Liu, X.H.; Li, C.L.; Ruan, F.T. Double etherification of corn starch to improve its adhesion to cotton and polyester fibers. *Int. J. Adhes. Adhes.* **2018**, *84*, 101–107. [CrossRef]
37. Chang, Y.J.; Choi, H.W.; Kim, H.S.; Lee, H.; Kim, W.; Kim, D.O.; Kim, B.Y.; Baik, M.Y. Physicochemical properties of granular and non-granular cationic starches prepared under ultra high pressure. *Carbohydr. Polym.* **2014**, *99*, 385–393. [CrossRef] [PubMed]
38. Li, W.; Zhu, Z.F. Electroneutral maize starch by quaterization and sulfosuccination for strong adhesion-to-viscose fibers and easy removal. *J. Adhes.* **2016**, *92*, 257–272. [CrossRef]
39. Li, W.; Xu, Z.Z.; Wang, Z.Q.; Xing, J. One-step quaternization/hydroxypropylsulfonation to improve paste stability, adhesion and film properties of oxidized starch. *Polymers* **2018**, *10*, 1110. [CrossRef] [PubMed]
40. Ge, W. Chemical analysis of starch-like mineral crystals to eliminate misidentification in ancient residue research. *Archaeometry* **2013**, *55*, 1122–1131. [CrossRef]
41. Yoshimura, T.; Yoshimura, R.; Seki, C.; Fujioka, R. Synthesis and characterization of biodegradable hydrogels based on starch and succinic anhydride. *Carbohydr. Polym.* **2006**, *64*, 345–349. [CrossRef]
42. Chang, P.R.; Qian, D.Y.; Anderson, D.P.; Ma, X.F. Preparation and properties of the succinic ester of porous starch. *Carbohydr. Polym.* **2012**, *88*, 604–608. [CrossRef]
43. Li, W.; Wu, J.; Guo, Y.; Lu, Y.H. Impact of hydroxypropylsulfonation on adhesion-to-fibers and film properties of corn starch. *J. Text. Inst.* **2019**, *110*, 1679–1686. [CrossRef]
44. Xie, W.L.; Wang, Y.B. Synthesis of high fatty acid starch esters with 1-butyl-3-methylimidazolium chloride as a reaction medium. *Starch-Stärke* **2011**, *63*, 190–197. [CrossRef]
45. Li, W.; Xu, W.Z.; Wei, A.F.; Xu, Z.Z.; Zhang, C.H. Quaternization/maleation of cornstarch to improve its adhesion and film properties for warp sizing. *Fibers Polym.* **2016**, *17*, 1589–1597. [CrossRef]
46. Wurzburg, O.; Szymanski, C. Modified starches for the food industry. *J. Agric. Food Chem.* **1970**, *18*, 997–1001.
47. Doublier, J.L.; Llamas, G.; Meur, M.L. A rheological investigation of cereal starch pastes and gels. Effect of pasting procedures. *Carbohydr. Polym.* **1987**, *7*, 251–275. [CrossRef]
48. Wong, R.B.K.; Lelievre, J. Rheological characteristics of wheat starch pastes measured under steady shear conditions. *J. Appl. Polym. Sci.* **1982**, *27*, 1433–1440. [CrossRef]
49. Wu, S.H. *Polymer Interface and Adhesion*; Marcel Dekker: New York, NY, USA, 1982; pp. 359–448.
50. Shen, S.Q.; Zhu, Z.F.; Liu, F.D. Introduction of poly [(2-acryloyloxyethyl trimethyl ammonium chloride)-*co*-(acrylic acid)] branches onto starch for cotton warp sizing. *Carbohydr. Polym.* **2016**, *138*, 280–289. [CrossRef] [PubMed]
51. Li, W.; Zhu, Z.F. Effect of sulfosuccinylation of corn starch on the adhesion to viscose fibres at lower temperature. *Indian J. Fibre Text. Res.* **2014**, *39*, 314–321.

52. Zhu, Z.F.; Zhang, L.Y.; Feng, X.M. Introduction of 3-(trimethylammonium chloride)-2-hydroxypropyls onto starch chains for improving the grafting efficiency and sizing property of starch-*g*-poly (acrylic acid). *Starch-Stärke* **2016**, *68*, 742–752. [CrossRef]
53. Zhang, K. *Interface Science of Polymers*; China Petrochemical Press: Beijing, China, 1996; p. 130.
54. Zhu, Z.F.; Liu, Z.J.; Li, M.L.; Xu, D.S.; Li, C.L. Mono-phosphorylation of cornstarch to improve the properties of wool yarns sized at reduced temperature. *J. Appl. Polym. Sci.* **2013**, *127*, 127–135. [CrossRef]
55. Zhu, Z.F. *Chemistry in Textile Engineering*; Donghua University Press Co. Ltd.: Shanghai, China, 2010; pp. 179–217.
56. Seydel, P.V.; Hunt, J.R. *Textile Warp Sizing*; Phoenix Printing Inc.: Atlanta, GA, USA, 1981; pp. 5–16, 247–267.
57. Thiré, R.M.S.; Simão, M.R.A.; Andrade, C.T. High resolution imaging of the microstructure of maize starch films. *Carbohydr. Polym.* **2003**, *54*, 149–158. [CrossRef]
58. Li, W.; Xu, Z.Z.; Wang, Z.Q.; Li, C.L.; Feng, Q.; Zhu, Y.N. Tertiary amination/hydroxypropylsulfonation of cornstarch to improve the adhesion-to-fibers and film properties for warp sizing. *Fibers Polym.* **2018**, *19*, 1386–1394. [CrossRef]

© 2019 by the authors. Licensee MDPI, Basel, Switzerland. This article is an open access article distributed under the terms and conditions of the Creative Commons Attribution (CC BY) license (http://creativecommons.org/licenses/by/4.0/).

Article

Low Dielectric Constant Polyimide Obtained by Four Kinds of Irradiation Sources

Hongxia Li [1], Jianqun Yang [1], Shangli Dong [1], Feng Tian [2] and Xingji Li [1,*]

[1] School of Material Science and Engineering, Harbin Institute of Technology, Harbin 150001, China; koalalhx@163.com (H.L.); yangjianqun@hit.edu.cn (J.Y.); sldong@hit.edu.cn (S.D.)
[2] Shanghai Synchrotron Radiation Facility, Zhangjiang Lab, Shanghai Advanced Research Institute, Chinese Academy of Sciences, Shanghai 201204, China; tianfeng@zjlab.org.cn
* Correspondence: lxj0218@hit.edu.cn

Received: 27 January 2020; Accepted: 12 March 2020; Published: 10 April 2020

Abstract: Irradiation is a good modification technique, which can be used to modify the electrical properties, mechanical properties, and thermal properties of polymer materials. The effects of irradiation on the electrical properties, mechanical properties, and structure of polyimide (PI) films were studied. PI films were irradiated by a 1 MeV electron, 3 MeV proton, 10 MeV proton, and 25 MeV carbon ion. Dielectric constant, dielectric loss, and resistance measurements were carried out to evaluate the changes in the electrical properties; moreover, the mechanical properties of the pristine and irradiated PI were analyzed by the tensile testing system. The irradiation induced chemical bonds and free radicals changes of the PI films were confirmed by the Fourier transform infrared (FTIR) spectra, X-ray photoelectron spectroscopy (XPS), and electron paramagnetic resonance (EPR). The dielectric constant of the PI films decreases with the increase of fluences by the four kinds of irradiation sources.

Keywords: low dielectric constant; PI; irradiation; dielectric loss

1. Introduction

In recent years, with the rapid development of microelectronics industry, the miniaturization of electronic components, the rapid growth of large scale integrated circuit chips, the increase of chip interconnect density, and the improvement the circuit connection of the resistance and capacitance, the signal delay of resistance production effect signal transmission speed and signal loss [1–3]. As a kind of functional material, low dielectric constant polymers have become an important research direction. Polyimide, as an important insulating and encapsulating material, is widely used in aerospace and microelectronic fields. Currently, many researchers have reported that polyimide (PI) thin films with low dielectric constant are obtained by doped fluorine-containing groups, doped porous, or other introduced functional groups methods. Goto et al. [4–7] successfully prepared a series of thermostable low dielectric constant polyimide (PI) by introducing diphenyl fluorine group into the main chain. The result shows that the lowest dielectric constant of the non-fluorinated PI is 2.77, while the lowest dielectric constant of the fluorinated PI is 2.35. Fujiwara et al. [8] prepared poly(norborneolin imide) with large fluorinated aromatic groups in the side chain and confirmed that the main chain structure of fluorinated poly(norborneolin imide) showed helical structure through molecular simulation. The results showed that the polymer had good heat resistance and glass transition temperature (T_g) available at above 400 °C, the spiral structure increased the free volume, and the dielectric constant was as low as 2.31 [9]. The reason for the decrease in the dielectric constant of the PI-PDMS polymer is that polydimethylsiloxane (PDMS) was added to the polyimide, and then the chain reaction produced carbon dioxide to form nanoporous films. The result showed that the copolymers had the lowest dielectric constant (2.58) when the mass fraction ω(PDMS) = 25% and ω(MDI) = 5%. Meanwhile, Lee et al. [10–12] prepared nanoporous polyimide films with polyoxyethylene (PEO) polyhedral oligomeric silsesquioxane (POSS) nanoparticles, and the dielectric constant of such porous hybrid films were 3.25–2.25. The method can be used to obtain low dielectric constant polymers by doping porous materials or introducing other functional groups.

Radiation treatment of polymers involves irradiation of the polymers, usually in a continuous mode, and modification of the polymers to improve their performance for industrial purposes. The irradiation processing of polymers mainly includes cross-linking, curing, grafting, and degradation [13–16]. Irradiation changes the structure and properties of polymer materials, and the dielectric constant of the irradiated polymer decreases when the content of polarized group decreases. Polyimide (PI) can be used as dielectric material for thin-film transistors and capacitors [17–20]. Therefore, it is of great significance to study the properties of irradiated polymers to evaluate the improvement or deterioration in their mechanical strength, molecular weight, dielectric properties, and thermal properties. At present, some researchers have reported the effects of irradiation on the dielectric properties of polymers. S. Raghu et al. [21] used electron beam and gamma ray irradiated polymer electrolyte films, the results showed that the dielectric constant and conductivity increased with the increase of irradiation fluences. Qureshi et al. [22] found that the dielectric constant and dielectric loss of polyimide increased with the increase of irradiation fluences after irradiation with 80 MeV O^{6+} ion. Quamara et al. [23] pointed out that the dielectric behavior of polyimide was a non-monotonic evolution law with the increase of irradiation fluences after 50 MeV Si^+ ion irradiation and that it was temperature-dependent. However, the irradiated polymers with low dielectric constant are rarely reported.

In this research work, PI films were exposed by irradiation sources including a 1 MeV electron, 3 MeV proton, 10 MeV proton, and 25 MeV carbon ion with different fluences. The dielectric properties of pristine and irradiated PI films were analyzed by dielectric spectroscopy, and it was found that the dielectric constant decreased to a minimum of 2.7. However, the mechanical properties did not change much compared with pristine PI. The four kinds of irradiation sources can decrease the dielectric constant of PI films.

2. Experimental Section

2.1. Materials and Equipment

Kapton-H polyimide (PI, Isophthalic anhydride diaminodiphenyl ether) films with the density of 1.34 g/cm^3 and 50 μm in thickness were purchased from Du Pont co., Ltd., with molecular formula $(C_{22}H_{10}N_2O_5)_n$. The high-energy electron irradiation test was performed by the DD 1.2 high-frequency high-voltage electronic accelerator in Heilongjiang Technical Physics Institute. High-energy proton and carbon ion irradiation experiments were carried out using the EN2 × 6 tandem electrostatic accelerator at Peking University.

2.2. Experimental Parameter

The PI films were cut into 5 × 5 cm^2 size and all the irradiation area was larger than this size. The electron energy is 1 MeV; particle radiation flux is 2×10^8 cm$^{-2} \cdot$s^{-1}; and the irradiation fluences are 3.8×10^{13}, 3.8×10^{14}, and 2.2×10^{16} e/cm^2, respectively. The 1 MeV electron penetration depth is ~100 μm calculated by Geant4. The proton energy is 3 MeV; the 3 MeV proton penetration depth is ~100 μm, as calculated by Stopping and Ranges of Ions in Matter (SRIM) software; and the irradiation fluences are 4.4×10^{11}, 1.1×10^{12}, and 2.2×10^{12} p/cm^2, respectively. The proton energy is 10 MeV; the 10 MeV proton penetration depth is ~1000 μm, as calculated by SRIM software; and the irradiation fluences are 5×10^{11}, 1×10^{12}, and 1×10^{13} p/cm^2, respectively. The carbon ion energy is 25 MeV, the 25 MeV carbon ion penetration depth is ~30 μm as calculated by SRIM software, and the films thickness is 50 μm, thus the 25 MeV carbon ion does not penetrate through the PI films, and the irradiation fluences are 2.8×10^{11}, 1.2×10^{12}, and 5×10^{12} ion/cm^2, respectively. The sizes for the dumbbell-shaped tensile samples are 12, 4, and 0.5 mm.

2.3. Characterizations

The dielectric constant, dielectric loss, and insulation resistance of the samples were measured by BDS 4000 Novocontrol broadband dielectric spectrometer/impedance, which was produced in Germany in a wide frequency range from 10^{-1} to 10^7 Hz at room temperature. Both sides of the PI samples were plated with a 25 mm diameter Al electrode by the vacuum evaporation machine before test. The equipment used in external stimulus is an electric field, E. The composition and chemical state of polyimide were characterized by the X-ray photoelectron spectroscopy (XPS) (K-Alpha) produced by the Thermal fisher scientific. The samples used for the tensile test are dumbbell-shaped samples with drawing rate of 2 µm/s at room temperature in air, which uses an MST810 material tensile testing system. Fourier infrared spectroscopy (FTIR) spectra of films have been obtained by Nicolet Magna-IR 560 spectrometer operated in transmission mode with the spectral region between 4000 and 700 cm^{-1} and scanned 32 times on average. Free radicals were tested by the A200 type Electron paramagnetic resonance (EPR) from the German Bruke company. In this experiment, the center of the magnetic field is 3580 G, the sweep width field is 90 G, the microwave frequency is 9.85 GHz, the time constant is 20 s, the spectral gain is 2×10^4, the modulation amplitude is 1 G, and the microwave power is 6.140 mW.

3. Results

3.1. Dielectric Constant Analysis

The dielectric constant (ε) of pristine and irradiated PI samples after 1 MeV electron with different fluences (3.8×10^{13} e/cm^2, 3.8×10^{14} e/cm^2, 2.2×10^{16} e/cm^2) at different frequency and room temperature are shown in Figure 1a. The dielectric constant of PI films decreases after the four kinds of irradiation sources; moreover, it decreases with the increase of the fluences. The lowest dielectric constant of PI irradiated by 1 MeV electron with 2.2×10^{16} e/cm^2 is ~2.7. The dielectric constant of polyimide decreases with the increase of irradiation fluences in the whole frequency range. It is worth noting that the degree of attenuation of the dielectric constant is not linear with the increase in irradiation fluences as shown in Figure 1e. All samples show normal dielectric dispersion behavior, and the dielectric constant is only slightly reduced. All of the dielectric constant curves are frequency dependent. In the higher frequency range (>10^6 Hz), the decrease in dielectric constant is more obvious. Under such high frequency conditions, the polar group with such orientation polarization motion cannot keep up with the change of electric field frequency, so the dielectric constant decreases.

Figure 1b,c shows the variation of the dielectric constant of PI samples before and after 3 MeV proton irradiation with different fluences (4.4×10^{11}, 1.1×10^{12}, and 2.2×10^{12} p/cm^2) and 10 MeV proton irradiation with different fluences (5×10^{11}, 1×10^{12}, and 1×10^{13} p/cm^2) as a function of the frequency at room temperature. Both dielectric constants of the irradiated PI decrease obviously with the increase of the fluences. All samples show normal dielectric dispersion behavior; the dielectric constant of PI samples decreases in the high-frequency range and the dielectric constant decreases with the increase of frequency. Figure 1d shows the variation of the dielectric constant of PI samples before and after 25 MeV carbon ion irradiated with different fluences (2.8×10^{11}, 1.2×10^{12}, and 5×10^{12} ions/cm^2) as a function of the frequency at room temperature. The dielectric constant of polyimide decreases with the increase of irradiation fluences in the whole frequency range. However, the dielectric constant of irradiated polyimide begins to decrease at lower frequency when compared with the dielectric constant of the pristine PI, as the Figure 1d shown. The dielectric constant of pristine PI begins to decrease at 10^6 Hz, the dielectric constant of irradiated PI with 2.8×10^{11} ions/cm^2 begins to decrease at 10^5 Hz, the dielectric constant of irradiated PI with 1.2×10^{12} ions/cm^2 begins to decrease at 10^4 Hz, and the dielectric constant of irradiated PI with 5×10^{12} ions/cm^2 begins to decrease at 10^3 Hz.

Figure 1e shows the variation in the dielectric constant of PI irradiated with different sources and fluences, the curve of 1 MeV electron uses bottom X-axis, and the curves of other irradiation sources use top X-axis at 100 Hz. The dielectric constant decreases nonlinearly with the increase of irradiation fluences. Figure 1f shows the measurement results with the rate of change of dielectric constant (at 100 Hz). The variation trend of Figure 4f is consistent with that of Figure 1e. These changes in the dielectric constant were calculated by using Equation (1).

$$\nabla \varepsilon = (\varepsilon_a - \varepsilon_b)\varepsilon_b \times 100 \quad \% \tag{1}$$

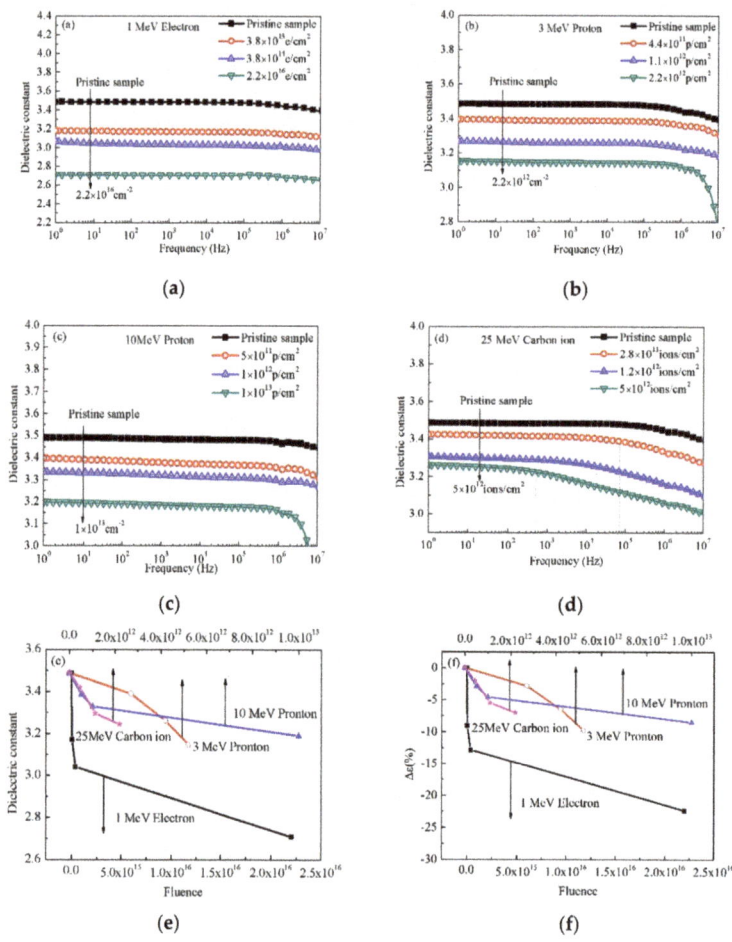

Figure 1. Frequency dependent dielectric constant of pristine and irradiated polyimide (PI) films with different sources and fluences at room temperature: (**a**) 1 MeV electron, (**b**) 3 MeV proton, (**c**) 10 MeV proton, (**d**) 25 MeV carbon ion, (**e**) dielectric constant at 100 Hz, and (**f**) the rate of change of dielectric constant at 100 Hz.

Here, ε_b is the dielectric constant of pristine PI and ε_a is the dielectric constant of irradiated PI.

According to the evolution characteristics of the dielectric properties of polymers, the dielectric constant of polymer is closely related to its characteristics of molecule, chain group structure, and the internal defects [21]. When the frequency is lower than 1×10^8 Hz, the dielectric behavior of PI films mainly depends on the orientation polarization, the interface polarization, and dipole polarization, as shown in Figure 9. The interface polarization mainly comes from the defects after irradiation, as shown in Figure 8. The orientation polarization mainly comes from the main chain of molecule. This experiment is tested at room temperature, so the polarization is weak. The results are different from the high dielectric constant induced by irradiation such as gamma, so there may be some other main polarization mechanism leading to the reduction of dielectric constant here.

3.2. Dielectric Loss and Resistance Analysis

The dielectric loss of pristine and irradiated PI with different sources and fluences is shown in Figure 2 at different frequency and room temperature. From Figure 2a, it can be seen that the dielectric loss of the irradiated PI with 1 MeV electron increases compared with the pristine sample, especially at low frequency, but the dielectric loss increases with the increase in frequency (>10^4 Hz) and the curves are almost the same after irradiation with different fluences. From Figure 2b,c, the data shows that the dielectric loss decreases a little with the increase in the frequency in the low frequency region, but the dielectric loss increases obviously with the increase of frequency in the high frequency region. It can be seen that the dielectric loss of the irradiated PI with 3 MeV proton shows a movement in the peak compared with the pristine sample at 10 Hz. The peak moves to 100 Hz after the 10 MeV proton. The dielectric loss of irradiated PI increases slightly compared with the pristine PI in the whole frequency. Figure 2d shows the variation of dielectric loss of PI irradiated by 25 MeV carbon ion in different frequency. The dielectric loss of PI irradiated by 25 MeV carbon ion increases with the increase of fluences in low frequency but the curves almost the same in high frequency compared with the pristine sample. The dielectric loss increases with the increase of the frequency for all irradiated and pristine PI samples. The dielectric loss depends on a number of factors, such as orientation polarization and dipole polarization. At low frequency, the dielectric loss is mainly caused by the main chain of the molecule movement polarization relaxation. There is a peak around 100 Hz that may be caused by the space charge polarization relaxation. The strong polarity in the molecular chain can produce unsaturated groups, such as carbonyl group, which have the polarization characteristics of space charge due to the movement ability limited, so the increase in the dielectric loss produces the peak. At high frequency, the dipole polarization relaxation produces the dielectric loss, and the dipole polarization cannot keep up with the change of frequency, thus the dielectric loss increase with the increase of the frequency.

The variation insulation resistance curves of PI irradiated with different fluences of four irradiation sources are shown in Figure 3 at the voltages of 500 V and 4000 V. It can be seen from the figure that the insulation resistance of PI decreases with the increase of fluences under different irradiation conditions. This indicates that all kinds of irradiation sources lead to the poor insulation of polyimide. This implies that the conductivity increases approximately with the increase of fluences. However, the resistance of the irradiated polyimide is 10^{12} Ω, which still maintains good insulation performance.

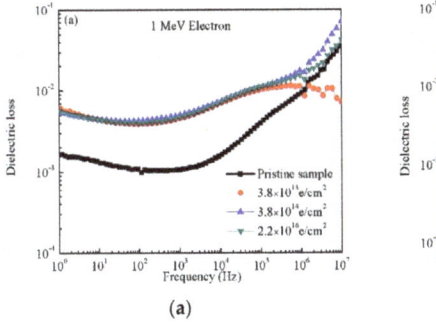
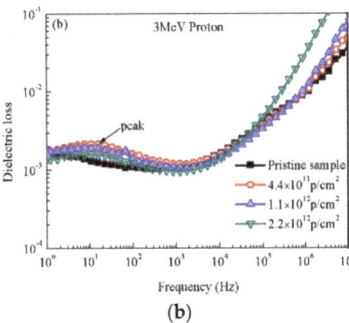

(a) (b)

Figure 2. *Cont.*

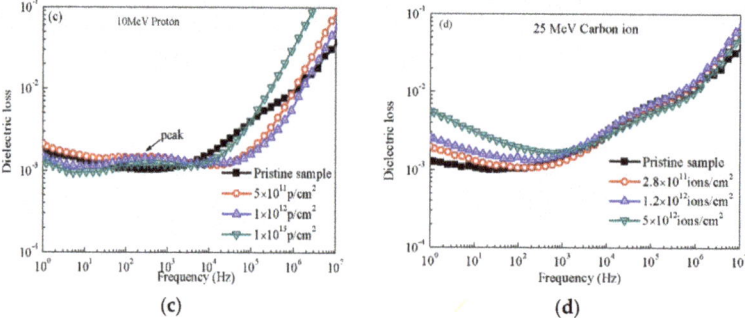

Figure 2. Frequency dependent dielectric loss of pristine and irradiated PI films with different sources and fluences at room temperature: (**a**) 1 MeV electron, (**b**) 3 MeV proton, (**c**) 10 MeV proton, and (**d**) 25 MeV carbon ion.

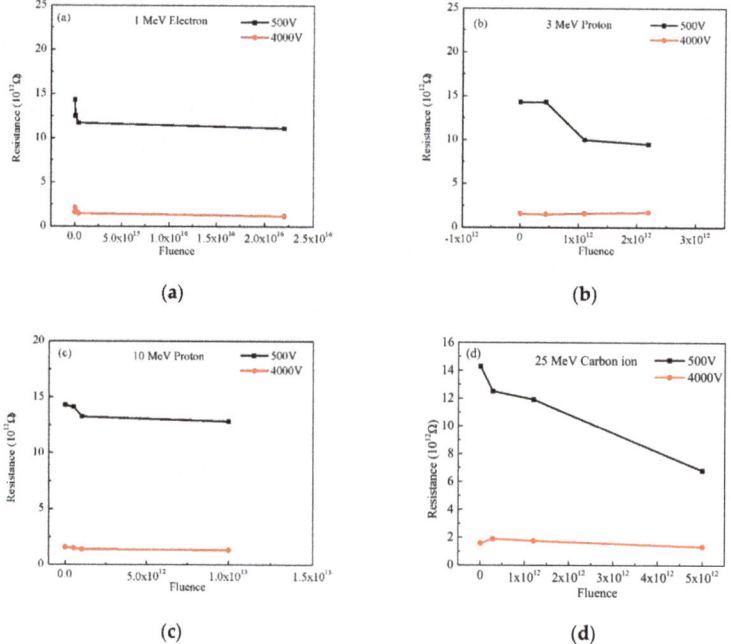

Figure 3. Frequency dependent resistance of pristine and irradiated PI films with different sources and fluences at room temperature: (**a**) 1 MeV electron, (**b**) 3 MeV proton, (**c**) 10 MeV proton, and (**d**) 25 MeV carbon ion.

3.3. XPS Analysis

To investigate the chemical bond states and content of PI samples after irradiation, the XPS fine spectrum of pristine polyimide C1 s and O1 s and the detailed relative component ratios of different bonds irradiated by 3 MeV proton with 2.2×10^{12} p/cm^2 and 25 MeV carbon ion with 5×10^{12} ions/cm^2 are studied, as shown in Figure 4. The two peaks at 286.3 eV and 288.4 eV are attributed to C–O bonds and carbonyl group C=O bonds, respectively. By analyzing the content of different chemical bonds in the C1 s fine spectrum of XPS, it can reflect the structural damage behavior of polyimide after irradiation. Compared with the pristine PI, the relative content of C=O bonds and C–N bonds reduces after irradiation by 3 MeV proton and 25 MeV carbon ion, whereas the relative content of C–C bonds and C–O bonds increases, as shown in Table 1. Moreover, the content of C–C bonds changes a little after 3 MeV proton irradiation. However, the content of C–C bonds increases significantly after 25 MeV carbon ion irradiation, because the carbon ion irradiation produces a large number of pyrolytic carbon free radicals, and then the compound of these free radicals generates a lot of C–C bonds. Compared with the pristine PI, the relative content of C=O bonds decreases after irradiation by the 3 MeV proton and 25 MeV carbon ion, whereas the relative content of C–O bonds increases in the O1 s spectrum, as shown in Table 2. Thus, the content of polar groups such as C–N bonds and C=O bonds decreases, whereas the content of nonpolar groups such as C–C bonds increases after irradiation. This reduces the dielectric constant of the material, and forms a PI film with a low dielectric constant.

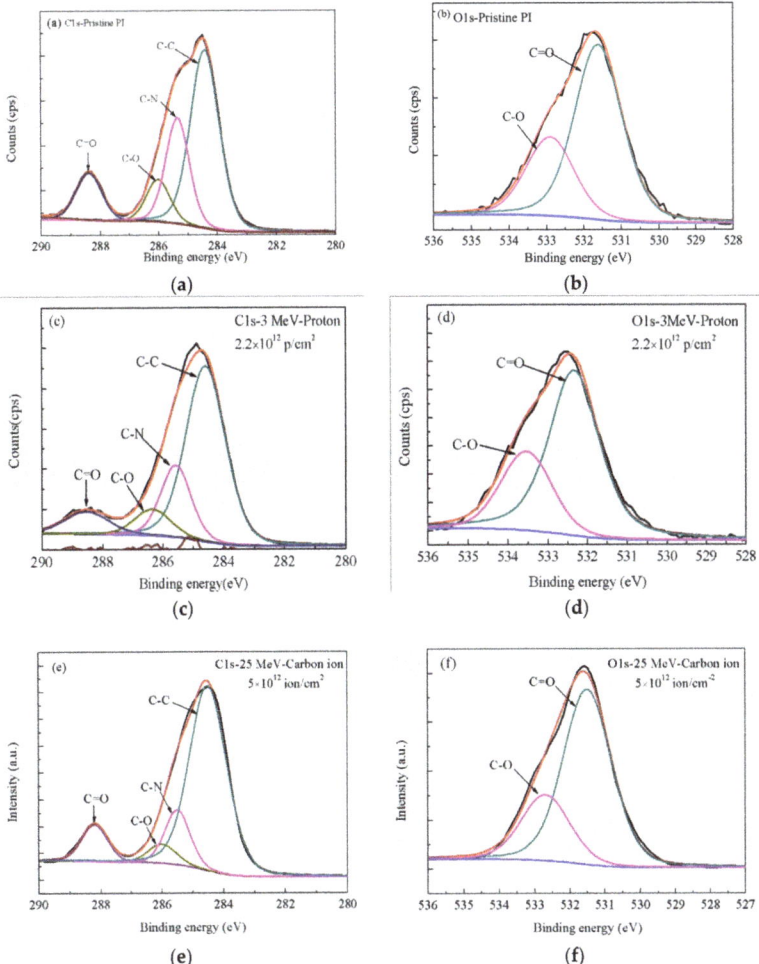

Figure 4. X-ray photoelectron spectroscopy (XPS) curves of pristine and irradiated PI films: (**a**) C1 s of pristine PI, (**b**) O1 s of pristine PI, (**c**) C1 s of irradiated PI with 3 MeV proton of 2.2×10^{12} p/cm^2, (**d**) O1 s of irradiated PI with 3 MeV proton of 2.2×10^{12} p/cm^2, (**e**) C1 s of irradiated PI with 25 MeV carbon ion of 5×10^{12} p/cm^2, and (**f**) O1 s of irradiated PI with 25 MeV carbon ion of 5×10^{12} p/cm^2.

Table 1. C element composition of PI with irradiation energy of 3 MeV proton and 25 MeV carbon ion.

Irradiation Energy	Fluence (cm^{-2})	Chemical Bond	Binding Energy (eV)	Proportion (%)
Pristine Polyimide		C–C	284.5	60.4
		C–N	285.5	20.1
		C–O	286.3	6.8
		C=O	288.4	12.7
3 MeV Proton	2.2×10^{12}	C–C	284.5	65.7
		C–N	285.5	16.8
		C–O	286.5	7.5
		C=O	288.5	10.0
25 MeV Carbon ion	5.0×10^{12}	C–C	284.5	71.6
		C–N	285.5	11.4
		C–O	286.5	7.3
		C=O	288.5	9.7

Table 2. O element composition of PI with irradiation energy of 3 MeV proton and 25 MeV carbon ion.

Irradiation Energy	Fluence (cm^{-2})	Chemical Bond	Binding Energy (eV)	Proportion (%)
Pristine Polyimide		C=O	532	74.1
		C–O	533.2	25.9
3 MeV Proton	2.2×10^{12}	C=O	532	71.1
		C–O	533.2	28.9
25 MeV Carbon ion	5.0×10^{12}	C=O	532	70.6
		C–O	533.2	29.4

3.4. FT-IR Analysis

The FT-IR spectra of the pristine and irradiated PI films with different sources and fluences as a function of the frequency are shown in Figure 5. All of the samples exhibit typical characteristic imide peaks at 1780 cm^{-1} (imide C=O asymmetric stretching), 1720 cm^{-1} (imide C=O symmetric stretching), 1246 cm^{-1} and 1168 cm^{-1} (aromatic ether (C–O–C) asymmetric stretching vibration), and 1380 cm^{-1} (C–N–C stretching vibration). The intensity of the bands at 1500 and 1116 cm^{-1} represents the C=C stretching vibration and C–C stretching vibration, respectively [21]. The peak positions and patterns of the absorption peaks of the irradiated characteristic groups are similar to the spectral lines of the pristine samples, and no new characteristic absorption peaks generate, which indicates that no new groups generate inside the irradiated materials. By comparing the strength of the characteristic absorption peaks of materials before and after the four irradiation sources, it can be seen that the strength of characteristic absorption peaks of each group decreases slightly with the increase of irradiation fluences. The reason for this change may be the increase of material surface roughness caused by irradiation [22], this improves the material surface infrared scattering and weakens the intensity of absorption.

3.5. EPR Analysis

The EPR spectra of the pristine and irradiated PI films with different sources and fluences are shown in Figure 6: (a) 1 MeV electron, (b) annealing curves after 1 MeV electron, (c) 25 MeV carbon ion, and (d) the change of the content of free radicals. The results show that a large number of free radicals appear in the PI films under the condition of charged particles irradiation. In Figure 6a, the content of free radicals increases with the increase in the electron irradiation fluences. The content of free radicals changes greatly after carbon ion irradiation, as shown in Figure 6c. In Figure 6b,d, the content of free radicals after 1 MeV electron irradiation decreases as the change of time, which corresponds to the results of XPS data. This means that the free radicals on the surface from irradiated PI films recombine with the surrounding environment active factors such as oxygen from air, which leads to the decreasing of the content of free radicals. Figure 6d shows that the content of free radicals changes with the change of fluences or time after irradiation. The data are calculated by Formula (2):

$$N = \frac{\iint S ds}{m} \quad (2)$$

where N is the content of free radicals, S is the measured EPR spectrum, and m is the quality of test sample here. As it can be seen from Figure 6d, the PI films produce a lot of free radicals and the content of the free radicals increases with the increase in the electron and carbon ions irradiation fluences. Moreover, more free radicals are produced after carbon ion irradiation than after electron irradiation. The g value is 2.0025 in this data, and the g value is stable and did not change with the irradiated particle energy and fluences. In the polymer material, the g values correspond to two kinds of free radical: one is a kind of pyrolytic carbon free radical, and the other is a hydroxyl superoxide radical. The irradiation experiment is under a vacuum environment, and therefore there is not enough oxygen reacting with free radicals, so it is impossible to generate such a large number of hydroxyl superoxide radicals. Moreover, combining the result in Figure 6c, the free radicals measured in experiment should be mainly pyrolytic carbon free radicals. The reason why the content of the free radicals after electron irradiation reduces in Figure 6d mainly because the free radicals on the surface recombine with active factors in the air, such as oxygen, as shown in Figure 8.

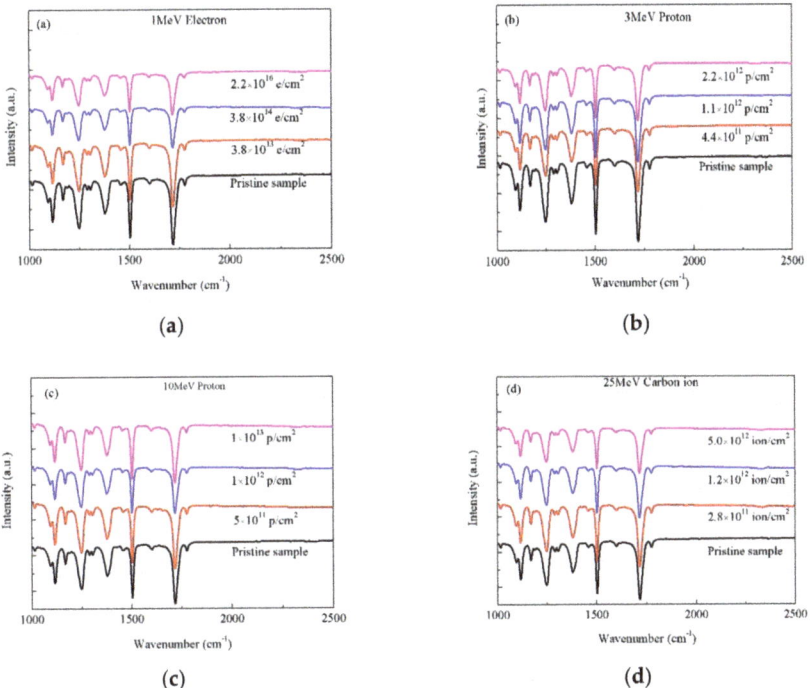

Figure 5. FT-IR curves of pristine and irradiated PI films with different sources and fluences: (**a**) 1 MeV electron, (**b**) 3 MeV proton, (**c**) 10 MeV proton, and (**d**) 25 MeV carbon ion.

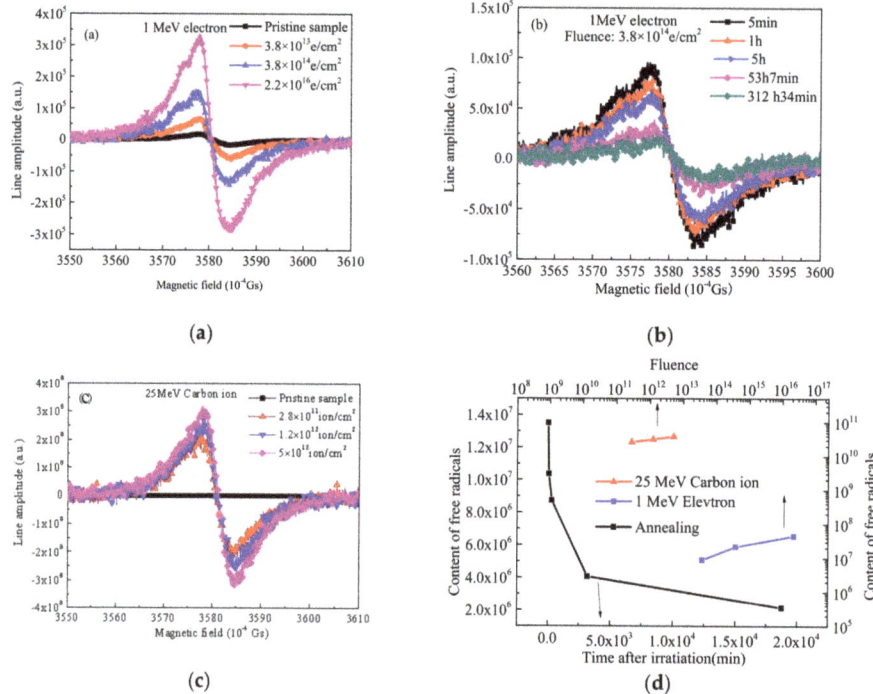

Figure 6. Electron paramagnetic resonance (EPR) curves of the pristine and irradiated PI films with different sources and fluences: (**a**) 1 MeV electron, (**b**) annealing after 1 MeV electron, (**c**) 25 MeV carbon ion, and (**d**) the change of the content of free radicals.

3.6. Mechanical Property Analysis

As we all know, stress is defined as the force per unit area of the samples and strain is the measure of the change in the sample length. The pristine and irradiated PI films with different sources and fluences exhibit typical stress–strain curves at room temperature, as shown in Figure 7. In the first stage of deformation, the stress gradually increases and reaches the yield strength about 110 Mpa, which shows the strain strengthening characteristics. With the obvious strain softening characteristics, it enters the second stage of deformation directly, and the tensile stress increases until the material breaks up. It is worth noting that the yield strength decreases slightly after the irradiation of four different sources, but the elongation at the break, at which the strain is ~60%, did not show an obvious change after irradiation. However, the elongation at the break decreased after the 10 MeV proton irradiation. The reason for this may be the degradation of PI after radiation, and then the reduction in molecular weight, resulting in the reduction in yield strength. Compared with other methods of introducing holes, which led to the mechanical properties of the materials decreasing sharply, the mechanical properties of the materials can still be maintained by the irradiation method. However, the disadvantage of this method is that a large number of fluences of irradiation for the material cannot be use; otherwise, the materials will degrade a lot and lose their available properties.

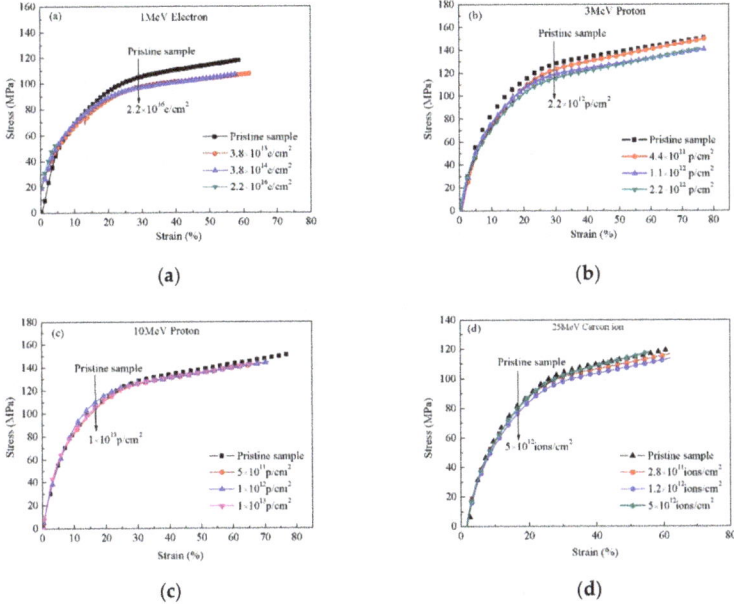

Figure 7. Stress–strain curves of pristine and irradiated PI films with different sources and fluences at room temperature: (**a**) 1 MeV electron, (**b**) 3 MeV proton, (**c**) 10 MeV proton, and (**d**) 25 MeV carbon ion.

4. Discussion

The schematic of the irradiation process and irradiation degradation process and results of XPS data of PI repeat unit is shown in Figure 8. The schematic shows that the PI films irradiated by the four kinds of irradiation sources, including 1 MeV electron, 3 MeV proton, 10 MeV proton, and 25 MeV carbon ion, may produce defects, and the free radicals on the surface of the irradiated PI films and the free radicals may combine with active factors such as oxygen. The accelerated 1 MeV electron, 3 MeV proton, 10 MeV proton, and 25 MeV carbon ion all have enough energy to break chemical bonds in organic materials. The irradiation sources may break the benzene ring, ether bonds, C–N bonds, and C=O bonds of the PI films. The most common result of chemical bond breakage is the formation of free radicals. From the EPR results, we know that the free radicals are the pyrolytic carbon free radicals. The irradiation processes can be classified according to the effects of formation of free radicals, including curing, cross-linking, degradation, and grafting [24]. The experimental XPS results show that the content of C=O bonds and C–N bonds decreases and the content of C–O bonds and C–C bonds increases. According to the results, we can know degradation is the main process and the fracture of chemical bonds often forms polymers with low molecular weight.

The sample schematic of orientation polarization, interface polarization, and dipole polarization in the PI films is shown in Figure 9. The dielectric constant section refers to these polarizations. The orientation polarization mainly regards the molecular main chain polarization and the ability of the polarization relaxation related to the movement of main chain molecules. The reason why the interface polarization happens is that the irradiated PI films produce defects on the surface [24]. The dipole polarization is closely related to the relaxation process of the corresponding groups within the molecular features, and highly polarized groups in the molecular chain of polyimide, which can dominate this dipole polarization process. Tables 3 and 4 show the typical molecular dipole moment and polarizability of the functional groups of the polyimide [25]. It can be seen that the groups C=O, C–O, and C–N have high polarizability and dipole moment, so electron density is concentrated around carbon, whereas oxygen and nitrogen are depleted in the condition of external field; these groups have strong polarization and corresponding relaxation phenomenon.

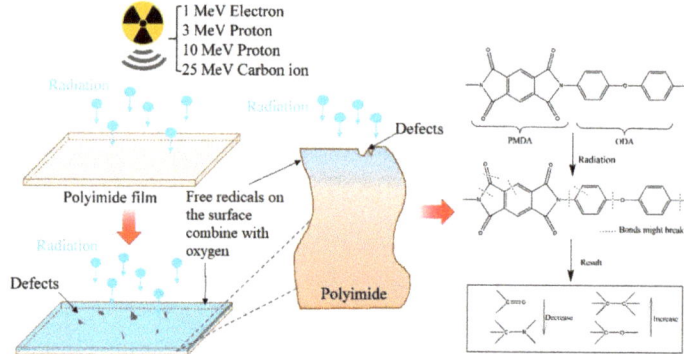

Figure 8. Schematic of the irradiation process and irradiation degradation process of PI repeat unit and the XPS results show that the content of C=O and C–N decrease and the content of C–C and C–O increase.

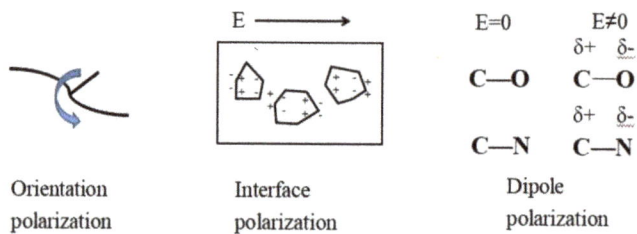

Figure 9. Simple schematic of orientation polarization, interface polarization and dipole polarization happened in PI films.

Table 3. Polarizabilities of the chemical bonds parallel and normal to the bond axis and the mean value for all three directions in space [24].

Bond	$\alpha_{//}$	$\alpha_{\perp}$	α_m
C–N	0.58	0.84	0.75
C–H	0.79	0.58	0.65
C–C	1.88	0.02	0.64
C–O	2.25	0.48	1.07
C=O	2.00	0.75	1.20

Table 4. Dipole moments of the chemical bonds of the relevant molecules expressed in Debye units [24].

Bond	$\mu\ (10^{-30}/CM)$
C–N	0.40
C–H	0.74
C–C	0.00
C–O	2.30
C=O	0.73

The dielectric constant of all samples decreases with the increase in the fluences and the rate of the decrease is not the same as all samples. Some researches have reported that polymer electrolyte films used electron beam and Gamma ray at 50 and 150 kGy, and the results showed that the dielectric constant increased with the increase of irradiation fluences and temperature because of the presence of an appreciable number of defects and dipoles in the form of chain scission [21]. Different irradiation sources, fluences, and irradiation rates have different effects on the structure, defects, and free radical damage of materials. The four different kinds of irradiation sources may have different mechanisms for low dielectric constant. The reasons why the four types of irradiation cause the decrease of dielectric constant are as follows. First, the molecular structure change is due to the increase of the free volume of the molecular structure and decrease of the density of the material. The large free volume reduces the number of polarizing groups per unit volume which can lead to the decrease of dielectric constant of the material. This should be the main reason for the decrease of the dielectric constant. Second, the irradiation may lead to defects of the PI films and the dielectric constant of air is 1, so the dielectric constant decreases. Third, the dipole polarization of the irradiated PI films decreases. The irradiation effect will directly lead to damage of the functional groups of the polymer. The polarity with large groups such as C=O, N–H, O–H, C–N, etc. is probably detached from the molecular main chain. Therefore, the decrease in the polarization ability of the groups inside the polymer material leads to the decrease in the dielectric constant.

5. Conclusions

The electron/proton/carbon ion beams of PI samples with different fluences have been systematically investigated, and the electrical and mechanical properties of the irradiated PI have been closely observed. The results show that the dielectric constant of PI irradiated by the four kinds of irradiation sources decrease with the increase of irradiation fluences. The dielectric loss of irradiated PI with the different fluences does not change obviously, but it increases significantly with the increase of frequency after 10^4 Hz. The XPS data implies that the polytetrachloric anhydride group in the polyimide degrades after irradiation, the content of polar groups such as C–N bonds and C=O bonds decreases, whereas the content of non-polar groups such as C–C bonds increases. The results of the EPR show that irradiation produces a lot of pyrolytic carbon free radicals which will compound when the irradiated PI is putted in air. The mechanical properties decrease slightly after irradiation.

Author Contributions: Conceptualization, H.L.; Data curation, H.L.; Investigation, H.L.; Writing – original draft, H.L.; Supervision, J.Y. and S.D.; Writing – review & editing, J.Y.; Formal analysis, X.L. and S.D.; Validation, S.D.; Visualization, S.D.; Funding acquisition, X.L.; Methodology, F.T.; Project administration, X.L.; Resources, X.L. All authors have read and agreed to the published version of the manuscript.

Funding: This research was funded by Science Challenge Project, grant number NO.TZ2018004 and the research was funded by the National Natural Science Foundation of China, grant number No. 51503053.

Conflicts of Interest: The authors declare no conflict of interest.

References

1. Vesely, D.; Finch, D.S.; Cooley, G.E. Electrical properties of polymers modified by electron beam irradiation. *Polymer* **1988**, 1402–1406. [CrossRef]
2. Yawson, B.N.; Noh, Y. Recent Progress on High-capacitance polymer gate dielectrics for flexible low-voltage transistors. *Adv. Funct. Mater.* **2018**, *28*, 1802201.
3. Constantinou, I.; Yi, X.; Shewmon, N.T.; Klump, E.D.; Peng, C.; Garakyaraghi, S.; Kin Lo, C.; Reynolds, J.R.; Castellano, F.N.; So, F. Effect of polymer-fullerene interaction on the dielectric properties of the blend. *Adv. Energy Mater.* **2017**, *7*, 1601947. [CrossRef]
4. Wu, T.; Dong, J.; Gan, F.; Fang, Y.; Zhao, X.; Zhang, Q. Low dielectric constant and moisture-resistant polyimide aerogels containing trifluoromethyl pendent groups. *Appl. Surf. Sci.* **2018**, *440*, 595–605. [CrossRef]
5. Yin, X.; Feng, Y.; Zhao, Q.; Li, Y.; Li, S.; Dong, H.; Hu, W.; Feng, W. Highly transparent, strong, and flexible fluorographene/fluorinated polyimide nanocomposite films with low dielectric constant. *J. Mater. Chem. C* **2018**, *6*, 6378. [CrossRef]

6. Chen, X.; Huang, H.; Shu, X.; Liu, S.; Zhao, J. Preparation and properties of a novel graphene fluoroxide/polyimide nanocomposite film with a low dielectric constant. *RSC Adv.* **2017**, *7*, 1956–1965. [CrossRef]
7. Lee, H.J.; Kim, J.; Kwon, O.; Lee, H.J.; Kwak, J.H.; Kim, J.M.; Lee, S.S.; Kim, Y.; Kim, D.-Y.; Jo, J.Y. Low coercive field of polymer ferroelectric via x-ray induced phase transition. *Appl. Phys. Lett.* **2015**, *107*, 262902. [CrossRef]
8. Bonardd, S.; Alegria, A.; Saldias, C.; Leiva, A.; Kortaberria, G. Polyitaconates: A new family of "all-polymer" dielectrics. *ACS Appl. Mater. Interfaces* **2018**, *10*, 38476–38492. [CrossRef]
9. Neese, B.; Chu, B.; Lu, S.; Wang, Y.; Furman, E.; Zhnag, Q.M. Large electrocaloric effect in ferroelectric polymers near room temperature. *Science* **2008**, *3211*, 821–823. [CrossRef]
10. Ismaiilova, R.S.; Magerramov, A.M.; Kuliev, M.M.; Akhundova, G.A. Electrical conductivity and dielectric permittivity of γ-irradiated nanocomposites based on ultrahigh-molecular-weight polyethylene filled with α-SiO2. *Surf. Eng. Appl. Electrochem.* **2018**, *54*, 6–11. [CrossRef]
11. Atta, A.; Lotfy, S.; Abdeltwab, E. Dielectric properties of irradiated polymer/multiwalled carbon nanotube and its amino functionalized form. *J. Appl. Polym. Sci.* **2018**, *135*, 46647.
12. Jaschin, P.W.; Bhimireddi, R.; Varma, K.B.R. Enhanced dielectric properties of LaNiO3/BaTiO3/PVDF: A three-phase percolative polymer nanocrystal composite. *ACS Appl. Mater. Interfaces* **2018**, *10*, 27278–27286.
13. Mackova, A.; Havránek, V.; Švorčík, V.; Djourelov, N.; Suzuki, T. Degradation of PET, PEEK and PI induced by irradiation with 150 keV Ar+ and 1.76 MeV He+ ions. *Nuclear Instrum. Methods Phys. Res. B* **2005**, *240*, 245–249.
14. Ismail, N.H.; Mustapha, M.; Ismail, H.; Kamarol Jamil, M.; Hairaldin, S.C. Effect of Electron Beam Irradiation on Dielectric Properties, Morphology and Melt Rheology of Linear Low Density Polyethylene/Silicone Rubber-Based Thermoplastic Elastomer Nanocomposites. *Polym. Eng. Sci.* **2018**, E135–E144.
15. Abdelrahman, M.M.; Osman, M.; Hashhash, A. Electrical properties of irradiated PVA film by using ion/electron beam. *Prog. Theor. Exp. Phys.* **2016**, *2016*, 023G01.
16. Mariania, M.; Ravasioa, U.; Varolia, V.; Consolati, G.; Faucitano, A.; Buttafava, A. Gamma irradiation of polyester films. *Radiat. Phys. Chem.* **2007**, *76*, 1385–1389.
17. Raghu, S.; Kilarkaje, S.; Sanjeev, G.; Nagaraja, G.K.; Devendrappa, H. Effect of electron beam irradiation on polymer electrolytes: Change in morphology, crystallinity, dielectric constant and AC conductivity with dose. *Radiat. Phys. Chem.* **2014**, *98*, 124–131.
18. Raghu, S.; Archana, K.; Sharanappa, C.; Ganesh, S.; Devendrappa, H. Electron beam and gamma ray irradiated polymer electrolyte films: Dielectric properties. *J. Radiat. Res. Appl. Sci.* **2016**, *9*, 117–124.
19. Raghu, S.; Subramanya, K.; Sharanappa, C. The Change in Dielectric Constant, AC Conductivity and Optical Band Gaps of Polymer Electrolyte Film: Gamma Irradiation. *Solid State Phys. AIP Conf. Proc.* **2014**, *1591*, 1272–1274.
20. Qureshi, A.; Singh, N.L.; Rakshit, A.K.; Singh, F.; Avasthi, D.K. Swift heavy ion induced modification in polyimide films. *Surf. Coat. Technol.* **2007**, *201*, 8308–8311.
21. Quamara, J.; Garg, M.; Prabhavathi, T. Effect of high-energy heavy ion irradiation on dielectric relaxation behaviour of kapton-H polyimide. *Thin Solid Films* **2004**, *449*, 242–247. [CrossRef]
22. Tkachenkoa, I.; Kononevichb, Y.; Kobzara, Y.; Purikova, O.; Yakovlev, Y.; Yakovlev, I.; Muzafarov, A.; Shevchenko, A. Low dielectric constant silica-containing cross-linked organic-inorganic materials based on fluorinated poly(arylene ether)s. *Polymer* **2018**, *157*, 131–138. [CrossRef]
23. Dutta, N.J.; Mohanty, S.R.; Buzarbaruah, N.; Ranjan, M.; Rawat, R.S. Self-organized nanostructure formation on the graphite surface induced by helium ion irradiation. *Phys. Lett. A* **2018**, *382*, 1601–1608. [CrossRef]
24. Naseem, S. Structural Optimization of SrMnO3 to Study Electro-Magnetic Characteristics. Master's Thesis, University of the Punjab, Lahore, Pakistan. No. SSP-1308 Session 2013–2015.
25. Yue, L.; Wu, Y.; Sun, C.; Xiao, J.; Shi, Y.; Ma, J.; He, S. Investigation on the radiation induced conductivity of space-applied polyimide under cyclic electron irradiation. *Nucl. Instrum. Methods Phys. Res., Sect. B* **2012**, *291*, 17–21. [CrossRef]

© 2020 by the authors. Licensee MDPI, Basel, Switzerland. This article is an open access article distributed under the terms and conditions of the Creative Commons Attribution (CC BY) license (http://creativecommons.org/licenses/by/4.0/).

Article

Preparation and Characterization of Waterborne UV Lacquer Product Modified by Zinc Oxide with Flower Shape

Yan Wu [1,2,*], Xinyu Wu [1,2], Feng Yang [3,*] and Jiaoyou Ye [4]

1. College of Furnishings and Industrial Design, Nanjing Forestry University, Nanjing 210037, China; aprilwu2019@163.com
2. Co-Innovation Center of Efficient Processing and Utilization of Forest Resources, Nanjing Forestry University, Nanjing 210037, China
3. Fashion Accessory Art and Engineering College, Beijing Institute of Fashion Technology, Beijing 100029, China
4. DeHua TB New Decoration Material Co. Ltd., Deqing 313200, China; yejiaoyou2020@163.com
* Correspondence: wuyan@njfu.edu.cn (Y.W.); yangfeng@bift.edu.cn (F.Y.)

Received: 3 March 2020; Accepted: 14 March 2020; Published: 17 March 2020

Abstract: In this paper, the waterborne UV lacquer product (WUV) was used as the main raw material, zinc oxide (ZnO) was used as the additive, and the stearic acid as the surface modifier. According to the method of spraying coating on the surface of poplar wood (*Populus tomentosa*), a simple and efficient preparation method was carried out to generate a super-hydrophobic surface and enhance the erosion resistance of the coating. By testing, the contact angle (CA) of water on the coating surface can reach 158.4°. The microstructure and chemical composition of the surface of coatings were studied by scanning electron microscope (SEM), Fourier transform infrared spectroscopy (FT-IR), and X-ray diffraction (XRD). The results showed that under acidic conditions, the non-polar long chain alkyl group of stearic acid vapor molecule reacted with the hydroxyl group in acetic acid, the metal ions of the ZnO were displaced to the stearic acid and generated globular zinc stearate ($C_{36}H_{70}O_4Zn$). The hydrophobic groups –CH_3 were grafted to the surface of zinc stearate (ZnSt2) particles and the micro/nano level of multistage flower zinc stearate coarse structure was successfully constructed on the surface of poplar wood, which endowed it with superhydrophobic properties. It is shown that the coating has good waterproof and erosion resistance.

Keywords: poplar wood; waterborne UV lacquer product; wood modification; contact angle; spectroscopy; super-hydrophobic coating

---

1. Introduction

Waterborne UV lacquer product (WUV) is a kind of coating which realizes crosslinking curable by UV irradiation [1]. It does not contain volatile toxic substances or irritating gases [2–4]. WUV combines UV-cured technology with waterborne polymer technology [5,6], which not only saves energy and protects environmental but also has the advantages of fast curing speed, pipeline production and high production efficiency [7,8]. The main component of WUV is waterborne acrylic resin [9]. It has the advantages of light color, high solid content, and strong adhesion, but its hardness, wear resistance, and mechanical properties limit its use [10–12]. However, due to the water-based acrylic resin with water as the dispersion medium, the phenomenon of incomplete curing is easy to occur during curing, and the residual water-based additives lead to poor water resistance of the coating. In addition, the evaporation of water requires more heat, which asks higher requirements during the drying process of the coating. In order to better play the application of UV cured coatings in the field of wood, the ultra-hydrophobic modification [13] and durability enhancement [14] are needed to improve the properties of WUV, so as to better meet the production and use requirements. Zhong et al. [15] used maleic anhydride and silicon modified waterborne alkyd resin, and prepared fluoro-acrylate resin by micro-emulsion polymerization without surfactant. The modified resin has better mechanical stability and corrosion resistance due to its larger contact angle with water. Saladino et al. prepared for the silica/PMMA nanocomposites with different silica quantities by a melt compounding method and systematically investigated it as a function of silica amount from 1 to 5 wt %. Results showed that silica nanoparticles are well dispersed in the polymeric matrix whose structure remains amorphous. The degradation of the polymer occurs at higher temperature in the presence of silica because of the interaction between the two components [16].

As a kind of inorganic material, zinc oxide nanoparticles have diversified morphology and excellent physicochemical properties [17]. Therefore, ZnO nanoparticles are a common packing, they are low cost [18–20], harmless to the environment [19], and have excellent photoelectric performance and rich form. Therefore, they have a broad application prospect in coatings [21–23], sensors [24,25], photoelectric material [26–28], medicine [20], and many other fields [29]. For example, Zahra et al. [30] used mixed ZnO/GO nanostructures to modify the surface of low carbon steel before acrylic resin coating. Through the structural properties and interactions between the oxygen-containing groups of ZnO and GO structures, the corrosion resistance of low carbon steel was improved. Zhou et al. [31] prepared the nano-hydroxyapatite/ZnO coating on biodegradable Mg-Zn-Ca block metallic glass by one-step hydrothermal method. Due to the presence of ZnO in the coating, the antibacterial rate of BMG in vitro was close to 100%. Guo et al. prepared two coatings on the surface of spruce panels, the first coating was coated by UV light absorbing ZnO and an additional hydrophobic layer of stearic acid, which endowed the wooden panels with water repellence as well as protected the ZnO coating from erosion due to rain. The second coating was based on a thin TiO_2 layer attached to the wood surface, which aimed to avoid a pronounced initial color change induced by the coating itself [32]. Nair et al. prepared a stable dispersion of nanoparticles of three metal oxides, zinc oxide (ZnO), cerium oxide (CeO_2), and titanium dioxide (TiO_2) by propylene glycol (PG) through ultrasonication. The stability test of the coating was measured by UV–vis absorption spectroscopy and an accelerated weathering tester. Results shown that the increase in concentration of nanoparticles in the dispersion imparted higher resistance to UV induced degradation [33].

In wood science, the utilization of wood can be improved by changing its dimensional stability, flammability, biodegradability, and other properties [34]. Tuong et al. treated the acacia hybrid wood with TiO_2 impregnation through the combination of pressure impregnation and hydrothermal post-treatment. Results showed that the color stability against UV irradiation of the TiO_2 impregnated wood was significantly improved than that of the untreated acacia hybrid wood. The TiO_2 nanoparticles were located on the inner surfaces of the wood vessels and it could improve the UV resistance of fabricated wood samples [35]. Yu et al. constructed ZnO nanostructures on the surface of solid wood via a simple two-step process consisting of generation of ZnO seeds on the wood surface followed by a solution treatment to promote crystal growth. Accelerated weathering was used to evaluate the photostability of treated wood. Results showed that the photostability of the treated wood was greatly enhanced [36].

The super-hydrophobic modification of WUV can make it have the properties of self-cleaning [37,38], antifouling [38] and water resistance [39] and improve the erosion resistance [3,40–43] of the coating, so that the wood material will possess a higher use value [44–46]. Yang et al. [47] soaked the poplar wood in a compound solution of maleic rosin and ethanol for 24 h. After being taken out and dried, the sample was then soaked again in the solution of TiO_2 and modifier to obtain a super-hydrophobic wood with a contact angle up to 157°. After soaking the modified poplar in water for a week, irradiation under the hot sun for a week or boiling at 100 °C, the wood surface still has super hydrophobicity. Gao et al. prepared superhydrophilic and underwater superoleophobic poly by salt-induced phase inversion method. This kind of poly can quickly and efficiently separate the oil–water mixture system from the emulsified oil–water system, and the separation property is stable. After repeated use, the separation efficiency of the oil–water mixture was above 98%, and that of the oil–water emulsion is above 91%, which can be widely used in the oil–water mixture system and emulsified oil–water system [48]. Wang et al. [49] impregnated wood into a mixture of zinc acetate dihydrate and triethylamine, constructed a ZnO coating with roughness on the wood surface, and modified the coating surface with stearic acid to obtain a super-hydrophobic coating with low surface energy. The superhydrophobic coating remained superhydrophobic after drying at 60 °C for a month or soaking in deionized water for a week, showing good air stability and erosion resistance, which retained the natural appearance of wood while minimizing maintenance intervals. Poplar wood is a kind of fast growing wood, which is sustainable, biodegradable, biocompatible [34], rich in natural resources and low in price. It is usually used in furniture manufacturing. In this paper, ZnO nanoparticles (ZnO NPs) were used as the modifier to prepare for a superhydrophobic WUV coating on poplar wood, which retained the natural appearance of wood while minimizing maintenance intervals [32,50]. The possibility of large-scale application of this low-cost and environmentally friendly superhydrophobic surface preparation technology is discussed theoretically. This study is of guiding significance to the application of functional materials with superhydrophobicity.

2. Experimental Part

2.1. Materials

The poplar wood (*Populus tomentosa*) cut from the longitudinal section with the size of 20 mm long × 20 mm wide × 2 mm thick was supplied by Yihua Lifestyle Technology Co., Ltd., Shantou, China. Its moisture content was 9.9% and the absolute dry density was 298 kg/m^3. The above data was tested at 10 a.m. in winter, in an environment of 25 °C and the relative humidity of 46%. The waterborne UV lacquer product (A185721006) was purchased from Huzhou Dazhou Polymer Material Co., Ltd., Huzhou, China. Ethyl acetate (99.5%) and acetic acid (CH_3COOH, 99.5%) were provided by Nanjing Chemical Reagent Co., Ltd., Nanjing, China. Zinc oxide (ZnO) was obtained from Xilong Scientific Co., Ltd.,Guangzhou, China. Anhydrous ethanol (99.7%), n-hexane (99%) and TBOT (98%) was purchased from Sinopharm Group, Shanghai, China. Stearic acid ($C_{18}H_{36}O_2$) was provided from Yonghua Chemical Technology Co., Ltd, Changshu, China.

2.2. Experimental Method

First, the poplar wood was dipped in anhydrous ethanol and distilled water respectively and experienced sonication by an ultrasonic crusher (Misonix, Inc., New York, USA) for 1 h, and then the treated wood was dried to absolute dry condition at 100 °C by a constant temperature blast drying oven (Shanghai Xinmiao Medical Devices Co., Ltd., Shanghai, China). 10 g WUV and 6 g ZnO powder was added into 13 mL ethyl acetate and the mixed solution was magnetically stirred by a magnetic stirrer (Yinyu High-tech Instrument factory, Gongyi, China) for 45 min, so that the WUV and ZnO powder were evenly dispersed in ethyl acetate. The obtained ZnO/WUV lacquer was sprayed onto dried poplar wood surface with a high-pressure electric spraying machine (Pritzker Power Tools, Ningbo, China) at a distance of about 15–20 cm. Two layers of UV lacquer product were applied to the poplar wood surface. The ZnO/WUV lacquer was cured under ultraviolet lamp (Wuxi Jinhua Test Equipment Co., Ltd, Wuxi, China) for 3.5 min and dried at 85 °C for 4 h to obtain ZnO/WUV coating.

Then the anhydrous ethanol was used as dispersant, 20 mL stearic acid and acetic acid were allocated as mixture under 70 °C. The solution was centrifuged at 4000 r/min with a centrifuge (Shanghai Anting Scientific Instrument Factory, Shanghai, China) for 5 min. The sonicated mixture was cooled to room temperature, and the ZnO/WUV sample was dipped into the mixture for 200 seconds before being removed. The multistage flower zinc stearate/waterborne UV lacquer super-hydrophobic coating (ZnSt2/WUV) was obtained after drying. The diagram of modification process of ZnO/WUV and ZnSt2/WUV coatings is shown in Figure 1.

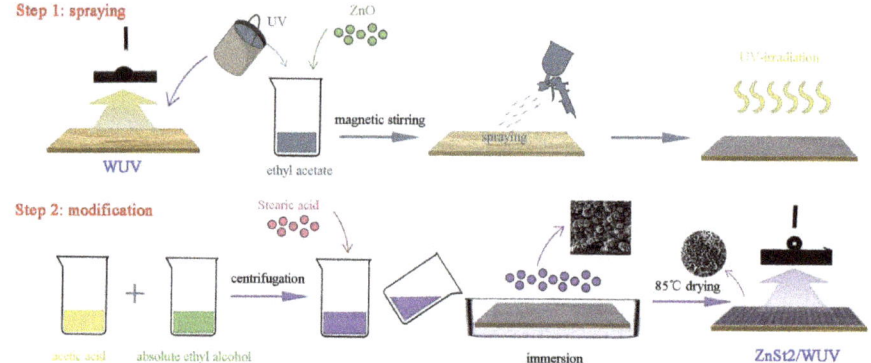

Figure 1. Diagram of modification process.

2.3. Contact Angle (CA) Test

The contact angle of water on the cross section of ZnSt2/WUV was measured by Theta t200 Optical contact goniometer (Sweden baiorin technology Co. Ltd., Gothenburg, Swedish). After drying, the WUV and ZnSt2/WUV samples were placed on the loading platform. About 2 μL of deionized water was dripped on the surface of the samples under test. The sample was tested three times in parallel at different locations.

2.4. SEM

After the prepared WUV and ZnSt2/WUV samples were dried, thin slices with a width of about 5 mm and a thickness of about 2 mm were cut and fixed to the conductive adhesive on which the samples were placed with tweezers. The samples were sprayed with gold for 30 s with a vacuum plating apparatus. The surface and internal morphology of the samples were observed with an environmental QUANTA 200 SEM (FEI Company, Hillsboro, OR, USA) at 3 kV voltage.

2.5. FTIR and XRD

A Vertex 80V infrared spectrum analyzer (Germany Bruker Co., Ltd., Karlsruhe, Germany) was used to determine the functional groups of the ZnSt2/WUV, and the wave number range was set at 500–4000 cm^{-1}. AXIS UltraDLD XRD (Nippon Koji Co. Ltd, Osaka, Japan) was used to characterize the internal molecular structure of the sample. The scanning speed was set from 5° to 70° [51], the acceleration voltage was 40 kV, and the impressed current was 30 mA.

2.6. Different pH Values of Solution and Organic Solvent Immersion Test

ZnSt2/WUV was soaked in the solutions with the pH values of 2, 4, 7, 9, 12, respectively and dried after a period of time [52]. Similarly, the samples were soaked in different organic solvents [53], dried after a period of time, and the contact angle of water on two samples in the same reagent were measured and the average value of them was obtained as the final data.

2.7. Water Resistance Test

In order to determine whether the ZnSt2/WUV coating had good water resistance, two samples for each of WUV and ZnSt2/WUV were impregnated in distilled water, and then which were taken out and weighed after a period of time, and the water absorption before and after modification was calculated by using Equation (1):

$$WA(\%) = (m_{ai} - m_{bi})/m_{bi} \times 100\% \tag{1}$$

where m_{bi} and m_{ai} represent the film weights before and after absorbed water, respectively [15].

3. Results and Discussion

3.1. Contact Angle (CA) Test

By testing, the contact angles of water on the WUV and ZnSt2/WUV (Table 1) coating samples were 68.2° and 158.4°, respectively. Compared with the WUV coating, the ZnSt2/WUV coating had the property of superhydrophobicity.

Table 1. Image of contact angle (A: WUV; B: ZnSt2/WUV)

Name of the Coating	Contact Angle of Water on the Coating Surface (°)	Image of Contact Angle
WUV	68.2	
ZnSt2/WUV	158.4	

3.2. SEM

The surface microstructure of WUV and ZnSt2/WUV coating samples was shown in Figure 2. It could be seen from Figure 2A and 2B that the surface of WUV was smooth and covered by a uniform and continuous WUV, with a few nanoparticles distributed on the surface. It could be seen from Figure 2C that, at a magnification of 250, the surface of ZnSt2/WUV coating was evenly arranged with micron-sized bulbous flower clusters, indicating that under the adhesive effect of WUV and the modification effect of stearic acid, ZnO NPs reacted with acetic acid solution to form a micro/nano flower-like structure. As could be seen from the Figure 2D, the surface of the micron-scale bulge was evenly arranged with the petal-like mastoid structure. The petal-like mastoid tip had a nanoscale folded structure, which enabled the gas to exist in the structure and lift the droplets when contacting with the water droplets, so that the water droplets cannot penetrate and spread. The micro/nano structure, similar to that of rose petals, was the key reason to the superhydrophobicity of the ZnSt2/WUV coating [29,54].

3.3. FTIR and XRD

Figure 3A shows the FTIR spectrum of ZnSt2/WUV. 2869.5 cm^{-1} corresponded to the symmetric tensile vibration peak of C–H bond. 1320.9 and 1164.7 cm^{-1} were the asymmetric stretching vibration peaks of C–O–C, while 1488.7 and 1432.8 cm^{-1} were the bending vibration peaks of –CH$_2$. 1376.9 cm^{-1} was the symmetric deformation vibration peak of –CH$_3$. The absorption peaks corresponding to the reaction products –CH$_2$ and –CH$_3$ were 2923.5 and 2850.2 cm^{-1}, respectively. It can be seen that the amount of C element and alkyl on the surface of ZnSt2/WUV increased significantly, indicating that the coating had been modified by stearic acid. The XRD pattern of the ZnSt2/WUV was shown in Figure 3B. It could be seen that the characteristic diffraction peak of zinc contained a small amount of C and O elements. In Figure 3B, the peaks at 16.2°, 22.6°, and 33.2° corresponded to the peaks of zinc stearate crystals. This indicated that the modification of stearic acid promotes the changes in topography of the surface and the formation of nano-particles, the carboxyl group in acetic acid reacted with ZnO particles, and the –CH$_3$ hydrophobic group was grafted onto the surface of ZnSt2 particles, successfully constructing the superhydrophobic coating on the surface of poplar wood.

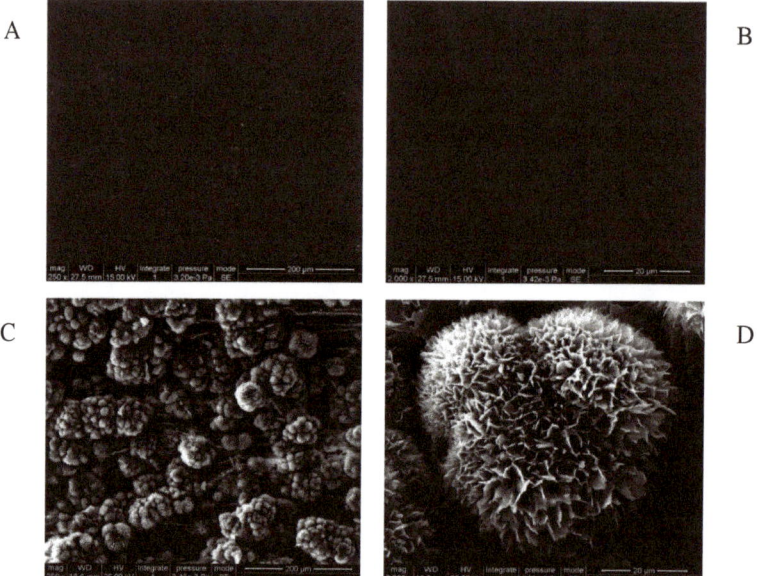

Figure 2. SEM images of WUV (**A**: at 250 magnification; **B**: at 2000 magnification) and ZnSt2/WUV (**C**: at 250 magnification; **D**: at 2500 magnification) coatings.

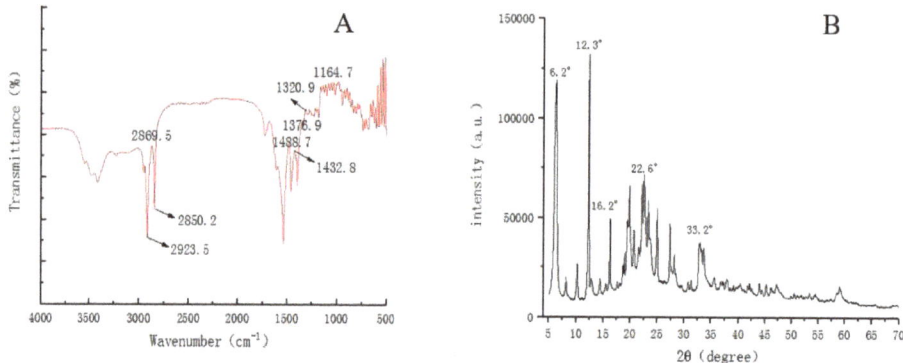

Figure 3. FTIR (**A**) and XRD (**B**) of ZnSt2/WUV.

3.4. Different pH Values of Solution and Organic Solvent Immersion Test

The water contact angle of the water on ZnSt2/WUV coating was measured after impregnated in different pH values of solution and organic solvent. As could be seen from Figure 4A, the CA of coating was greater than 150° after being soaked in the solutions with pH values of 2, 4, and 7 for 50 h. After soaking in the solutions with pH values of 2, 4, and 7 for 100 h, the CA was greater than 140°, this was owing to the ZnSt2/WUV coating decompose into stearic acid and corresponding salt when it encounters acid, and its hydrophobicity would not lose, indicating its ability to maintain hydrophobicity in acid solution and neutral solution. The superhydrophobic property of the ZnSt2/WUV coating lost after impregnating in the solution with pH values of 9 for 15 h and 12 for 5 h, respectively.

This indicated that it is difficult for ZnSt2/WUV coating to maintain hydrophobicity in alkaline solution for a long time. The contact angle was above 140° after the ZnSt2/WUV coating was impregnated for 60 h in the solution with pH value of 9 and the contact angle was above 140° after the coating was impregnated in the solution with pH value of 9 for 3 h. The contact angle of the coating was above 130° after immersion for 100 h in the solution with pH values of 9 and 12. The results showed that the surface of the sample was still hydrophobic after dipping with the alkaline solution, the hydrophobicity of ZnSt2/WUV can be well maintained in acidic medium. As shown in Figure 4B, because the ethyl alcohol and n-hexane used for impregnation was neutral and acetic acid was said to be acidic, the ZnSt2/WUV coating remained super-hydrophobic after being impregnated in the solution of n-ethane, anhydrous ethanol, and acetic acid for 50 h, respectively. The stability of ZnSt2/WUV coating in ethyl acetate and tetrabutyl titanate solutions—which all present alkaline after hydrolysis—was relatively poor, which might be due to the reason that the low surface energy material on the coating surface was easy to dissolve in alkaline organic solvent, resulting in the loss of superhydrophobicity of the coating. In Figure 4A, the R2 coefficient of the contact angle of water on ZnSt2/WUV with the immersion time is −0.93, and in Figure 4B, the R2 coefficient of the contact angle of water on ZnSt2/WUV with the immersion time is −0.94. It can be seen that there is a negative correlation between the contact angle of water on ZnSt2/WUV and the immersion time. In a word, the coating could be better to maintain its hydrophobicity in acid organic solvents. Therefore, the application of this superhydrophobic coating onto wood surface could protect wood from erosion when it is exposed to an acid or alkaline environment, and extend the use of poplar wood.

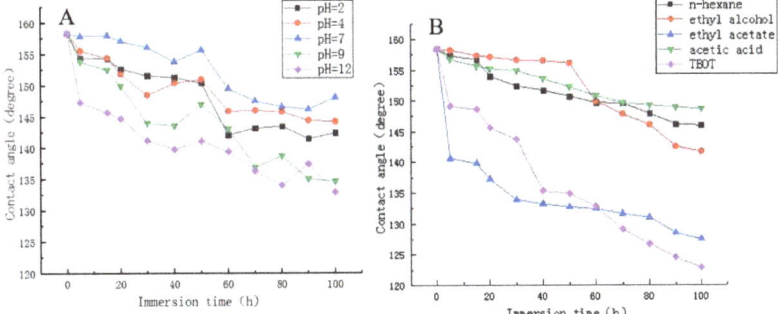

Figure 4. Different pH values of solution and organic solvent immersion test of ZnSt2/WUV (**A**: immersion in different pH solutions; **B**: immersion in different organic solvent).

3.5. Water Resistance Test

The water resistance test results of WUV and ZnSt2/WUV coatings are shown in Figure 5. WUV and ZnSt2/WUV was soaked in distilled water for 45 days. The water absorption rate increased with the increase of days and the water absorption of WUV went up even more and faster. When the samples were immersed in water for 45 days, the absorption rate of the WUV coating increased from 6.1% to 88.9%, while that of the ZnSt2/WUV increased from 0.9% to 67.8%. It could be seen that the ZnSt2/WUV showed good water resistance. This is because the modification of stearic acid constructed a super-hydrophobic structure of multigrade ZnO on the surface of poplar wood, so that the wood surface containing hygroscopicity components was covered by ZnSt2, which reduced the hygroscopicity of poplar wood. According to the test, after 50 days of immersion in distilled water, the contact angle of water on the surface of the sample was 154.4°, which showed a persistent superhydrophobicity [53]. Therefore, this method of constructing superhydrophobic coatings on the surface of wood substrates was expected to expand the use of poplar wood in waterproof and stain resistant [55].

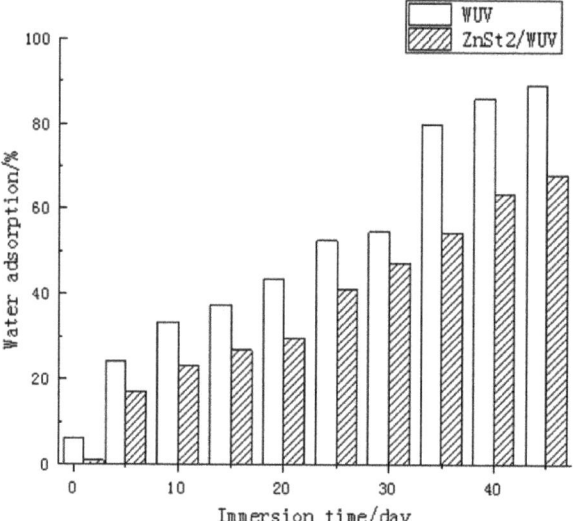

Figure 5. Water resistance test of WUV and ZnSt2/WUV.

4. Conclusions

A superhydrophobic coating was constructed on the surface of poplar wood with a contact angle of up to 158.4° through the water-based UV-cured wood coating which was modified by ZnO and stearic acid. The results showed that under acid condition, the nonpolar long chain alkyl group of stearic acid vapor molecule reacted with the hydroxyl group in acetic acid, the metal ions of the ZnO were displaced to the stearic acid and generated globular zinc stearate ($C_{36}H_{70}O_4Zn$). $-CH_3$ hydrophobic groups were grafted to the surface of ZnSt2 particles and micro/nano level of multistage flower ZnSt2 coarse structure was successfully constructed on the surface of poplar wood, which was the key to the superhydrophobic property of the coating. The pH, corrosion resistance, and water resistance tests revealed that the ZnSt2/WUV coating had good resistance to acid medium and some organic solvent corrosion ability. Compared with WUV, the water resistance of ZnSt2/WUV was stronger, which was conducive to prepare superhydrophobic coatings in an easy and environmentally friendly way and expand the application scope of poplar products in waterproof field.

Author Contributions: Conceptualization, Y.W.; Data curation, X.W.; Funding acquisition, Y.W., F.Y. and J.Y.; Methodology, X.W.; Resources, F.Y.; Writing—original draft, X.W.; Writing—review & editing, Y.W. All authors have read and agreed to the published version of the manuscript.

Funding: The authors gratefully acknowledgement the financial support from the project funded by "Nan Taihu Lake elite plan" project ([2018] no. 2), the Special Scientific Research Fund of Construction of High-level teachers Project of Beijing Institute of Fashion Technology (BIFTQG201805), the Project of Science and Technology Plan of Beijing Municipal Education Commission (KM202010012001), and Yihua Lifestyle Technology Co., Ltd. Projects funded (YH-NL-201507 and YH-JS-JSKF-201904003).

Acknowledgments: The authors gratefully acknowledgement the financial support from China and would like to thank Shang Huang for helping us prepare the samples.

Conflicts of Interest: The authors declare no conflict of interest.

References

1. Yan, X.; Qian, X.; Chang, Y. Preparation and characterization of urea formaldehyde@ epoxy resin microcapsule on waterborne wood coatings. *Coatings* **2019**, *9*, 475. [CrossRef]
2. Wang, N.; Zhang, Y.; Chen, J.; Zhang, J.; Fang, Q. Dopamine modified metal-organic frameworks on anti-corrosion properties of waterborne epoxy coatings. *Prog. Org. Coat.* **2017**, *109*, 126–134. [CrossRef]

3. Xu, W.; Fang, X.Y.; Han, J.T.; Wu, Z.H.; Zhang, J.L. Effect of coating thickness on sound absorption property of four wood species commonly used for piano soundboards. *Wood Fiber Sci.* **2020**, *52*, 28–43. [CrossRef]
4. Yan, X.; Qian, X.; Lu, R.; Miyakoshi, T. Comparison and optimization of reactive dyes and coating performance on fraxinus mandshurica veneer. *Polymers* **2018**, *10*, 1302. [CrossRef]
5. Shahriari, L.; Mohseni, M.; Yahyaei, H. The effect of cross-linking density on water vapor and oxygen permeability of hybrid UV cured nano coatings. *Prog. Org. Coat.* **2019**, *134*, 66–77. [CrossRef]
6. Aizpurua, J.; Martin, L.; Fernández, M.; González, A.; Irusta, L. Recyclable, remendable and healing polyurethane/acrylic coatings from UV curable waterborne dispersions containing Diels-Alder moieties. *Prog. Org. Coat.* **2020**, *139*, 105460. [CrossRef]
7. Wang, S.; Wu, Y.; Dai, J.; Teng, N.; Peng, Y.; Cao, L.; Liu, X. Making organic coatings greener: Renewable resource, solvent-free synthesis, UV curing and repairability. *Eur. Polym. J.* **2020**, *123*, 109439. [CrossRef]
8. Yan, X.; Qian, X.; Chang, Y.; Lu, R.; Miyakoshi, T. The effect of glass fiber powder on the properties of waterborne coatings with thermochromic ink on a Chinese Fir surface. *Polymers* **2019**, *11*, 1733. [CrossRef]
9. Rawat, R.S.; Chouhan, N.; Talwar, M.; Diwan, R.K.; Tyagi, A.K. UV coatings for wooden surfaces. *Prog. Org. Coat.* **2019**, *135*, 490–495. [CrossRef]
10. Xiong, X.Q.; Yuan, Y.Y.; Niu, Y.T.; Zhang, L.T.; Wu, Z.H. Effects of different treatments on surface activity of rice straw particleboard. *Sci. Adv. Mater.* **2020**, *12*, 289–295. [CrossRef]
11. Nosrati, R.; Olad, A.; Maryami, F. Visible-light induced anti-bacterial and self-cleaning waterborne polyacrylic coating modified with TiO_2/polypyrrole nanocomposite; preparation and characterization. *J. Mol. Struct.* **2018**, *1163*, 174–184. [CrossRef]
12. Wu, Y.; Sun, Y.; Yang, F.; Zhang, H.; Wang, Y. The implication of benzene-ethanol extractive on mechanical properties of waterborne coating and wood cell wall by nanoindentation. *Coatings* **2019**, *9*, 449. [CrossRef]
13. Hang, T.T.X.; Anh, N.T.; Truc, T.A.; Van Truoc, B.; Hoang, T.; Thanh, D.T.M.; Daopiset, S. Synthesis of 3-glycidoxypropyltrimethoxysilane modified hydrotalcite bearing molybdate as corrosion inhibitor for waterborne epoxy coating. *J. Coat. Technol. Res.* **2016**, *13*, 805–813. [CrossRef]
14. Xu, R.; He, T. Corrosion of self-curing waterborne zinc oxide-potassium silicate coating modified with aluminium powder. *J. Alloy. Compd.* **2019**, *811*, 152008. [CrossRef]
15. Zhong, S.; Li, J.; Yi, L.; Cai, Y.; Zhou, W. Cross-linked waterborne alkyd hybrid resin coatings modified by fluorinated acrylate-siloxane with high waterproof and anticorrosive performance. *Polym. Adv. Technol.* **2019**, *30*, 292–303. [CrossRef]
16. Saladino, M.L.; Motaung, T.E.; Luyt, A.S.; Spinella, A.; Nasillo, G.; Caponetti, E. The effect of silica nanoparticles on the morphology, mechanical properties and thermal degradation kinetics of PMMA. *Polym. Degrad. Stab.* **2012**, *97*, 452–459. [CrossRef]
17. Zhu, W.; Wu, Y.; Zhang, Y. Synthesis and characterisation of superhydrophobic CNC/ZnO nanocomposites by using stearic acid. *Micro Nano Lett.* **2019**, *14*, 1317–1321. [CrossRef]
18. Zhong, W.L. Structure distortion, optical and electrical properties of ZnO thin films co-doped with Al and Sb by sol-gel spin coating. *Chin. Geogr.* **2010**, *19*, 515–519.
19. Ismail, A.; Alahmad, M.; Alsabagh, M.; Abdallah, B. Effect of low dose-rate industrial Co-60 gamma irradiation on ZnO thin films: Structural and optical study. *Microelectron. Reliab.* **2020**, *104*, 113556. [CrossRef]
20. Shalu, C.; Shukla, M.; Tiwari, A.; Agrawal, J.; Bilgaiyan, A.; Singh, V. Role of solvent used to cast P3HT thin films on the performance of ZnO/P3HT hybrid photo detector. *Phys. E Low-Dimens. Syst. Nanostruct.* **2020**, *15*, 113694. [CrossRef]
21. Wang, Z.; Liu, F.; Han, E.; Ke, W.; Luo, S. Effect of ZnO nanoparticles on anti-aging properties of polyurethane coating. *Sci. Bull.* **2009**, *54*, 3464–3472. [CrossRef]
22. Qu, J.E.; Ascencio, M.; Jiang, L.M.; Omanovic, S.; Yang, L.X. Improvement in corrosion resistance of WE43 magnesium alloy by the electrophoretic formation of a ZnO surface coating. *J. Coat. Technol. Res.* **2019**, *16*, 1559–1570. [CrossRef]
23. Lin, J.; Chen, C.; Lin, C. Influence of sol–gel-derived ZnO:Al coating on luminescent properties of Y_2O_3:Eu^{3+} phosphor. *J. Sol-Gel Sci. Technol.* **2019**, *92*, 562–574. [CrossRef]
24. Yadav, A.B.; Pandey, A.; Jit, S. Effects of annealing temperature on the structural, optical, and electrical properties of ZnO thin films grown on n-Si⟨100⟩substrates by the sol–gel spin coating method. *Acta Metall. Sin.* **2014**, *27*, 682–688. [CrossRef]

25. Hussain, F.; Imran, M.; Khalil, R.M.A.; Niaz, N.A.; Rana, A.M.; Sattar, M.A.; Ismail, M.; Majid, A.; Kim, S.; Iqbal, F.; et al. An insight of Mg doped ZnO thin films: A comparative experimental and first-principle investigations. *Phys. E Low-Dimens. Syst. Nanostruct.* **2020**, *115*, 113658. [CrossRef]
26. Saleem, M.; Farooq, W.A.; Khan, M.I.; Akhtar, M.N.; Rehman, S.U.; Ahmad, N.; Khalid, M.; Atif, M.; AlMutairi, M.A.; Irfan, M. Effect of ZnO nanoparticles coating layers on top of ZnO Nanowires for morphological, optical, and photovoltaic properties of dye-sensitized solar cells. *Micromachines* **2019**, *10*, 819. [CrossRef]
27. Kumar, V.; Ntwaeaborwa, O.M.; Swart, H.C. Effect of oxygen partial pressure during pulsed laser deposition on the emission of Eu doped ZnO thin films. *Phys. B Condens. Matter* **2020**, *576*, 411713. [CrossRef]
28. Srisuai, N.; Boonruang, S.; Horprathum, M.; Sarapukdee, P.; Denchitcharoen, S. Growth of highly uniform size-distribution ZnO NR arrays on sputtered ZnO thin film via hydrothermal with PMMA template assisted. *Mater. Sci. Semicond. Process.* **2020**, *105*, 104736. [CrossRef]
29. Zhu, W.; Wu, Y.; Zhang, Y. Fabrication and characterization of superhydrophobicity ZnO nanoparticles with two morphologies by using stearic acid. *Mater. Res. Express* **2019**, *6*, 1150d1. [CrossRef]
30. Sharifalhoseini, Z.; Entezari, M.H.; Davoodi, A.; Shahidi, M. Surface modification of mild steel before acrylic resin coating by hybrid ZnO/GO nanostructures to improve the corrosion protection. *J. Ind. Eng. Chem.* **2020**, *83*, 333–342. [CrossRef]
31. Zhou, J.; Li, K.; Wang, B.; Ai, F. Nano-hydroxyapatite/ZnO coating prepared on a biodegradable Mg–Zn–Ca bulk metallic glass by one-step hydrothermal method in acid situation. *Ceram. Int.* **2020**, *46*, 6958–6964. [CrossRef]
32. Guo, H.; Michen, B.; Burgert, I. Estudios reales en banco de pruebas de la Casa ETH de Recursos Naturales – protección superficial de la madera para aplicaciones exteriores. *Inf. Constr.* **2017**, *69*, 220. [CrossRef]
33. Nair, S.; Nagarajappa, G.B.; Pandey, K.K. UV stabilization of wood by nano metal oxides dispersed in propylene glycol. *J. Photochem. Photobiol. B Biol.* **2018**, *183*, 1–10. [CrossRef] [PubMed]
34. Mishra, P.K.; Giagli, K.; Tsalagkas, D.; Mishra, H.; Talegaonkar, S.; Gryc, V.; Wimmer, R. Changing Face of Wood Science in Modern Era: Contribution of Nanotechnology. *Recent Pat. Nanotechnol.* **2018**, *12*, 13–21. [CrossRef] [PubMed]
35. Tuong, V.M.; Chu, T.V. Improvement of color stability of acacia hybrid wood by TiO_2 nano sol impregnation. *BioResources* **2015**, *10*, 5417–5425. [CrossRef]
36. Yu, Y.; Jiang, Z.; Wang, G.; Song, Y. Growth of ZnO nanofilms on wood with improved photostability. *Holzforschung* **2010**, *64*, 385–390. [CrossRef]
37. Zhang, J.; Zhang, W.; Lu, J.; Zhu, C.; Lin, W.; Feng, J. Aqueous epoxy-based superhydrophobic coatings: Fabrication and stability in water. *Prog. Org. Coat.* **2018**, *121*, 201–208. [CrossRef]
38. Jia, S.; Lu, X.; Luo, S.; Qing, Y.; Yan, N.; Wu, Y. Efficiently texturing hierarchical epoxy layer for smart superhydrophobic surfaces with excellent durability and exceptional stability exposed to fire. *Chem. Eng. J.* **2018**, *348*, 212–223. [CrossRef]
39. Jeong, J.H.; Han, Y.C.; Yang, J.H.; Kwak, D.S.; Jeong, H.M. Waterborne polyurethane modified with poly(ethylene glycol) macromer for waterproof breathable coating. *Prog. Org. Coat.* **2017**, *103*, 69–75. [CrossRef]
40. Grüneberger, F.; Künniger, T.; Huch, A.; Zimmermann, T.; Arnold, M. Nanofibrillated cellulose in wood coatings: Dispersion and stabilization of ZnO as UV absorber. *Prog. Org. Coat.* **2015**, *87*, 112–121. [CrossRef]
41. Miklečić, J.; Turkulin, H.; Jirouš-Rajković, V. Weathering performance of surface of thermally modified wood finished with nanoparticles-modified waterborne polyacrylate coatings. *Appl. Surf. Sci.* **2017**, *408*, 103–109. [CrossRef]
42. Yu, D.; Wen, S.; Yang, J.; Wang, J.; Chen, Y.; Luo, J.; Wu, Y. RGO modified ZnAl-LDH as epoxy nanostructure filler: A novel synthetic approach to anticorrosive waterborne coating. *Surf. Coat. Technol.* **2017**, *326*, 207–215. [CrossRef]
43. Wang, S.; Hu, Z.; Shi, J.; Chen, G.; Zhang, Q.; Weng, Z.; Wu, K.; Lu, M. Green synthesis of graphene with the assistance of modified lignin and its application in anticorrosive waterborne epoxy coatings. *Appl. Surf. Sci.* **2019**, *484*, 759–770. [CrossRef]
44. Dong, H.; Zheng, L.; Yu, P.; Jiang, Q.; Wu, Y.; Huang, C.; Yin, B. Characterization and Application of Lignin–Carbohydrate Complexes from Lignocellulosic Materials as Antioxidants for Scavenging In Vitro and In Vivo Reactive Oxygen Species. *ACS Sustain. Chem. Eng.* **2020**, *8*, 256–266. [CrossRef]

45. Sun, S.; Zhao, Z.; Umemura, K. Further Exploration of Sucrose-Citric Acid Adhesive: Synthesis and Application on Plywood. *Polymers* **2019**, *11*, 1875. [CrossRef] [PubMed]
46. Zhao, Z.; Sun, S.; Wu, D.; Zhang, M.; Huang, C.; Umemura, K.; Yong, Q. Synthesis and characterization of sucrose and ammonium dihydrogen phosphate (SADP) adhesive for plywood. *Polymers* **2019**, *11*, 1909. [CrossRef] [PubMed]
47. Yang, M.; Chen, X.; Lin, H.; Han, C.; Zhang, S. A simple fabrication of superhydrophobic wood surface by natural rosin based compound via impregnation at room temperature. *Eur. J. Wood Wood Prod.* **2018**, *76*, 1417–1425. [CrossRef]
48. Gao, H.; Duan, Y.; Yuan, Z. Preparation and oil–water separation performance of ultra-hydrophilic subaqueous ultra-hydrophobic pvdf-g-paa porous membrane. *Acta Chem. Sin.* **2016**, *37*, 1208–1215.
49. Wang, S.; Shi, J.; Liu, C.; Xie, C.; Wang, C. Fabrication of a superhydrophobic surface on a wood substrate. *Appl. Surf. Sci.* **2011**, *257*, 9362–9365. [CrossRef]
50. Zhao, Z.; Sakai, S.; Wu, D.; Chen, Z.; Zhu, N.; Gui, C.; Zhang, M.; Umemura, K.; Yong, Q. Investigation of synthesis mechanism, optimal hot-pressing conditions, and curing behavior of sucrose and ammonium dihydrogen phosphate adhesive. *Polymers* **2020**, *12*, 216. [CrossRef]
51. Wei, X.; Li, Q.; Hao, H.; Yang, H.; Li, Y.; Sun, T.; Li, X. Preparation, physicochemical and preservation properties of Ti/ZnO in situ SiOx chitosan composite coatings. *J. Sci. Food Agric.* **2019**, *100*, 570–577. [CrossRef] [PubMed]
52. He, Y.; Wan, M.; Wang, Z.; Zhang, X.; Zhao, Y.; Sun, L. Fabrication and characterization of degradable and durable fluoride-free super-hydrophobic cotton fabrics for oil/water separation. *Surf. Coat. Technol.* **2019**, *378*, 125079. [CrossRef]
53. Pandey, S.; Singh, S.P. Organic solvent tolerance of an α-Amylase from haloalkaliphilic bacteria as a function of pH, temperature, and salt concentrations. *Appl. Biochem. Biotechnol.* **2012**, *166*, 1747–1757. [CrossRef]
54. Wang, C.; Tzeng, F.; Chen, H.; Chang, C. Ultraviolet-durable superhydrophobic zinc oxide-coated mesh films for surface and underwater–oil capture and transportation. *Langmuir* **2012**, *28*, 10015–10019. [CrossRef] [PubMed]
55. Cai, P.; Bai, N.; Xu, L.; Tan, C.; Li, Q. Fabrication of superhydrophobic wood surface with enhanced environmental adaptability through a solution-immersion process. *Surf. Coat. Technol.* **2015**, *277*, 262–269. [CrossRef]

© 2020 by the authors. Licensee MDPI, Basel, Switzerland. This article is an open access article distributed under the terms and conditions of the Creative Commons Attribution (CC BY) license (http://creativecommons.org/licenses/by/4.0/).

Article

Preparation, Thermal Analysis, and Mechanical Properties of Basalt Fiber/Epoxy Composites

Konstantinos Karvanis [1,*], Soňa Rusnáková [1], Ondřej Krejčí [2] and Milan Žaludek [1]

[1] Department of Production Engineering, Faculty of Technology, Tomas Bata University in Zlín, Vavrečkova 275, 760 01 Zlín, Czech Republic; rusnakova@utb.cz (S.R.); zaludek@utb.cz (M.Ž.)
[2] Department of Polymer Engineering, Faculty of Technology, Tomas Bata University in Zlin, Vavrečkova 275, 760 01 Zlín, Czech Republic; okrejci@utb.cz
* Correspondence: karvanis@utb.cz

Received: 14 May 2020; Accepted: 6 August 2020; Published: 10 August 2020

Abstract: In this study, basalt fiber-reinforced polymer (BFRP) composites with epoxy matrix, 20 layers, and volume fraction of fibers V_f = 53.66%, were prepared by a hand lay-up compression molding combined method. The fabric of the basalt fibers is in twill 2/2 weave. Through dynamic mechanical analysis (DMA), their viscoelastic behavior at elevated temperatures and in various frequencies was explored, whereas thermomechanical analysis (TMA) took part in terms of creep recovery and stress-relaxation tests. Moreover, the glass transition temperature (T_g) of the BFRP composites was determined through the peak of the tanδ curves while the decomposition of the BFRP composites and basalt fibers, in air or nitrogen atmosphere, was explored through thermogravimetric analysis (TGA). The mechanical behavior of the BFRP composites was investigated by tensile and three-point bending experiments. The results showed that as the frequency is raised, the BFRP composites can achieve slightly higher T_g while, under the same circumstances, the storage modulus curve obtains a less steep decrease in the middle transition region. Moreover, the hand lay-up compression molding hybrid technique can be characterized as efficient for the preparation of polymer matrix composites with a relatively high V_f of over 50%. Remarkably, through the TGA experiments, the excellent thermal resistance of the basalt fibers, in the temperature range 30–900 °C, was revealed.

Keywords: basalt fiber; epoxy composite; glass transition temperature; DMA; TMA; creep recovery; stress-relaxation

1. Introduction

Nowadays, fiber-reinforced polymer (FRP) composites are broadly used, with the most use in critical applications. These composites have proved their significant mechanical behavior in various applications whereas research on them, using various materials and methods for their production, is still ongoing. In particular, FRP composites consist of a polymer matrix reinforced with fibers. The usual applications of the FRPs are in the aerospace, marine, automotive, and construction industries [1]. Additionally, due to their high strength-to-weight ratio, high stiffness-to-weight ratio, corrosion resistance, and light weight, the FRP composites are appealing in civil engineering applications [2]. It must be noted that these materials have generally remarkable costs, so the applications must verify them and the parameters such as fibers' architecture and composites' production method affect the properties of the FRP composites to a great extent.

The most commonly used fibers in the FRP composites are the carbon, glass, aramid, and basalt; boron and silicon carbide fibers are also used, but in limited amounts. When long fibers are applied as reinforcement phase, the composite obtains characteristics familiar for structural applications because of the remarkable ability of these fibers to carry loads.

A promising material for usage in various applications is the basalt, which is solidified volcano lava. The basalt fibers are attractive to be used as reinforcement phase in composites because they combine superior properties with low price. In particular, basalt fibers can be fabricated through the use of basalt rocks [3]. When these fibers are embedded in a polymer matrix, the named basalt fiber-reinforced polymer (BFRP) composites are formed. The stiff and brittle nature of the basalt fibers is associated with their disadvantages [4].

During the production of the basalt fibers, shattered basalt rocks are melted at 1400 °C and the molten material is drawn [5]. The continuous basalt fibers are fabricated with technology similar to the E-glass, with the primary discrepancy that the latter are produced through a complex group of materials, whereas the basalt filament is produced through the melting of basalt rocks without other additives [6]. Corresponding to the degree of the contained SiO_2, the basalt materials are categorized to alkaline basalts (up to 42% SiO_2), midly acidic basalts (43% to 46% SiO_2), and acidic basalts (more than 46% SiO_2) [7].

One of the most common polymers, which is used as a matrix in the FRP composites, is epoxy. Epoxy is primarily used for aerospace composites, but its long curing time makes it not the first choice in automotive applications where polyester, vinyl ester, or polyurethane polymer matrices are preferred due to their lower curing time than the epoxy [8].

The FRP composites production methods are mainly divided to autoclave and out of autoclave (OOA) methods. The former include the use of autoclave, with which high quality FRP composites are produced, but with significant costs for operation and energy, whereas the latter methods are more economic and not so high level equipment needed. The lay-up techniques are those in which layers are stacked, and the most familiar technique of them is the hand lay-up. In this method, the laminate, usually with fiber fabric as reinforcement phase, is prepared by placing ply over ply by hand until the desired thickness is achieved by the use of a roller as the resin is applied on the laminate and as the excessive amount of it is removed. The hand lay-up technique has two main disadvantages; it is difficult to achieve uniform distribution of the matrix, resulting in non-uniform percentage of fibers/matrix in the composite laminate, and that during the curing, usually there is no pressure on the laminate, which in turn increases the porosity of the composite plate. In this research, after the preparation of the lay-up of the BFRP composite plate and during its curing, compression molding was used to eliminate its porosity.

One of the main disadvantages of the FRP composites is their low thermal resistance and, generally, their weakness at high temperatures, something which is mainly caused by the low thermal resistance of the polymer matrix. The thermal behavior of the FRP composites is investigated through thermal analysis, which is the study of the materials' properties in accordance with the temperature; measurement of properties under a specific temperature range or in an isothermal situation. Familiar experiments of this sector are the dynamic mechanical analysis (DMA), thermomechanical analysis (TMA), and thermogravimetric analysis (TGA). Generally, the materials' behaviors are influenced by high temperatures, so it is necessary for these to be tested and verified under these circumstances.

DMA is used for the investigation of the rheology of the materials. By this technique, materials' properties such as storage modulus, loss modulus, tanδ, and glass transition temperature (T_g), under various temperatures, frequencies, forces, and deformation modes can be determined. Moreover, by DMA, the properties of viscoelastic materials, like the polymers, can be explored. With the term viscoelastic, materials which exhibit both elastic and viscous characteristics, when they are deformed, are described. Additionally, another characteristic of the DMA experiments is that the deformation can be applied on the specimens through various ways, depending on the instrument, such as three-point bending, single cantilever and dual cantilever bending, tension, compression, and shear mode. It must be noted that, in DMA tests, it is important the measurements to take part in the linear viscoelastic region of the materials.

Furthermore, through the DMA, the glass transition temperature (T_g) can be specified. The T_g is an important point in the mechanical behavior of the materials and usually, at this temperature, the polymers change phase from a solid state to a rubbery. However, T_g is not very accurate defined whereas there are many methods for its determination such as peak of loss modulus or tanδ curves, so it is essential the method which was followed for its determination to be described. Furthermore, the T_g is quite different from the melting temperature (T_m), because the melting point is the beginning of the materials' melting whereas at T_g the materials becomes softer [9].

During the real life applications of the materials, forces are exerted on them for long periods of time so it is essential their properties to be studied in dependence with the time. Through stress-relaxation and creep recovery tests the materials' time dependent behavior can be investigated [10]. It should be noted that, because the polymer matrix has viscoelastic behavior, polymer matrix composites show viscoelastic behavior with the stress and strain to be dependent on time and temperature [11]. In particular, the long molecular chains of the polymer matrix cause the viscoelastic phenomenon of the polymer composites [11]. Contrastively with the solid materials, the viscoelastic, when they are under a constant load, do not exhibit a constant deformation but they continue to flow with the time—the phenomenon named "creep" [12]. Additionally, the viscoelastic materials have time-dependent behavior and permanent deformation [13].

During a creep experiment, constant stress is applied on the material at isothermal temperature, and the arising strain is recorded in accordance with the time [14]. The parameters which affect the creep behavior of FRP composites are temperature, frequency, applied stress, and the interface between polymer-fibers. In the stress-relaxation experiment, a constant extension is applied on the sample and the reducing force is recorded as the time is increased.

Various properties of BFRP composites have been investigated by researchers. Huaian Zhang et al. [15] studied the effects of strain rate and temperature on the tensile properties of BFRP composites which were produced through vacuum-assisted resin infusion technique.

Using two different matrices, vinylester resin and epoxy resin, and the same type of basalt fibers as reinforcement, C. Colombo et al. [16] produced two types of basalt fiber-reinforced composites, and they explored them experimentally through static tensile tests, static compression tests, static delamination tests, fatigue tests, and stepwise tests. The results showed that the basalt fiber-reinforced composites with epoxy matrix had higher ultimate tensile strength and ultimate compressive strength than the one with vinylester matrix [16].

T. Bhat et al. [17] studied the fire structural properties of a basalt fiber-reinforced polymer laminate under compressive loading and compared them with the properties of an E-glass fiber composite which had the same fiber content, ply orientation, and polymer matrix. Dynamic mechanical analysis was performed on BFRP plates by Zhongyu Lu et al. [18].

V. Lopresto et al. [19] fabricated, through vacuum bag technology, E-glass and basalt fiber-reinforced plastic laminates, and they investigated them in terms of tensile, bending, shear, compression, and impact tests. A study about the tensile, shear, and impact strengths of basalt fiber-reinforced unsaturated polyester composites with and without acid and alkali treatments of the fabrics was conducted by V. Manikandan et al. [20].

The influence of elevated temperatures on the flexural fatigue behavior of a pultruded basalt fiber-reinforced epoxy plate was studied by Zike Wang et al. [21]. In particular, the BFRP specimens were subjected to elevated temperature treatment in an oven at 150 °C and 250 °C for 0.5, 1, and 2 h [21].

Farzin Azimpour Shishevan et al. [22] investigated the low-velocity impact behavior of BFRP composites. In particular, in this study, BFRP and carbon fiber-reinforced polymer (CFRP) composites were produced through vacuum-assisted resin transfer molding (VARTM) method and they were tested to low-velocity impact experiments with different energy magnitudes [22].

Salvatore Carmisciano et al. [23] produced basalt and E-glass woven fabric reinforced composites. Particularly, these composite laminates were fabricated by a resin transfer molding (RTM) system, and through various tests, it was revealed that the basalt fiber composites exhibited higher flexural modulus and interlaminar shear strength but lower flexural strength compared to the composites containing E-glass fibers [23].

P. Amuthakkannan and V. Manikandan [24] studied the free vibration and dynamic mechanical properties of surface modified basalt fiber-reinforced polymer composites. In particular, by hand lay-up method, they fabricated untreated, acid-treated, and base-treated basalt fiber composite and investigated them through DMA [24].

However, a research study of BFRP composites, which have volume fraction of fibers (V_f) over 50%, representing enough experimental results, both for the mechanical and the thermal behavior of these compounds, appears to be missing. At this point, it must be noted that in the FRP composites' industry, the introduction of natural fibers, in percentage of 50% and over in a composite, while maintaining good mechanical behavior of it, is of great interest as much from an ecological point of view as from mechanical investigation aspects. In order to gain a better understanding of an FRP composite, due to its viscoelastic matrix, several thermal analysis techniques must be employed to study and characterize it.

In the present study, BFRP composites, consisting of 46.4% epoxy and 53.6% of basalt fibers, volume fraction percentages determined through TGA, in 20 layers, were successfully prepared by a hand lay-up compression molding combined technique and their dynamic mechanical properties, in terms of storage modulus, loss modulus, tanδ, and glass transition temperature were determined. Through the maximum values of the tanδ curves, the glass transition temperatures of them were defined while TMA was performed in the modes of creep recovery and stress-relaxation experiments. Moreover, TGA was used for the exploration of the decomposition of basalt fibers and BFRP composites, as well as for the determination of the weight fraction of reinforcement phase and matrix, whereas the mechanical behavior of the BFRP composites was investigated through tension and three-point bending experiments.

2. Experimental

2.1. Materials

For the preparation of the BFRP composites, a mixture of Epoxy resin Epidian® 652 CIECH Sarzyna S.A (Cieszyn, Poland) and hardener TFF CIECH Sarzyna S.A (Cieszyn, Poland), in a mixing ratio of 100:27 parts per weight, was used as matrix. A fabric of basalt fibers 235 g/m², in twill 2/2 weave, supplied by Havel Composites (Cieszyn, Poland) was used as reinforcement phase (Figure 1). The fact that the specific weight of the used basalt fibers is 2.67 g/cm³ must be stressed. For the production of the BFRP composites, the steps described below were followed. By hand lay-up technique, a laminate composed of 20 layers of polymer fibers was prepared, and then it was placed between two rectangular metal plates forming a mold. It must be noted that the polymer matrix was applied on the fibers fabric through a roller. Consequently, this system was placed in a compression machine, where it was pressed under 20 MPa for 24 h, at a laboratory temperature of 24 °C. Then, the composite plate was left for curing at room temperature for a week, and finally, specimens were cut in the desired dimensions through water jet and mechanical cutting.

2.2. General Experimental Conditions

In all of the experiments, both of thermal analysis and mechanical behavior, the specimens were in storage in a laboratory environment at 24 °C for not less than 40 h prior of the tests. During the experiments, the conditions in the laboratory were at a temperature of 23 °C and humidity of approximately 50%.

Figure 1. The used basalt fibers. The photo was taken with the Carl Zeiss Stemi 2000C Microscope (Jena, Germany) from the edges of the mold.

2.3. Dynamic Mechanical Analysis (DMA)

The DMA experiments were performed by the DMA 1 instrument, from METTLER TOLEDO (Schwerzenbach, Switzerland), with using the STARe Software and under single cantilever configuration (Figure 2). The dimensions of the rectangular shape specimens were 25 mm × 5.7 mm × 2.1 mm (length × width × thickness).

Figure 2. DMA specimen on the single cantilever mode.

Firstly, for the determination of the linear viscoelastic region of the BFRP composites, strain sweep tests over the range 1–31 μm were performed at 25 °C and 1 Hz frequency.

Next, temperature sweep tests were conducted over the temperature range of 30–180 °C, with a heating rate of 2 K/min under three different frequencies: 1, 5, and 10 Hz. The displacement amplitude was adjusted to be 8 μm. Moreover, the T_g temperatures of the BFRP composites was determined through the peak of the tanδ and they are presented in Table 1.

Table 1. Glass transition temperature of the BFRP composites.

Frequency	1 Hz	5 Hz	10 Hz
T_g (peak of tanδ)	67.1 °C	72.1 °C	75.4 °C

2.4. Thermomechanical Analysis (TMA)

The creep recovery and stress-relaxation tests were carried out by the METTLER TOLEDO DMA 1 (Schwerzenbach, Switzerland) instrument, with the STARe Software and under TMA mode and three-point bending configuration. For both types of the experiments, creep recovery and stress-relaxation, the specimens were in rectangular shape with dimensions 40 mm × 5.7 mm × 2.1 mm (length × width × thickness) with the span length between the supports to be 30 mm. Specifically, the particular specifications of these tests are described below.

2.4.1. Creep Recovery Tests

In the first type of the creep recovery experiments, at isothermal 25 °C, a force of 1, 3, or 5 Newton was applied on the specimens for 30 min and then the recovery behavior of them was recorded under 0 N for 120 min.

The second type of the creep recovery tests took part under three different temperatures. In detail, 1 N was applied on the specimens whereas the temperature was 25, 50, or 75 °C and the recovery of them, under 0 N, was measured for 120 min.

2.4.2. Stress-Relaxation Tests

The stress-relaxation tests were performed at three different temperatures: 25, 50, and 60 °C. In detail, during these experiments, 20 μm extension was applied on the specimens and their time-dependent stress was being measured for 60 min. It should be noted that experiments were performed at 75 °C, but they were failed due to the small resistance of the BFRP sample in deformation at this temperature; the force was always falling in negative values.

2.5. Thermogravimetric Analysis (TGA)

The TGA was performed with the TGA Q50 Thermogravimetric Analyzer, from the TA Instruments (New Castle, DE, USA), in the temperature range 30–900 °C, with heating rate 10 °C/min, in air or nitrogen atmosphere for two different separate runs. It must be noted that the TGA tests were setting and performed with the Thermal Advantage Release 5.4.0 software and the results were evaluated with the TA Instruments Universal Analysis 2000 version 4.5A program. In particular, during these experiments, samples with weights of approximately 60 mg were placed in an alumina crucible, and the total flow was set at 100 mL/min; balance purge flow at 40 mL/min, and sample purge flow 60 mL/min. It must be noted that the complex structure of the BFRP composites requires enough mass during the TGA experiments, so as to be excluded reliable conclusions for its decomposition and that is why the weight of the TGA samples was relative big.

2.6. Mechanical Properties

The flexural strength of the BFRP composites was determined at room temperature using the testing machine Zwick/Roell 1456 (Ulm, Germany) with the software testXpert® II V2.1. In particular, the three-point bending experiments were exhibited with specimens of dimensions 75 mm × 10 mm × 2.1 mm (length × width × thickness) with the span length-to-depth ratio of 20:1, whereas the crosshead speed was set to 1 mm/min. For reliability in the measurements, four specimens were tested, and the average values of the flexural characteristics are presented in Table 2.

Table 2. Average values of the three-point bending results.

	F at 0.2% Plastic Deformation [N]	Upper Yield Point [N]	σ_{fsmax} [MPa]
BFRP composite	187.75	195	282

The tensile tests were carried out, at room temperature, with the testing machine Vibrophore 100, from Zwick/Roell Company (Ulm, Germany) with the software TestExpert III. In particular, the tensile strength of the BFRP composites was measured with specimens of dimensions 150 mm × 13 mm × 2.1 mm (Figure 3) with the gripping sections to be each 40 mm, giving a clear tension testing length of 70 mm (Figure 4). Furthermore, the crosshead speed was set at 1 mm/min. For this kind of experiment, five samples were tested, and the average values of them are presented in Table 3.

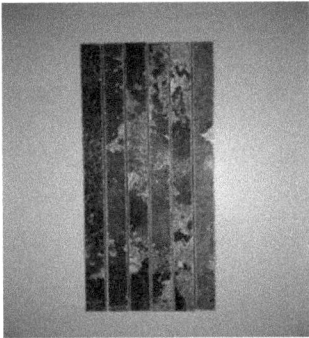

Figure 3. Tension samples.

Figure 4. Tension specimen on the grips.

Table 3. Average values of the tension experiments results.

	F at 0.2% Plastic Strain [N]	σ_{tsmax} [MPa]	Strain at Breakage [%]
BFRP composite	2580	494.4	4.68

3. Results

3.1. DMA Tests

3.1.1. Displacement Sweep Test

Figure 5 depicts the graph of a displacement sweep test on BFRP compound specimen. At the chosen 8 μm, the initial storage modulus value does not change remarkably so it is assumed that the DMA experiments take part in the linear viscoelastic region of the materials.

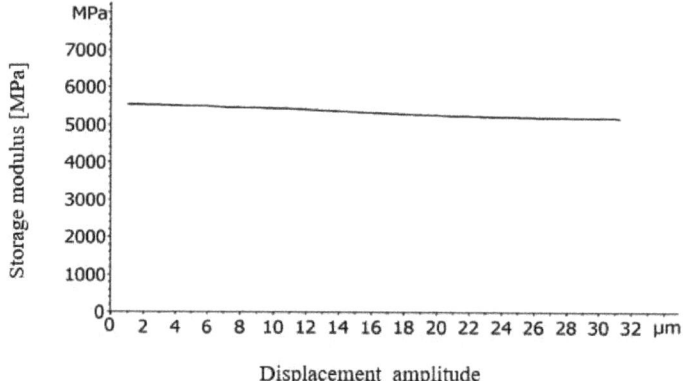

Figure 5. Displacement sweep test of the BFRP specimen; storage modulus versus displacement.

3.1.2. DMA Experiments

The storage modulus (E') is associated with the stored energy of the materials. Figure 6 shows the storage modulus of the BFRP composites, as the temperature is increased, at three different frequencies; 1, 5 or 10 Hz. From this graph, it is observed that as the temperature is raised the storage modulus of the BFRP composites is reduced, with high rate, up to the 90 °C. Remarkably, after about the 55 °C, a steep drop in the values is observed. Generally, this steep fall in the storage modulus curve specifies the maximum working temperature of the materials and it is associated with the T_g. However, it should be noted that as the frequency is increased, this transition region becomes less abrupt and with longer duration. Furthermore, after about the 90 °C, the storage modulus values are in a steady low value line until the final temperature.

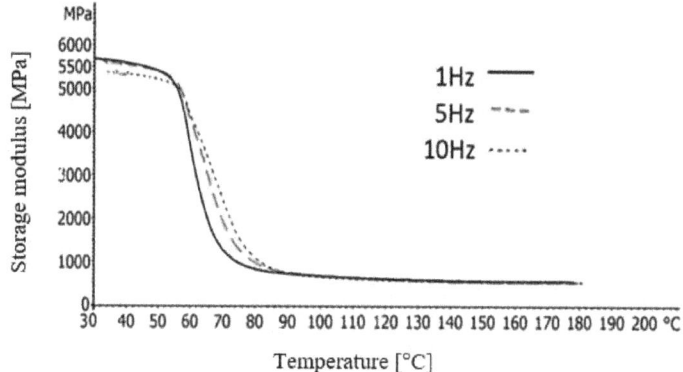

Figure 6. Storage modulus of the BFRP composites, as a function of the temperature, at 1, 5 and 10 Hz.

The loss modulus (E") is correlated with the lost energy, as form of heat, under DMA tests. High values of loss modulus point out viscous behavior therefore remarkable damping properties [25]. The loss moduli of the BFRP composites, at three different frequencies and as a function of temperature, is presented in Figure 7. As it can be seen, at low temperatures, the curves follow a steep increase achieving peak values whereas then a steep fall is observed until almost zero values which are maintained up to the final 180 °C. It is worth mentioning that higher frequency seems to have a negative impact in the peak values of the loss modulus.

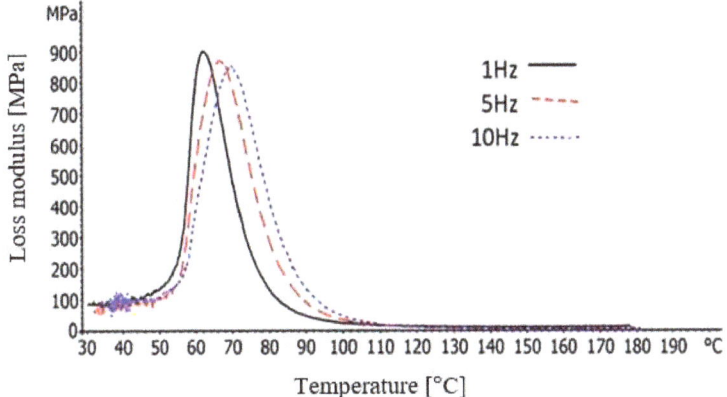

Figure 7. Loss modulus of the BFRP composites, as a function of the temperature, at 1, 5 and 10 Hz.

The tanδ is the ratio of loss modulus to storage modulus

$$\tan\delta = E''/E' \quad (1)$$

and it is sometimes called damping factor.

The tanδ curves of the BFRP composites, over the temperature range of 30–180 °C, whereas the frequency is 1, 5, or 10 Hz, are shown in Figure 8. As it can be seen, at low temperatures the values are almost stable, whereas after approximately the 50 °C, they follow a step upward trend, achieving a peak value, which can be used for the identification of the materials' glass transition temperature, and then they follow a steep decrease.

Table 1 depicts the T_g of the BFRP composites which were acquired through the corresponding temperature of the peak of the tanδ curves. It can be revealed that as the frequency is increased the BFRP can obtain higher T_g. Daohai Zhang et al. [26] determined, through tanδ peak, the T_g of long glass fiber reinforced thermoplastic polyurethane/poly(butylene terephthalate) composites, and they found the T_g of these composites to shift in higher temperatures as the frequency was increased. Moreover, the T_g of auto polymerized hard direct denture reline resins, determinate through dynamic mechanical analysis, was found to be higher in greater frequencies [27].

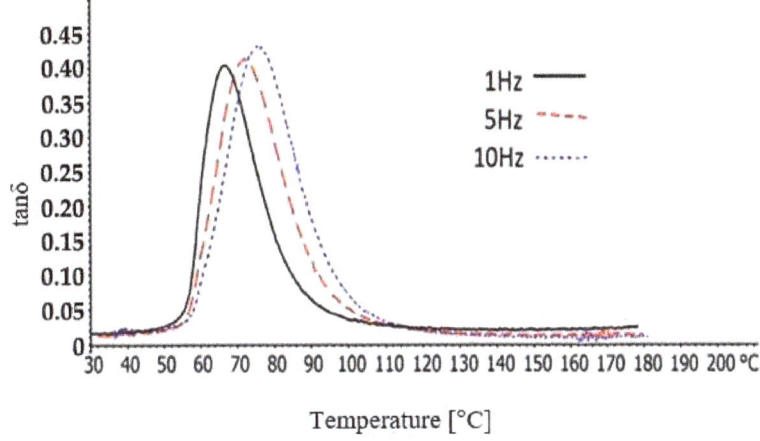

Figure 8. tanδ of the BFRP composites, as a function of the temperature, at 1, 5, and 10 Hz.

3.2. Thermomechanical Analysis (TMA)

3.2.1. Creep Recovery Tests

Figure 9 depicts the creep recovery behavior of the BFRP composites with an initial force of 1, 3 or 5 Newton at temperature 25 °C. In this graph, it can be noticed that the creep behavior of these composites is almost stable or very slightly increases during the first 30 min of the experiments whereas their recovery took part immediately after the release of the force. The final value of these curves is the permanent deformation which was caused on the composites and as it was expected the 1 Newton initial force caused the less.

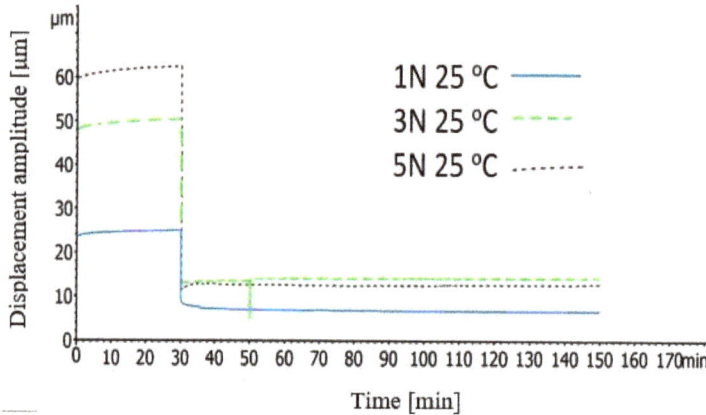

Figure 9. Creep recovery isothermal curves of the BFRP composites under various initial forces.

Figure 10 illustrates the effect of the temperature on the creep recovery behavior of the BFRP composites. In this graph, a remarkable point is that, in higher temperatures, the resistance of the BFPR composites in deformation is reduced. Additionally, whereas at 25 °C, the deformation on the samples was almost steady during the first 15 min of the force application, at 50 and 75 °C, the deformation is rising rapidly during this time. This can be attributed to the fact that, at high temperatures, the structure of the composites becomes more compliant, thus the viscoelasticity becomes more evident and, as a consequence, this growing in deformation can be noticed.

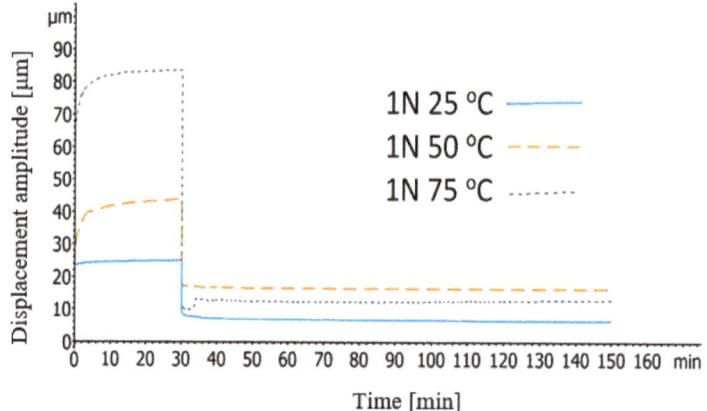

Figure 10. Creep recovery curves of the BFRP composites at various temperatures under initial force of 1 Newton.

3.2.2. Stress-Relaxation Tests

Figure 11 depicts the stress-relaxation of the BFRP composites at 25, 50, or 60 °C. Remarkably, the viscoelastic nature of the BFRP composites is visible in the first minutes of these curves; initially, the composites' structure require a considerable high force which then is reduced, finally reaching almost stable values.

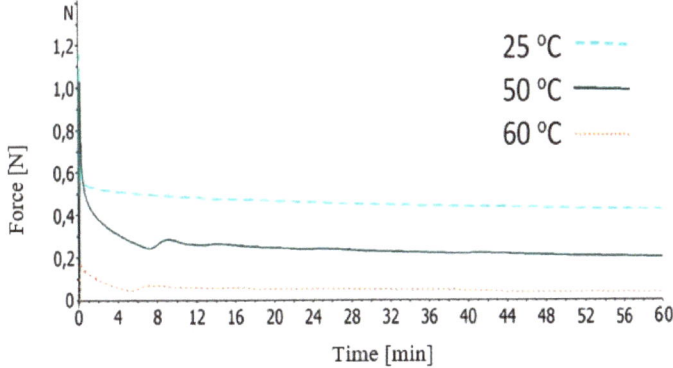

Figure 11. Stress-relaxation curves of the BFRP composites at various temperatures.

3.3. Flexural Experiments

Strength of the materials is their ability to withstand forces without breakage. The determination of the flexural strength is very useful, especially in structural applications, whereas its identification also helps for the evaluation of the composites' matrix–fiber interface. In engineering, the yield strength is the point at which the materials change from elastic to plastic deformation, and usually, it determines the maximum working strain of the materials, as after this point, there is permanent deformation on their structure. Table 2 shows the three-point bending experiments results of the BFRP composites, whereas Figure 12 represents the flexural curves of them in detail.

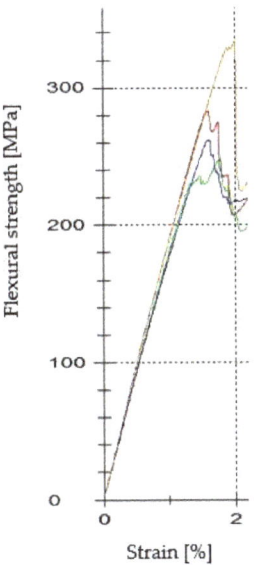

Figure 12. Flexural strength curves of the BFRP composites.

3.4. Tension Experiments

The tension strength is the force which a material can withstand, due to pulling, before its breakage and it is one of the most important properties of the materials. The curves of the five tension experiments, are presented in Figure 13. As it can be revealed, a relative good reliability, for the hand lay-up fabrication method and composite heterogeneous structure, can be obtained, as three of them have almost identical curves, whereas the other two curves, despite the fact that they exhibited lower maximum values, follow the same general trend. Additionally, the BFRP composites are characterized as brittle materials because they break suddenly without showing plastic deformation.

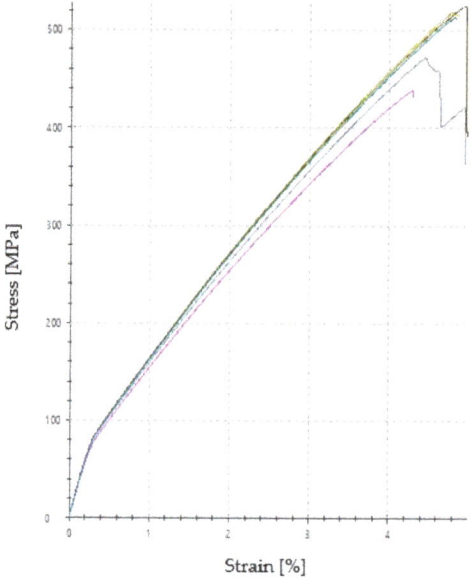

Figure 13. Tensile stress-strain curves of the five specimens.

3.5. TGA

Figures 14 and 15 show the weight of the BFRP composite as a function of the temperature, in the range 30–900 °C, in air or nitrogen atmosphere. As it can be seen, during the decomposition of the BFRP composite in both atmospheres, a sharp decline starts at approximately 260–300 °C, and this is attributed to the main weight loss of the epoxy matrix. Comparing the decomposition of the BRFP compound in air with its decomposition in nitrogen atmosphere, it can be revealed that the degradation of its matrix, as the basalt fibers are not thermally affected in this temperature, in the latter atmosphere takes part with a lower rate and continues up to the final 900 °C whereas in air the epoxy has completely decomposed up to the 550 °C. A remarkable point in the DTG graphs, is the maximum peaks which correspond to the maximum degradation rate of the investigated materials. In the case of the BFRP composite, TGA in N_2, one peak is appeared at 341 °C, which is dedicated to degradation of its matrix, whereas in air, two maximum heating rates are presented at 336.4 and 458.6 °C, respectively.

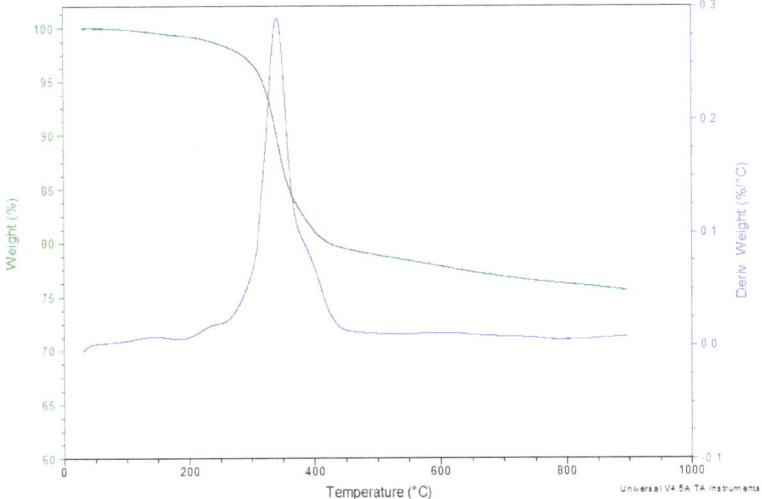

Figure 14. TGA and DTG curves of the BFRP composite in Nitrogen atmosphere.

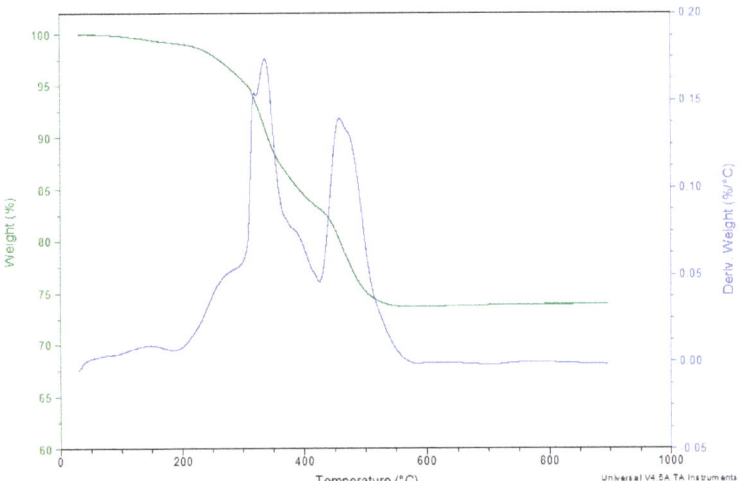

Figure 15. TGA and DTG curves of the BFRP composite in air atmosphere.

Figures 16 and 17 depict the decomposition of the basalt fibers in air or N_2 atmosphere. As it can be revealed, these fibers show excellent thermal resistance; they are not thermally influenced under both atmospheres up to the final 900 °C.

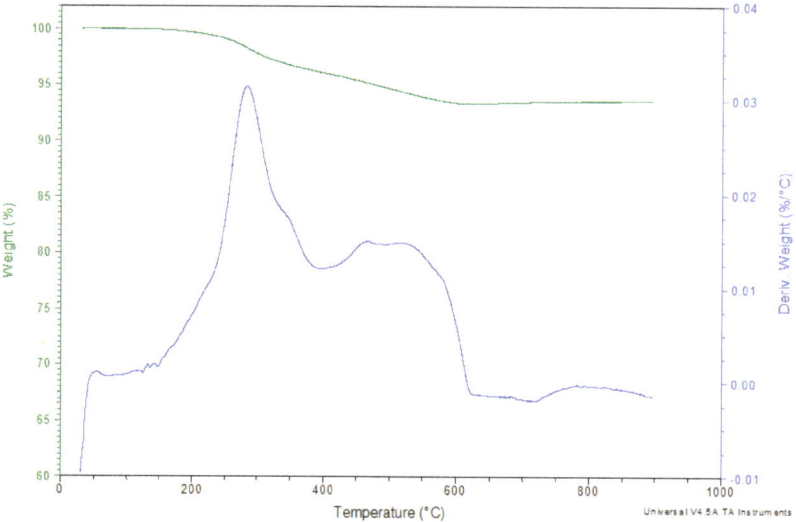

Figure 16. TGA and DTG curves of the basalt fibers in Nitrogen atmosphere.

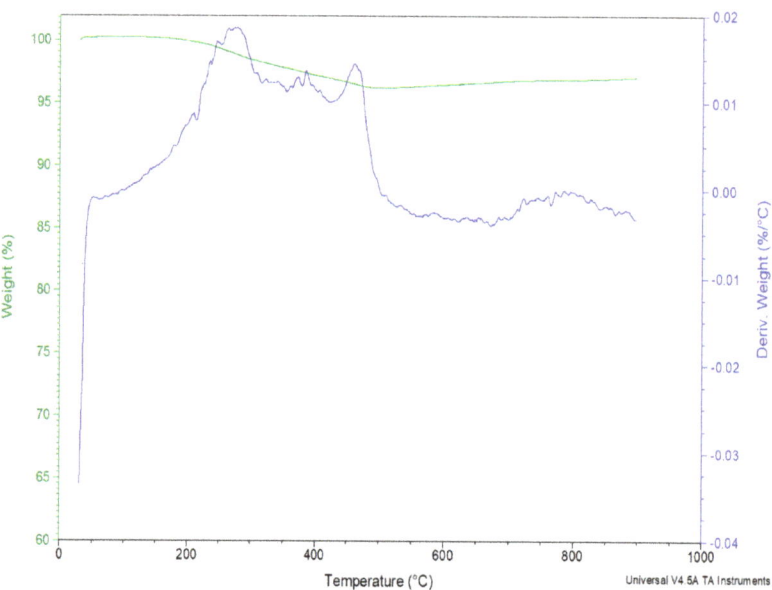

Figure 17. TGA and DTG curves of the basalt fibers in air atmosphere.

The used basalt fibers, in this TGA experiment, were taken from over the edges of the hand lay-up mold so they were not absolutely pure. The approximately 6% weight reduction, in the temperature range 200–600 °C, is attributed in decomposition, of a very small part of epoxy which was stuck on the basalt fibers during the composites' production. This is also verified from the fact that after the approximately the 630 °C the basalt fibers are not thermally affected.

The used basalt fibers, in this TGA experiment, were taken from over the edges of the hand lay-up composite plate so they were not absolutely pure. The approximately 4% weight reduction, in the temperature range 200–500 °C is attributed in decomposition, of a very small part of epoxy which was stuck on the basalt fiber during the composites' production. This is also verified from the fact that after approximately the 500 °C the basalt fibers are not thermally affected.

3.6. Calculation of the Fibers' Volume Fraction (V_f) and Volume of the Matrix (V_m)

Based on the TGA experiments, and using the weight fraction of the reinforcement (W_f), the weight fraction of the matrix (W_m) and the densities of the basalt fibers and epoxy matrix, the V_f of the reinforcement in the BFRP composite was calculated. In detail, the decomposition curve of the BFRP composite, in air atmosphere, after approximately the 550 °C became flat and steady, thus meaning that the epoxy matrix was completely decomposed where at the same time the basalt fibers were not thermally affected. At this temperature, the weight percentage is 73.76% and so this includes only basalt fibers; thus, W_f = 73.76% and W_m = 26.24 %.

The densities of the basalt fibers and epoxy matrix are known, 2.67 and 1.10 g/cm^3 respectively. So, the volume fraction of the reinforcement phase, in terms of weight fraction, was found according to the equation [28]:

$$V_f = \frac{W_f * \rho_m}{W_f * \rho_m + W_m * \rho_f} \tag{2}$$

where:

V_f = volume of the reinforcement
V_m = volume of the matrix
ρ_m = density of matrix
ρ_f = density of reinforcement
So, finally, V_f = 53.66% and V_m = 46.34%.

4. Conclusions

During this research study, BFRP composites with relatively high V_f over 50% were successfully fabricated and were experimentally investigated through various thermal analysis and mechanical behavior tests. At this point, it must be noted that the basalt fibers, which have natural origin, when they are used in polymer matrix composites, especially with high V_f, give them biodegradable characteristics, an advantage which nowadays is becoming more and more important due to environmental concerns. Based on the experimental results, the following conclusions can be drawn. Through the creep recovery and stress-relaxation tests, the significant impact of the high temperatures on the materials' structure was revealed. Especially near and over the T_g, the BFRP polymer showed remarkable reduction in resistance in deformation forces. Thus, it can be concluded that the T_g is a critical point in the polymer matrix composites' behavior. Through the DMA, it was revealed that as the frequency increases, the peak values of the tanδ of the BFRP composite move to higher temperatures, leading to a greater T_g. The T_g is affected by the frequency.

One of the significant revelations in this research is the thermal resistance of the basalt fibers. In particular, in the TGA runs, these fibers showed no decomposition, in the temperature range 30–900 °C, in nitrogen or air atmosphere. Due to this characteristic, the basalt fibers can be classified as potential material for usage in a very broad range of applications, as the resistance to high temperatures is always a critical factor. These could be as thermal insulations against fire dangers, and also in the surrounding regions of exhaust systems of airplanes, cars, and vehicles. Moreover, due to the basalt fibers' low density, the combination of them in polymer matrix results in a composite with low weight, which turns to very important materials' characteristics such as high strength-to-weight and high stiffness-to-weight. These features make the BFRP composites ideal for commercial aircraft applications where the BFRP composite' low weight will turn in reduction in airplanes' fuel consumption, whereas also in the case of fire, these fibers will show remarkable thermal resistance and will continue to provide structural support up to high temperatures.

From an overall point of view, the BFRP composites showed very good tensile and flexural strength, something which means that the epoxy matrix formed a very good interfacial bond with the basalt fibers. Generally, the FRP composites demonstrate low mechanical behavior in case of a poor bond between matrix and fibers. Moreover, based on the overall results, especially on those of the mechanical behavior of the BFRP composites, as in those in which a various number of specimens were used for reliability, and due to the fact that the composite plate has remarkably stable thickness (±0.1 mm) through its structure, the hybrid hand lay-up compression molding technique can be characterized as efficient and reliable for the production of FRP composites, with a high V_f of fibers of over 50% as the produced BFRP composites.

Author Contributions: Investigation, K.K., S.R., O.K. and M.Ž.; Methodology, K.K., S.R., O.K. and M.Ž.; Supervision, K.K. and S.R.; Validation, K.K., S.R., O.K. and M.Ž.; Writing—original draft, K.K.; Writing—review & editing, K.K. and S.R. All authors have read and agreed to the published version of the manuscript.

Funding: This research received no external funding.

Acknowledgments: This work and the project is realized with the financial support of the internal grant of TBU in Zlin No. IGA/FT/2020/004 funded from the resources of specific university research.

Conflicts of Interest: The authors declare no conflicts of interest.

References

1. Martin, A. Introduction of Fibre-Reinforced Polymers—Polymers and Composites: Concepts, Properties and Processes. In *Fiber Reinforced Polymers*; IntechOpen: Rijeka, Croatia, 2013; Available online: https://www.intechopen.com/books/fiber-reinforced-polymers-the-technology-applied-for-concrete-repair/introduction-of-fibre-reinforced-polymers-polymers-and-composites-concepts-properties-and-processes (accessed on 1 May 2020). [CrossRef]
2. Cao, S.; Wu, Z. Tensile Properties of FRP Composites at Elevated and High Temperatures. *J. Appl. Mech.* **2008**, *11*, 963–970. [CrossRef]
3. Vikas, G.; Sudheer, M. A Review on Properties of Basalt Fiber Reinforced Polymer Composites. *Am. J. Mater. Sci.* **2017**, *7*, 156–165. [CrossRef]
4. Zhang, Y.; Yu, C.; Chu, P.K.; Lv, F.; Zhang, C.; Ji, J.; Zhang, R.; Wang, H. Mechanical and thermal properties of basalt fiber reinforced poly(butylene succinate) composites. *Mater. Chem. Phys.* **2012**, *133*, 845–849. [CrossRef]
5. Maxineasa, S.G.; Taranu, N. Life cycle analysis of strengthening concrete beams with FRP. In *Eco-Efficient Repair and Rehabilitation of Concrete Infrastructures*; Woodhead Publishing: Sawston Cambridge, UK, 2018; pp. 673–721. ISBN 9780081021811. [CrossRef]
6. Fiore, V.; di Bella, G.; Valenza, A. Glass–basalt/epoxy hybrid composites for marine applications. *Mater. Des.* **2011**, *32*, 2091–2099, ISSN 0261-3069. [CrossRef]
7. Tamas, D.; Tibor, C. Chemical Composition and Mechanical Properties of Basalt and Glass Fibers: A Comparison. *Text. Res. J.* **2009**, *79*, 645–651. [CrossRef]
8. Mallick, P.K. *Fiber-Reinforced Composites: Materials Manufacturing and Design*, 3rd ed.; Mallick, P.K., Ed.; CRC Press: Boca Raton, FL, USA, 2007.
9. Saba, N.; Jawaid, M.; Alothman, O.Y.; Paridah, M.T. A review on dynamic mechanical properties of natural fibre reinforced polymer composites. *Constr. Build. Mater.* **2016**, *106*, 149–159. [CrossRef]
10. Papanicolaou, G.C.; Zaoutsos, S.P. Viscoelastic constitutive modeling of creep and stress relaxation in polymers and polymer matrix composites. In *Creep and Fatigue in Polymer Matrix Composites*; Woodhead Publishing: Sawston Cambridge, UK, 2011; pp. 3–47. [CrossRef]
11. Tang, T.; Felicelli, S.D. Computational evaluation of effective stress relaxation behavior of polymer composites. *Int. J. Eng. Sci.* **2015**, *90*, 76–85. [CrossRef]
12. Epaarachchi, J.; Guedes, R.; Eparachchi, J.A. The effect of viscoelasticity on fatigue behaviour of polymer matrix composites. In *Creep and Fatigue in Polymer Matrix Composites*; Guedes, R.M., Ed.; Woodhead Publishing: Sawston Cambridge, UK, 2011; pp. 4913–4925.
13. Roy, S.; Reddy, J. *Computational Modeling of Polymer Composites: A Study of Creep and Environmental Effects*; CRC Press: Boca Raton, FL, USA, 2013.

14. Daver, F.; Kajtaz, M.; Brandt, M.; Shanks, R.A. Creep and Recovery Behaviour of Polyolefin-Rubber Nanocomposites Developed for Additive Manufacturing. *Polymers* **2016**, *8*, 437. [CrossRef] [PubMed]
15. Zhang, H.; Yao, Y.; Zhu, D.; Mobasher, B.; Huang, L. Tensile mechanical properties of basalt fiber reinforced polymer composite under varying strain rates and temperatures. *Polym. Test.* **2016**, *51*, 29–39. [CrossRef]
16. Colombo, C.; Vergani, L.; Burman, M. Static and fatigue characterisation of new basalt fibre reinforced composites. *Compos. Struct.* **2012**, *94*, 1165–1174. [CrossRef]
17. Bhat, T.; Kandare, E.; Gibson, A.; Di Modica, P.; Mouritz, A. Compressive softening and failure of basalt fibre composites in fire: Modelling and experimentation. *Compos. Struct.* **2017**, *165*, 15–24. [CrossRef]
18. Lu, Z.; Xian, G.; Li, H. Effects of elevated temperatures on the mechanical properties of basalt fibers and BFRP plates. *Constr. Build. Mater.* **2016**, *127*, 1029–1036. [CrossRef]
19. LoPresto, V.; Leone, C.; De Iorio, I. Mechanical characterisation of basalt fibre reinforced plastic. *Compos. Part B* **2011**, *42*, 717–723. [CrossRef]
20. Manikandan, V.; Jappes, J.W.; Kumar, S.S.; Amuthakkannan, P. Investigation of the effect of surface modifications on the mechanical properties of basalt fibre reinforced polymer composites. *Compos. Part B* **2012**, *43*, 812–818. [CrossRef]
21. Wang, Z.; Yang, Z.; Yang, Y.; Xian, G. Flexural fatigue behavior of a pultruded basalt fiber reinforced epoxy plate subjected to elevated temperatures exposure. *Polym. Compos.* **2016**, *39*, 1731–1741. [CrossRef]
22. Shishevan, F.A.; Akbulut, H.; Mohtadi-Bonab, M.A. Low Velocity Impact Behavior of Basalt Fiber-Reinforced Polymer Composites. *J. Mater. Eng. Perform.* **2017**, *26*, 2890–2900. [CrossRef]
23. Carmisciano, S.; De Rosa, I.M.; Sarasini, F.; Tamburrano, A.; Valente, M. Basalt woven fiber reinforced vinylester composites: Flexural and electrical properties. *Mater. Des.* **2011**, *32*, 337–342. [CrossRef]
24. Amuthakkannan, P.; Manikandan, V. Free vibration and dynamic mechanical properties of basalt fiber reinforced polymer composites. *Indian J. Eng. Mater. Sci.* **2018**, *25*, 265–270. Available online: http://nopr.niscair.res.in/handle/123456789/44932 (accessed on 30 September 2018).
25. Wagner, M. *Thermal Analysis in Practice-Fundamental Aspects*; Carl Hanser Verlag GmbH Co KG: Munich, Germany, 2018.
26. Zhang, D.; He, M.; Qin, S.; Yu, J.; Guo, J.; Xu, G. Study on dynamic mechanical, thermal, and mechanical properties of long glass fiber reinforced thermoplastic polyurethane/poly(butylene terephthalate) composites. *Polym. Compos.* **2018**, *39*, 63–72. [CrossRef]
27. Takase, K.; Watanabe, I.; Kurogi, T.; Murata, H. Evaluation of glass transition temperature and dynamic mechanical properties of autopolymerized hard direct denture reline resins. *Dent. Mater. J.* **2015**, *34*, 211–218. [CrossRef]
28. Sheikh-Ahmad, J.Y. *Machining of Polymer Composites*; Springer: New York, NY, USA, 2009; ISBN 978-0-387-35539-9. [CrossRef]

© 2020 by the authors. Licensee MDPI, Basel, Switzerland. This article is an open access article distributed under the terms and conditions of the Creative Commons Attribution (CC BY) license (http://creativecommons.org/licenses/by/4.0/).

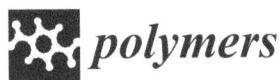

Article

Pressure Field Assisted Polycondensation Nonaqueous Precipitation Synthesis of Mullite Whiskers and Their Application as Epoxy Resin Reinforcement

Guo Feng [1,*], Feng Jiang [2,*], Zi Hu [3,*], Weihui Jiang [1,2,*], Jianmin Liu [1], Quan Zhang [1], Qing Hu [2], Lifeng Miao [1], Qian Wu [1] and Jian Liang [1]

1. National Engineering Research Center for Domestic & Builing Ceramics, Jingdezhen Ceramic Institute, Jingdezhen 333000, China; liujianmin@jci.edu.cn (J.L.); zhangquan@jci.edu.cn (Q.Z.); miaolifeng@jci.edu.cn (L.M.); wuqian@jci.edu.cn (Q.W.); liangjian@jci.edu.cn (J.L.)
2. Department of Material Science and Engineering, Jingdezhen Ceramic Institute, Jingdezhen 333000, China; huqing@jci.edu.cn
3. Jiangxi Ceramic Research Institute, Jingdezhen 333000, China
* Correspondence: fengguo@jci.edu.cn (G.F.); jiangfeng@jci.edu.cn (F.J.); fg19840421@163.com (Z.H.); jiangweihui@jci.edu.cn (W.J.)

Received: 6 November 2019; Accepted: 3 December 2019; Published: 4 December 2019

Abstract: Mullite whiskers were novelty prepared via pressure field assisted polycondensation nonaqueous precipitation method. The precipitate phase transition in heating process, phase compositions and microstructure of samples calcined at different temperatures, effect of pressure field on precursors polycondensation and AlF_3 amount on sample morphology, the structure and the growth mechanism of whiskers were investigated. The results indicate that pressure field caused by kettle treatment promotes the polycondensation reaction between AlF_3 and tetraethyl orthosilicate (TEOS), the excess aluminum fluoride coordinates with the precipitate skeleton of the =Al–O–Si≡, which brings about the low mullitization temperature (900 °C). The sample prepared with the optimal amount of aluminum fluoride (1.3 of the theoretical amount) calcined at 1100 °C presents high yield and aspect ratio (>15, 100 nm in diameter) of mullite whiskers. Growth of whiskers prepared via pressure field assisted polycondensation nonaqueous precipitation method is attributed to a vapor-solid (VS) mechanism with the inducement of screw. These mullite whiskers with the structure of multi-needle whiskers connected in the same center can be distributed evenly in epoxy resin, which greatly improves the mechanical properties of epoxy resin.

Keywords: mullite; whiskers; nonaqueous precipitation method; aluminum fluoride; polar transformation; screw

1. Introduction

Mullite ($3Al_2O_3·2SiO_2$) materials are widely applied as high-temperature engineering and refractory materials, due to their unique excellent properties of high-temperature strength, low thermal conductivity, high creep resistance, relatively low thermal expansion coefficient, excellent chemical stability and creep resistance [1,2]. Mullite has a stable crystal structure of orthorhombic. Its lattice constants (a, b and c) are of 7.545 Å, 7.689 Å and 2.884 Å (JCPDS Card # 15-0776). The crystal growth of mullite is generally more quickly in the c-axis direction than any other direction, which brings a high orientation degree to form mullite whiskers. Mullite whiskers have attracted much attention as the reinforcement for high-temperature materials [3–5].

Various processing techniques have been utilized to prepare mullite whiskers, including the mineral decomposition method [6], vapor-phase reaction method [7], molten salt method [8], hydrolytic sol–gel method [9], nonhydrolytic sol–gel method [10], nonhydrolytic sol–gel combined with the molten salt method [11–14], etc. The nonaqueous precipitation method is a novel materials synthesis method, and it holds the merits of simple process, short cycle, non-aggregation of products [15–17]. However, in order to give full play to the advantages of nonaqueous precipitation method, the precursor materials must undergo nonhydrolytic polycondensation. In our previous researches, it was found that the polarity of precursor material should not be too large, the ion bond percentage should not be more than 50%, in order to make the precursor material directly undergo nonhydrolytic polycondensation. At the same time, these polar compounds with a large percentage of ionic bond are often characterized by low cost, low toxicity and environmental protection. However, it is a traditional problem in the field of nonhydrolytic polycondensation that how to make compounds with more than 50% ionic bond participate in nonhydrolytic polycondensation. If the compounds with high ionic bond percentage more than 50% can also participate in nonhydrolytic polycondensation, the range of raw materials for nonhydrolytic polycondensation can be greatly increased. In this work, a method of pressure field assisted polycondensationis developed to make the compound with more than 50% ionic bond percentage participate in nonhydrolytic polycondensation. This method is proposed to be used in mullite whiskers in-situ synthesis via a facile nonaqueous precipitation process with high ionic bond percentage. In comparison with traditional mullite whisker preparation methods, this in-situ nonaqueous precipitation method has the superiorities of low mullitization temperature, high homogeneity and efficiency, simple operation. The present work studies the phase transition process and the structure of precipitate. Effects of aluminum fluoride amount on whisker preparation are also investigated. The whiskers growth mechanism is discussed and their application as epoxy resin reinforcement is also investigated.

2. Materials and Methods

Analytical grade of anhydrous aluminum fluoride (AlF_3), tetraethoxysilane ($Si(OC_2H_5)_4$) and anhydrous ethanol (C_2H_5OH) were produced by China Medicine (Group) Shanghai Chemical Reagent, Co., Ltd. They were directly used without further purification.

In the glove-box, 13.5 mL tetraethyl orthosilicate (TEOS) was dissolved in 120 mL anhydrous ethanol with the formation of TEOS-ethanol solution (0.5 mol/L). Then 15.116 g (theoretical amount (TA) for mullite synthesis, A_0), 16.628 g (1.1 of TA, A_1), 18.139 g (1.2 of TA, A_2), 19.651 g (1.3 of TA, A_3) and 21.162 g (1.4 of TA, A_4) anhydrous aluminum fluoride was added to the ethanol solution of TEOS, respectively. Mixtures were transformed to a kettle with the nominal volume of 200 mL, and then held at 130 °C for 12 h. After washed repeatedly with ethanol and filtered, the precipitate powders were obtained. They were finally dried at 110 °C for 2 h, and then calcined to a temperature scale from 900 to 1100 °C for 4h to get the final samples.

As to the epoxy resin samples, 5 g A_3 mullite whisker ($M^\#$) or without whisker ($E^\#$), and 6 g diethanolamine were weighed and added to 5 0g epoxy resin with continuous and fierce stirring. After the mixture was stirred evenly, they were poured into tin paper mold. The samples were then degassed for 1 h in air and 30 min in negative pressure of 0.06 MPa, and cured at 80 °C for 12 h to get the final sample.

Crystal phases of the samples calcined at different temperatures were tested via XRD (X-ray diffractometer, D8, Bruker, Karlsruhe, Germany) with radiation of CuK_α operated at 30 mA and 40 kV. The bonds contained in the precipitates were determined via FT-IR (Fourier transform infrared spectroscopy, Nicolet 5700, Thermo, Boston, MA, USA) in the wavenumber of 4000–400 cm^{-1}. The samples morphology with different aluminum fluoride amounts was characterized by FE-SEM (field-emission scanning electron microscopy, SU-8010, JEOL, Tokyo, Japan). The whiskers structure was determined via TEM (transmission electron microscopy, JEM-2010, JEOL, Tokyo, Japan).

The mechanical properties of pure epoxy resin and mullite whisker-epoxy resin composite were measured by universal testing machine (TH-8203, Suzhou, China). Two-body abrasive wear test was determined by pin-on-disc machine (MPX-2000A, Zhangjiakou, China) under multi-load conditions. The specimens with the size of 10 mm × 10 mm × 3 mm and surface rubbed were glued on the steel sample clip. They were rubbed with the abrasive paper of SiC, which was pasted on the disc through the adhesive.

The specific abrasive wear rate (W_S) was calculated by the following relation:

$$W_s = K \frac{V_S V_C}{E H \varepsilon_f \mu_a F_N}, \quad (1)$$

where the K is proportionality constant, H is the hardness, E is the elastic modulus, ε_f is the failure strain, μ_a is the friction coefficient, F_N is the normal load, V_S is the sliding speed and V_C is the crack growth speed. In present work, the V_S (150 rpm) and the F_N (5, 10 and 15 N) are small. It is applicable of Equation (1).

3. Results and Discussion

3.1. Precipitate Phase Transition Analysis

Figure 1 presents the XRD patterns of A3 precipitate calcined in the temperature range of 800–1100 °C with intervals of 100 °C. As can be seen from Figure 1, the mullitization occurred at 900 °C, which could be inferred from the XRD pattern of the sample heated at 800 °C had no diffraction peak and sample heated at 900 °C presented a unique mullite phase without any other diffraction peak. It is worth noting the mullitization temperature is much lower than the traditional one [18] generally generated via the solid phase reaction between SiO$_2$ and Al$_2$O$_3$ according to formula (2), which is benefited from the precipitate mullitization is generated by the =Al–O–Si≡ bonds rearrangement confirmed by the later FT-IR analysis.

$$3Al_2O_3 + 2SiO_2 \rightarrow Al_6Si_2O_{13}. \quad (2)$$

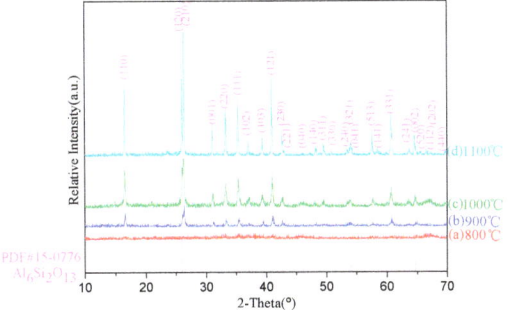

Figure 1. XRD patterns of the samples calcined at different temperatures.

As the temperature further increased to 1100 °C, it is also noteworthy that no impurity phase peak, such as Si-Al spinel or γ-Al$_2$O$_3$, could be observed in the XRD patterns from 900 to 1100 °C. Mullite is a unique phase appeared in the whole heating process. These indicate TEOS had completely reacted with AlF$_3$, and the excessive AlF$_3$ wasto the point.

3.2. Effect of Pressure Field Caused by Kettle Treatment

Figure 2 shows the XRD patterns of the samples without kettle treatment (a) and with kettle treatment (b), both of them were calcined at 900 °C. No crystal phase diffraction peak existed in the sample prepared without kettle treatment (a), indicating that the sample was amorphous without the formation of mullite phase. In stark contrast, only mullite phase diffraction peaks were detected in the sample prepared with kettle treatment (b), suggesting that the crystalline phase of the sample was the pure mullite phase. Figure 2 also presents the FT-IR spectra of precipitates without kettle treatment (a) and with kettle treatment (b) to study chemical root for low mullitization temperature of the sample prepared with kettle treatment. The FT-IR spectrum of the sample without kettle treatment (a) shows a typical FT-IR spectrum of TEOS, the characteristic vibrations of Si–O–C in TEOS are shown in it. However, no aluminum related vibration was detected, indicating that AlF_3 didnot participate in reaction. It was mainly because the ionic character of Al–F bond calculated according to formula (3) was 64.89%. It indicates an obvious ionic character of Al–F bond. AlF_3 preferred to exist in ions form in precursors mixtures liquid theoretically in conventional conditions. In sharp contrast, the FT-IR spectrum of precipitates with kettle treatment shown in Figure 2 presents typical absorbance peaks of =Al–O–Si≡. It is the intermediate product of reaction between AlF_3 and TEOS. The vibrations at 492 cm^{-1} and 1044 cm^{-1} ascribed to $\delta(SiO_4)$ and $\nu(SiO_4)$ indicate the silica tetrahedron formation. This certificates that molecules TEOS had reacted with (partial of) AlF_3 molecules completely. The vibrations at 810 and 854 cm^{-1} assigned to $\nu(AlO_4)$ also show that the three Fs bonded with Al in partial of AlF_3 molecules were completely replaced by the groups of Si–O. While the appearance of Al–F bond at 607 cm^{-1} was caused by the excess of AlF_3. In addition, the C=O bonds in sample (b) were caused by adsorption of carbon dioxide in air, which indicates that the precipitate skeleton had higher coordination ability and polycondensation.

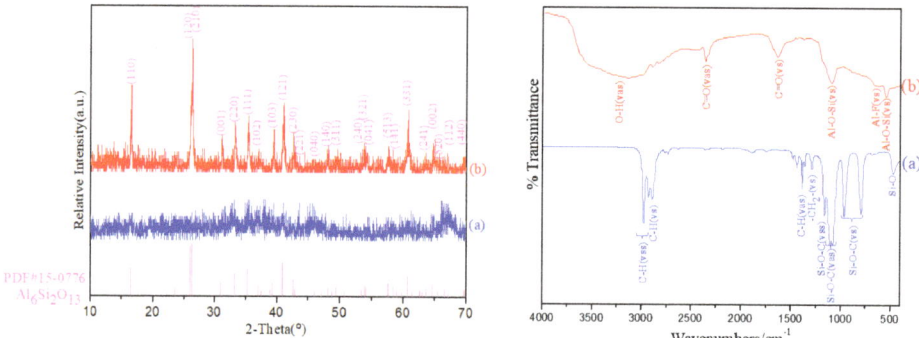

Figure 2. XRD patterns and FT-IR spectra of the samples without kettle treatment (**a**) and with kettle treatment (**b**).

These results indicate the reaction between AlF_3 and TEOS shown as formula (3) and the structure of precipitate with kettle treatment is deduced as shown in Figure 3.

$$\text{Ionic character percentage (\%)} = 1 - \exp[-(X_A - X_B)^2/4]. \tag{3}$$

In formula (3), X_A and X_B are the electronegativities of the two elements in the compound AB.

$$\begin{array}{c}
\text{C}_2\text{H}_5\text{O-Si-OC}_2\text{H}_5 + \text{F-Al} \xrightarrow{130\ °C} \text{C}_2\text{H}_5\text{O-Si-OC}_2\text{H}_5\ \text{F-Al} \xrightarrow{130\ °C} \text{C}_2\text{H}_5\text{O-Si-O-Al} + \text{C}_2\text{H}_5\text{F}
\end{array} \tag{4}$$

The polarity of the compound can be reversed in organic chemical synthesis reactions. The concept of "polar transformation" in the reaction has been paid attention to in recent years. It is generally believed that the change of entropy (isothermal) and temperature (adiabatic) will induce the change of dipole state when the molecular dipole of the material changes from a disordered state to an ordered state. If the temperature change and the entropy change are large, it is called the electrothermal effect of the material. Electric dipoles undergo heating fluctuations without external pressure, and their orientations are random, similar to those of water molecules. When molecules are subjected to external pressure, the dipoles may turn. When the external pressure becomes high enough, the dipole can even be completely symmetrical, and the material is polarize-saturated to form polar covalent bonds with high covalent bond percentage. Polarity conversion can broaden the selections of raw materials in the organic related synthesis process. Therefore, when designing the reaction process, it should consider not only the nature of the raw material itself and the inherent performance in the reaction but also the possible transformation of raw material in the reaction process.

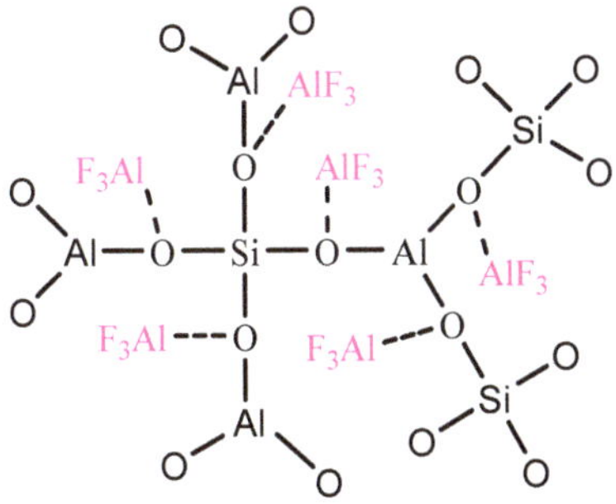

Figure 3. Deduced structure of precipitate with kettle treatment.

"Polar transformation" can further enrich the contents of nonhydrolytic reactions and discussion on the polarity of raw materials from the perspective of environment temperature and pressure. It plays an important role in mastering the organic synthesis and provides a simpler way for the synthesis of new compounds. It also enriches raw materials selection for nonhydrolytic reactions.

3.3. Effect of AlF$_3$ Amount on Sample Morphology

Figure 4 presents the FE-SEM graphs of the samples prepared with different AlF$_3$ amounts, A_0 (a; 1.0 theoretical amount (TA)), A_1 (b; 1.1TA), A_2 (c; 1.2TA), A_3 (d; 1.3TA) and A_4 (e; 1.4TA). Figure 4a shows characteristic powder morphology; it is similar to the powder morphology prepared via a non-hydrolytic sol–gel method [19]. The sample presents rice-like grains with the one-dimensional growth trend. In comparison Figure 4a–d, the latter three samples show the characteristic whisker-like morphology. Figure 4b–d) show that the particles mingled gradually disappeared in whiskers with the increase of AlF$_3$ amount. The whiskers diameter further increased with the AlF$_3$ amount increase. The sample finally shows cluster-like-structured whisker morphology. However, when the amount of AlF$_3$ further increased to 1.4TA (A_4; e), mullite whisker had disappeared. Sample morphology develops into a sheet-like structure with a thickness of about 0.1–0.25 µm. When AlF$_3$ was used as a vapor catalyst for crystal growth, vapor saturation was the decisive factor for the final products. Only appropriate low vapor saturation could generate high-quality whiskers. In this work, when AlF$_3$ amount equaled to the theoretical amount for mullite synthesis, the whole system had no AlF$_3$ vapor phase. Mullite crystal could hardly grow preferentially and finally formed a particulate. For sample A_1 and sample A_2, they were in low vapor concentration, their mullitization and mullite whiskers growth were limited. The limited mullitization and mullite whiskers growth led to the mullite whiskers being mixed with mullite particles. However, when the AlF$_3$ amount further changed to 1.4TA, the vapor supersaturation concentration of AlF$_3$ was at an over high degree, AlF$_3$ preferentially reacted with H$_2$O vapor to generate alumina phase. This process is shown in formula (5) and formula (6). Al$_2$O$_3$ preferred to grow into platelet-shaped corundum in AlF$_3$ vapor. Consequently, the optimal AlF$_3$ amount was 1.3TA, namely 1.3 of the theoretical amount for mullite synthesis.

$$AlF_3 + H_2O = AlOF + 2HF. \quad (5)$$

$$2AlOF + H_2O = Al_2O_3 + 2HF. \quad (6)$$

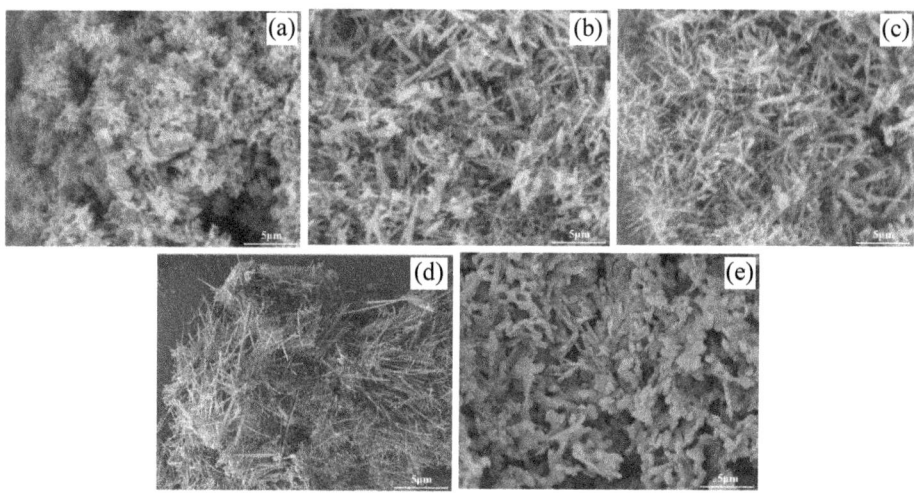

Figure 4. Field-emission (FE)-SEM images of the samples with different AlF$_3$ amounts; (**a**) A$_0$, (**b**) A$_1$, (**c**) A$_2$, (**d**) A$_3$ and (**e**) A$_4$.

3.4. Structure Analysis of Whiskers

Figure 5 shows the TEM graph (a), SAED pattern (b) and HR-TEM graph (c) of the optimal A$_3$ sample. TEM (a) graph shows the typical morphology of the as-prepared A$_3$ mullite whisker. It demonstrated that the whisker had a relatively uniform microstructure, which shows the well-distributed whiskers had formed. The mullite whiskers were less than 100 nm in diameter and more than 15 in the aspect ratio. It was consistent with the results of FE-SEM. There was dark bands in the whisker center, which indicates that it was a solid whisker rather than a tube. SAED pattern (b) indicates the single mullite phase diffraction pattern. The whisker SAED pattern revealed a single diffraction pattern of the mullite phase. In the SAED pattern, the cell constants were measured a = 0.757 nm, b = 0.769 nm and c = 0.289 nm. They were in excellent agreement with the theoretical mullite values (JCPDF file no. 15-0776) and those calculated from XRD pattern. It was also deduced the SAED pattern that [1$\bar{1}$0] was the crystal band axis of mullite whiskers, and the axial diffraction spots along the whisker corresponded to [001]. The mullite whiskers growth direction was parallel to [001] direction. It was along the c axis direction. It was also proved by the HR-TEM graph shown in Figure 5c. The HR-TEM of sample also clearly revealed that the whisker was a perfect mullite monocrystal, without the crystal defects, such as low angle tilt grain-boundary or dislocation in the whisker.

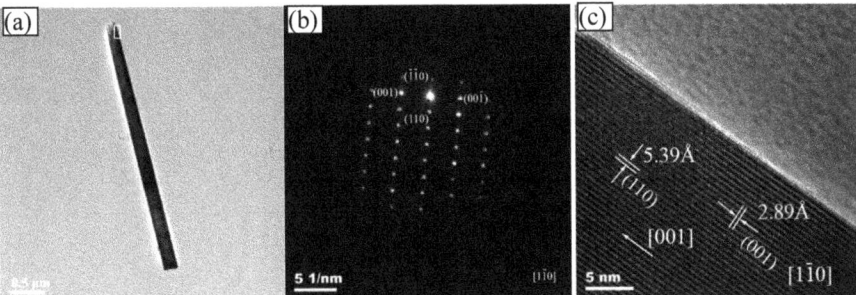

Figure 5. TEM, SAED and HR-TEM graphs of whisker. (**a**) TEM graph; (**b**) SAED pattern and (**c**) HR-TEM graph.

3.5. Whiskers Growth Mechanism Analysis

To confirm the mullite whiskers growth mechanism prepared via nonaqueous precipitation in-situ synthesis method, Figure 6 shows FE-SEM (a) and TEM (b) graphs of the sample A$_3$ held at 1100 °C for 0.5 h. Figure 6a presents many rod-like crystals on the gel skeleton surface. These crystals show an evident anisotropic growth trend. There are also many tiny dots on the particles. Figure 6b shows the high magnification TEM graph of the short-rod-like crystals shown in Figure 6a. A mass transport path appears in the direction shown by the white arrow in Figure 6b, combining with non-circular tip, which is generally thought to be relative to screw, these indicate the vapor–solid (VS) model with the inducement of screw for mullite whiskers prepared in this work. Based on the results and analysis above, a possible growth mechanism for mullite whiskers synthesized via nonaqueous precipitation method is schematically illustrated in Figure 7. Firstly, heterogeneous polycondensation reaction between the precursors of tetraethyl orthosilicate (TEOS) and aluminum fluoride occurred with the aid of kettle treatment. The heterogeneous polycondensation product =Al–O–Si≡ generated low-temperature nucleation of mullite, which was quite in favor to the further anisotropic growth of mullite. This low mullitization temperature also ensured enough three-dimensional growth dynamics differences, and it was also slightly lower than the volatilization temperature of aluminum fluoride. Secondly, it is shown in Figure 6b that slender whiskers had formed by calcining at 1100 °C for 0.5 h, which indicates vapor sediments on the surface of the whisker with sharp-pointed top, which indicates the formation of screw. These sediments with markedly surface diffusion sign were caused by the deposition of vapor molecule on the whisker surface. Figure 6b also shows the sediments diffusion sign and process along the whisker surface to the growth point. It is generally known that mullite whiskers growth is a dynamic physicochemical process, in which AlF$_3$ is widely regarded as the most effective auxiliary for mullite whiskers growth [20]. The vapor phase diffusion is known as the main mass transport mechanism during the anisotropic growth of mullite whiskers, it can also induce the formation of screw, which is also beneficial to the mass transfer and whisker growth. The AlF$_3$ vapor accelerates the mullite whiskers growth, because the mass transport for crystal growth is enhanced in the presence of vapor transport. The large mass transport accelerated by the vapor phase promotes the grains growth near the surfaces and the formation of whiskers (Figure 7). The growth structure at the end of whisker shown in Figure 6 was related to the enrichment of vapor nucleating particles during calcining process. At this time, the aggregates formed in the nucleating process could not crystallize along the orientation of whisker lattice, thus forming polycrystalline aggregation shown in Figure 6b. The further growth process could only occur on the small crystal surfaces with different orientations, that is, secondary growth at the end of the whisker. It is believed that the secondary growth phenomenon of mullite whisker may be due to the local wave of AlOF and SiF$_4$ [2]. This is because when the top of the whisker enters a region with high vapor supersaturation, the vapor phase reactants with high supersaturation will nucleate rapidly at the top of the whisker (possibly with the help of many dislocation points at the top of the whisker), and form a polycrystalline pile-up pattern at the top of the whisker. With the formation of mullite crystal nucleus, the supersaturation of gas-phase reactants decreased rapidly and restored to the appropriate supersaturation for mullite whisker growth. The growth started from the different orientation of the microcrystalline surfaces in the polycrystalline aggregation surface, and it grew outwards continuously, forming the secondary growth of mullite whisker. There was avisible terminal-bottleneck phenomenon at the end of the whisker shown in Figure 6b. The terminal-bottleneck phenomenon detected in this work was generally thought as the most typical characteristic of vapor–solid (VS) whisker growth mechanism.

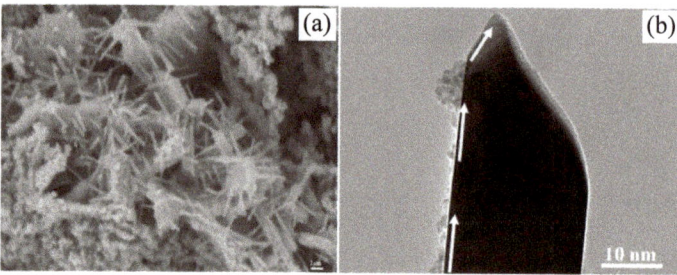

Figure 6. FE-SEM and TEM of the sample A$_3$ holding at 1100 °C for 0.5 h; (a) FE-SEM and (b) TEM.

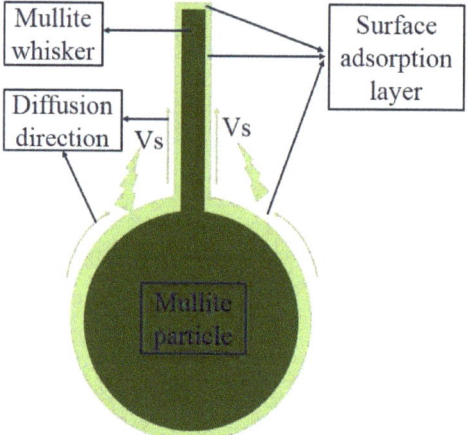

Figure 7. Schematic diagram illustrating whiskers growth mechanism.

According to the results of this work, our previous researches [11–15] and reference [21], Figure 7 schematically illustrates the growth mechanism of mullite whiskers prepared by the nonaqueous precipitation method. Due to non-hydrolytic heterogeneous polycondensation reaction between ethyl silicate and aluminum fluoride shown in formula (3) with the help of kettle treatment, and the formation of =Al–O–Si≡ bonds in precipitate, the mullitization temperature is 900 °C. It was slightly lower than the volatilization temperature of aluminum fluoride. These could ensure when aluminum fluoride begins to volatilize, there are a large number of mullite crystal nucleus. Low mullitization temperature also ensures the appropriate three-dimensional growth dynamic differences. Meanwhile, residual aluminum fluoride in precipitate, which uniformly coordinates with the gel skeleton, volatilizes and reacts with O$_2$ in the air according to formula (7) to form AlOF and F, and then =Al–O–Si≡ reacts with F to form AlOF and SiF$_4$ [10,19]. They can further react with O$_2$ to form mullite (shown in formula (8)). The newly formed mullite are generally on the surface of the particle, it plays the role of raw material for whisker growth and continuously diffuses to the top of the whisker with the help of gas-phase mass transport and concentration gradient. With the continuous reaction and mass transport process, precipitates particles eventually grow into mullite whiskers.

$$= Al-O-Si \equiv (amorphous) + 5F \rightarrow AlOF + SiF_4 \quad (7)$$

$$6AlOF + 2SiF_4 + 3.5O_2 \rightarrow 3Al_2O_3 \cdot 2SiO_2(whisker) + 14F. \quad (8)$$

3.6. Application of Whiskers in Epoxy Resin Reinforcement

Figure 8a shows the flexural strength of epoxy resin without mullite whisker ($E^\#$) and mullite whisker-epoxy resin composite ($M^\#$). For the epoxy resin, its flexural strength was 4.2 MPa (Figure 8 $E^\#$). The flexural strength for mullite whisker-epoxy resin composite ($M^\#$) dramatically increased to 47.6 MPa, which was 11.3 times of pure epoxy resin. The mullite whisker-epoxy resin composite ($M^\#$) also shows a lower relative flexural strength deviation, which might be due to the addition of mullite whiskers reduces the influence of defects on the materials. Specific abrasive wear rates of pure epoxy resin and mullite whisker-epoxy resin composite as a function of normal load are shown in Figure 8b. The abrasive wear rate of mullite whisker-epoxy resin composite was much smaller than that of pure epoxy resin. Both of their abrasive wear rates have presented a sharp decreasing trend with the load increasing, which is in excellent agreement with Lhymn's mathematical model [22]. It was also obvious that effects of load on epoxy resin without mullite whisker ($E^\#$) were much larger than that of mullite whisker-epoxy resin composite ($M^\#$), which shows a strong reinforcement of mullite whiskers to epoxy resin.

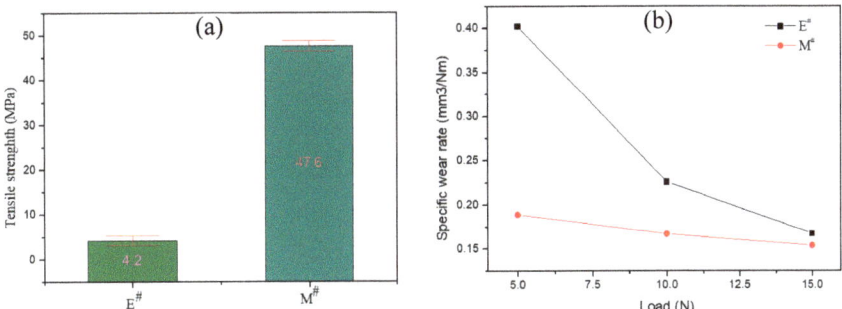

Figure 8. Mechanical properties of the samples. (a) The flexural strength of epoxy resin without mullite whisker ($E^\#$) and mullite whisker-epoxy resin composite ($M^\#$); (b) Specific abrasive wear rates of pure epoxy resin and mullite whisker-epoxy resin composite as a function of normal load.

To confirm the reinforcement mechanism of mullite whisker to epoxy resin, the morphology of mullite whisker-epoxy resin composite is presented in Figure 9. Compared with the original epoxy resin shown in Figure 9a, the morphology of mullite whisker-epoxy resin composite clearly indicates that mullite whiskers were three-dimensionally distributed in epoxy resin. The multi-needle whiskers were connected in the same center, and there were whiskers in all directions, which could play a better synergistic role. They could prevent the generation and development of cracks and resist the damage of epoxy resin caused by friction. The unique structure of the multi-needle whiskers connected to the same center guarantees that when a whisker is exposed to shear and pressure, other whiskers will disperse the force of the whisker, thus eliminating stress concentration, preventing cracks and reducing the probability of damage.

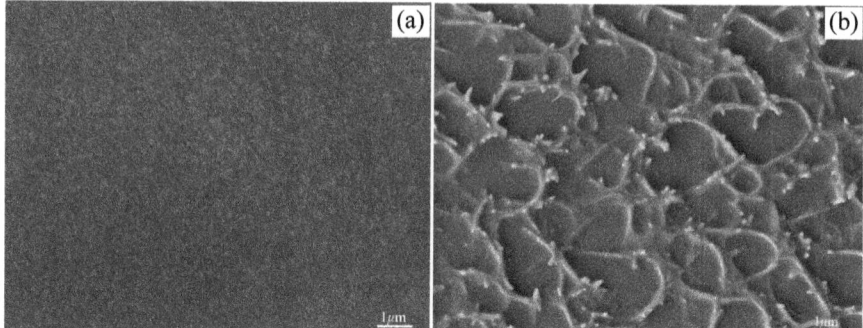

Figure 9. Of samples, (**a**) original epoxy resin and (**b**) mullite whisker-epoxy resin composite.

4. Conclusions

Well-developed mullite whiskers were prepared via nonaqueous precipitation in-situ synthesis method at 1100 °C taking tetraethoxysilane (TEOS) as silicon source, anhydrous AlF_3 as aluminum source and growth auxiliary for whisker. The precipitate was composed of =Al–O–Si≡ bonds and coordinated excessive AlF_3. Kettle treatment facilitated the formation of =Al–O–Si≡ bonds due to it changing the polarity of aluminum fluoride. The mullitization temperature of the precipitates was 900 °C, and they grew into mullite whiskers at 1100 °C. The whiskers grew preferentially along the direction parallel to the c-axis, resulting in an orthorhombic-type crystallographic structure. XRD and FE-SEM results show that the whiskers were in high purity with a high yield. The whiskers had the aspect ratio of >15 (100 nm in diameter). The growth process of mullite whiskers wasdominated by a vapor–solid (VS) mechanism combined with the inducement of =Al–O–Si≡ bonds formed in the precipitates were beneficial for the low mullitization temperature and whiskers growth. The AlF_3 amount was optimized to be 1.3 of the theoretical amount, which ensured appropriate AlF_3 vapor supersaturation concentration. With the help of the vapor promoted mass transport process, precipitates particles eventually grew into mullite whiskers.

Author Contributions: G.F., F.J. and W.J., funding acquisition; Z.H. and Q.Z., experiment investigation; J.L. (Jianmin Liu) and Q.H., data curation; L.M. and Q.W., formal analysis; J.L. (Jian Liang), project administration.

Acknowledgments: This work was supported by the National Natural Science Foundation of China [grant numbers 51662016, 51962014]; Key Research and Development Program of Jiangxi Province [grant number 20192BBEL50022]; the Key Science Foundation of Jiangxi Provincial Department of Education, China [grant number GJJ180699]; Scientific Research Fund of Jiangxi Provincial Education Department (grant number GJJ160881); the Youth Science Foundation of Jiangxi Provincial Department of Education, China [grant number GJJ180740]; and the Jingdezhen Science and technology program [grant number 20161GYZD011-007].

Conflicts of Interest: The authors declare no conflict of interest.

References

1. Li, S.; Du, H. Preparation of self-reinforcement of porous mullite ceramics through in situ synthesis of mullite whisker in flyash body. *Ceram. Int.* **2011**, *38*, 1027–1032. [CrossRef]
2. Xu, L.; Xi, X.; Zhu, W. Preparation of mullite whisker skeleton porous ceramic. *Ceram. Int.* **2015**, *41*, 11576–11579. [CrossRef]
3. She, J.H.; Ohji, T. Fabrication and characterization of highly porous mullite ceramics. *Mater. Chem. Phys.* **2003**, *80*, 610–614. [CrossRef]
4. Wang, M.; Xie, C. Polymer-derived silicon nitride ceramics by digital light processing based additive manufacturing. *J. Am. Ceram. Soc.* **2019**, *102*, 5117–5126. [CrossRef]
5. Zhang, K.Q.; Xie, C. High solid loading, low viscosity photosensitive Al_2O_3 slurry for stereolithography based additive manufacturing. *Ceram. Int.* **2019**, *45*, 203–208. [CrossRef]

6. Kim, B.M.; Cho, Y.K. Mullite whiskers derived from kaolin. *Ceram. Int.* **2009**, *35*, 579–583. [CrossRef]
7. Okada, K.; Otsuka, N. Synthesis of mullite whiskers by vapour-phase process. *J. Mater. Sci. Lett.* **1989**, *8*, 1052–1054. [CrossRef]
8. Nie, J.; Zhang, H. Synthesis of mullite whiskers using molten salt method. *Rare Metal. Mat. Eng.* **2009**, *38*, 1154–1157.
9. Xiang, W.; Li, J.H. Phase evolution and dynamics of cerium-doped mullite whiskers synthesized by sol–gel process. *Ceram. Int.* **2013**, *39*, 9677–9681.
10. Jiang, W.H.; Peng, Y.F. Preparation of mullite whisker via non-hydrolytic sol-gel route. *J. Inorg. Mater.* **2010**, *25*, 532–536. [CrossRef]
11. Fu, K.L.; Jiang, W.H. Low temperature preparation of mullite whisker via non-hydrolytic sol-gel process combined with molten salt method. *Adv. Mater. Res.* **2014**, *936*, 975–980. [CrossRef]
12. Fu, K.L.; Jiang, W.H. Study on mullite whiskers preparation via non-hydrolytic sol-gel process combined with molten salt method. *Mater. Sci. Forum.* **2016**, *848*, 295–300. [CrossRef]
13. Liu, J.M.; Fu, K.L. Preparation of mullite whiskers using lithium molybdate as molten salt. *Chin. Ceram.* **2017**, *53*, 7–12.
14. Wu, Q.; Jiang, W.H. Preparation of mullite whiskers by non-hydrolytic sol-gel process combined with molten salt method. *Chin. Ceram.* **2016**, *52*, 12–16.
15. Feng, G.; Jiang, F. Novel facile nonaqueous precipitation in-situ synthesis of mullite whisker skeleton porous materials. *Ceram. Int.* **2018**, *44*, 22904–22910. [CrossRef]
16. Feng, G.; Jiang, W.H. Novel nonaqueous precipitation synthesis of alumina powders. *Ceram. Int.* **2017**, *43*, 13461–13468. [CrossRef]
17. Feng, G.; Jiang, F. Effect of oxygen donor alcohol on nonaqueous precipitation synthesis of alumina powders. *Ceram. Int.* **2019**, *45*, 354–360. [CrossRef]
18. Vieira, S.C.; Ramos, A.S. Mullitization kinetics from silica- and alumina-rich wastes. *Ceram. Int.* **2007**, *33*, 59–66. [CrossRef]
19. Jiang, W.H.; Peng, Y.F. Low temperature synthesis of high purity mullite powder via non-hydrolytic sol-gel method. *Chin. Ceram.* **2010**, *25*, 532–536.
20. Hua, K.; Xi, X. Effects of AlF_3 and MoO_3 on properties of mullite whisker reinforced porous ceramics fabricated from construction waste. *Ceram. Int.* **2016**, *42*, 17179–171840. [CrossRef]
21. Yang, F.; Shi, Y. Analysis of whisker growth on a surface of revolution. *Phys. Lett. A* **2017**, *381*, 1–5. [CrossRef]
22. Lhymn, C.; Tempelmeyer, K.E. The abrasive wear of short fibre composites. *Composites* **1985**, *16*, 127–136. [CrossRef]

© 2019 by the authors. Licensee MDPI, Basel, Switzerland. This article is an open access article distributed under the terms and conditions of the Creative Commons Attribution (CC BY) license (http://creativecommons.org/licenses/by/4.0/).

Article

Tailored Crosslinking Process and Protective Efficiency of Epoxy Coatings Containing Glycidyl-POSS

Mirjana Rodošek, Mohor Mihelčič, Marija Čolović *, Ervin Šest, Matic Šobak, Ivan Jerman * and Angelja K. Surca *

National Institute of Chemistry, Hajdrihova 19, 1000 Ljubljana, Slovenia; mirjana.rodosek@gmail.com (M.R.); mohor.mihelcic@fs.uni-lj.si (M.M.); ervin.sest@gmail.com (E.Š.); matic.sobak@ki.si (M.Š.)
* Correspondence: marija.colovic@ki.si (M.Č.); ivan.jerman@ki.si (I.J.); angelja.k.surca@ki.si (A.K.S.); Tel.: +386-1-4760-200 (A.K.S.)

Received: 10 February 2020; Accepted: 3 March 2020; Published: 5 March 2020

Abstract: Versatile product protective coatings that deliver faster drying times and shorter minimum overcoat intervals that enable curing at faster line speeds and though lower energy consumption are often desired by coating manufacturers. Product protective coatings, based on silsesquioxane-modified diglycidyl ether of bisphenol-A (DGEBA) epoxy resin, are prepared through a glycidyl ring-opening polymerization using dicyandiamide (DICY) as a curing agent. As silsesquioxane modifier serves the octaglycidyl-polyhedral oligomeric silsesquioxane (GlyPOSS). To decrease the operational temperature of the curing processes, three different accelerators for crosslinking are tested, i.e., N,N-benzyl dimethylamine, 2-methylimidazole, and commercial Curezol 2MZ-A. Differential scanning calorimetry, temperature-dependent FT-IR spectroscopy, and rheology allow differentiation among accelerators' effectiveness according to their structure. The former only contributed to epoxy ring-opening, while the latter two, besides participate in crosslinking. The surface roughness of the protective coatings on aluminum alloy substrate decreases when the accelerators are applied. The scanning electron microscopy (SEM) confirms that coatings with accelerators are more homogeneous. The protective efficiency is tested with a potentiodynamic polarization technique in 0.5 M NaCl electrolyte. All coatings containing GlyPOSS, either without or with accelerators, reveal superior protective efficiency compared to neat DGEBA/DICY coating.

Keywords: polymers; octaglycidyl-POSS; DGEBA; dicyandiamide; accelerators; corrosion; protective coatings; infrared spectroscopy; rheology

---

1. Introduction

Typically, alloys corrode merely from exposure to moisture and pollutants in the air atmosphere. According to NACE International (National Association of Corrosion Engineers), the global cost of corrosion is estimated to €2.2 billion, which is equivalent to 3.4% of global gross domestic product (GDP) (<€640 billion or 3.8% GDP in Europe) [1]. Just by available proper corrosion control practices, it is estimated that savings of between 15–35% of the cost of corrosion can be realized (globally €340–800 billion per year) [1]. Commonly used anticorrosion approaches to slow down the corrosion rate are cathodic protection with sacrificial anodes, deposition of protective coatings [2], the addition of inhibitors directly into corrosive environments, or in the structure of protective coatings [3]. Protective coatings are one of the most prospective and widespread methods for the corrosion prevention of metals, which makes it one of the main critical technologies underpinning the competitiveness of the European industry. The deposited coatings isolate the metal surface from the atmosphere or any other corrosive media.

As a corrosion barrier, the epoxy coatings [4] have been a subject of research and commercial applications for a long time [5]. The well-established routes of their preparation use curing of epoxy-precursors with amino groups-containing compounds; for example, a reaction of diglycidyl ether of bisphenol-A (DGEBA) with dicyandiamide (DICY) curing agent (Figure 1). Despite that, epoxy-based coatings continued to remain a promising topic of investigations through the formation of various composite materials [6–8] and advanced metal/polymer laminates [9,10]. For instance, as a challenging issue has remained the development of nanocomposite polymers, in which nanosized reinforcement is applied to obtain improved performance of such protective coatings. Most commonly, nano reinforcement compounds of <100 nm size have been added to the polymeric matrix, and the uniform dispersion is the crucial key for achieving the desired properties. Nanoparticles are commonly introduced as fillers [11] but can also be bound directly into the polymeric matrix as pendant groups.

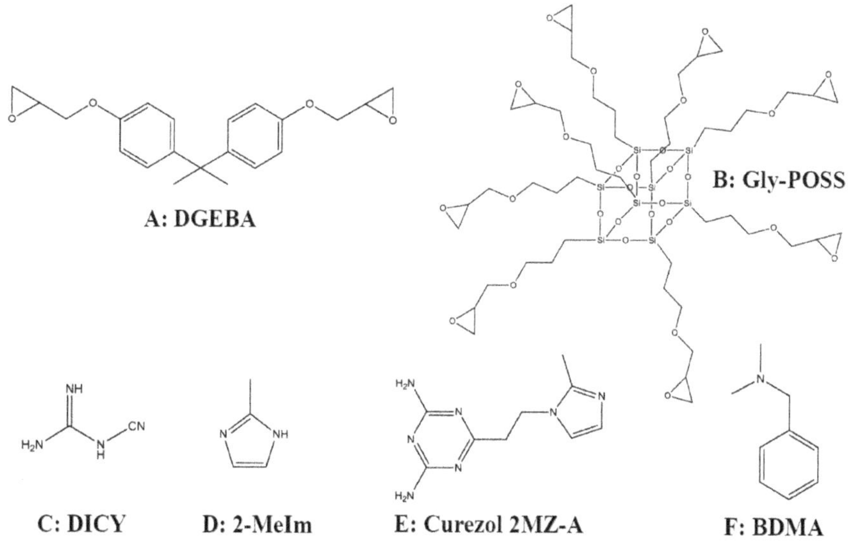

Figure 1. Structures of the precursors: (**A**) diglycidyl ether of bisphenol-A (DGEBA), (**B**) octaglycidyl-POSS (GlyPOSS), (**C**) dicyandiamide (DICY), (**D**) 2-methylimidazole (2-MeIm), (**E**) 2,4-diamino-6-[2′-methylimidazolyl-(1′)]-ethyl-s-triazine (Curezol) and (**F**) N,N-benzyl dimethylamine (BDMA).

Nanoparticles that can, for example, be bound in the epoxy networks are polyhedral oligomeric silsesquioxanes (POSS) [12]. They belong to a group of organic-inorganic hybrids [13]. Specifically, the core is composed of an inorganic silsesquioxane, while different organic pendant groups can be attached to each of the eight corners of the T_8 cage. Homoleptic POSS nanoparticles have the same organic groups in their shells, while organic groups in heteroleptic POSS can differ. Consequently, POSS can incorporate in different modes into novel materials. (i) When no reactive organic group is present in the shell, POSS simply behaves as a nanofiller. (ii) On the other hand, the presence of a suitable reactive organic group can lead to the bounding of POSS into the epoxy network. POSS can enter the epoxy curing reaction either as: (ii-a) glycidoxy-group containing precursor or (ii-b) an amine-POSS hardener compound. (ii-a) The examples of the former approach are epoxy networks based on DGEBA that contained different fractions of monoglycidyl-heptaisobutyl-POSS [14–17] and were cured with different hardeners. Strachota et al. [18], on the other hand, used POSS with 1, 2, 4, or 8 glycidyl groups while remaining groups to T_8 cage being either phenyl, isooctyl, or cyclopentyl. (ii-b) The latter approach was demonstrated by partial exchange of hardener with either monoamino-functionalized POSS [17,19–21] or octaamino-functionalised POSS [22]. (iii) Nevertheless, the open-cube trisilanol-heptaphenyl-POSS was studied as a promotor of the curing reaction between glycidyl and amino groups, i.e., the influence of silanol groups on epoxy curing kinetics [23].

By incorporation of POSS nanoparticles into polymer matrices, the composite materials with superior functional properties can be obtained. For example, a combination of epoxy and various glycidoxy-POSS nanoparticles have been studied from mechanical [14] and viscoelastic [18] perspective. It was found that monoglycidoxy-POSS does not contribute to the deformation process of the network while enhancing the thermal properties [14]. The thermal stability of epoxy/glycidyl-POSS materials is improved, but higher loadings tend to decrease this beneficial influence [18]. The behavior of DGEBA with monoglycidoxy-heptaisobutyl-POSS cured with short aromatic amines was studied from the kinetic perspective by differential scanning calorimetry (DSC) [16]. In a similar material, local thermal analysis (LTA) and DSC gave evidence of amorphous POSS-rich domains which can eventually arise from phase separation [15]. As the main reason for that, the incompatibility of the isobutyl groups of monoglycidoxy-heptaisobutyl-POSS and the aromatic epoxy-amine network was suggested. Anyhow, when octaglycidoxy-functionalized POSS (Figure 1B) was introduced into the DGEBA-based nanocomposite system, it was found to accelerate the rate of opening of glycidyl epoxy rings of DGEBA [24].

POSS molecules also have a robust resistance to environmental degradation factors, such as moisture, oxidation, corrosion, and UV radiation. It is therefore not surprising that various POSS nanoparticles have already been tested as an additive in polymeric protective coatings for alloys, for example, aminopropyl-heptaisooctyl-POSS in epoxy coatings [20,21]. However, although epoxy coatings are used worldwide as corrosion protective coatings [7], we haven't found any research report on the corrosion topic where the glycidyl-POSS molecules are added to improve the protective efficiency of epoxy coatings. Much more obvious are reports on the addition of different silanes [8] or the application of silane primers [5]. Since the addition of monoglycidyl-heptaisobutyl-POSS nanoparticles can cause phase separation, as hydrophobic isobutyl groups are not compatible with the epoxy network [15], the much easier is to incorporate homoleptic octaglycidyl-POSS (abbreviated GlyPOSS in Figure 1) molecules in the epoxy matrix [24]. Such molecule can be represented by formula $(R–SiO_{1.5})n$ ($n = 8$, $R = –(CH_2)_3–O–CH_2–[C_2H_3O]$). Under this premise, it is expected that the developed coatings shall have lower curing and glass transition temperature (T_g), lower roughness, dense structure, and consequently, also the outstanding protective properties.

To achieve the dense coating structure and other above-listed properties, special attention has to be given to the selection of the amine hardener. In reports on epoxy composites with glycidyl-POSS nanoparticles as hardeners, different aromatic amines [15,16] and polyetheramines (Jeffamines) [14,18] were studied. However, the small aliphatic molecule of dicyandiamide (DICY in Figure 1C) also suggests that dense crosslinking of the epoxy matrix is possible [7].

In order to decrease the production cost of protective coatings, the lowering of curing temperature, is desired. This can be achieved through the acceleration of curing reactions, i.e., the addition of suitable accelerators. The lowering of the activation energy for glycidoxy ring opening can be achieved by the presence of proton donors, for example, alcohols or hydroxyl groups emerging from previous reactions [25]. Further lowering of the ring-opening temperature is achieved by the addition of tertiary amines [25]. Different tertiary amines have been tested for DGEBA-based systems, ranging from benzyl dimethylamine (BDMA in Figure 1F) to various imidazolium-based structures [26,27]. The results confirmed the influence of tertiary amine accelerators on the curing dynamics and resulting materials. This suggests that specific studies should include different accelerators.

Herein, we report on a successful preparation of composite epoxy protective coatings for aluminum alloy AA 2024. Part of DGEBA precursor is exchanged by GlyPOSS to decrease the influence of bisphenol-A on public health, to decrease the production costs, and to achieve better protective efficiency of coatings. Three different amine groups-containing accelerators, i.e., N,N-benzyl dimethylamine (BDMA), 2-methylimidazole (2-MeIm), and commercial accelerator Curezol (Figure 1D–F), are compared regarding the triggering of the curing reaction. They are studied regarding their capacity for the opening of glycidoxy rings but also their eventual contribution to crosslinking. The influence of accelerators on the curing is proved via thermal- and time-dependent FT-IR spectroscopy and rheological examination. Differential scanning calorimetry (DSC) is used to determine the thermal properties of epoxy-octaglycidyl-POSS composites. Morphology of the coatings is checked using atomic force microscopy (AFM) and scanning electron microscopy (SEM). The electrochemical technique, i.e., potentiodynamic polarization, gives a clear answer on improved protective efficiency of the developed epoxy-GlyPOSS coatings.

2. Materials and Methods

2.1. Materials

Diglycidyl ether of bisphenol-A (DGEBA) was obtained from ABCR (Karlsruhe, Germany), as well as solvent 2-butanone (ACS, 99%). Commercial octaglycidyl-POSS (abbreviated GlyPOSS) nanoparticles were purchased from Hybrid Plastics (Hattiesburg, MS, USA). Curing agent dicyandiamide (DICY) and accelerators 2-methylimidazole (2-MeIm, 99%) and *N,N*-benzyl dimethylamine (BDMA, 99%) were purchased from Sigma-Aldrich (St. Louis, MO, USA). Commercial accelerator Curezol 2MZ-A (abbreviated Curezol) was obtained from Air Products (Allentown, PA, USA). Dimethyl sulfoxide (DMSO) was purchased from Merck (Darmstadt, Germany). All chemicals were used as supplied.

2.2. Preparation of Coatings

The preparation procedure of coatings is depicted in Figure 2. DGEBA (1.6 g) and GlyPOSS (0.4 g) were dissolved in butanone (2 g) and stirred for 30 min. For coating without GlyPOSS, only 2 g of DGEBA was dissolved in butanone (2 g). Separately, DICY (0.25 g) was dissolved in DMSO (1 g). Into the latter solution, if appropriate, one of three accelerators was introduced in molar ratio DICY:accelerator = 1:0.1. Both solutions were finally mixed. Before dip-coating deposition on aluminum alloy AA 2024 (Aviometal, Italy) with a pulling velocity of 10 cm/min, mixtures were left to stir 5 min. The substrates (dimensions: 2×5 cm^2) were polished using 3 M Perfect-IT III paste and subsequently sonicated in hexane, acetone, and methanol for 15 min. Specifically, for this research five types of coatings were prepared:

- DGEBA/DICY (D coating) without the accelerator,
- DGEBA/GlyPOSS/DICY (D-P coating) without the accelerator,
- DGEBA/GlyPOSS/DICY/BDMA (D-P-BDMA) coating,
- DGEBA/GlyPOSS/DICY/2-MeIm (D-P-2MeIm) coating,
- DGEBA/GlyPOSS/DICY/Curezol (D-P-Curezol) coating.

In parenthesis, the abbreviations that are depicted on graphs are shown. The curing process was performed at 150 °C for 1 h for coatings without accelerators and at 120 °C for 1 h for coatings with accelerators.

2.3. Methods

FT-IR absorbance measurements were made on a Bruker spectrometer, model IFS 66/S (Bruker, Billerica, MA, USA). All samples, i.e., either precursors or formulations, were deposited on silicon wafers. Temperature-dependent FT-IR absorbance measurements, from room temperature to either 120 or 150 °C, were performed in Spectra-Tech heated demountable cell with a controller. When the desired temperature was achieved, the spectrum was recorded. The spectra were gathered every 10 min during measurements at 120 and 150 °C. The resolution was 4 cm^{-1}.

DSC measurements were performed on a Mettler-Toledo DSC-1 (Columbus, OH, USA) calorimeter under a nitrogen atmosphere with a flow rate of 50 mL/min. Samples were sealed in 40 µL alumina crucibles with the lids. The mass of the samples was around 10 mg for all lyophilized mixtures. The analysis was performed with a heating rate of 5 K/min from −30 to 300 °C.

The rheological behavior of the epoxy resin was observed by nonisothermal dynamic oscillation with a rotational controlled rate rheometer (Physica MCR301, Anton Paar, Graz, Austria), equipped with a parallel geometry (PP-25). The epoxy samples were heated by convection. A solvent trap was used to minimize solvent evaporation, while a temperature-controlled hood was applied to prevent heat dissipation. The measurements were performed under a constant shear strain (10–20%) with a gap of 0.5 mm. Dynamic measurements were performed using a heating rate of 2 °C/min from 23 to 180 °C under a constant flow of dry nitrogen to eliminate any oxidative processes during heating.

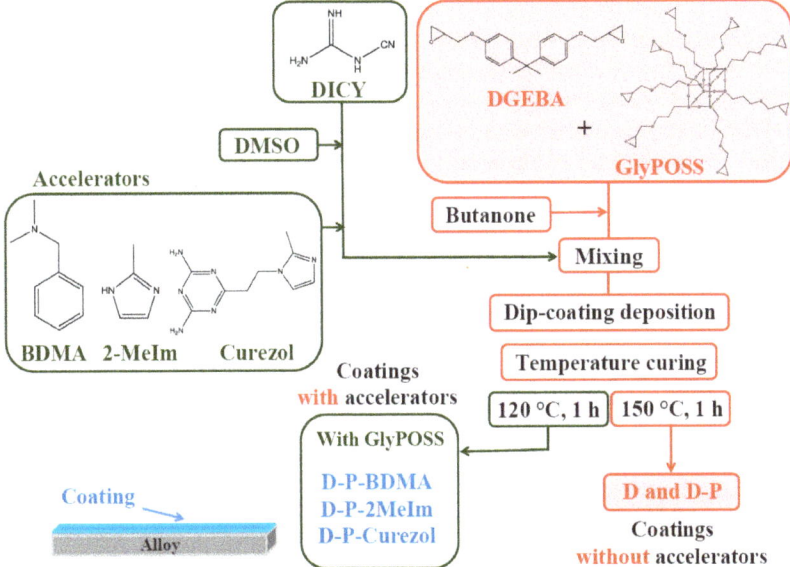

Figure 2. The preparation procedure of the protective coatings.

Taylor Hobson Series II profilometer (Leicester, United Kingdom) was used for coatings thickness determination. Coatings were characterized by a scanning electron microscope FE-SEM Zeiss SUPRA 35VP (Zeiss, Oberkochen, Germany). Atomic force microscopy (AFM) images were made on an AFM attachment of WITec alpha 300 confocal Raman spectrometer (Ulm, Germany). The images were recorded on areas of 10×10 µm^2 of the prepared coatings deposited on AA 2024 coupons. The images are presented in the two-dimensional representation without filtering. According to the scale bars alongside the images, the brighter the color, the higher the spot on the surface. Consequently, surface roughness (SR) was calculated. Samples were measured at room atmosphere and temperature.

An Autolab PGSTAT30 potentiostat-galvanostat (Metrohm Autolab, Utrecht, The Netherlands) was used to perform electrochemical measurements. Potentiodynamic polarization measurements were made in a K0235 flat cell (Ametek Scientific Instruments, Oak Ridge, USA) with a built-in Pt grid counter electrode. The cell was filled with a 0.5 M NaCl electrolyte. The coating on AA 2024 was mounted as a working electrode, while Ag/AgCl/KCl$_{sat}$ served as a reference electrode. The coating was held at an open circuit potential for 30 min before the measurement. Then linear sweep voltammetry was swept from 1.0 to 0.0 V using a scan rate of 1 mV/s. The corrosion current density (j_{corr}) was extrapolated with Tafel slopes from the measured potentiodynamic polarization curves.

3. Results

3.1. Characteristics of Formulations and Curing Process

Epoxy-based formulations belong to thermosetting materials that cure upon heating. Insight into curing dynamics can be obtained using different analytical techniques. DSC analysis, for example, provides information on glass transition temperature (T_g), start and completion of the curing process and the enthalpy of the curing process (Figure 3). Time-dependent FT-IR absorbance measurements during temperature treatment of coatings (Figures 4–6) enables insight into the crosslinking by observation of decreasing intensity of the epoxy mode at 915 cm^{-1} [28]. Rheology (Figures 7 and 8), on the other hand, gives quite a piece of information on changes in consistency during the transformation of the formulation into a crosslinked structure [18,29]. When the obtained results are observed in parallel experiments, designed to approach to the same conditions as much as possible, they can throw light to the curing processes that lead to the formation of highly crosslinked protective coatings. The important point of this study is to follow the influence of accelerators on the curing process. We look forward to connecting their structural properties with the behavior during the curing of formulations.

3.1.1. Thermal Properties

DSC analysis showed that for all curves of D-P-based formulations, the characteristic exothermic peaks appear (Figure 3). The curing of the D-P formulation without accelerator took place in the range of 130–280 °C and the enthalpy of the reaction reached 506 J/g (Table 1). As expected, the addition of any of the three accelerators speeded up the curing. Consequently, the curing maximums were found at approximately 30–40 °C lower temperatures (Table 1). The enthalpy detected for formulations with accelerators was lower for 100–160 J/g, meaning that some reactions already started during the lyophilization preparation procedure. In the case of BDMA and 2-MeIm accelerators, this phenomenon was more prominent as the crosslinking reaction started at lower onset temperatures (T_{onset}: 2-MeIm < BDMA < Curezol). The shapes of DSC curves of D-P and D-P-Curezol formulations were the same indicating that the behavior of Curezol is similar to DICY. Namely, Curezol acts as an epoxy ring-opening initiator (2-methyl imidazole part of its molecule accelerates and shifts the curing reactions to lower temperature) and as a crosslinking agent (through amino groups on the triazine ring). However, the Curezol accelerator shifted the D-P-Curezol curve to lower onset temperature. As Curezol, also 2-MeIm can act in both roles, i.e., as ring-opener and crosslinking agent. The crosslinking role of imidazoles was confirmed already long ago by NMR studies [30]. In addition, it was proposed that imidazole ring in the polymer matrix positively influences its physical and chemical properties [30].

It is commonly accepted that accelerators in epoxy resins decrease the curing temperature and enhance the curing rate. As described above, DSC curves (Figure 3, Table 1) confirmed that for all three tested accelerators in our D-P system. The addition of any accelerator also led to a change in T_g value [30–32]. However, these changes vary in both directions with regard to the T_g value of 116 °C obtained for D-P formulation without accelerator (Table 1). Interestingly, the T_g value increased in case of BDMA and Curezol, but decreased for accelerator 2-MeIm. It is reported for the DGEBA+DICY system that the increasing amount of accelerator 1-ethyl-3-methyl-imidazolium dicyanamide consistently decrease the T_{peak} temperature. All described formulations exhibit T_g above 120 °C and the values increased slightly with the increase in the concentration of the accelerator 1-ethyl-3-methyl-imidazolium dicyanamide. On the other hand, in a similar DGEBA+DICY system with either BDMA or 2-MeIm accelerators showed similar values of T_g to ours, i.e., 115–116 °C for the 2-MeIm accelerator and 139–147 °C for the BDMA accelerator depending on cure characteristics [32]. More detailed dependency studies of T_g were performed for the BDMA accelerator [32]. Specifically, the T_g value is dependent on the concentration of the BDMA accelerator and ratio DICY/BDMA, which was confirmed with the maximal T_g value achieved in the middle of the tested concentration profile. Moreover, dependency of T_g on amine/epoxy ratio in systems with BDMA was also demonstrated [32].

It can be concluded though that the T_g value considerably depends on the systems and should be more detailed studied for each system separately. This will be made in our future work.

Some studies observed the appearance of the second peak in DSC curves, as was noted for accelerator 1-ethyl-3-methyl-imidazolium dicyanamide [31]. Such peaks became more dominant with the increasing amount of the imidazolium salt accelerator. It was suggested that the second peaks can be tentatively attributed to the more pronounced reaction of DICY (since the addition of this accelerator decreased the amount of the unreacted DICY [31]. In addition, this aspect remains for our future concentration-dependent investigation of our system.

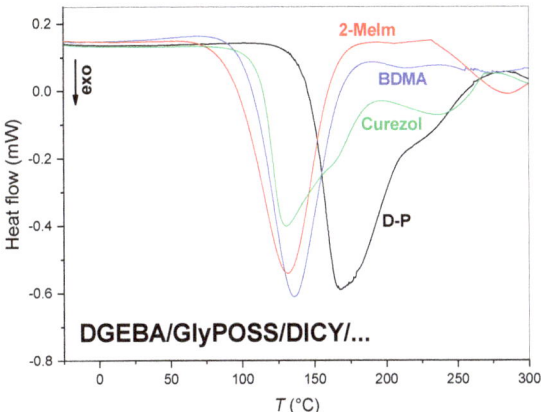

Figure 3. Normalized differential scanning calorimetry (DSC) curves of the D-P-based formulations.

Table 1. Thermal properties of D-P-based formulations used for the preparation of coatings.

Sample	T_{onset} [1] (°C)	T_{peak} [2] (°C)	ΔH [3] (J/g)	T_g [4] (°C)
D-P	112	168	−506	116
D-P-BDMA	76	125	−345	128
D-P-2MeIm	62	131	−363	113
D-P-Curezol	78	130	−405	120

[1] T_{onset}—onset temperature; [2] T_{peak}—temperature of the peak; [3] ΔH—enthalpy of the curing reactions; [4] T_g—glass temperature.

3.1.2. Temperature-Dependent FT-IR Spectra

The characteristic bands in the FT-IR absorbance spectra of precursors (DGEBA, GlyPOSS) and formulations D and D-P are evident from Figure 4 and Figure S1, the latter in Supporting Information. The epoxy modes appear in the spectrum of DGEBA at 3056 (ν_s(C–H)$_{epoxy\ ring}$), 1132 (ν(C–O–C)$_{epoxy\ ether}$) and 915 (ν(C–O)$_{epoxy\ ring}$) cm^{-1} [5,16,28]. The first two bands are of low intensity and in the proximity of other modes. The intensity of the band at 915 cm^{-1} is moderate and can serve as a measure of the extent of the curing reaction, i.e., opening of glycidoxy rings and their crosslinking into polymeric materials [28]. The spectrum of D formulation, in addition to vibrations of DGEBA, shows the stretching of primary and secondary amines (ν(NH$_2$), ν(NH)) between 3430 and 3150 cm^{-1} and cyano group at 2208 and 2162 cm^{-1} [33] (Figure 4A). When DGEBA is in part exchanged with GlyPOSS, i.e., for D-P formulation, the spectrum retains the basic characteristics of amino and cyano groups (Figure 4A). As well, the epoxy bonds at 3056 and 915 cm^{-1} remained visible in this spectrum, but the low-intensity epoxy band at 1132 cm^{-1} became overlapped with the broad bands in the spectral region 1200–1080 cm^{-1}. They belong to the stretching vibrations of ν(Si–O–Si) originating from the silsesquioxane cage of GlyPOSS. Such intensive ν(Si–O–Si) band was also noted in the spectrum of GlyPOSS/4,4'-(1,3-phenylenediisopropylidene) epoxy composites [24]. Importantly, in that study, the appearance of the glycidoxy groups of GlyPOSS is reported at ~745 cm^{-1} [24]. In our spectrum of GlyPOSS, the nearest band that can be assigned to glycidoxy groups is at 762 cm^{-1} (Figure 4B). However, it overlaps with the glycidoxy groups of DGEBA. Therefore, it only marginally contributes to the intensity increase of the 762 cm^{-1} band in D-P formulation regarding the D formulation. For this reason, it cannot be used to follow the curing of GlyPOSS glycidoxy groups.

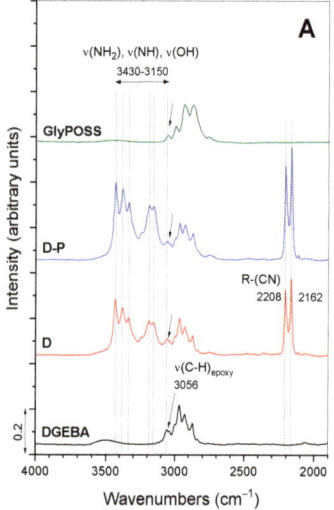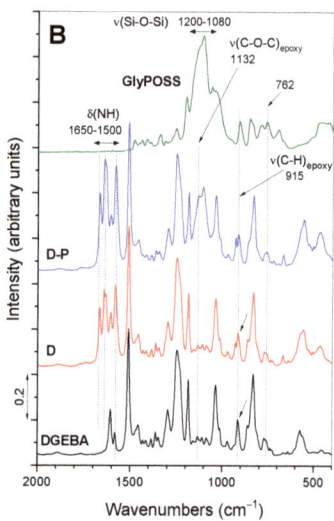

Figure 4. FT-IR absorbance spectra of precursors DGEBA and GlyPOSS, and initial D and D-P formulations in the spectral ranges: (**A**) 4000–1900 cm^{-1} and (**B**) 2000–400 cm^{-1}.

The FT-IR absorbance spectra obtained during the curing of D-P formulation with temperature is evident from Figure 5 and, for comparison, for D formulation in Figure 6. The whole spectra are shown in Figures S2 and S3 in Supporting Information. The spectrum recorded at 100 °C for D-P formulation already showed a notable change (Figure 5). Specifically, the reactions between the opened glycidoxy rings of DGEBA, GlyPOSS and primary and secondary amine groups of DICY started to occur. The result was also the formation of hydroxyl groups [5,25], while the opening of glycidoxy rings of DGEBA was indicated through the decrease in the intensity of the above-mentioned epoxy stretching bands at 3056 and 915 cm^{-1}. Regarding the low intensity of the 3056 cm^{-1} band, the 915 cm^{-1} band (Figure 5) remains the most appropriate one for the following of this curing reaction through a calculation of its integral intensity [28]. The uncertainties can occur in the final curing stages when the concentration of glycidoxy rings became very low [28]. As the reference band for normalization, 1509 cm^{-1} band of C–C stretching of the aromatic ring was taken.

The same trend, i.e., decrease in the intensity during curing (Figure 5), was observed for the bands associated with amines of DICY in the spectral region 3430–3150 cm^{-1} (ν(NH$_2$), ν(NH)) and 1650–1500 cm^{-1} (δ(NH)). According to the literature, cyano–CN groups (2208 and 2162 cm^{-1}) can also provide crosslinking at higher temperatures [31]. Such network crosslinking in the formulation is evident from the appearance of the stretching ν(C–N) band at 1084 cm^{-1} at temperatures around 190 °C. Since this band is overlapped by ν(Si–O–Si) bands of silsesquioxane GlyPOSS in Figure 5, its evolution can only be followed in the spectra of D formulation (Figure 6; Figure S3 in Supporting Information). The curing process of D-P formulation was finished after 80 min at 200 °C (Figure 5). The main characteristics of the FT-IR bands with curing were similar, in the case of D formulation (Figure 6). The spectra revealed the decrease in intensity of epoxy, amino and cyano bands at 200 °C.

As mentioned above, due to the absence of ν(Si–O–Si) bands, the increase in the intensity of stretching ν(C–N) band can be observed (Figure 6B).

The spectra of D-P-accelerator formulations revealed similar features than the described D-P one (Figure 5) due to the low concentration of accelerators. Consequently, the time-dependent FT-IR spectra during thermal curing are not shown for the formulations with accelerators. Anyhow, the influence of accelerators on the course of the curing reactions is shown through the presentation of the integral intensity decrease in the 915 cm^{-1} band for either of the formulations (Figure 7). The curing was performed up to either 120 or 150 °C, at which the process was followed for a certain period, before the continuation with the temperature increase. The spectra of formulations without accelerators (D, D-P) were only cured with the 150 °C temperature profile (Figure 7A).

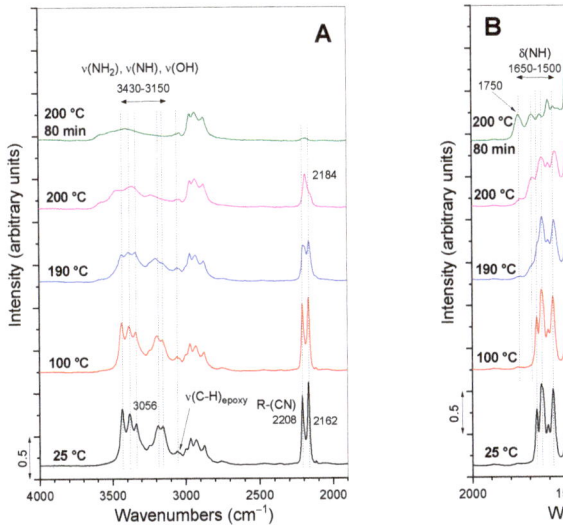
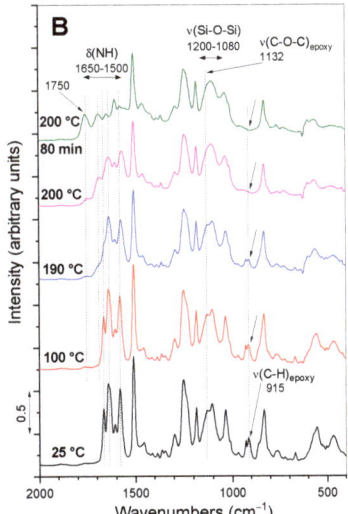

Figure 5. Time-dependent FT-IR absorbance spectra during the thermal curing process of formulation D-P in the spectral ranges: (**A**) 4000–1900 cm^{-1} and (**B**) 2000–400 cm^{-1}.

The curing of D and D-P formulations occurred quite gradually (Figure 7A). For D formulation two slopes can be discerned. When GlyPOSS was added to the coating formulation (D-P), the integral intensity of the 915 cm^{-1} band decreased slower during the initial 20 min, but then somewhat more abruptly around 60 °C. After reaching the isothermal temperature of 150 °C, the initial increase in slope is followed by some relaxation and then a steady decrease. The integral intensity approached zero after ~2 h of isothermal treatment for either of formulations, D or D-P. The observed differences in the intensity decrease reflect the composition of both formulations. The incorporation of GlyPOSS molecules into the epoxy matrix of D-P formulation induced the sterical hindrances. Specifically, such molecules are equipped with eight reactive groups extending into all directions. When only DGEBA and DICY are the reactive species in the D formulation (Figure 6; Figure S3 in Supporting Information), the formation of the epoxy matrix demands the approaching of the reactive amine groups to glycidoxy rings. The addition of GlyPOSS with eight glycidoxy rings in the corners of the cube silsesquioxane, however, enabled their bonding without any necessary preceding orientation of the silsesquioxane. Consequently, the crosslinking occurred somewhat faster at certain temperatures, which becomes apparent at 60 °C. After that temperature, the inclinations of the slopes are similar for both formulations, i.e., D and D-P, throughout the whole isothermal treatment.

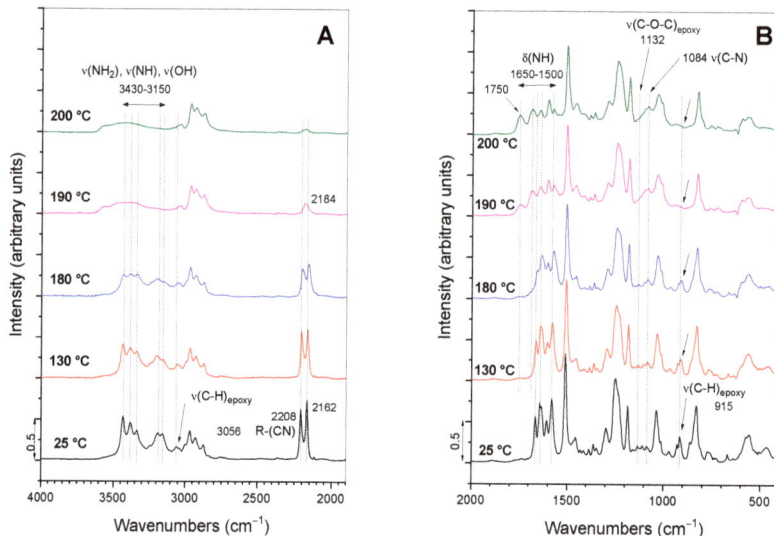

Figure 6. Time-dependent FT-IR absorbance spectra during the thermal curing process of formulation D in the spectral ranges: (**A**) 4000–1900 cm^{-1} and (**B**) 2000–400 cm^{-1}.

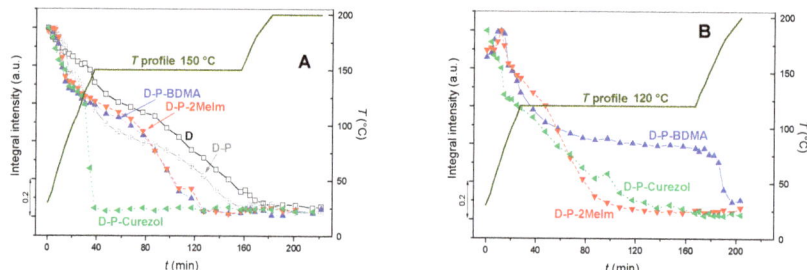

Figure 7. Time-dependent integral intensity changes in the characteristic epoxy band of 915 cm^{-1} with thermal curing. The temperature profile that was followed with time is depicted, as well: (**A**) 150 °C and (**B**) 120 °C.

The addition of accelerators to D-P formulation changed the course of the integral intensity curves (Figure 7A). During the temperature increase up to 150 °C, the integral intensity of the 915 cm^{-1} band decreased quicker compared to that of the neat D-P formulation. Interestingly, the accelerators BDMA and 2-MeIm behaved similarly. Such behavior was expected since at this temperature these two accelerators are characterized as compounds that catalyze the opening of glycidoxy rings [32]. When accelerator Curezol was added, the integral intensity decreased abruptly after reaching 130–140 °C (after 30–40 min in Figure 7A). This is understandable since the Curezol—in addition to the opening of the glycidoxy rings—can also contribute to the crosslinking reactions (post-curing) through primary amino groups at the triazine ring (Figure 1E).

The inspection of the integral intensity behavior of the 915 cm^{-1} band during a 120 °C temperature profile (Figure 7B) revealed the difference among both accelerators with the predominantly catalytic activity. Initially, during heating to 120 °C (40 min in Figure 7B), the intensity decrease was similar for BDMA and 2-MeIm accelerators. However, during isothermal treatment at 120 °C, the integral intensity remained considerably constant when BDMA was added. It did not continue to decrease until 150 °C (190 min) was reached, showing that BDMA possesses only the ring-opening properties. In contrast, formulation with 2-MeIm showed a considerable decrease in the intensity of the 915 cm^{-1} band already during the isothermal treatment. The described discrepancy pointed out that the 2-MeIm as aromatic amine is an epoxy ring-opener which can be after initiation ingrained in the polymer structure [31]. Curezol remained the quickest in its action that occurred through an opening of glycidoxy rings and crosslinking up to 120 °C, while during isothermal treatment, it approached the intensity behavior of formulation with 2-MeIm.

3.1.3. Rheological Characteristics

With the same aim, i.e., to investigate the influence of different accelerators on the curing of D-P-based formulations, rheological characterization was made. Figure 8 shows the dynamic storage modulus (G') and loss modulus (G'') versus temperature during heating at a heating rate of 2 °C/min. These rheograms confirm that curing processes are time- and temperature-dependent [18,29]. Although the shapes of the curves obtained for elastic and viscous modulus are similar for all formulations, the curing reactions start at considerably different temperatures (i.e., times). However, G' or G'' tend to final values of the order of magnitude 3×10^5 Pa (G') and 10^4 Pa (G'') for all formulations, respectively. The evolution of the elastic G' modulus, which is proportional to the rigidity of the cured epoxy, indicate the formation of molecular networks through the formation of chemical bonds (crosslinking) in the either of the investigated epoxy systems [29].

The curves of the basic formulation D (Figure 8, Table 2) reveal that up to approximately 154 °C when the viscous modulus G'' dominated the elastic G' one, this epoxy system was in a viscous state ($G'' > G'$). This gradual increase in the G'', viscous modulus shows that the reaction processes have already started in the formulation. The crosslinking, however, occurred at 66 min and the temperature of 155 °C, when G' equals G'' (Table 2). Afterward, the gel state occurred, but the values of elastic and viscous moduli continue to increase. The elastic G' portion prevails the viscous G'' one thereupon. The final stage of the curing process is the formation of the plateaus for both G' or G'', indicating the cured epoxy system. The more significant is the difference between both moduli in the plateau region, the stronger is the cured epoxy system (i.e., more rigid solid matter).

The processes that are described for the D formulation also occurred in other investigated systems but at a different temperature-scale (Figure 8). In D-P formulation, the crosslinking happened already after 62 min and the temperature of 146 °C (Table 2). Other characteristics of the curing processes remained similar as for D formulation. However, the absolute value of elastic modulus G' reached higher values for the formulation with GlyPOSS compared to neat D formulation. This indicated that more crosslinked and stiff structure formed when GlyPOSS with eight glycidoxy groups steaming from the siloxane cube center was added.

The addition of any accelerator (BDMA, 2MeIm, Curezol) to D-P formulation contributed to even faster crosslinking (~50 min) and at approximately 22 °C lower temperatures (Figure 8A, Table 2). Differences among accelerators were small. Interestingly, the differences between plateau G' and G'' values were similar for all three accelerators, but when BDMA was added, both plateau were shifted to somewhat lower values. This indicates that the extent of the crosslinking of the internal matrix was the lowest in the case of BDMA accelerator (as it did not take part in crosslinking), which is in the correlation with FT-IR results (Figure 7B) as more or less sufficient crosslinking was achieved at 120 °C. When 2-MeIm or Curezol were used, the achieved moduli G' and G'' were similar to the ones obtained by D-P formulation, resulting in a similar crosslinking. These two accelerators could also collaborate in crosslinking, they are not only epoxy ring-openers. BDMA only acts as the epoxy ring-opener.

As described above, rheology was used to check the influence of accelerators on the D-P system (Figure 8A, Table 2). Besides, we evaluated their influence on sole D formulation (Figure 8B, Table 2). At first glance, the temperature of crosslinking is quite similar for either of the D-accelerator formulations. However, if an average value is calculated out of temperatures of crosslinking for formulations with accelerators, the average temperature obtained for D-accelerator formulation is about 1 °C lower (Table 2). This implements that accelerators can exert a slightly larger influence on D formulation compared to D-P formulation. Such an effect can be understandable since the introduction of GlyPOSS brings about some sterical hindrances and tensions during the packaging of the compounds into the crosslinked solid material.

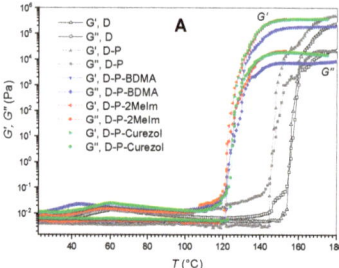 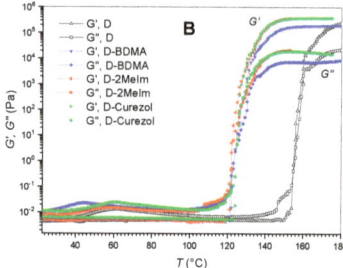

Figure 8. Rheological characterization formulations during thermal curing: (**A**) D-P-based and (**B**) D-based formulations.

Table 2. Rheological characterization of D-P-accelerator formulations used for the preparation of coatings.

Sample	t [1] (min)	T [2] (°C)	$T_{average}$ [3] (°C)
D	66	155	
D-BDMA	50	123	
D-2MeIm	49	119	122
D-Curezol	51	125	
D-P	62	146	
D-P-BDMA	51	124	
D-P-2MeIm	50	122	124
D-P-Curezol	51	125	

[1] t—time, at which $G' = G''$; [2] T—temperature, at which $G' = G''$; [3] $T_{average}$—average temperature, at which $G' = G''$ for all three formulations with accelerators.

The typical characteristics of the crosslinking processes in the investigated epoxy formulations can also be viewed through the behavior of phase angle (Figure 9). This parameter is defined as a ratio between the lost and the stored deformation energy, which means the ratio between the viscous and the elastic contribution to the viscoelastic behavior. Namely, the proper balance between the viscous and the elastic contribution in the viscoelastic formulation before the deposition of the coatings is of particular importance when our aim is the deposition of the protective coatings with excellent barrier properties [18,29].

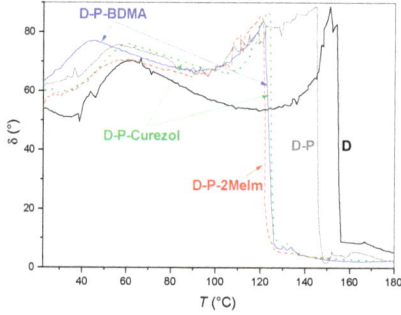

Figure 9. Phase angles of D, D-P and D-P-accelerator formulations during curing.

Specifically, the phase angle δ shows the time delay between the determined sinusoidal parameters and the measured characteristics G' and G''. These differences occur due to the viscosity changes of the measured formulations. The values of phase angle are 45° to 90° for liquids, at δ = 45° crosslinking occurs, while below 45° to 0° G' dominates over G'' indicating formation of solid materials. An ideal elastic deformation would result in δ = 0°, an ideal viscous deformation in δ = 90°. These characteristics can also be viewed from Figure 9. Namely, in the beginning, when formulations are in a liquid state, but their viscosity increase, the values of phase angle are between 90° and 45°. When crosslinking occurs for $G' = G''$, the phase angle equals 45°. Then the decrease in the values of the phase angle towards 0° indicates the solid-state of the measured material. In this range, the epoxy formulations become highly crosslinked, molecules with larger and larger molar mass form, and their mobility is significantly decreased. The rigidity of the material increase, but during all stiffening processes also some reactive components may be present.

3.2. Characterstics of Protective Coatings

3.2.1. Surface Properties

Protective efficiency of coatings is usually affected also by their surface morphology, originating from the intensity of the interface between the coating surface and the corrosive media. Consequently, SEM analyses of the surface and cross-cut of D-P and D-P-BDMA coatings are shown in Figure 10. The surface of the D-P coating (Figure 10A,B) reveals tiny inhomogeneities. On the contrary, the surface of the coating with the BDMA accelerator (Figure 10C,D) shows the presence of some elevated particles, but the surrounding surface is more homogeneous compared to the D-P coating. This tentatively suggests that when the accelerator was added the surface became more homogeneous. Nevertheless, some particle formation was observed also in this coating with the accelerator. The SEM of the cross-cut examples showed a compact coating structure for either coating without or with accelerator which is a prerequisite for sufficient protective efficiency (Figure 10E,F).

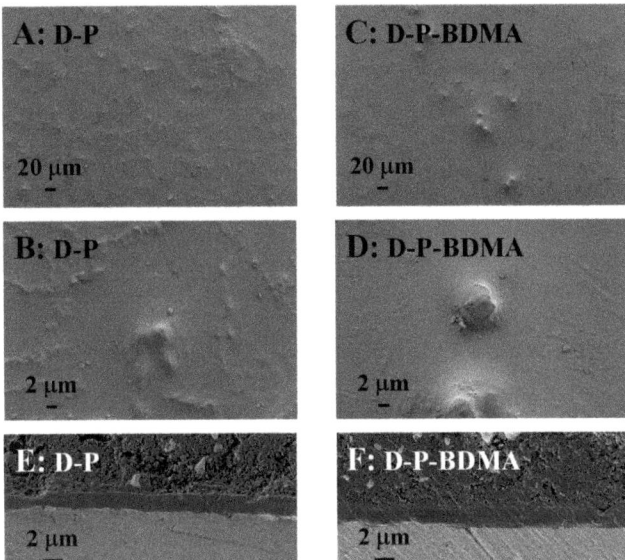

Figure 10. SEM micrographs of protective coatings at different magnifications: (**A,B**) D-P coating and (**C,D**) D-P-BDMA coatings. (**E,F**) Cross-cut SEM micrographs of D-P (**E**) and D-P-BDMA (**F**) coatings.

Besides, AFM images were recorded for D, D-P and D-P-accelerator coatings (Figure S4 in Supporting Information). The surface roughness is reported in Table 3. The AFM image of D coating shows areas with spherical particles in the coating composition with the size of a few hundreds of nanometers to the size of a micron. The measurements show random particle distribution. Moreover, similar morphology with some spherical particles can be observed in the D-P coating (Figure S4). Surface roughness values (SR), calculated from images demonstrate a considerably higher value for the coating composed of D compared to the D-P coating.

Table 3. Morphological and electrochemical characterization of the protective coatings.

Protective Coating	d [1] (µm)	SR [2] (nm)	E_{corr} [3] (V)	j_{corr} [4] (A/cm^2)	Protective Coating	E_{corr} [3] (V)	j_{corr} [4] (A/cm^2)
D	1.2	426	−0.336	1.8×10^{-11}			
D-P	1.6	148	−0.549	2.8×10^{-11}	D	−0.336	1.8×10^{-11}
D-P-BDMA	1.5	91	−0.695	9.4×10^{-12}	D-BDMA	−0.695	9.4×10^{-12}
D-P-2MeIm	3.3	80	−0.138	2.6×10^{-12}	D-2MeIm	−0.138	2.6×10^{-12}
D-P-Curezol	3.5	58	−0.721	1.7×10^{-11}	D-Curezol	−0.721	1.7×10^{-11}

[1] d—thickness; [2] SR—surface roughness; [3] E_{corr}—corrosion potential; [4] j_{corr}—corrosion current density.

As already shown by SEM (Figure 10), the addition of accelerators causes a more uniform and homogenous morphology of the coatings (Figure S4C–E in Supporting Information). As well, the addition of accelerators decreased the average roughness values (Table 3), ranging from 91 nm for the coating with BDMA to the smallest value of 58 nm obtained for the coating with Curezol (Figure S4C,E). Such a difference can originate from added accelerators. Specifically, all three accelerators stimulate the opening of the glycidoxy rings, which mitigated the reactions of crosslinking. Despite the reactions start at lower temperatures in the presence of accelerators, the support in ring-opening enabled more uniform crosslinking. Except in surface roughness, AFM images do not reveal any significant differences among accelerators (Figure S4C–E). The eventual bonding of Curezol or 2-MeIm in the epoxy coating does not influence their images.

3.2.2. Electrochemical Characteristics

Potentiodynamic polarization is a quick test of the protective efficiency of investigated coatings. The test is relative and the measurements should be performed in the same way and conditions. Figure 11 reports the potentiodynamic curves for D-P- and D-based coatings without and with accelerators in comparison. The difference in the current density between the measurement of the uncovered AA 2024 and the substrates covered with coatings reveals their protective efficiency. It is obvious that the shape of the potentiodynamic curves obtained for either of the D-P-based coatings shows its considerably improved performance with regard compared to D coating (Figure 11A). Even when either of accelerators was added into the D formulation (D-accelerator coatings in Figure 11B), the shape of the curves did not improve but resembled that of the D coating. Namely, the anodic current density for the D and D-accelerator coatings increased significantly with the increase in the potential. On the other hand, the current density of the D-P-based coatings remained low for all examined potentials. This showed the excellent protective efficiency of the D-P and D-P-accelerator coatings. Their compact structure (Figure 10E,F) prevents the formation of corrosion products at the interface coating substrate.

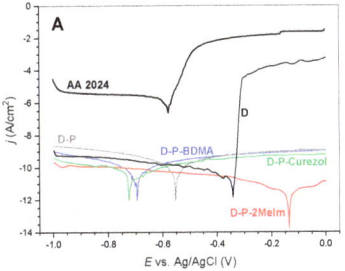
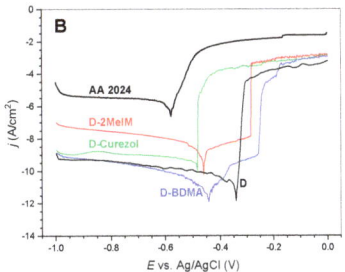

Figure 11. Potentiodynamic polarization curves of protective coatings; (**A**) D-P-based formulations and (**B**) D-based formulations.

4. Summary and Conclusions

Epoxy coatings are known for their excellent protective efficiency in corrosion applications. Mostly, increased protective efficiency can be achieved through the formation of advanced epoxy nanocomposites [5–8]. The addition of suitable nanoparticles can result in improved temperature, chemical, and electrochemical resistance. An important parameter for the formation of nanocomposite materials is the appropriate incorporation of nanoparticles in the polymer matrix. Although in some cases reinforcement can be achieved by the simple addition of nanoparticles, their eventual bonding into the epoxy matrix can prevent leakage, migration, and agglomeration. Consequently, homoleptic octaglycidyl-POSS was used in this study. Specifically, 20 wt.% of DGEBA precursor was exchanged by GlyPOSS in reaction with DICY hardener. Three different accelerators that promote the opening of the glycidoxy rings were also tested.

DSC (Figure 3), temperature-dependent FT-IR absorbance (Figures 4–7) and rheology (Figure 8) measurements confirmed that the addition of accelerators induced the lowering of the temperature at which the crosslinking processes occur. Such an effect is desired from the industrial viewpoint, significantly simplifying the production procedures and reducing the production costs. DSC (Figure 3) revealed a similar lowering of crosslinking temperature for 2-MeIm (131 °C) and Curezol (130 °C), while somewhat lower peak temperature was obtained by BDMA (136 °C). Temperature-dependent FT-IR absorbance measurements further differentiated among the accelerators. When isothermal treatment was performed at 150 °C (Figure 7A), the Curezol was the quickest in its action. BDMA was found the slowest during isothermal treatment at 120 °C (Figure 7B). Rheologically, the lower plateau values of elastic G' and viscous G'' moduli for D-P-BDMA formulation showed that the BDMA accelerator, as it did not take part in crosslinking, led to less strong internal structure compared to 2-MeIm and Curezol (Figure 8). Temperatures at which $G' = G''$ were similar for all three accelerators, although slightly lower (122 °C) when accelerator 2-MeIm was used. The described findings reflect the basic characteristics of the chosen accelerators. According to their structure, all three accelerators function as glycidoxy ring openers. However, Curezol can also contribute to crosslinking processes via primary amino groups on the triazine ring and 2-MeIm via aromatic secondary amine. These differences in the structures of accelerators reflect in the crosslinking behavior of the D-P-based formulations. All three applied measurement techniques differentiate BDMA accelerator from 2-MeIm and Curezol ones. The double function of the latter two, being openers for the glycidoxy rings and their possibility to collaborate in the crosslinking, prevented any final decision on the superior action of either of them. Interestingly, the rheological time sweep experiments performed to determine gel time ($G' = G''$) based on viscoelastic parameters showed even its decrease when 1-benzyl-2-methylimidazole catalyst was added to DGEBA/triethylene-tetraamine formulation [29]. The G' and G'' crossovers were determined isothermally at five different temperatures from 60 to 100 °C and only in case of some diluted formulations an accelerative effect of the catalyst was noted. This pointed to the importance of the presence of aromatic secondary amine in the structure of 2-MeIm (Figure 1). Anyhow, the slight differences in our DSC (Figure 3), FT-IR absorbance (Figures 4–7) and rheology (Figure 8) measurements do not allow the decision on the preference. The preference may, however, be set via experimental demands. Namely, while the admixture of BDMA or 2-MeIm is straightforward, the addition of Curezol might occasionally result in the formation of an opaque formulation. This is a consequence of the low solubility of Curezol in organic solvents and water.

The surface roughness values obtained from AFM (Figure S4 in Supporting Information) showed distinct changes among the surfaces. When accelerators were not applied, i.e., neat D and D-P coatings, the surfaces revealed the presence of spherical particles and inhomogeneities in AFM images. The addition of either of accelerators resulted in lower values of the surface roughness (Table 3). SEM micrographs (Figure 10) confirmed the rougher surface of the D-P coating concerning the D-P-BDMA coating with the BDMA accelerator. It is worth mentioning, that also the coating with BDMA contained certain elevated areas, but the surrounding surface was much flatter and more homogeneous (Figure 10C,D). The SEM measurements of the cross-cut samples (Figure 10E,F) revealed the compact inner structures of D-P and D-P-BDMA coatings. Such a compact structure also resulted in the extremely good protective efficiency of D-P and D-P-accelerator coatings with regard to the neat D coating (Figure 11).

In conclusion, we can say that the time-dependent FT-IR absorbance measurements showed that partial exchange of DGEBA with GlyPOSS resulted in curing at a somewhat lower temperature. The reason probably lies in the eight glycidoxy groups that are positioned in the corners of this homoleptic cube-shaped GlyPOSS. The further lowering of the temperature of curing was achieved by the addition of various accelerators. It was found that the action of the accelerators considerably depends on their structure. Although all three accelerators are capable of the opening of glycidoxy rings, only Curezol and 2-MeIm can collaborate in crosslinking reactions. Specifically, Curezol can bind via two primary amino groups and 2-MeIm through aromatic secondary amine. Potentiodynamic polarization tests showed that all coatings comprising GlyPOSS show better protective efficiency compared to neat DGEBA/DICY (D) coatings. Consequently, our material is a promising candidate for a wide range of applications, such as coatings for food cans, white goods, etc.

Supplementary Materials: The following are available online at http://www.mdpi.com/2073-4360/12/3/591/s1, Figure S1: FT-IR absorbance spectra of precursors DGEBA and GlyPOSS, and initial D and D-P formulations, Figure S2: Time-dependent FT-IR absorbance spectra during the thermal curing process of D-P formulation, Figure S3: Time-dependent FT-IR absorbance spectra during the thermal curing process of D formulation, Figure S4: AFM images of protective coatings: (A) D, (B) D-P, (C) D-P-BDMA, (D) D-P-2MeIm, (E) D-P-Curezol.

Author Contributions: Conceptualization: M.R., M.Č., I.J.; Methodology: M.R., M.Č., I.J., A.K.S.; Software: M.R., E.Š., M.Š., M.Č., A.K.S.; Formal analysis: M.R., M.M., E.Š., M.Š.; Investigation: M.R., M.M., E.Š., M.Š., A.K.S.; Data curation: M.R., M.M., E.Š., M.Š., M.Č., A.K.S.; Writing—Original Draft Preparation: M.R., A.K.S.; Writing—Review & Editing: A.K.S., M.Č., I.J.; Supervision: I.J., M.Č., A.K.S.; Project Administration: I.J.; Funding Acquisition: I.J. All authors have read and agreed to the published version of the manuscript.

Funding: This project was financed by the Slovenian Research Agency (Programme P2-0393).

Acknowledgments: Helena Spreizer is acknowledged for infrared measurements.

Conflicts of Interest: The authors declare no conflict of interest.

References

1. Koch, G.; Varney, J.; Thompson, N.; Moghissi, O.; Gould, M.; Payer, J. International measures of prevention, application, and economics of corrosion technologies study. In *International Measures of Prevention, Application and Economics of Corrosion Technologies Study*; Jacobson, G., Ed.; NACE International: Huston, TX, USA, 2016.
2. Olajire, A.A. Recent advances on organic coating system technologies for corrosion protection of offshore metallic structures. *J. Mol. Liq.* **2018**, *269*, 572–606. [CrossRef]
3. Zhang, F.; Ju, P.; Pan, M.; Zhang, D.; Huang, Y.; Li, G.; Li, X. Self-healing mechanisms in smart protective coatings: A review. *Corros. Sci.* **2018**, *144*, 74–88. [CrossRef]
4. Jin, F.; Li, X.; Park, S. Journal of Industrial and Engineering Chemistry Synthesis and Application of Epoxy Resins: A Review. *J. Ind. Eng. Chem.* **2015**, *29*, 1–11. [CrossRef]

5. Saliba, P.A.; Mansur, A.A.; Santos, D.B.; Mansur, H.S. Fusion-bonded epoxy composite coatings on chemically functionalized API steel surfaces for potential deep water petroleum exploration. *Appl. Adhes. Sci.* **2015**, *3*, 22. [CrossRef]
6. Navarchian, A.H.; Joulazadeh, M.; Karimi, F. Investigation of corrosion protection performance of epoxy coatings modified by polyaniline/clay nanocomposites on steel surfaces. *Prog. Org. Coat.* **2014**, *77*, 347–353. [CrossRef]
7. Miyauchi, K.; Takita, Y.; Yamabe, H.; Yuasa, M. A study of adhesion on stainless steel in an epoxy/dicyandiamide coating system: Influence of glass transition temperature on wet adhesion. *Prog. Org. Coat.* **2016**, *99*, 302–307. [CrossRef]
8. Kongparakul, S.; Kornprasert, S.; Suriya, P.; Le, D. Self-healing hybrid nanocomposite anticorrosive coating from epoxy/modified nanosilica/perfluorooctyl triethoxysilane. *Prog. Org. Coat.* **2017**, *104*, 173–179. [CrossRef]
9. Hader-Kregl, L.; Wallner, G.M.; Kralovec, C.; Eyssell, C. Effect of inter-plies on the short beam shear delamination of steel/composite hybrid laminates. *J. Adhes.* **2019**, *95*, 1088–1100. [CrossRef]
10. Pugstaller, R.; Wallner, G.M.; Strauss, B.; Fluch, R. Advanced characterization of laminated electrical steel structures under shear loading. *J. Adhes.* **2019**, *95*, 834–848. [CrossRef]
11. Sprenger, S. Epoxy resins modified with elastomers and surface-modified silica nanoparticles. *Polymer* **2013**, *54*, 4790–4797. [CrossRef]
12. Ayandele, E.; Sarkar, B.; Alexandridis, P. Polyhedral oligomeric silsesquioxane (POSS)-containing polymer nanocomposites. *Nanomaterials* **2012**, *2*, 445–475. [CrossRef]
13. Li, G.; Wang, L.; Ni, H.; Pittman, C.U. Polyhedral oligomeric silsesquioxane (POSS) polymers and copolymers: A review. *J. Inorg. Organomet. Polym.* **2002**, *11*, 123–154. [CrossRef]
14. Lee, A.; Lichtenhan, J.D. Viscoelastic responses of polyhedral oligosilsesquioxane reinforced epoxy systems. *Macromolecules* **1998**, *9297*, 4970–4974. [CrossRef] [PubMed]
15. Abad, J.; Barral, L.; Fasce, D.P.; Williams, R.J.J. Epoxy networks containing large mass fractions of a monofunctional polyhedral oligomeric silsesquioxane (POSS). *Macromolecules* **2003**, *36*, 3128–3135. [CrossRef]
16. Ramírez, C.; Abad, M.J.; Barral, L.; Cano, J.; Díez, F.J.; López, J.; Montes, R.; Polo, J. Thermal behaviour of a polyhedral oligomeric silsesquioxane with epoxy resin cured by diamines. *J. Therm. Anal. Calorim.* **2003**, *72*, 421–429. [CrossRef]
17. Constantin, F.; Gârea, S.A.; Iovu, H. The influence of organic substituents of polyhedral oligomeric silsesquioxane on the properties of epoxy-based hybrid nanomaterials. *Compos. B* **2013**, *44*, 558–564. [CrossRef]
18. Strachota, A.; Kroutilova, I.; Kova, J.; Mate, L. Epoxy networks reinforced with polyhedral oligomeric silsesquioxanes (POSS). Thermomechanical Properties. *Macromolecules* **2004**, *37*, 9457–9464. [CrossRef]
19. Frank, K.L.; Exley, S.E.; Thornell, T.L.; Morgan, S.E.; Wiggins, J.S. Investigation of pre-reaction and cure temperature on multiscale dispersion in POSS-epoxy nanocomposites. *Polymer* **2012**, *53*, 4643–4651. [CrossRef]
20. Kumar, S.A.; Sasikumar, A. Studies on novel silicone/phosphorus/sulphur containing nano-hybrid epoxy anticorrosive and antifouling coatings. *Prog. Org. Coat.* **2010**, *68*, 189–200. [CrossRef]
21. Rodošek, M.; Rauter, A.; Slemenik Perše, L.; Merl Kek, D.; Šurca Vuk, A. Vibrational and corrosion properties of poly(dimethylsiloxane)-based protective coatings for AA 2024 modified with nanosized polyhedral oligomeric silsesquioxane. *Corros. Sci.* **2014**, *85*, 193–203. [CrossRef]
22. Zhang, Z.; Gu, A.; Liang, G.; Ren, P. Thermo-oxygen degradation mechanisms of POSS/epoxy nanocomposites. *Polym. Degrad. Stab.* **2007**, *92*, 1986–1993. [CrossRef]
23. Fu, B.X.; Namani, M.; Lee, A. Influence of phenyl-trisilanol polyhedral silsesquioxane on properties of epoxy network glasses. *Polymer* **2003**, *44*, 7739–7747. [CrossRef]
24. Ramírez, C.; Rico, M.; Torres, A.; Barral, L.; López, J.; Montero, B. Epoxy/POSS organic–inorganic hybrids: ATR-FTIR and DSC studies. *Eur. Polym. J.* **2008**, *44*, 3035–3045. [CrossRef]
25. McCoy, J.D.; Ancipink, W.B.; Clarkson, C.M.; Kropka, J.M.; Celina, M.C.; Giron, N.H.; Hailesilassie, L.; Fredj, N. Cure mechanisms of diglycidyl ether of bisphenol A (DGEBA) epoxy with diethanolamine. *Polymer* **2016**, *105*, 243–254. [CrossRef]
26. Tseng, C.-C.; Chen, K.-L.; Lee, K.-W.; Takayam, H.; Lin, C.-Y. Soluble PEG600-imidazole derivatives as the thermal latent catalysts for epoxy-phenolic resins. *Prog. Org. Coat.* **2019**, *127*, 385–393. [CrossRef]

27. Chun, H.; Kim, Y.-J.; Park, S.-Y.; Park, S.-J. Curing mechanism of alkoxysilyl-functionalized epoxy (II): Effect of catalyst on the epoxy chemistry. *Polymer* **2019**, *172*, 272–282. [CrossRef]
28. Cholake, S.T.; Mada, M.R.; Raman, R.K.S.; Bai, Y.; Zhao, X.L.; Rizkalla, S.; Bandyopadhyay, S. Quantitative analysis of curing mechanisms of epoxy resin by Mid- and Near- Fourier Transform infrared spectroscopy. *Def. Sci. J.* **2014**, *64*, 314–321. [CrossRef]
29. Wang, Y.; Science, M. Effect of Additives on the rheological properties of fast curing epoxy resins. *J. Silic. Based Compos. Mater.* **2015**, *67*, 25–27. [CrossRef]
30. Farkas, A.; Strohm, P.F. Imidazole catalysis in the curing of epoxy resins. *J. Appl. Polym. Sci.* **1968**, *12*, 159–168. [CrossRef]
31. Neumeyer, T.; Staudigel, C.; Bonotto, G.; Altstaedt, V. Influence of an imidazolium salt on the curing behaviour of an epoxy-based hot-melt prepreg system for non-structural aircraft applications. *CEAS Aeronaut. J.* **2015**, *6*, 31–37. [CrossRef]
32. Hayaty, M.; Honarkar, H.; Hosain, B.M. Curing behavior of dicyandiamide/epoxy resin system using different accelerators. *Iran. Polym. J.* **2013**, *22*, 591–598. [CrossRef]
33. Fedtke, M.; Domaratius, F.; Walter, K.; Pfitzmann, A. Curing of epoxy resins with dicyandiamide. *Polym. Bull.* **1993**, *31*, 429–435. [CrossRef]

© 2020 by the authors. Licensee MDPI, Basel, Switzerland. This article is an open access article distributed under the terms and conditions of the Creative Commons Attribution (CC BY) license (http://creativecommons.org/licenses/by/4.0/).

Article

The Effect of the Preparation Method and the Dispersion and Aspect Ratio of CNTs on the Mechanical and Electrical Properties of Bio-Based Polyamide-4,10/CNT Nanocomposites

Itziar Otaegi [1], Nora Aranburu [1], Maider Iturrondobeitia [2], Julen Ibarretxe [2] and Gonzalo Guerrica-Echevarría [1,*]

[1] POLYMAT and Polymer Science and Technology Department, Faculty of Chemistry, University of the Basque Country UPV/EHU, Paseo Manuel de Lardizabal 3, 20018 Donostia-San Sebastián, Spain; itziar.otaegi@ehu.eus (I.O.); nora.aramburu@ehu.eus (N.A.)

[2] eMERG, School of Engineering of Bilbao, building II-I, University of the Basque Country UPV/EHU, Rafael Moreno Pitxitxi 3, 48013 Bilbao, Spain; maider.iturrondobeitia@ehu.eus (M.I.); julen.ibarretxe@ehu.eus (J.I.)

* Correspondence: gonzalo.gerrika@ehu.eus; Tel.: +34-943-01-5443

Received: 14 November 2019; Accepted: 10 December 2019; Published: 11 December 2019

Abstract: Bio-based polymeric nanocomposites (NCs) with enhanced electrical conductivity and rigidity were obtained by adding multi-walled carbon nanotubes (CNTs) to a commercial bio-based polyamide 4,10 (PA410). Two different types of commercial CNTs (Cheap Tubes and Nanocyl NC7000^{TM}) and two different preparation methods (using CNTs in powder form and a PA6-based masterbatch, respectively) were used to obtain melt-mixed PA410/CNT NCs. The effect of the preparation method as well as the degree of dispersion and aspect ratio of the CNTs on the electrical and mechanical properties of the processed NCs was studied. Superior electrical and mechanical behavior was observed in the Nanocyl CNTs-based NCs due to the enhanced dispersion and higher aspect ratio of the nanotubes. A much more significant reduction in aspect ratio was observed in the Cheap Tubes CNTs than in the Nanocyl CNTs. This was attributed to the fact that the shear stress applied during melt processing reduced the length of the CNTs to similar lengths in all cases, which pointed to the diameter of the CNTs as the key factor determing the properties of the NCs. The PA6 in the ternary PA410/PA6/CNT system led to improved Young's modulus values because the reinforcing effect of CNTs was greater in PA6 than in PA410.

Keywords: aspect ratio; carbon nanotube; dispersion; masterbatch; nanocomposite; polyamide

1. Introduction

The study of polymer nanocomposites (NCs), based on carbon nanotubes (CNTs), has attracted significant academic and industrial interest in recent years [1]. Due to their outstanding conductive properties, CNTs make it possible to obtain electrically-conductive thermoplastic/CNT NCs [2], with conductivity values ranging from 10^{-8} to 10^3 S/cm. Conductivity is achieved when the electrical percolation concentration (p_c) is reached—i.e., when a three-dimensional network of interconnected nanotubes is formed within the matrix. Furthermore, the addition of CNTs has been widely linked to improvements in the mechanical, thermal, and barrier properties of thermoplastic materials. Therefore, polymer/CNT nanocomposites would seem to be materials of great interest and can potentially be used in leading technological applications such as electrostatic dissipation (ESD) and electromagnetic interference shielding (EMI) [3].

Polyamides (PA) have been widely studied in the field of matrices. As they have excellent thermo-mechanical properties, these engineering polymers are widely used in the automotive, packaging, electrical, and electronics industries [4]. Moreover, polyamides from renewable resources (bio-PAs) have recently come onto the market [5–7], adding to the wide variety of commercial polyamides [6] already available. Indeed, there has been a growing demand in recent years to replace conventional petrochemical polymers with plastics derived from renewable resources. The production of bio-based polymers is expected to triple from 5.1 million tons in 2013 to 17 million tons in 2020 (Figure 1), and moderate growth is predicted in the bio-polyamides market [6]. The PA410 used in this study is similar to conventional technical polyamides such as PA6 or PA66 in terms of properties, but it is made from bio-based sebacic acid obtained from castor oil [7]. Therefore, its carbon content is approximately 70% renewable, reducing its carbon footprint to almost zero compared with other polyamides. While the butanediamine counterpart used in this study is derived from petroleum, it can also be derived from commercially available bio-based succinic acid or from the direct fermentation of sugars [5].

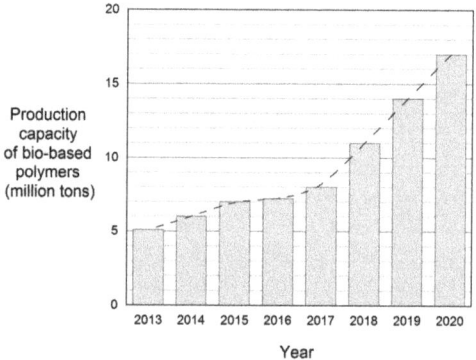

Figure 1. Evolution of the production capacity of bio-based polymers.

In the extensive literature on PA/CNT NCs, PA6 is the commercial polyamide most frequently used in the production of this type of nanocomposite [8–10]. However, other PAs, such as PA66 [11,12], PA12 [13], PA610 [14], PA1010 [15], and PA11 [16], and to a lesser extent PA46 [17] and PA1212 [18], have also been studied. It is worth mentioning that articles on bio-based PAs are rare [5,19,20] and studies on bio-PA/CNT NCs are practically non-existent.

Regarding the preparation methods of CNT-based nanocomposites, melt mixing is a well-established and cost-effective way of developing new polymeric materials with diverse properties as it is fast, simple, solvent-free, and does not require specific equipment [21]. Therefore, melt mixing is the preparation method of choice in the industry. It is also widely recognized that the orientation, dispersion, and aspect ratio, i.e., the main factors that define the end properties of CNT nanocomposites, are determined by the shear during melt processing [22]. Electrical measurements are commonly used to obtain information about the dispersion of CNTs and the formation of CNT networks, with higher conductivity values indicating well-formed conductive nanotube networks [22]. Accordingly, well-dispersed masterbatches (typically in the range of 10–20 wt% CNT content) are often used as they enable homogeneous dispersions to be obtained. However, a single nanotube parameter is not enough to describe the overall effect on electrical and mechanical properties. Several other parameters need to be considered, such as the length, dispersion, surface quality, and nature of the nanotubes (single, double, or multi-walled carbon nanotubes) [23]. Haggenmueller et al. [24] observed that the single-walled carbon nanotubes composites (SWCNTs) that maintained their length, but exhibited poor dispersion, showed enhanced elastic modulus and electrical conductivity. Finally, the functionalized

SWCNTs were better dispersed, but had shorter, more separated nanotubes (caused by the presence of functional groups) and limited mechanical and electrical properties.

The aspect ratios of CNTs play an important role in the percolation threshold for polymer/CNT composites. According to continuum percolation theory, for randomly oriented ideal monodisperse penetrable rods with aspect ratios much greater than one, the following equation can be used to predict the critical percolation (volume) concentration [25–27]:

$$\varnothing_p = \frac{D}{2L} \quad (1)$$

where $\varnothing_p$ is the volume percolation threshold, D is the diameter, and L is the length of the CNT.

This equation has been found to reasonably describe the experimental data for the electrical conductivity obtained from different multi-walled carbon nanotubes (MWCNTs), as reported by Castillo et al. [26]. In their work, the lowest percolation thresholds recorded for different polycarbonate/MWCNT NCs were attributed to the higher aspect ratios of the corresponding CNTs. Similarly, Guo et al. [25] reported that the electrical percolation threshold of their PC/MWCNT NCs decreased as the CNT aspect ratio increased. Caamaño el al. [27] found that, in their PA66/MWCNT NCs, the percolation thresholds of the larger diameter tubes increased as the L/D ratio decreased. Haggenmueller et al. [24] demonstrated that, at a given CNT concentration, the electrical conductivity of PA66/SWCNT NCs increased with the length of the nanotubes, showing that longer nanotubes are more likely to build percolating paths than shorter nanotubes at a fixed nanotube loading.

Mechanical properties are also affected by the aspect ratio of CNTs. Castillo et al. [26] reported that the storage modulus of PC/MWCNT NCs at 180 and 200 °C increased as the aspect ratio of the MWCNTs increased. Haggenmueller et al. [24] provided evidence that the elastic moduli of nanotube/nylon composites were enhanced at higher nanotube aspect ratios. Other studies have also indicated that the diameter and length, as well as the degree of functionality, of either single-walled or multi-walled CNTs, exert a significant influence on the electrical conductivity, and mechanical and thermal properties of polymer/CNT composites [23,24,27,28].

Therefore, effective mechanical reinforcement and improved electrical properties in CNT-polymer nanocomposites can be achieved with nanotubes with large aspect ratios and good dispersion levels [24]. In the present work, three different commercial bio-PA410/CNT-based systems were prepared and characterized using different CNTs (Cheap Tubes and NC7000™) and different preparation methods (involving CNTs in powder form and a PA6-based CNT masterbatch). Two of the systems shared the same preparation method (CNTs in powder form was melt-blended with the PA410), which allowed for the influence of the different natures of the CNTs to be studied, while the other two shared the same CNTs (NC7000™), which enabled the influence of the preparation method to be examined. Thus, an exhaustive characterization of the aspect ratio of the CNTs employed, comparing the before and after-processing situations, allowed us to evaluate the effect of this parameter on the final properties of the NCs. Furthermore, using a masterbatch is more innovative than using CNTs in powder form because, firstly, masterbatches promote homogeneous CNT dispersion, and secondly, as they are readily available commercially, so the need to use harmful CNT powder is removed from the process, a highly attractive prospect for the polymer industry. This is why the effect of the preparation method on the electrical and mechanical properties of the processed NCs was also studied.

2. Materials and Methods

2.1. Materials

The PA410 used in this work was EcoPaXX® Q150-D, kindly provided by DSM (Genk, Belgium). Two kinds of commercially available pristine MWCNTs were used: Cheap Tubes (20–30 nm) (Grafton, VT, USA) and Nanocyl NC7000™ (Sambreville, Belgium). Their properties are summarized in Table 1. Additionally, a PA6-based MWCNT masterbatch—also manufactured by Nanocyl (Sambreville,

Belgium) and commercialized as Plasticyl™ PA 1503—containing 15 wt % of the aforementioned NC7000™ was also used to obtain the PA410-based NCs.

Table 1. Properties of the multi-walled carbon nanotubes (MWCNTs) used, according to their respective data sheets.

Sample	Diameter (nm)	Length (μm)	Estimated Aspect Ratio (L/D)	Surface Area (m^2/g)	Carbon Purity (%)
Cheap Tubes [1]	20–30	10–30	333–1500	110	>95
Nanocyl [2]	9.5	1.5	158	250–300	90

[1] Cheap Tubes. Description Multi Walled Carbon Nanotubes 20–30 nm; [2] Nanocyl. Technical Data Sheet: NC7000™, 12th July 2016, V08.

2.2. Composite Processing

All the composites were prepared using the same processing techniques and conditions described in a previous study [29]—it is available as an open access article in this journal. In order to prevent moisture-induced degradation reactions during processing, the PA410 and Plasticyl™ PA 1503 masterbatch were dried in a dry air dehumidifier (Wittmann Drymax, Kottingbrunn, Austria) for 48–72 h at 80 °C. PA410/PA6/CNT NCs were obtained by melt-mixing, using a Collin ZK25 co-rotating twin screw extruder-kneader (Ebersberg, Baviera, Germany) at 270 °C with a screw rotation speed of 200 rpm and diluting the masterbatch with the correct amount of neat PA410. The diameter and length-to-diameter ratios of the screws were 25 mm and 30, respectively. Directly melt-mixed PA410/CNT binary NCs were also obtained under the same conditions, using Cheap Tubes and NC7000™ CNTs. Table 2 shows the wt% content of each component for all the compositions studied.

Table 2. PA410, PA6, and carbon nanotubes (CNT) wt % content in the compositions studied.

	Nomenclature	PA410 (wt %)	PA6 (wt %)	Cheap Tubes (wt %)	Nanocyl NC7000™ (wt %)
PA410/CNT(1) binary NCs	100/0	100	-	0	-
	99/1	99	-	1	-
	98/2	98	-	2	-
	97/3	97	-	3	-
	96/4	96	-	4	-
	95/5	95	-	5	-
	94/6	94	-	6	-
PA410/CNT(2) binary NCs	99.5/0.5	99.5	-	-	0.5
	99/1	99	-	-	1
	98/2	98	-	-	2
	97/3	97	-	-	3
	96/4	96	-	-	4
	95/5	95	-	-	5
	94/6	94	-	-	6
PA410/PA6/CNT ternary NCs	96.6/2.9/0.5	96.6	2.9	-	0.5
	93/6/1	93	6	-	1
	87/11/2	87	11	-	2
	80/17/3	80	17	-	3
	73/23/4	73	23	-	4

As described previously [29], the extrudates were cooled in a water bath, pelletized, and dried again. Subsequent injection molding of dried pellets was carried out in a Battenfeld PLUS 350/75 reciprocating screw injection molding machine (Kottingbrunn, Austria) with a press closing force of 350 kN to obtain tensile (ASTM D-638, type IV, thickness 2 mm) specimens. The diameter of the screw in the plasticizing unit was 25 mm and the L/D ratio was 14. The melt and mold temperatures were 270 and 85 °C, respectively. The injection speed, pressure-holding, and cooling times were

42 cm^3/s, 3 s, and 15 s, respectively. All specimens were kept in a desiccator to prevent post-processing humidity absorption.

Standard circular sheets for electrical conductivity measurements (diameter and thickness: 70 mm and 1 mm, respectively) were obtained by hot pressing in a Collin P200E hydraulic press (Ebersberg, Baviera, Germany). The molding process was carried out at a temperature of 270 °C and a pressure of 130 bar in three stages: Preheating or plasticizing (closure without pressure, 2 min), compression (closure under pressure, 3 min), and cooling under pressure (6 min). Therefore, the only property of the materials in the study that was not measured using injection molded tensile specimens was the conductivity. While injection molding and hot pressing techniques can be expected to produce different CNT dispersion levels, the methodology is still valid when the aim of the measurements is to compare the electrical conductivity of the materials developed using different CNTs and preparation methods.

2.3. Phase Structure

The phase behavior was studied by dynamic mechanical analysis (DMA) using a TA Q800 viscoelastometer (New Castle, DE, USA) that provided the loss tangent (tanδ) against temperature. The scans were carried out in single cantilever bending mode at a constant heating rate of 4 °C/min and a frequency of 1 Hz, from −100 to 150 °C. The melting and crystallization behavior of the materials was studied by DSC using a Perkin–Elmer DSC-7 calorimeter (Waltham, MA, USA), which was calibrated using an indium standard as a reference. The samples were first heated from 30 to 300 °C at 20 °C/min and then cooled at the same rate. The melting and crystallization temperatures (T_m, T_c) were determined, respectively, from the maxima of the corresponding peaks during the heating and cooling scans, and the melting and crystallization enthalpies were determined from the areas of each of these peaks. The degree of crystallization of PA410 was calculated from the melting and cold crystallization enthalpies, taking the enthalpy of a 100% crystalline (ΔH_f^∞) PA410 as 269 J/g [5].

2.4. Morphology

The nanostructure was analyzed by transmission electron microscopy (TEM). The samples were obtained from injection-molded specimens and ultrathin-sectioned at ~100 nm using a Leica EMFC 6 ultramicrotome (Wetzlar, Germany) equipped with a diamond knife. The micrographs were obtained in a Tecnai G2 20 twin apparatus (FEI, Waltham, MA, USA), operating at an accelerating voltage of 200 kV.

2.5. Characterization of Nanotube Diameter and Length Distributions

The nanotube length and diameter distribution in the composites after processing was analyzed in the three systems under study with 2 wt% CNT content (98/2 in the case of the binary PA410/CNT NCs and 87/11/2 in the case of the ternary PA410/PA6/CNT NCs). The CNTs were extracted as described by Krause et al. [30]. The samples for this characterization were pellets, as received from the extruder after cooling. They were dissolved in formic acid at room temperature for 1 h and the dispersions were treated afterwards for 3 min in an ultrasonic bath. The nanotube concentration in the dispersions was 0.1 g/l.

For TEM observations, a drop of the newly prepared dispersion was placed onto a glow-discharged, carbon-coated, TEM copper grid (300 mesh) and dried at ambient temperature. TEM micrographs were used for the quantitative characterization of the geometry (length and diameter) of the CNTs. Image analysis was carried out using the free Fiji software [31]. For the characterization of the lengths, TEM micrographs with a pixel size of 2.1 nm/pixel were used. For each sample, 5 TEM micrographs (the number of quantified nanotubes ranged from 223 to 1281, depending on the sample) were analyzed following the procedure described here:

1. The background of the micrographs was homogenized by subtracting the background and applying Gaussian filters.

2. The micrograph was binarized (the objects of interest, the CNTs, were separated from the background). A thresholding method was used for the binarization (Figure 2).
3. Artifacts were removed based on the particle size (Figure 2).
4. Morphological transformations were carried out to improve the quality of the binarized particles.
5. The particles were skeletonized.
6. A quantitative analysis of the skeleton was performed by computing the skeleton length for each particle.

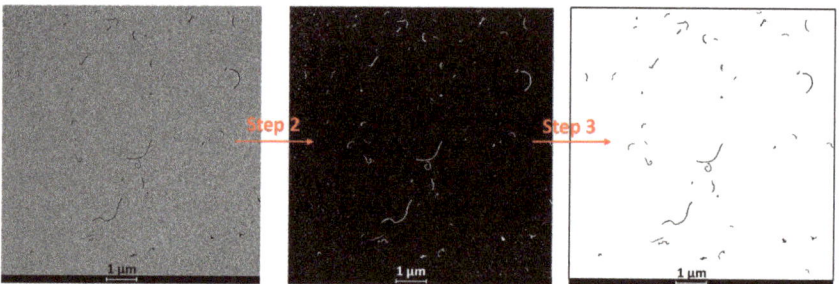

Figure 2. Graphical diagram of Steps 2 and 3 of the image analysis procedure.

The thickness of the nanotubes was measured using higher magnification micrographs with a pixel size of 0.75 nm/pixel (Figure 3). For each sample, 60–70 measurements were performed, using a total of 5 micrographs per sample. Due to their irregular shape, 3–4 measurements were carried out on each nanotube. These measurements were made manually on raw TEM micrographs.

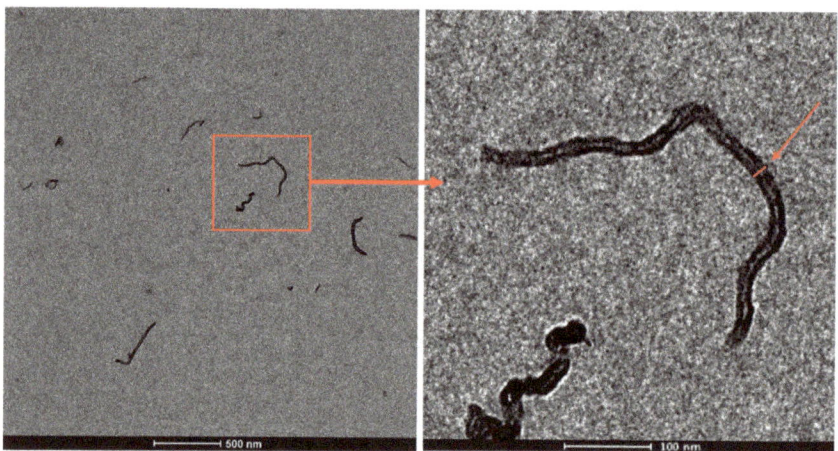

Figure 3. Procedure for nanotube diameter measurements.

2.6. Mechanical Properties

The tensile tests were carried out in an Instron 5569 tensile tester (Instron, Norwood, MA, USA). An extensometer at a crosshead speed of 1 mm/min was used to measure the Young's modulus. Yield stress and ductility, measured as the break strain (ε_b), were determined from the load-displacement curves at a crosshead speed of 10 mm/min. A minimum of five tensile specimens were tested for each reported value.

2.7. Electrical Properties

The electrical resistivity was determined according to the ASTM D4496-87 standard. Volume resistances were measured and converted to conductivity values. Measurements were taken at 1 V using a Keithley 6487 picoammeter (Cleveland, OH, USA) and a Keithley 8009 Resistivity Test Fixture. Three measurements were performed for each reported value.

3. Results and Discussion

3.1. Phase Structure

As previously reported [29], the PA410/CNT(1) NCs showed slight T_g increases when compared with the neat PA410, due to both the characteristic hindering effect of the CNTs in the movement of polymeric chains in the amorphous phase and the reduction of the free volume. With respect to the crystalline phase, the CNTs did not affect the melting behavior of PA410, and exerted a nucleating effect during cooling, enhancing the overall non-isothermal crystallization rate of the PA410 from the melt. The PA410/CNT(2) NCs, which were not studied in the previous work, showed very similar phase behavior to that of the PA410/CNT(1) NCs (slight increases in T_g, no effect on the melting behavior and a nucleating effect during cooling).

As described in the previous work [29], since the T_g observed for the ternary PA410/PA6/CNT NCs was similar to that of neat PA410 – caused by the CNTs and PA6 offsetting each other, it can be concluded that these ternary NCs are completely miscible in the amorphous phase. With respect to the crystalline phase, a progressive decrease in T_m was observed at increasing CNT contents, due to the increase in the PA6 contents. During cooling, the nucleating effect of the CNTs was still evident despite the drop in T_c caused by the presence of PA6 in the PA410 matrix. A fully detailed explanation of this result can be found in our previous work [29].

3.2. Characterization of Nanotube Diameter and Length Distributions

Figures 4 and 5 show, respectively, the length and diameter distributions of the CNTs in the three systems, recovered from the melt-extruded NCs after processing, measured from TEM micrographs and following the image analysis procedure described in the experimental part. Table 3 summarizes the experimentally measured average length and diameter values of the different CNTs and the calculated aspect ratio (L/D).

As can be inferred from Figure 4 and seen in Table 3, the average length of the nanotubes in the three systems studied was similar. This result is particularly relevant considering that, before processing, the Cheap Tubes CNTs in the PA410/CNT(1) system were 7–20 times longer (10–30 μm, Table 1) than the Nanocyl CNTs in the PA410/CNT(2) and PA410/PA6/CNT systems (1.5 μm, Table 1). It has previously been demonstrated in the literature that melt processing CNT-based NCs causes the nanotubes to break [22,32] due to the characteristic shear stress in this method of processing. Moreover, as the three systems in the study were obtained under the same processing conditions and therefore subjected to the same shear stress, this result suggests that a certain shear stress reduces the average length of all the CNTs to a similar one, regardless of their initial length. Similar results have previously been obtained in PC-based NCs [25].

With respect to the diameter of the nanotubes, as Figure 5 suggests and Table 3 shows, the values measured by TEM were consistent with those reported in the data sheets (Table 1). Thus, it can be stated that, as expected and unlike its effect on the length, the shear stress applied during melt processing barely affects the diameter of the CNTs.

The aspect ratio of the nanotubes of the three systems, calculated from the length and diameter measurements, is shown in Table 3. As can be seen, the values for the two systems containing Nanocyl CNTs are similar because the length of the CNTs is similar and the diameter is practically identical. However, the aspect ratio of the system containing the Cheap Tubes CNTs is significantly lower than that of the Nanocyl CNT systems, because the length of the Cheap Tubes nanotubes is similar but the diameter is greater. When the experimental L/D values are compared with the values on the data sheets (Table 1), the decrease in the three systems is very significant, but much more so in the PA410/CNT(1) system (95%–99%) than in the systems containing Nanocyl CNTs (82%–84%). Therefore, the high aspect ratio of the Cheap Tubes CNTs before processing was compared with the Nanocyl CNTs drops after processing.

It has been widely reported in the literature and will be demonstrated later in this work that the aspect ratio of carbon nanotubes affects the mechanical and electrical properties of NCs. Many previous studies have also shown that higher aspect ratios result in lower electrical percolation thresholds [25,26], as well as improved mechanical performance.

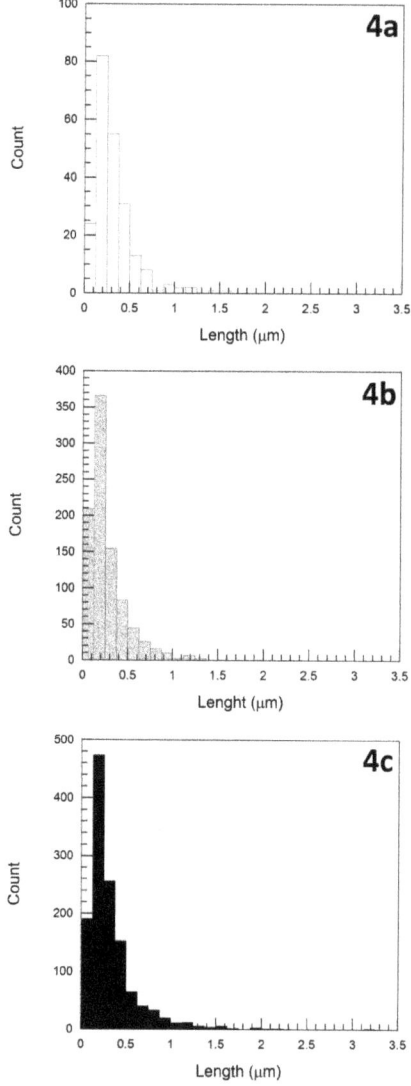

Figure 4. Nanotube length distributions after processing, as recovered from melt extruded PA410/CNT(1) 98/2 (**a**), PA410/CNT(2) 98/2 (**b**), and PA410/PA6/CNT 87/11/2 (**c**) NCs. Total number of particles: 223, 925, and 1281, respectively.

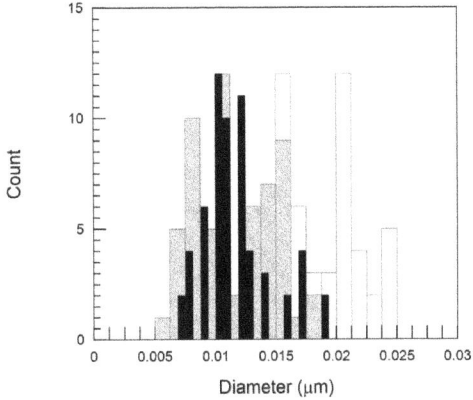

Figure 5. Comparison of diameter distributions of Cheap Tubes (white bars), Nanocyl NC7000™ (grey bars) and Plasticyl™ PA 1503 masterbatch (black bars) after processing, as recovered from melt extruded PA410/CNT(1) 98/2, PA410/CNT(2) 98/2, and PA410/PA6/CNT 87/11/2 NCs. Total number of particles: 65, 60, and 62, respectively (with 3–4 measurements for each nanotube).

Table 3. Average values of the diameter and length, measured from transmission electron microscopy (TEM) micrographs, and estimated aspect ratio of the CNTs after processing the NCs.

NC	Diameter (nm)	Length (μm)	Aspect ratio (L/D)
PA410/CNT(1)	18	0.32	18
PA410/CNT(2)	11	0.27	25
PA410/PA6/CNT	12	0.33	28

3.3. Morphology

Figure 6 shows the TEM micrographs of the PA410/CNT(1), PA410/CNT(2), and PA410/PA6/CNT NCs at a CNT concentration of 1 wt%. As can be seen, the nanotube dispersion levels are qualitatively good in the three NCs under study. Moreover, while difficult to appreciate in the images in Figure 6, a systematic observation of all the TEM micrographs revealed that the nanotube dispersion level of the systems that contained Nanocyl CNTs, i.e., PA410/CNT(2) and PA410/PA6/CNT, was better than the system with the Cheap Tubes CNTs (PA410/CNT(1)). In the former, the nanotubes appeared uncurled and individually dispersed, while in the latter, they were more curled, and there were more small aggregates.

As mentioned in the previous section, the nanotube aspect ratio was significantly higher in the Nanocyl CNTs-based NCs than in the Cheap Tubes CNTs-based ones, due to the similar length but smaller diameter of the nanotubes. Moreover, the density of the pristine Nanocyl CNTs was also lower than that of the Cheap Tubes CNTs. Therefore, given a certain nanofiller wt% content, there were more nanotubes per volume unit in the Nanocyl CNTs-based NCs than in the Cheap Tubes CNTs-based ones.

In conclusion, the dramatic decrease in length—and consequently, in the aspect ratio—of the Cheap Tubes gave rise to lower after-processing aspect ratio values than those of both Nanocyl-based systems. This, along with the greater number and better dispersion of the nanotubes in PA410/CNT(2) and PA410/PA6/CNT NCs, influences both the electrical and mechanical properties of the resulting materials, as will be seen in the following sections.

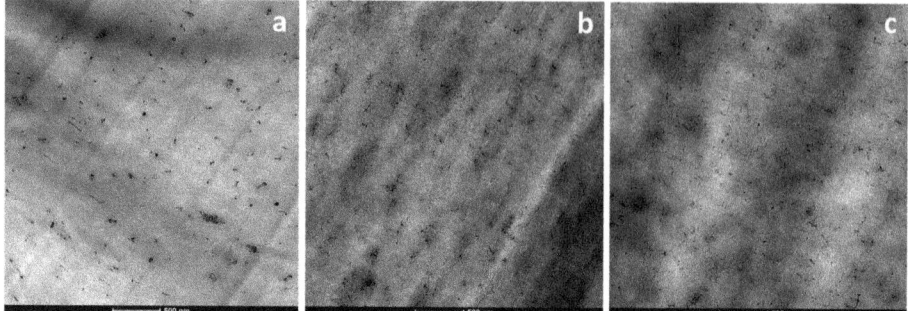

Figure 6. TEM micrographs of the PA410/CNT(1) 99/1 (**a**), PA410/CNT(2) 99/1 (**b**) and PA410/PA6/CNT 93/6/1 (**c**) NCs at x14500 magnification.

3.4. Electrical Properties

Figure 7 shows the electrical conductivity of the PA410/CNT(1), PA410/CNT(2), and PA410/PA6/CNT NCs vs. the CNT content. As mentioned in the experimental section, the specimens for the electrical conductivity measurements were obtained by hot pressing, rather than by injection molding, which was used for the characterization of the other properties in the study. Therefore, as discussed in a previous work [29], different CNT dispersion levels, orientations, and/or even aspect ratios to the injection-molded samples were to be expected. However, given that the samples used for taking the electrical conductivity measurements were all processed in the same way (extrusion followed by compression molding) and under the same conditions, the conductivity results are still valid for the purposes of comparison.

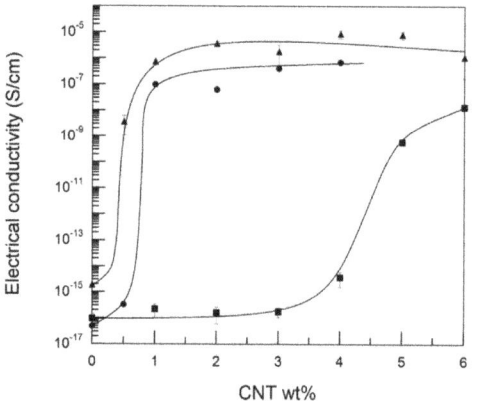

Figure 7. Electrical conductivity of the PA410/CNT(1) NCs (■), the PA410/CNT(2) NCs (▲), and the PA410/PA6/CNT NCs (●). When not visible, the error bars are smaller than the symbols.

As can be seen in Figure 7, the electrical conductivity increased by over seven decades in the three systems studied, indicating that electrical percolation of the carbon nanofiller occurred. The maximum conductivity values obtained were similar in the PA410/CNT(2) and in the PA410/PA6/CNT systems, but were higher than the figures for the PA410/CNT(1) system. The electrical percolation threshold was calculated from the experimental data using a simple power law Equation (2) [33]:

$$\sigma(p) = B(p - p_c)^t \qquad (2)$$

where $\sigma(p)$ is the experimental electrical conductivity, B is the proportionality constant, t is the critical exponent, p is the CNT concentration, and p_c is the percolation concentration; where p is always higher than p_c ($p > p_c$). The results obtained are stated in Table 4. As can be inferred from Figure 7 and shown in Table 4, the electrical percolation of the carbon nanofiller occurred at similar, significantly lower values in the two Nanocyl CNTs-based NCs than in the Cheap Tubes CNTs-based ones.

The three main factors that affect the electrical behavior of a CNT-reinforced polymeric system are: (1) Orientation, (2) geometry, usually characterized by the aspect ratio, and (3) dispersion level, which encompasses the number of individual CNTs, the aggregates, and the curled or uncurled geometry. The first factor can be excluded because TEM observations clearly show randomly oriented CNTs in the three systems of the study. With respect to the other two, it was previously demonstrated that the aspect ratio and the dispersion level were similar in the PA410/CNT(2) and PA410/PA6/CNT NCs, but higher and better, respectively, than in the PA410/CNT(1) NCs. Thus, the superior electrical performance of the Nanocyl CNT-based NCs compared with the Cheap Tubes CNT-based ones can be attributed to their good dispersion level and higher aspect ratio.

Table 4. Experimental percolation thresholds of the NCs.

NC.	Percolation Threshold (wt%)
PA410/CNT(1)	3.98
PA410/CNT(2)	0.50
PA410/PA6/CNT	0.65

When these experimental percolation concentrations are compared with the theoretical values obtained using Equation (1), calculated using the L and D values provided in the data sheets, it can be seen that the experimental percolation thresholds (Table 4) are much higher than the theoretical values (0.06–0.29 wt%) in the PA410/CNT(1) NCs, and similar or slightly higher than the latter results (0.51 wt%) in the PA410/CNT(2) and PA410/PA6/CNT NCs. The difference in values can thus be attributed to the reduction in the aspect ratio of the different CNTs during the processing referred to earlier. Consistent with the aforementioned data regarding the reduction in the aspect ratio, the maximum difference between the experimental and theoretical percolation values was observed in the PA410/CNT(1) NCs with the Cheap Tubes CNTs.

However, when the theoretical p_c values from Equation (1) were calculated using the experimentally measured L and D values, i.e., after processing, the results were higher (5,17, 3,31, and 3,29 wt%, respectively, for PA410/CNT(1), PA410/CNT(2), and PA410/PA6/CNT NCs) than the experimental p_c-s shown in Table 4. Given that Equation (1) assumes randomly oriented ideal monodisperse penetrable rods, the lower experimental values obtained in this work are unexpected and point to the experimental methodology used as underestimating the real L/D ratios of the CNTs. It is notable that Equation (1) is consistent with experimentally measured p_c values in certain polymer/CNT systems [25,26], however, as calculations for these systems were performed with the initial L/D values of the CNTs, the effect of the melt-mixing processing on the aspect ratio was not considered in any of these studies.

Finally, it is notable that careful selection of the CNTs brought about some of the lowest p_c values reported in the literature for melt-processed PA-based NCs, with obtained values ranging from 0.4% [34] to 6% [35], depending on the PA, processing method, and conditions used. As previously demonstrated, given that the shear stress applied during processing reduced the initial dissimilar lengths of the CNTs to similar lengths, the initial diameter of the CNTs, which was unchanged by processing, plays a key role in the final aspect ratio of the CNTs, and therefore in the electrical behavior of the NCs.

3.5. Mechanical Properties

Table 5 shows the Young's modulus, tensile strength, and ductility values of the PA410/CNT(1), PA410/CNT(2), and PA410/PA6/CNT NCs. Young's modulus vs. the CNT content is also shown in

Figure 8. As expected in well-dispersed CNT-filled NCs, the stiffness increased at increasing CNT contents in all the NC systems. The overall increase in Young's modulus is caused by a decrease in the molecular mobility of the polymeric chains, which is also supported by the T_g increase shown by DMTA. The large interfacial area-to-dispersed phase volume ratio characteristic of well-dispersed CNTs facilitates the presence of interactions between the polymer and the filler [36]. When the three systems were compared, the maximum modulus increase was observed in the PA410/PA6/CNT system, where a 10.3% increase was observed with a 4 wt% CNT content, followed by the PA410/CNT(2) system (an increase of 11% with a 6 wt% CNT content), and the PA410/CNT(1) system, where a 4.5% increase with a 6 wt% CNT content was observed. Therefore, the reinforcing efficiency of the Nanocyl CNTs is clearly greater than that of the Cheap Tubes CNTs, and can be attributed to the higher aspect ratio and better dispersion rate of the former, as previously discussed.

Table 5. Young's modulus, tensile strength and ductility values of the PA410/CNT(1), PA410/CNT(2), and PA410/PA6/CNT NCs.

	Composition	Young's Modulus (MPa)	Tensile Strength (MPa)	Strain at Break (%)
PA410/CNT(1) NCs	100/0	2880 ± 30	75.5 ± 0.4	100 ± 30
	99/1	2850 ± 100	74.3 ± 1.2	7 ± 4
	98/2	2860 ± 60	75.6 ± 0.2	7 ± 1
	97/3	2870 ± 40	74.3 ± 1.2	5 ± 1
	96/4	2930 ± 40	75.1 ± 0.7	7 ± 1
	95/5	3010 ± 10	74.5 ± 0.8	5 ± 1
	94/6	3010 ± 50	76.3 ± 0.5	6 ± 0
PA410/CNT(2) NCs	99.5/0.5	2630 ± 10	75.8 ± 0.3	14 ± 6
	99/1	2680 ± 20	76.5 ± 3.4	5 ± 2
	98/2	2730± 30	79.2 ± 0.4	6 ± 4
	97/3	2880 ± 20	83.2 ± 0.4	9 ± 5
	96/4	2950 ± 60	82.6 ± 1.8	7 ± 4
	95/5	3010 ± 20	85.8 ± 0.9	8 ± 1
	94/6	3160 ± 40	84.5 ± 1.0	4 ± 1
PA410/PA6/CNT NCs	96.6/2.9/0.5	2930 ± 90	74.3 ± 0.3	8 ± 4
	93/6/1	2850 ± 40	73.5 ± 0.5	5 ± 1
	87/11/2	2930 ± 40	73.8 ± 1.0	6 ± 2
	80/17/3	3040 ± 80	71.8 ± 3.6	4 ± 1
	73/23/4	3180 ± 70	74.1 ± 2.0	3 ± 0

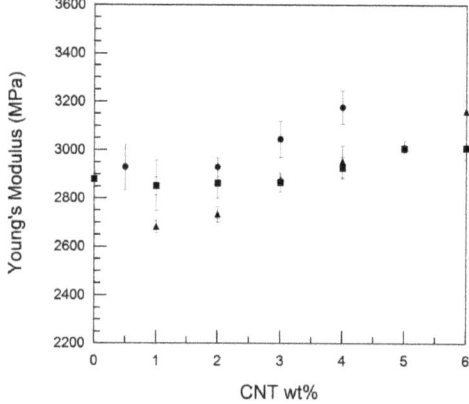

Figure 8. Young's modulus of the PA410/CNT(1) NCs (■), PA410/CNT(2) NCs (▲), and PA410/PA6/CNT NCs(●).

When comparing the two Nanocyl CNTs-based NCs, the superior mechanical performance of the PA410/PA6/CNT NCs is particularly noteworthy, considering that the Young's modulus of PA6 ($\approx$2400 MPa) is significantly lower than that of neat PA410 ($\approx$2900 MPa). As mentioned before regarding conductivity, the most important factors affecting the mechanical behavior of CNT-filled NCs are orientation, aspect ratio, and dispersion level. However, these parameters were found to be similar in both the Nanocyl CNTs-based NCs in this study. In fact, the only difference between the two systems is the presence of a certain amount of PA6 in the PA410/PA6/CNT NCs. In the literature, the reinforcing capacity of the Nanocyl CNTs was better in a PA6 matrix [36] than in the PA410 matrix of the present work. This is because the PA6/CNT masterbatch used here was used in a pure PA6 matrix, and a 36% increase in modulus was reported after 5% CNTs were added. In the present work, the addition of the same CNT content to PA410 led to a modulus increase of only 4.5%. As already explained in a previous work [29], the PA410/PA6/CNT NCs studied in the present work were prepared by diluting the aforementioned PA6/CNT masterbatch with the appropriate amounts of pure PA410, while the PA410/CNT(2) NCs were prepared by direct melt-mixing pristine Nanocyl CNTs and PA410. Consequently, the presence of a certain amount of PA6 in the ternary system produces a synergistic reinforcing effect of the CNTs, which, in turn, produces higher Young's modulus values in PA410/PA6/CNT NCs.

As can be seen in Table 5, the behavior of the yield stress was similar to that of Young's modulus in the binary PA410/CNT(1) and PA410/CNT(2) NCs, showing slight and moderate increases as the CNT content increased. Similar behavior has been observed in other CNT-filled NCs [9,13,37], although the local nature of the yielding process—unlike the Young's modulus where the whole section of the specimen is involved—often leads to constant [8,38] or even decreasing [8,39] values at increasing CNT contents. This seems to be the case of the ternary PA410/PA6/CNT system, where the presence of PA6, with its lower yield stress ($\approx$66 MPa), coupled with the local nature of the yielding process lead to values similar to that of neat PA410.

The ductility of all the NC systems in the study decreased sharply even at the lowest CNT contents, as expected in CNT-reinforced NCs. This decrease is attributed to restrictions in the mobility of the matrix chains caused by the single CNTs, which promote fracture, and to the intrinsic crack-sensitive nature of PAs [40–42]. No relevant differences were observed among them, and all the filled compositions broke just after yielding.

4. Conclusions

Bio-based polymeric NCs with enhanced electrical conductivity and rigidity were obtained by adding MWCNTs to PA410. Superior electrical and mechanical behavior was observed in the Nanocyl CNTs-based NCs due to the improved dispersion and higher aspect ratio of the nanotubes. A much more significant reduction in aspect ratio was observed in the Cheap Tubes CNTs compared to the Nanocyl CNTs. This is attributed to the fact that the shear applied during melt processing reduced the length of the CNTs to a similar value in all cases, pointing to the diameter of the CNTs as the key factor determining the properties of the NCs. The presence of PA6 in the ternary PA410/PA6/CNT system led to improved Young's modulus values because the CNTs had greater reinforcing ability in PA6 than in PA410.

Author Contributions: Investigation, I.O., N.A., M.I., J.I., G.G.-E.; Writing—original draft, I.O., N.A., M.I., J.I., G.G.-E; Writing—review & editing, I.O., N.A., M.I., J.I., G.G.-E.

Funding: This research was funded by the Basque Government through project IT309-19 and through the grant awarded to I. Otaegi.

Acknowledgments: The authors are grateful for the technical and human support provided by SGIker (UPV/EHU/ ERDF, EU).

Conflicts of Interest: The authors declare no conflict of interest.

References

1. Grady, B.P. *Carbon Nanotube-Polymer Composites: Manufacture, Properties and Applications*; American Chemical Society (ACS): Washington, DC, USA; John Wiley & Sons: Hoboken, NJ, USA, 2011; p. 352. [CrossRef]
2. Alig, I.; Pötschke, P.; Lellinger, D.; Skipa, T.; Pegel, S.; Kasaliwal, G.R.; Villmow, T. Establishment, morphology and properties of carbon nanotube networks in polymer melts. *Polymer* **2012**, *53*, 4–28. [CrossRef]
3. Chen, J.; Yan, L. Effect of carbon nanotube aspect ratio on the thermal and electrical properties of epoxy nanocomposites. *Fuller. Nanotub. Carb. Nanostruct.* **2018**, *26*, 697–704. [CrossRef]
4. Arboleda-Clemente, L.; Ares-Pernas, A.; Garcia, X.; Dopico, S.; Abad, M.J. Influence of polyamide ratio on the CNT dispersion in polyamide 66/6 blends by dilution of PA66 or PA6-MWCNT masterbatches. *Synth. Met.* **2016**, *221*, 134–141. [CrossRef]
5. Moran, C.S.; Barthelon, A.; Pearsall, A.; Mittal, V.; Dorgan, J.R. Biorenewable blends of polyamide-4, 10 and polyamide-6, 10. *J. Appl. Polym. Sci.* **2016**, *133*. [CrossRef]
6. Aeschelmann, F.; Carus, M. 2nd edition: Bio-based Building Blocks and Polymers in the World: Capacities, Production, and Applications –Status Quo and Trends Towards 2020. *Ind. Biotechnol.* **2015**, *11*, 154–159. [CrossRef]
7. Niaounakis, M. Introduction to Biopolymers. In *Biopolymers Reuse, Recycling, and Disposal*; Niaounakis, M., Ed.; William Andrew Publishing: Oxford, UK, 2013; pp. 1–75.
8. Meng, H.; Sui, G.X.; Fang, P.F.; Yang, R. Effects of acid- and diamine-modified MWNTs on the mechanical properties and crystallization behavior of polyamide 6. *Polymer* **2008**, *49*, 610–620. [CrossRef]
9. Mahmood, N.; Islam, M.; Hameed, A.; Saeed, S.; Khan, A.N. Polyamide-6-based composites reinforced with pristine or functionalized multi-walled carbon nanotubes produced using melt extrusion technique. *J. Compos. Mater.* **2014**, *48*, 1197–1207. [CrossRef]
10. Li, J.; Ke, C.; Fang, K.; Fan, X.; Guo, Z.; Fang, Z. Crystallization and Rheological Behaviors of Amino-functionalized Multiwalled Carbon Nanotubes Filled Polyamide 6 Composites. *J. Macromol. Sci. Part B Phys.* **2010**, *49*, 405–418. [CrossRef]
11. Qiu, L.; Chen, Y.; Yang, Y.; Xu, L.; Liu, X. A study of surface modifications of carbon nanotubes on the properties of polyamide 66/multiwalled carbon nanotube composites. *J. Nanomater.* **2013**, *2013*, 252417. [CrossRef]
12. Lin, S.-Y.; Chen, E.-C.; Liu, K.-Y.; Wu, T.-M. Isothermal crystallization behavior of polyamide 6, 6/multiwalled carbon nanotube nanocomposites. *Polym. Eng. Sci.* **2009**, *49*, 2447–2453. [CrossRef]
13. Chatterjee, S.; Nuesch, F.A.; Chu, B.T.T. Comparing carbon nanotubes and graphene nanoplatelets as reinforcements in polyamide 12 composites. *Nanotechnology* **2011**, *22*, 275714. [CrossRef] [PubMed]
14. Jiang, J.; Zhang, D.; Zhang, Y.; Zhang, K.; Wu, G. Influences of Carbon Nanotube Networking on the Conductive, Crystallization, and Thermal Expansion Behaviors of PA610-Based Nanocomposites. *J. Macromol. Sci. Part B Phys.* **2013**, *52*, 910–923. [CrossRef]
15. Wang, B.; Sun, G.; Liu, J.; He, X.; Li, J. Crystallization behavior of carbon nanotubes-filled polyamide 1010. *J. Appl. Polym. Sci.* **2006**, *100*, 3794–3800. [CrossRef]
16. Huang, Y.; Tan, L.; Zheng, S.; Liu, Z.; Feng, J.; Yang, M.-B. Enhanced dielectric properties of polyamide 11/multi-walled carbon nanotubes composites. *J. Appl. Polym. Sci.* **2015**, *132*, 42642. [CrossRef]
17. Chiu, F.-C.; Kao, G.-F. Polyamide 46/multi-walled carbon nanotube nanocomposites with enhanced thermal, electrical, and mechanical properties. *Compos. Part A* **2012**, *43*, 208–218. [CrossRef]
18. Sun, X.; Yuan, X.; Zhang, X.; Liu, W.; Zhou, Q. Crystal structure study of nylon1212/carbon nanotube composites under tension force. *Qingdao Keji Daxue Xuebao Ziran Kexueban* **2006**, *27*, 423–426.
19. Leszczynska, A.; Kicilinski, P.; Pielichowski, K. Biocomposites of polyamide 4, 10 and surface modified microfibrillated cellulose (MFC): Influence of processing parameters on structure and thermomechanical properties. *Cellulose* **2015**, *22*, 2551–2569. [CrossRef]
20. Pagacz, J.; Raftopoulos, K.N.; Leszczynska, A.; Pielichowski, K. Bio-polyamides based on renewable raw materials–glass transition and crystallinity studies. *J. Therm. Anal. Calorim.* **2016**, *123*, 1225–1237. [CrossRef]
21. Breuer, O.; Sundararaj, U. Big returns from small fibers: A review of polymer/carbon nanotube composites. *Polym. Compos.* **2004**, *25*, 630–645. [CrossRef]

22. Krause, B.; Potschke, P.; Haeussler, L. Influence of small scale melt mixing conditions on electrical resistivity of carbon nanotube-polyamide composites. *Compos. Sci. Technol.* **2009**, *69*, 1505–1515. [CrossRef]
23. Krause, B.; Ritschel, M.; Taeschner, C.; Oswald, S.; Gruner, W.; Leonhardt, A.; Poetschke, P. Comparison of nanotubes produced by fixed bed and aerosol-CVD methods and their electrical percolation behaviour in melt mixed polyamide 6.6 composites. *Compos. Sci. Technol.* **2009**, *70*, 151–160. [CrossRef]
24. Haggenmueller, R.; Du, F.; Fischer, J.E.; Winey, K.I. Interfacial in situ polymerization of single wall carbon nanotube/nylon 6, 6 nanocomposites. *Polymer* **2006**, *47*, 2381–2388. [CrossRef]
25. Guo, J.; Liu, Y.; Prada-Silvy, R.; Tan, Y.; Azad, S.; Krause, B.; Poetschke, P.; Grady, B.P. Aspect ratio effects of multi-walled carbon nanotubes on electrical, mechanical, and thermal properties of polycarbonate/MWCNT composites. *J. Polym. Sci. Part B Polym. Phys.* **2014**, *52*, 73–83. [CrossRef]
26. Castillo, F.Y.; Socher, R.; Krause, B.; Headrick, R.; Grady, B.P.; Prada-Silvy, R.; Poetschke, P. Electrical, mechanical, and glass transition behavior of polycarbonate-based nanocomposites with different multi-walled carbon nanotubes. *Polymer* **2011**, *52*, 3835–3845. [CrossRef]
27. Caamano, C.; Grady, B.; Resasco, D.E. Influence of nanotube characteristics on electrical and thermal properties of MWCNT/polyamide 6, 6 composites prepared by melt mixing. *Carbon* **2012**, *50*, 3694–3707. [CrossRef]
28. Bose, S.; Bhattacharyya, A.R.; Bondre, A.P.; Kulkarni, A.R.; Potschke, P. Rheology, electrical conductivity, and the phase behavior of co-continuous PA6/ABS blends with MWNT: Correlating the aspect ratio of MWNT with the percolation threshold. *J. Polym. Sci. Part B Polym. Phys.* **2008**, *46*, 1619–1631. [CrossRef]
29. Otaegi, I.; Aramburu, N.; Muller, A.J.; Guerrica-Echevarria, G. Novel Biobased Polyamide 410/Polyamide 6/CNT Nanocomposites. *Polymers* **2018**, *10*, 986. [CrossRef]
30. Krause, B.; Boldt, R.; Poetschke, P. A method for determination of length distributions of multiwalled carbon nanotubes before and after melt processing. *Carbon* **2011**, *49*, 1243–1247. [CrossRef]
31. Schindelin, J.; Arganda-Carreras, I.; Frise, E.; Kaynig, V.; Longair, M.; Pietzsch, T.; Preibisch, S.; Rueden, C.; Saalfeld, S.; Schmid, B.; et al. Fiji: An open-source platform for biological-image analysis. *Nat. Method.* **2012**, *9*, 676–682. [CrossRef]
32. Socher, R.; Krause, B.; Mueller, M.T.; Boldt, R.; Poetschke, P. The influence of matrix viscosity on MWCNT dispersion and electrical properties in different thermoplastic nanocomposites. *Polymer* **2012**, *53*, 495–504. [CrossRef]
33. Weber, M.; Kamal, M.R. Estimation of the volume resistivity of electrically conductive composites. *Polym. Compos.* **1997**, *18*, 711–725. [CrossRef]
34. Gorrasi, G.; Bredeau, S.; Di Candia, C.; Patimo, G.; De Pasquale, S.; Dubois, P. Electroconductive Polyamide 6/MWNT Nanocomposites: Effect of Nanotube Surface-Coating by in situ Catalyzed Polymerization. *Macromol. Mater. Eng.* **2011**, *296*, 408–413. [CrossRef]
35. Meincke, O.; Kaempfer, D.; Weickmann, H.; Friedrich, C.; Vathauer, M.; Warth, H. Mechanical properties and electrical conductivity of carbon-nanotube filled polyamide-6 and its blends with acrylonitrile/butadiene/styrene. *Polymer* **2004**, *45*, 739–748. [CrossRef]
36. Gonzalez, I.; Eguiazabal, J.I.; Nazabal, J. Attaining high electrical conductivity and toughness in PA6 by combined addition of MWCNT and rubber. *Compos. Part A* **2012**, *43*, 1482–1489. [CrossRef]
37. Dintcheva, N.T.; Arrigo, R.; Nasillo, G.; Caponetti, E.; La Mantia, F.P. Effect of the nanotube aspect ratio and surface functionalization on the morphology and properties of multiwalled carbon nanotube polyamide-based fibers. *J. Appl. Polym. Sci.* **2013**, *129*, 2479–2489. [CrossRef]
38. Puch, F.; Hopmann, C. Morphology and tensile properties of unreinforced and short carbon fibre reinforced Nylon 6/multiwalled carbon nanotube-composites. *Polymer* **2014**, *55*, 3015–3025. [CrossRef]
39. Puch, F.; Hopmann, C. Nylon 6/multiwalled carbon nanotube composites: Effect of the melt-compounding conditions and nanotube content on the morphology, mechanical properties, and rheology. *J. Appl. Polym. Sci.* **2014**, *131*, 40893. [CrossRef]
40. Aramburu, N.; Eguiazabal, J.I. Compatible blends of polypropylene with an amorphous polyamide. *J. Appl. Polym. Sci.* **2013**, *127*, 5007–5013. [CrossRef]

41. Wang, Y.; Wang, W.; Peng, F.; Liu, M.; Zhao, Q.; Fu, P.-F. Morphology of Nylon 1212 toughened with a maleated EPDM rubber. *Polym. Int.* **2009**, *58*, 190–197. [CrossRef]
42. Gonzalez, I.; Eguiazabal, J.I.; Nazabal, J. Toughening and brittle-tough transition in blends of an amorphous polyamide with a modified styrene/ethylene-butylene/styrene triblock copolymer. *Polym. Eng. Sci.* **2009**, *49*, 1350–1356. [CrossRef]

© 2019 by the authors. Licensee MDPI, Basel, Switzerland. This article is an open access article distributed under the terms and conditions of the Creative Commons Attribution (CC BY) license (http://creativecommons.org/licenses/by/4.0/).

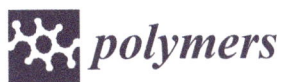

Article

The Covalent Tethering of Poly(ethylene glycol) to Nylon 6 Surface via *N,N'*-Disuccinimidyl Carbonate Conjugation: A New Approach in the Fight against Pathogenic Bacteria

Sumita Swar, Veronika Máková * and Ivan Stibor

Department of Nanochemistry, Institute for Nanomaterials, Advanced Technologies and Innovation, Technical University of Liberec, Studentská 1402/2, 46117 Liberec 1, Czech Republic; dearsumita@gmail.com (S.S.); ivan.stibor@tul.cz (I.S.)
* Correspondence: veronika.makova@tul.cz; Tel.: +420-48-5353863

Received: 27 August 2020; Accepted: 22 September 2020; Published: 24 September 2020

Abstract: Different forms of unmodified and modified Poly(ethylene glycols) (PEGs) are widely used as antifouling and antibacterial agents for biomedical industries and Nylon 6 is one of the polymers used for biomedical textiles. Our recent study focused on an efficient approach to PEG immobilization on a reduced Nylon 6 surface via *N,N'*–disuccinimidyl carbonate (DSC) conjugation. The conversion of amide functional groups to secondary amines on the Nylon 6 polymer surface was achieved by the reducing agent borane-tetrahydrofuran (BH_3–THF) complex, before binding the PEG. Various techniques, including water contact angle and free surface energy measurements, atomic force microscopy, scanning electron microscopy, X-ray photoelectron spectroscopy, and Fourier-transform infrared spectroscopy, were used to confirm the desired surface immobilization. Our findings indicated that PEG may be efficiently tethered to the Nylon 6 surface via DSC, having an enormous future potential for antifouling biomedical materials. The bacterial adhesion performances against *S. aureus* and *P. aeruginosa* were examined. In vitro cytocompatibility was successfully tested on pure, reduced, and PEG immobilized samples.

Keywords: poly(ethylene glycol) (PEG); conjugation; *N,N'*-disuccinimidyl carbonate (DSC); immobilization; surface modification

1. Introduction

The continuous increase and spread of infections caused by pathogenic bacteria and viruses in healthcare facilities worldwide also leads to an increase in the development of microbial resistance to antibiotics, antivirotics, and disinfectants [1,2]. From this point of view, many research groups worldwide are intensively working on the development of effective solutions to this problem [3–5]. Most biomaterials used in healthcare facilities, which are often in direct contact with patients, are made of polymers, particularly polyamides [6–8]. The possibility of modifying these materials via chemical treatments provides these materials with added value and may thus improve patient care in the fight against infections caused by pathogenic bacterial strains and/or viruses.

The most commonly used materials for fouling resistance are poly(ethylene glycol) (PEG) and its derivatives, which are widely used to engineer the surface of various polymers by enabling them to be hydrophilic, non-toxic, and biocompatible [9]. PEG also has the effect of minimizing the ability of bacteria to colonize and form biofilms on a material's surface [10]. According to the studies by Jeon and co-workers, protein resistance to PEG-functionalized surfaces in water is caused by Van der Waals attractions and steric repulsion between a solid substrate and proteins [11]. Recently, various different surface modification methods have been applied to graft PEG molecules onto various substrates such as polymers, metals and composites [12–17]. Several experiments have shown that longer PEG chains (e.g., M_W > 1000 g/mol) grafted to polymer surfaces are much more effective in minimizing bacterial adhesion. It is usually considered that the benefit of using long polymer chains is related to more efficient surface coverage. However, this statement closely relates to the type of tested bacterial strain [18]. The stability of a PEG coating has been examined and was found to be stable under exposure to phosphate-buffered saline (PBS) as well as sterilization conditions [19]. Below 37 °C, the PEG chain configurations are not affected. Therefore, PEG–protein interactions and the antifouling effect are not much altered [9].

The biomedical industry widely utilizes polyamides, including Nylon 6, for various purposes such as wound dressings, suture material, dialysis membranes, catheters, angioplasty balloons and porous bone scaffolds, etc. [20]. The recent worldwide crisis due to novel COVID-19 has enhanced the importance of polymers for producing effective personal protective equipment (PPE) to create a barrier between humans and germs. PPE comprises protective gloves, masks, protective eye shields, and clothing (gowns, aprons, head covering, and shoe covers). Nylon is used as a PPE material [21]. Suitable surface modifications may improve the performance of polyamide as a PPE biomaterial [22]. The surface of nylon is commonly modified by plasma treatment to immobilize the PEG chains [23]. Nevertheless, physical processes for polymeric immobilization have their own disadvantages, mainly instability issues.

Conjugation chemistry is vastly used for modification reactions. N,N'-Disuccinimidyl carbonate (DSC) is the smallest homo-bifunctional crosslinking reagent and contains only one carbonyl group with two N-hydroxysuccinimide (NHS) esters [24,25]. This compound is highly reactive toward nucleophiles [26]. DSC is widely used to produce modified polymer surfaces and to introduce PEG to the definite sites of proteins or nucleic acids. The advantages of DSC are a longer lifespan, wide range of suitable solvents and higher reactivity towards the activated hydroxyl groups [27]. Several studies have shown that DSC-mediated amination techniques provide higher coupling yields [28,29]. Antibiotics have been activated with DSC to conjugate with poly(catechin) aiming to achieve antimicrobial efficacy using polyurethane and a double-lumen silicone catheter [30]. DSC has also been employed in peptide and protein bonding with PEG or its derivatives for drug delivery and in anti-fouling polymer materials [31–33]. In the context of surface chemistry, no examples of Nylon 6/polyamide surface modification with DSC activated PEG or PEG derivatives have been reported to date.

In our previous studies, we explored various different ways of modifying the surface of Nylon 6 to alter the surface functional groups to graft PEG derivatives via chemical modification [20,34]. Our recent studies were chiefly focused on the efficient immobilization of PEG chains on a reduced Nylon 6 surface using a conjugation reaction with the carbonyl group of DSC. The surface characterization was examined by water contact angle (WCA) and free surface energy (FSE) measurements. Furthermore, surface topography and morphology were analyzed by atomic force microscopy and scanning electron microscopy. Surface modification was also verified by X-ray photoelectron spectroscopy and Fourier-transform infrared spectroscopy to confirm PEG chain tethering to the Nylon 6 surface.

Bacterial adhesion inhibition was examined against *S. aureus* and *P. aeruginosa*. In vitro cytocompatibility tests were performed to assess the cytotoxicity of the pure and modified samples.

2. Materials and Methods

2.1. Materials

Nylon 6 film (thickness: 15 μm) was supplied by Goodfellow Cambridge Ltd., (Huntingdon, UK). Methanol (99.8%), 2-propanol (99.8%), hexane (99.9%), and sodium hydroxide (NaOH) were supplied by PENTA Ltd. (Praha, Czech Republic). Borane-Tetrahydrofuran complex (1 M, BH_3-THF) and solvents tetrahydrofuran (99.95% THF), dimethyl sulfoxide (99% DMSO), acetone (99.5%), ethanol (96%), and hydrochloric acid (HCl 35%) were purchased from Lach:Ner, s.r.o., (Neratovice,

Czech Republic). Polyethylene glycol (M_W = 1450 g/mol, PEG) was purchased from Sigma-Aldrich Co., CZ. N,N′-diisopropylethylamine (≥99%), 4-(dimethylamino) pyridine (≥98%), N,N′-Disuccinimidyl carbonate (≥95%, DSC), anhydrous tetrahydrofuran (≥99.9%, inhibitor-free THF), and anhydrous acetone (≥99.5%) were obtained from Merck(Praha, Czech Republic). All modified and unmodified sample washings were performed using deionized water.

Bacterial adhesion tests were performed using Gram positive *Staphylococcus aureus*-CCM 3953 and Gram negative *Pseudomonas aeruginosa*-CCM 3955 (ALE-G18, CSNI, collection of microorganisms, Masaryk University, Brno, Czech Republic). Soyabean Casein Digest Medium-HIMEDIA®REF (HiMedia, Einhausen, Germany) was used to prepare agar plates, while Luria-Bertani (LB) broth-MILLER (Sigma-Aldrich-Merck, Praha, Czech Republic) was used to prepare the nutrient solution.

Cell experiments were performed using a 3T3 clone A31 mouse fibroblast cell line. Dulbecco's Modified Eagle's Medium (DMEM), a penicillin/streptomycin antibiotic mixture and 3-(4,5-dimethylthiazol-2-yl)-2,5-diphenyltetrazolium bromide (MTT) were supplied by Sigma-Aldrich Merck. Fetal bovine serum (Biosera, Loire Valley, France) and newborn calf serum (GibcoVR, Thermo Fisher Scientific, Praha, Czech Republic) were used as protein supplements. Positive (PM-A) and negative (RM-C) cytotoxicity controls were supplied by Hatano Research Institute (FDSC, Hatano, Japan). Staining of cell nuclei and cytoplasm was performed using 3,3′dihexyloxacarbocyanine iodide (DiOC6(3)), Triton X-100 and propidium iodide (PI) supplied by Sigma-Aldrich Merck. PBS, glutaraldehyde, Phalloidin–Fluorescein Isothiocyanate (phalloidin–FITC) and 4′,6-Diamidine-2′-phenylindole dihydrochloride (DAPI), which were used for cell adhesion tests, were also purchased form Sigma-Aldrich Merck.

2.2. Methods

2.2.1. Preparation of Samples

The Nylon 6 films (thickness: 15 µm, size: 6 × 6 cm) were thoroughly rinsed with the following solvents: deionized water, ethanol, 2-propanol, acetone, THF, and hexane. Samples were sonicated with each solvent for 3 min. The washed samples were dried at 50 °C/3 h in a vacuum and stored in a desiccator over silica gel until used.

2.2.2. Reduction of Nylon 6 with BH_3-THF

The amide functional groups present in Nylon 6 were reduced to a secondary amine according to a modified procedure described by Jia et al. [35]. Under an inert (argon) atmosphere, dry THF (50 mL) was introduced into a Schlenk flask (250 mL), containing one weighed dry sample (~90 mg). A total of 8 mL of BH_3-THF solution [1 M (8 mmol)] was added at 0 °C with stirring at 150 rpm. The reaction mixture was stirred for 1 h at r.t. and left overnight (18 h) at 50 °C. After cooling down to r.t., the modified samples were sonicated for 3 min with each of the following solvents: THF, 1 M HCl, deionized water, 1 M NaOH, deionized water, THF, ethanol, acetone, and hexane. The reduced samples were dried at 50 °C/3 h in a vacuum and stored in the desiccator until further treatment. For long-term storage, the 1 M NaOH solution was skipped and the modified Nylon 6 was stored as ammonium chloride salt prepared by HCl washing [34]. The reduced sample was referred to as Nylon 6-NH. All of the reduced samples (size 6 × 6 cm) were cut into 2 × 2 cm size pieces before further functionalization.

2.2.3. Tethering of PEG to the Nylon 6-NH Surface by Conjugating DSC

A verified procedure was followed for the conjugation reaction, after the required modification [26]. N,N′-diisopropylethylamine (5 mL) and DSC (0.3843 g, 1.5 mmol) were stirred in 60 mL of dry acetone at r.t./500 rpm/1 h. This suspension was added into the Schlenk flask (250 mL) containing five Nylon 6-NH samples under argon and stirred at r.t./200 rpm/3 h. The reaction medium was carefully removed using a syringe and the samples were washed twice with dry acetone in a closed system. Meanwhile, polyethylene glycol (2.9 g, 2 mmol PEG) was dried at r.t./2 h in a cold trap using liquid nitrogen (N_2). Anhydrous acetone (35 mL) was added to a 100-mL round bottom flask containing dry PEG under an inert atmosphere and stirred at r.t./400 rpm/2 h. In another flask, 4-(dimethylamino) pyridine (1.466 g, 12 mmol) was dissolved in 20 mL of anhydrous acetone and stirred at r.t./200 rpm/1 h. First, the PEG solution was added to the washed Nylon 6-NH samples after reaction with DSC. Then, the 4-(dimethylamino) pyridine solution was introduced, stirred at r.t./200 rpm, and allowed to react overnight. Finally, the samples were removed and rinsed with acetone, ethanol, and hexane in a sonication bath, then dried at r.t./5 h in a vacuum and stored in a desiccator. The modified samples were referred to as Nylon 6-N-PEG.

2.2.4. Surface Characterization

The WCAs and FSEs were analyzed using a computer-based portable instrument having special software that follows the ISO 27448:2009 test method (See System E, Advex Instruments s.r.o., Brno, Czech Republic). WCA measurement is a rapid, easy and useful surface analytical technique. Contact angle measurements were performed by vertically positioning 10 droplets (3.5 µL per droplet) of deionized water on each sample. The droplets were allowed to equilibrate for 5 s before the measurement. The mean values were taken for plotting WCA bar graphs including standard deviations (±SD). The FSEs related to the mean values of the WCAs were directly measured by the software using the Kwok–Neumann model.

The topography of the pure and modified Nylon 6 films was studied in the air at atmospheric pressure with NanoWizard® 3 NanoScience AFM (JPK Instruments, Berlin, Germany). For more accurate scanning, a contact mode with Cantilever NANOSENSORSTM PPP-CONTSCR (resonance frequency = 23 kHz; contact force = 0.2 Nm^{-1}; tip radius <10 nm; tip height 10–15 µm) was used. Sample scans and subsequent surface roughness (Ra) evaluation were performed on areas of 10×10 µm and 1×1 µm. The obtained data were processed using the freeware Gwyddion and JPK Data Processing.

The surface morphology changes were examined by the SEM (ZEISS, Sigma Family, Jena, Germany). Pure and modified Nylon 6 films were sputtered with platinum to from a 2-nm thin conductive layer, and subsequently viewed as secondary electron images (1 kV).

X-ray photoelectron spectroscopic (XPS) measurements investigating C_{1s} and N_{1s} binding energies (eV) before and after Nylon 6 modification were performed using a Thermo Scientific K-Alpha X-ray Photoelectron Spectrometer (Thermo Fisher Scientific, Waltham, MA, USA) with monochromatic Al Kα radiation (h$_\gamma$ = 1486.6 eV).

The Nicolet™ iS™10 FT-IR Spectrometer (Thermo Scientific™) was employed to investigate the modified Nylon 6 samples compared to the pure ones.

2.2.5. Bacterial Adhesion Tests

To analyze the bacterial adhesion on the surface of each sample (Nylon 6, Nylon 6-NH and Nylon 6-N-PEG), the below-mentioned protocol, with the necessary modifications described below, was followed according to the reference [36]. All of the samples were incubated in the prepared bacterial inoculums of *Pseudomonas aeruginosa* and *Staphylococcus aureus* separately.

Bacterial adhesion tests were performed using Gram positive *Staphylococcus aureus*-CCM 3953 and Gram-negative *Pseudomonas aeruginosa*-CCM 3955 (ALE-G18, CSNI, collection of microorganisms, Masaryk University, Brno, Czech Republic). The tested bacterial strains were firstly revived and leave it to grow on the agar plates made of Soyabean Casein Digest Medium-HIMEDIA®REF for 48 h at 37 °C in the incubator. After this time, the bacterial inoculums were prepared in the following manner.

Both grown bacterial strains were firstly diluted in 20 mL of Luria-Bertani (LB) broth MILLER and then were centrifuged for 3 min/5000 rpm to remove supernatant, further were washed with PBS at pH 7.2 twice, and resuspended in a sterile physiological saline solution (0.15 M NaCl, pH 7.0, 20 mM NaHCO$_3$) with the aim to reach the initial optical cell density at 600 nm 0.15 ± 0.08 for both bacterial inoculums. Subsequently, each sample was immersed in the bacterial inoculum for 1 h at 37 °C. After this time, the samples were rinsed with distilled water three times, put on the slide glass, and covered with a Live/Dead Backlight, 1 Kit 30× diluted solution containing 1.67 mM of SYTO9-A and 18.3 mM propidium iodide—B, molar ratio 1:1. Finally, the samples were kept in the dark for 15 min and further analyzed at 630 nm using 44 FITC (green) and 43 cy3 (red) filters.

Finally, the live (green) and dead (red) bacteria (*S. aureus* and *P. aeruginosa*), attached on the six (in total) analyzed samples, were imaged using a fluorescent microscope (ZEISS Axio Imager 2).

2.2.6. Cytotoxicity Assessment

Cytotoxicity was assessed by direct contact cytotoxicity tests as well as cell adhesion and proliferation analyses. Prior to the in vitro experiment, the samples (Nylon 6, Nylon 6-NH and Nylon 6-N-PEG) were sterilized by immersion in 70% ethanol for 30 min, followed by washing three times in a PBS solution. Mouse 3T3-SA fibroblasts (passage 8) were used for the cytotoxicity assessment of the films according to ISO 10993-5:2009 (Biological evaluation of medical devices—Part 5: Tests for in vitro cytotoxicity).

For direct contact cytotoxicity analysis, the fibroblasts were seeded in 12 well plates in a concentration of 5×10^4/well. After reaching subconfluency (24 h), the tested samples measuring approximately 5×5 mm were added into the wells ($n = 4$ per each sample). After 24 h of incubation, the cells were observed by optical microscopy and cell metabolic activity was determined by the colorimetric MTT test. The cells in the complete media served as a NC: negative control (DMEM). The complete medium with the addition of the cytotoxic agent Triton X-100 was used as a PC: positive control (DMEM+Triton X-100). After 24 h of incubation, the cell metabolic activity was examined by a metabolic MTT assay ($n = 12$ per each group). The measured absorbance of the negative control (NC) was considered as 100% viability. Recalculated values of cell viability were plotted in the results section, showing the cytotoxic effect of the samples. Direct contact cytotoxicity was determined by a decrease in viability below 70% of the control cells (marked with a red line in the graph).

Next, cell adhesion and proliferation tests were conducted. Staining of the cells that adhered to the tested films was used to visualize cell spreading and adhesion after seeding the cells on the top of the samples. Sterile samples measuring approximately 2×2 cm were placed on the bottom of six well plates and seeded with fibroblasts. After 1 and 7 days of incubation, the films with the adhered cells were washed twice in PBS and fixed in 2.5% glutaraldehyde. Then, the cells were stained by phalloidin-FITC, which binds to actin filaments in the cytoplasm, and DAPI that visualizes the cell nuclei in blue. A fluorescence microscope was used to capture cell spreading (magnification 200×) and to evaluate the cell morphology. Fluorescent imaging was carried out on a Zeiss Axio Imager M2 microscope using an EC Plan-Neofluor 20× objective lens. The density of the adhered cells was determined by counting the cell nuclei on the fluorescent images of each sample using ImageJ software.

3. Results

3.1. Tethering of PEG to the Nylon 6-NH Surface by Conjugating DSC

It is well known that DSC is a homo-bifunctional NHS ester crosslinking reagent that is highly reactive towards nucleophiles [26]. DSC can activate both hydroxyl (–OH) and amine (–NH$_2$ and >NH) functional groups. DSC undergoes rapid hydrolysis in an aqueous solution and therefore anhydrous organic solvents are required to carry out the treatments. Scheme 1 shows the three-step preparation protocol. The first step indicates the reduction of the Nylon 6 surface functional groups, amides (–CONH–), to secondary amines (–CH$_2$NH–) with the help of a borane-tetrahydrofuran (BH$_3$-THF) complex, producing the modified Nylon 6-NH surface. The next reaction shows the activation of secondary amine groups (>NH) on the surface via DSC. The third step shows the immobilization of PEG on the DSC activated Nylon 6-NH surface. The amine activated surface and the PEG tethered final surface are referred to as Nylon 6-N-SC and Nylon 6-N-PEG, respectively.

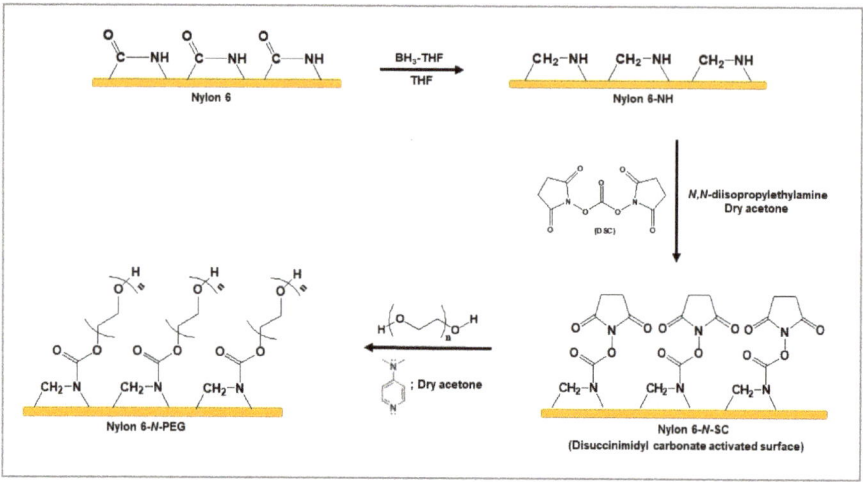

Scheme 1. Preparation of the PEG tethered Nylon 6-NH surface (Nylon 6-N-PEG) via DSC.

3.2. Characterization

3.2.1. WCA and FSE Analyses of Pure and Modified Nylon 6 Samples

Figure 1a,b shows the WCAs and FSE measurements on the pure and modified films, before and after PEG immobilization. The initial surface contact angle (mean value) was 70.8° ± 4.3° whereas after reduction the surface contact angle increased to 81.2° ± 3.1° (Figure 1a), making the surface more hydrophobic than the pure one [34]. The DSC activated Nylon 6-N-PEG surface (mean WCA—51.8° ± 2.9°) made the Nylon 6-NH films hydrophilic, as expected, by tethering PEG chains [37]. The anti-fouling properties of PEG chains on the surface depend on their high chain mobility, large exclusion volume, and steric hindrance effect of the highly hydrated layer. Therefore, the WCA analysis exhibits the initial indication of the successful tethering of PEG chains to the Nylon 6 surface [34].

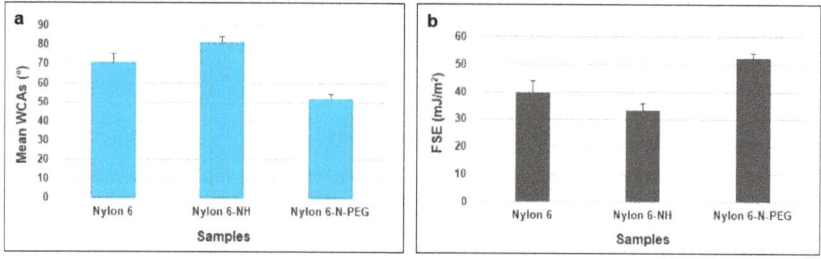

Figure 1. Mean WCA data (**a**) and FSE values (**b**), for pure and modified Nylon 6 films.

The FSE values decreased after Nylon 6 reduction to Nylon 6-NH and further increased after tethering PEG to the Nylon 6-NH surface via DSC to form Nylon 6-N-PEG (Figure 1b). The FSE of pure Nylon 6 was found to be 39.9 mJ/m^2. The FSE was significantly enhanced from 33.3 mJ/m^2 to 52.4 mJ/m^2, as the PEG immobilization transformed the Nylon 6-NH surface to a more hydrophilic one. Wide scientific studies have reported that an FSE of between 23 and 30 mJ/m^2 is related to the lowest bacterial adhesion [38]. In our previous work, we exhibited the remarkable antibacterial property of the Nylon 6-NH surface and the FSE may have been one of the significant factors [20].

3.2.2. AFM Analyses

AFM was used to study surface topography alterations caused by the chemical treatments. Table 1 shows the surface roughness (Ra) values of two different analyzed areas, (1 × 1) μm² and (10 × 10) μm², obtained for pure Nylon 6, modified Nylon 6-NH, and Nylon 6-N-PEG. The results indicated significant changes after chemical modifications and immobilization of the PEG chains. The Ra value decreased noticeably after Nylon 6 reduction from 35.7 nm to 7.5 nm in the case of the 10 × 10 μm² examined area. On the contrary, the Nylon 6-N-PEG film demonstrated a rougher surface (18.4 nm) than Nylon 6-NH for the same analyzed area. The change in Ra values implies the successful modification. The Ra values were supported by AFM analyses (Figure 2) and SEM micrographs (Figure 3).

Table 1. Changes in the surface roughness (Ra) observed before and after immobilization of the Nylon 6 surface.

Samples	Surface Area	
	(1 × 1) μm²	(10 × 10) μm²
Nylon 6	5.3 ± 0.4 nm	35.7 ± 4.6 nm
Nylon 6-NH	1.4 ± 0.3 nm	7.5 ± 0.2 nm
Nylon 6-N-PEG	3.8 ± 0.2 nm	18.4 ± 1.1 nm

The AFM micrographs (3D) of pure and modified Nylon 6 samples are shown for a visual understanding of the surface topography changes after various treatments (Figure 2a–c). The (10 × 10) μm² surface areas were investigated to compare the pure Nylon 6 with the modified Nylon 6-NH and Nylon 6-N-PEG samples. Significant differences can be seen. Nylon 6 appears to have the roughest surface (Figure 2a) while Nylon 6-NH shows the significant decrease of the Ra value compared to the unmodified Nylon 6 (Figure 2b). Subsequently, in the case of the sample Nylon 6-N-PEG (Figure 2c), a noticeable increase in the surface roughness occurs due to PEG immobilization.

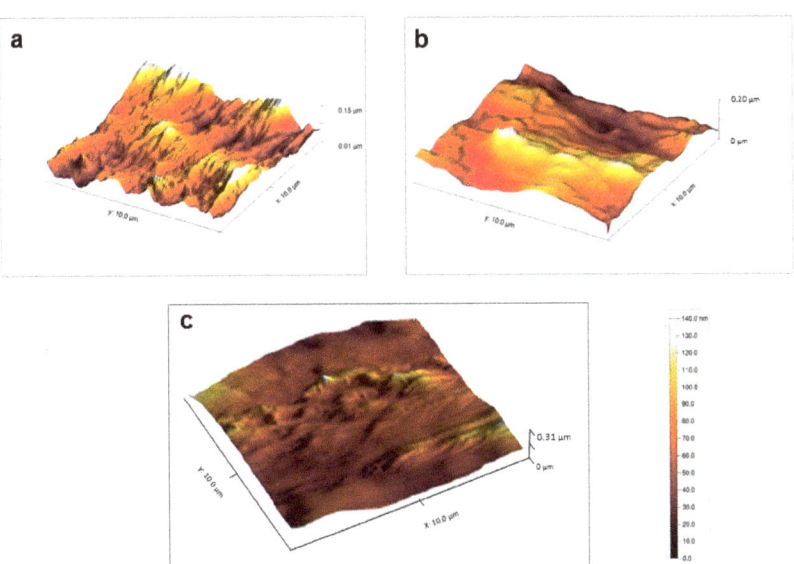

Figure 2. 3D (10 × 10) μm² AFM images of Nylon 6 (**a**); Nylon 6-NH (**b**) and Nylon 6-N-PEG (**c**).

3.2.3. SEM Analyses

The surface morphologies were also examined by SEM analyses. The SEM micrographs for the Nylon 6 and chemically treated samples revealed that the surface of Nylon 6 was uneven compared to Nylon 6-NH (Figure 3a,b). SEM analyses demonstrated the successful grafting, as the formed layers could be clearly identified on the Nylon 6-N-PEG surface (Figure 3c,d). Nylon 6-N-PEG showed densely grafted PEG chain layers on the surface by exhibiting the alteration in Nylon 6-N-PEG surface morphology compared to Nylon 6-NH. The SEM images supported the AFM micrographs discussed earlier.

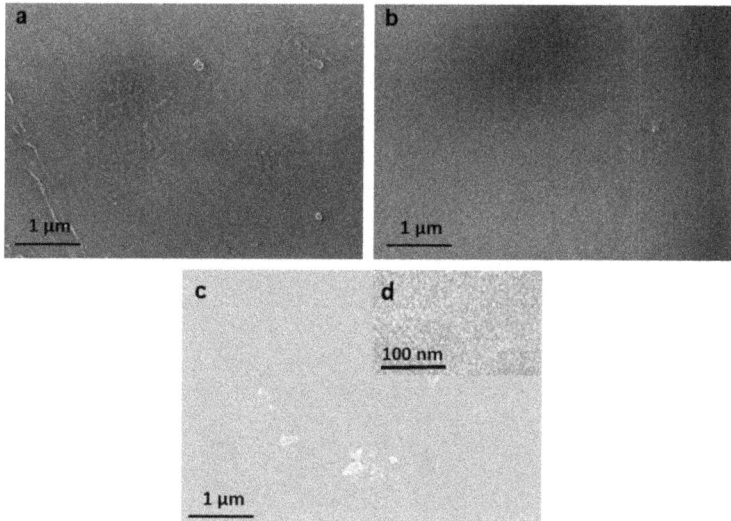

Figure 3. SEM micrographs showing Nylon 6 (**a**); Nylon 6-NH (**b**); Nylon 6-N-PEG (**c**) with an inset image of the PEG-modified surface in detail (**d**).

3.2.4. XPS Analyses

The XPS analyses of pure Nylon 6 and reduced Nylon 6 samples, marked as Nylon 6 and Nylon 6-NH, were explained in detail in our previously published article [34], which has been focused on an effective reductive modification of the Nylon 6 surface. In the above-mentioned article, we published the XPS C_{1s} and N_{1s} spectra comparing Nylon 6 surface before and after reduction. In this context, we have decided to add C_{1s} XPS spectra comparing unmodified and modified Nylon 6 samples (Figure 4).

Yields of converted amine groups in % were calculated according to the reference [35] using the following equation: yield (%) = $[(A^{C=O})_t - (A^{C=O})_{t0}]/(A^{C=O})_{t0}$, where the area of the signal at a given time $(A^{C=O})_t$ was compared to the value for pure Nylon 6 $(A^{C=O})_{t0}$. Based on this relationship, the yields of secondary amine groups formed after an overnight (18 h) reaction with BH_3-THF were moved in the interval 62–65% in our experiments. On the contrary to our observation, Jia. X. et al. [35] declare maximum yields of converted amine groups on the surface of Nylon 6/6 69% after 10 h of the reduction.

The elemental composition of the three sample surfaces, investigated by XPS analyses, is given in Table 2. The results showed that the oxygen (O) percentage decreased in the Nylon 6-NH sample compared to pure Nylon 6 from 12.74% to 7.08%. On the contrary, the PEG chain tethering enhanced the oxygen (O) percentage of Nylon 6-N-PEG notably from 7.08% to 14.59%. The significant increase in the oxygen (O) percentage is a very strong indicator of successful PEG chain tethering to the DSC activated PEG grafted surface. Comparative elemental composition (%) studies of oxygen (O) among the Nylon 6-NH and modified samples with other PEG derivatives exhibited similar results in our previous work [20]. Furthermore, the immobilization was confirmed by FT-IR spectroscopy.

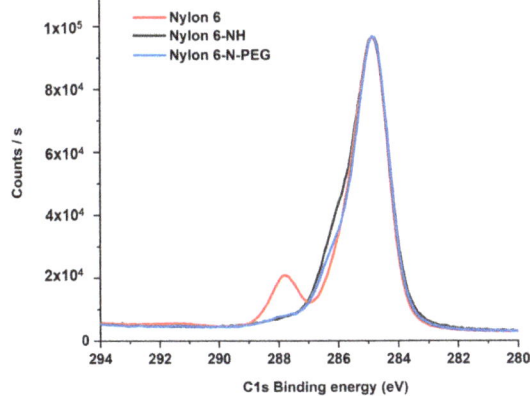

Figure 4. The XPS C_{1s} spectra of the unmodified and modified Nylon 6 samples.

Table 2. XPS data for the unmodified and modified Nylon 6 films.

Sample	Elemental Composition of Sample Surface (%)		
	C	N	O
Nylon 6	76.11	11.15	12.74
Nylon 6-NH	81.98	10.94	7.08
Nylon 6-N-PEG	74.88	10.53	14.59

3.2.5. FT-IR Analyses

Figure 5 compares the FT-IR spectra of pure Nylon 6 and the modified samples (Nylon 6-NH and Nylon 6-N-PEG). The characteristic stretching vibrations are seen for amide I and amide II groups (1660 cm^{-1} and 1541 cm^{-1}). Further we can observe, the secondary amine groups (3290 cm^{-1}; stretching vibration of -NH). These groups are visible for both samples before and after the tethering of PEG to Nylon 6-NH. It is important to note that all of the above-mentioned stretching bands are also seen in the case of the pure Nylon 6 samples [34]. For Nylon 6-NH, the band that appeared at 2436–2200 cm^{-1} corresponds to the imine groups introduced via chemical treatment (BH$_3$-THF reduction), indicating a step towards the conversion of amides to amines. The trace –NH$^+$ salts forms (their stretching and deformation vibrations) can also be identified in this interval.

The aliphatic groups coming from PEG chains can be observed at the following peaks; 1421 cm^{-1} (deformation vibration of –C–H bonds), 2860 cm^{-1} (symmetric stretching vibration of -CH$_2$ groups) and 2931 cm^{-1} (asymmetric stretching vibration of –CH$_2$ groups). In the case of Nylon 6-N-PEG, the characteristic stretching vibrations at 1084 cm^{-1} and 1117 cm^{-1} (–C–O–C– ether bond) may be attributed to the PEG chains tethered by conjugation with DSC [29]. The additional peaks that appeared at 1205 cm^{-1} (stretching vibration of –C–O–) and 1737 cm^{-1} (stretching vibration of –C=O) represent the ester bond formation during conjugation of PEG with DSC and aliphatic ketones on the surface.

There can be find also stretching vibration of –C=O ester bonds at 1757 cm^{-1} and at 1800 cm^{-1} [39]. These vibrations can be also found for bonds which are present close to the 4–5 cycle rings. This observation may support the presence of DCS traces in the PEG-modified samples. In the case of Nylon 6-N-PEG sample, the imine stretching band present in Nylon 6-NH (2420–2250 cm^{-1}) disappears completely, further confirming the successful immobilization of PEG.

Polymers **2020**, *12*, 2181

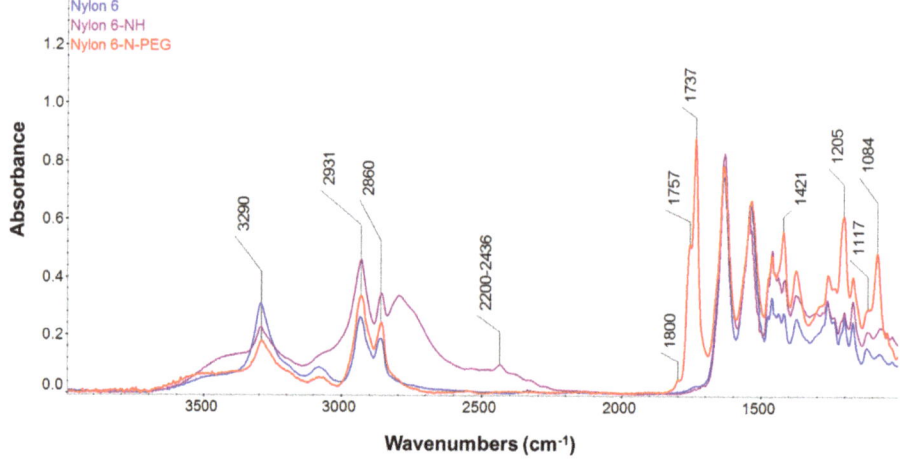

Figure 5. The FTIR spectra comparing pure and modified Nylon 6 samples.

3.3. Bacterial Adhesion

The bacterial attachment on the surfaces of the pure and modified Nylon 6 samples against *S. aureus* and *P. aeruginosa* was assessed by fluorescence imaging (Figures 6 and 7). After 1 h incubation of the Nylon 6, Nylon 6-NH, and Nylon 6-N-PEG samples in bacterial suspensions, the green and red fluorescence in the images indicated the live and dead bacteria on the surfaces, respectively.

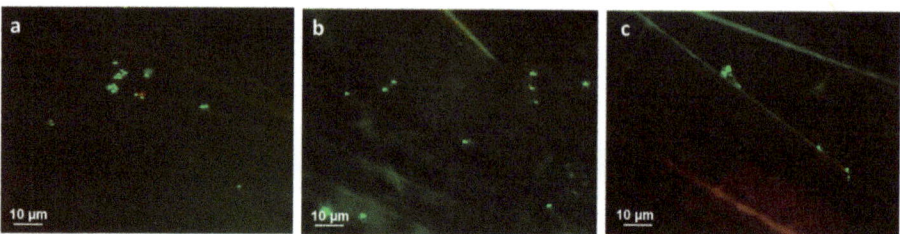

Figure 6. Fluorescence microscopy images of *S. aureus* on the surfaces of Nylon 6 (**a**); Nylon 6-NH (**b**) and Nylon 6-N-PEG (**c**).

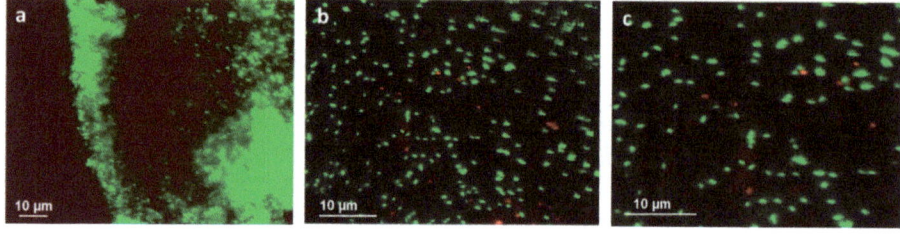

Figure 7. Fluorescence microscopy images of *P. aeruginosa* on the surfaces of Nylon 6 (**a**); Nylon 6-NH (**b**) and Nylon 6-N-PEG (**c**).

The pure sample (Nylon 6) and Nylon 6-NH exhibited more green fluorescence spots dispersed on the surface, suggesting the presence of live *S. aureus* bacteria (Figure 6a,b). The images clearly indicate that both samples did not allow significant attachment of *S. aureus* onto their surfaces. Moreover, this result is in conformity with the previous literature [40,41]. At the same time, red fluorescence was almost absent on both samples, which indicates that both samples lack any antibacterial properties. The PEG immobilized Nylon 6-N-PEG material exhibited antifouling properties due to the presence of hydrophilic PEG chains on the surface (Figure 6c). This phenomenon has been reported in previously published studies, suggesting that *S. aureus* bacteria have a tendency to adhere less on hydrophilic surfaces [4].

The adhesion of *P. aeruginosa* on the three tested sample surfaces provided very interesting results. The Nylon 6 surface image exhibits intense green fluorescence spots (Figure 7a). This phenomenon indicates the presence of viable *P. aeruginosa* bacterial predominance on the Nylon 6 surface and confirms a massive adhesion of this bacterial strain. It is assumed that the surface adhesion property of *P. aeruginosa* is different from *S. aureus* due to the hydrophobic cell walls around the *P. aeruginosa* [23]. The second sample (Nylon 6-NH) exhibited less green fluorescence along with a few red dots (Figure 7b) compare to the first sample. The image analysis significantly confirms the decrease in the attachment of viable *P. aeruginosa* bacterial cells to the surface of the reduced sample compares to the unmodified sample [20]. The Nylon 6-N-PEG sample seems to have the best antifouling properties against *P. aeruginosa* comparing to the previous two samples (Figure 7c). This sample shows a significant reduction of the bacterial adhesion compared to the unmodified Nylon 6.

Green and weak red fluorescence observed on the surface of the PEG tethered sample (Nylon 6-N-PEG) indicated both viable and dead *P. aeruginosa* bacterial cells, respectively. It is well known that PEG polymer chains (often with a brush-like structure) increase mobility inhibit protein adhesion and support antifouling effect, but there are many other factors that play an important role in the bacterial adhesion inhibition [23]. In Figure 6c, a relatively large amount of viable bacterial cells (green patch) present on the sample surface indicates that Nylon 6-N-PEG is not as effective for bacterial inhibition against *P. aeruginosa* as it is against *S. aureus* (Figure 6c).

3.4. Cytotoxicity Assessment

3.4.1. Direct Contact Cytotoxicity

Direct contact with the samples was assessed by measurement of cell metabolic activity after 24 h of incubation. The films floated in the media with subtle contact with the bottom where the cells were grown. Therefore, slightly limited metabolic activity in the wells containing the material was observed due to mechanical damage during the removal of the films after 24 h of incubation. Therefore, slightly limited metabolic activity does not indicate the cytotoxic effect of the material. Moreover, assessment of cell appearance under an optical microscope did not indicate any cell damage. Figure 8 shows the results of the cell viability (%) tests for the pure and modified Nylon 6 samples compared to the controls. The measured viability of the cells (%) in contact with the films reached 78–92% of the metabolic activity of the cells cultured in the complete medium. The viability of fibroblasts upon contact with Nylon 6 was 78.39%. In the case of Nylon 6-NH, the cell viability reached 79.65%. Nylon 6-N-PEG appeared to have the maximum cell viability (91.78%) of all of the tested samples. All of the measured values exceeded the 70% limit of cytotoxicity. Therefore, all of the samples may be considered cytocompatible with 3T3 mouse fibroblasts upon direct contact.

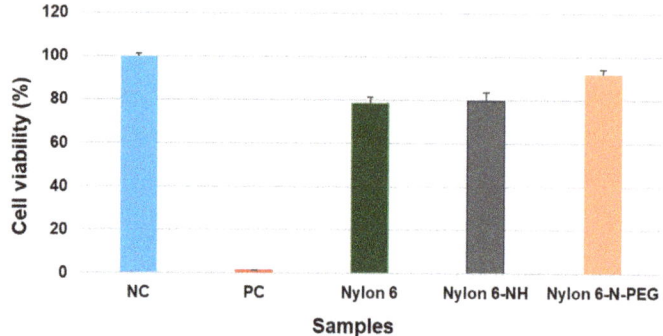

Figure 8. Metabolic activity of fibroblasts after 24 h of incubation in a complete medium (NC), medium + Triton X-100 (PC), in a complete medium with the presence of the tested samples. The red line refers to the cytotoxic level according to ISO 10993-5:2009.

3.4.2. Cell Adhesion and Proliferation

The cell proliferation on the pure and modified samples is clearly seen in Figure 9. Table 3 shows the results of 3T3 mouse fibroblast cell counts on all of the three tested samples. The highest cell adhesion rate was observed for the pure Nylon 6 material, after one day and seven days of culturing. Nylon 6-NH also exhibited similar cell adhesion profiles (both for one day and seven days) as the pure Nylon 6 sample, although the cell adhesion was slightly less than the pure sample. Nylon 6-N-PEG adhered a lesser number of cells after one day of culturing, compared to the other two tested samples. PEG polymer brushes have been shown to inhibit protein adsorption and as a result cell adhesion to surfaces [42]. Interestingly, the cell count of the Nylon 6-N-PEG sample increased after seven days, showing better proliferation compared to day 1 of culturing, but the cell morphology was affected.

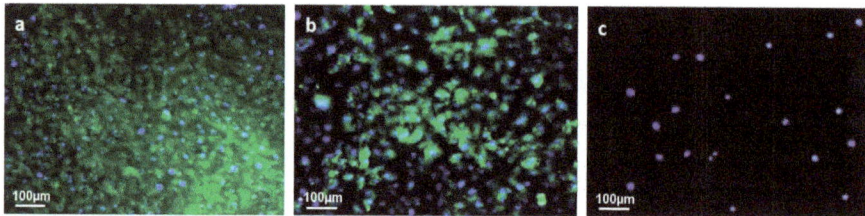

Figure 9. Fluorescence microscopy images of samples Nylon 6 (**a**); Nylon 6-NH (**b**) and Nylon 6-N-PEG (**c**), seeded with fibroblasts after 7 days of culturing.

Table 3. Results of cell count after 1 day and 7 days of culturing.

Sample	Cell Count (Number of Cells/cm^2)			
	1 Day		7 Days	
	Mean	SD	Mean	SD
Nylon 6	15,911	±3970	uncountable	-
Nylon 6-NH	12,254	±2350	318,202	±40,285
Nylon 6-N-PEG	493	±110	61,912	±1405

Figure 9a–c shows the cell morphologies on Nylon 6, Nylon 6-NH and Nylon 6-N-PEG samples, respectively, after one week of culturing. In the case of the pure Nylon 6 and reduced Nylon 6-NH samples, cell monolayers were observed on the surfaces after 7 days. Pure Nylon 6 supported a high cell adhesion and proliferation rate with normal cell morphology (Figure 9a). Nylon 6-NH also

exhibited significant cell adhesion and proliferation on the surface, although a few cell nuclei were bigger than normal cells. Overall, both of these tested samples exhibited acceptable cell adhesion rates followed by high proliferation rates. The fibroblast cells had an unusual morphology in the case of the PEG immobilized sample (Nylon 6-N-PEG), where the cells on the surface were rounded and small (Figure 9c). It is evident that the PEG allowed a limited number of cells to adhere with rounded morphology. Therefore, it may be assumed that the PEG tethered sample was somewhat non-adherent for the fibroblast cell line [3].

4. Conclusions

Our studies show the successful tethering of biocompatible PEG chains to a reduced Nylon 6 surface using unique chemical treatments, based on the utilization N,N'-Disuccinimidyl carbonate. This opens up the possibility to introduce antifouling coatings onto nylon as well as other polyamide materials, especially those used for biomedical applications. As it is well known that the antifouling efficacy of PEG is correlated to the chain length of the PEG polymer, the immobilization of longer PEG chains using a conjugation technique may serve as a useful method. The two-step modification procedure for PEG immobilization on Nylon 6 was confirmed by various spectroscopic and non-spectroscopic techniques. Cytocompatible Nylon 6-N-PEG exhibited significant inhibition of bacterial adhesion against *S. aureus* and *P. aeruginosa*. The disposable polyamide materials used in medicine are subject to ever-increasing demand due to the current worldwide crisis. The present COVID-19 pandemic has given us the opportunity and challenge to emphasize the importance of personal protection and the use of effective materials that have a multifunctional protective potential.

Author Contributions: Conceptualization, I.S. and S.S.; methodology, I.S. and V.M.; validation, S.S., V.M., and I.S.; investigation, S.S.; resources, V.M.; data curation, I.S.; writing—original draft preparation, S.S.; writing—review and editing, V.M.; visualization, S.S.; supervision, I.S.; project administration, I.S.; All authors have read and agreed to the published version of the manuscript.

Funding: This research was funded by the Ministry of Education, Youth and Sports of the Czech Republic and the European Union—European Structural and Investment Funds in the framework of the Operational Programme Research, Development and Education project "Modular Platform for Autonomous Chassis of Specialized Electric Vehicles for Freight and Equipment Transportation", Reg. No. CZ.02.1.01/0.0/0.0/16_025/0007293.

Conflicts of Interest: The authors declare no conflict of interest.

References

1. Adlhart, C.; Verran, J.; Azevedo, N.F.; Olmez, H.; Keinänen-Toivola, M.M.; Gouveia, I.; Melo, L.F.; Crijns, F. Surface modifications for antimicrobial effects in the healthcare setting: A critical overview. *J. Hosp. Infect.* **2018**, *99*, 239–249. [CrossRef] [PubMed]
2. Jamal, M.; Ahmad, W.; Andleeb, S.; Jalil, F.; Imran, M.; Nawaz, M.A.; Hussain, T.; Ali, M.; Rafiq, M.; Kamil, M.A. Bacterial biofilm and associated infections. *J. Chin. Med. Assoc.* **2018**, *81*, 7–11. [CrossRef] [PubMed]
3. Goor, O.J.; Brouns, J.E.; Dankers, P.Y. Introduction of anti-fouling coatings at the surface of supramolecular elastomeric materials via post-modification of reactive supramolecular additives. *Polym. Chem.* **2017**, *8*, 5228–5238. [CrossRef]
4. Krasowska, A.; Sigler, K. How microorganisms use hydrophobicity and what does this mean for human needs? *Front. Cell. Infect. Microbiol.* **2014**, *4*, 112. [CrossRef] [PubMed]
5. Zhang, X.; Wang, L.; Levänen, E. Superhydrophobic surfaces for the reduction of bacterial adhesion. *Rsc Adv.* **2013**, *3*, 12003–12020. [CrossRef]
6. Abedini, F.; Ahmadi, A.; Yavari, A.; Hosseini, V.; Mousavi, S. Comparison of silver nylon wound dressing and silver sulfadiazine in partial burn wound therapy. *Int. Wound J.* **2013**, *10*, 573–578. [CrossRef]
7. Chu, C.C.; Tsai, W.C.; Yao, J.Y.; Chiu, S.S. Newly made antibacterial braided nylon sutures. I. In vitro qualitative and in vivo preliminary biocompatibility study. *J. Biomed. Mater. Res.* **1987**, *21*, 1281–1300. [CrossRef]

8. Thokala, N.; Kealey, C.; Kennedy, J.; Brady, D.B.; Farrell, J.B. Characterisation of polyamide 11/copper antimicrobial composites for medical device applications. *Mater. Sci. Eng. C Mater. Biol. Appl.* **2017**, *78*, 1179–1186. [CrossRef]
9. Zhang, X.; Brodus, D.S.; Hollimon, V.; Hu, H. A brief review of recent developments in the designs that prevent bio-fouling on silicon and silicon-based materials. *Chem. Cent. J.* **2017**, *11*, 1. [CrossRef]
10. Ozcelik, B.; Ho, K.K.K.; Glattauer, V.; Willcox, M.; Kumar, N.; Thissen, H. Poly(ethylene glycol)-based coatings combining low-biofouling and quorum-sensing inhibiting properties to reduce bacterial colonization. *ASC Biomater. Sci. Eng.* **2017**, *3*, 78–87. [CrossRef]
11. Jeon, S.I.; Lee, J.H.; Andrade, J.D.; De Gennes, P. Protein Surface Interactions in the Presence of Polyethylene Oxide. *J. Colloid Interface Sci.* **1991**, *142*, 149–158. [CrossRef]
12. Zhang, T.-D.; Zhang, X.; Deng, X. Applications of protein-resistant polymer and hydrogel coatings on biosensors and biomaterials. *Ann. Biotechnol.* **2018**, *2*, 1–7. [CrossRef]
13. Sarita. Research on polyethylene glycol, crosslinked polyethylene glycol & polyethylene glycol chitosan conjugate coating for biomedical application. *Int. J. Recent Technol. Eng.* **2019**, *8*, 2921–2925. [CrossRef]
14. Francolini, I.; Silvestro, I.; Di Lisio, V.; Martinelli, A.; Piozzi, A. Synthesis, characterization, and bacterial fouling-resistance properties of polyethylene glycol-grafted polyurethane elastomers. *Int. J. Mol. Sci.* **2019**, *20*, 1011. [CrossRef] [PubMed]
15. Lowe, S.; O'Brien-Simpson, N.M.; Connal, L.A. Antibiofouling polymer interfaces: Poly(ethylene glycol) and other promising candidates. *Polym. Chem.* **2015**, *6*, 198–212. [CrossRef]
16. Damodaran, V.B.; Murthy, S.N. Bio-inspired strategies for designing antifouling biomaterials. *Biomater. Res.* **2016**, *20*, 1–11. [CrossRef]
17. Peng, L.; Chang, L.; Liu, X.; Lin, J.; Liu, H.; Han, B.; Wang, S. Antibacterial Property of a Polyethylene Glycol-Grafted Dental Material. *ACS Appl. Mater. Interfaces* **2017**, *9*, 17688–17692. [CrossRef]
18. Francolini, I.; Vuotto, C.; Piozzi, A.; Donelli, G. Antifouling and antimicrobial biomaterials: An overview. *Apmis* **2017**, *125*, 392–417. [CrossRef]
19. Branch, D.W.; Wheeler, B.C.; Brewer, G.J.; Leckband, D.E. Long-term stability of grafted polyethylene glycol surfaces for use with microstamped substrates in neuronal cell culture. *Biomaterials* **2001**, *22*, 1035–1047. [CrossRef]
20. Swar, S.; Máková, V.; Horáková, J.; Kejzlar, P.; Parma, P.; Stibor, I. A comparative study between chemically modified and copper nanoparticle immobilized Nylon 6 films to explore their efficiency in fighting against two types of pathogenic bacteria. *Eur. Polym. J.* **2020**, *122*, 109392. [CrossRef]
21. Garrigou, A.; Laurent, C.; Berthet, A.; Colosio, C.; Jas, N.; Daubas-Letourneux, V.; Jackson Filho, J.M.; Jouzel, J.N.; Samuel, O.; Baldi, I.; et al. Critical review of the role of PPE in the prevention of risks related to agricultural pesticide use. *Saf. Sci.* **2020**, *123*, 104527. [CrossRef]
22. Irzmańska, E.; Brochocka, A. Modified polymer materials for use in selected personal protective equipment products. *Autex Res. J.* **2017**, *17*, 35–47. [CrossRef]
23. Dong, B.; Jiang, H.; Manolache, S.; Wong, A.C.L.; Denes, F.S. Plasma-mediated grafting of poly(ethylene glycol) on polyamide and polyester surfaces and evaluation of antifouling ability of modified substrates. *Langmuir* **2007**, *23*, 7306–7313. [CrossRef] [PubMed]
24. Wulf, K.; Teske, M.; Löbler, M.; Luderer, F.; Schmitz, K.P.; Sternberg, K. Surface functionalization of poly (ε-caprolactone) improves its biocompatibility as scaffold material for bioartificial vessel prostheses. *J. Biomed. Mater. Res.-Part. B Appl. Biomater.* **2011**, *98*, 89–100. [CrossRef]
25. Yang, M.; Tsang, E.M.W.; Wang, Y.A.; Peng, X.; Yu, H.Z. Bioreactive surfaces prepared via the self-assembly of dendron thiols and subsequent dendrimer bridging reactions. *Langmuir* **2005**, *21*, 1858–1865. [CrossRef] [PubMed]
26. Hermanson, G.T. *Bioconjugate Techniques*; Elsevier: Amsterdam, The Netherlands, 2008; ISBN 9780123705013.
27. Meschaninova, M.I.; Novopashina, D.S.; Semikolenova, O.A.; Silnikov, V.N.; Venyaminova, A.G. Novel convenient approach to the solid-phase synthesis of oligonucleotide conjugates. *Molecules* **2019**, *24*, 4266. [CrossRef]
28. Ghosh, A.K.; Doung, T.T.; McKee, S.P.; Thompson, W.J. N, N′-dissuccinimidyl carbonate: A useful reagent for alkoxycarbonylation of amines. *Tetrahedron Lett.* **1992**, *33*, 2781–2784. [CrossRef]

29. Diamanti, S.; Arifuzzaman, S.; Elsen, A.; Genzer, J.; Vaia, R.A. Reactive patterning via post-functionalization of polymer brushes utilizing disuccinimidyl carbonate activation to couple primary amines. *Polymer* **2008**, *49*, 3770–3779. [CrossRef]
30. Gonçalves, I.; Abreu, A.S.; Matamá, T.; Ribeiro, A.; Gomes, A.C.; Silva, C.; Cavaco-Paulo, A. Enzymatic synthesis of poly(catechin)-antibiotic conjugates: An antimicrobial approach for indwelling catheters. *Appl. Microbiol. Biotechnol.* **2015**, *99*, 637–651. [CrossRef]
31. Lee, B.S.; Chi, Y.S.; Lee, K.B.; Kim, Y.G.; Choi, I.S. Functionalization of poly(oligo(ethylene glycol) methacrylate) films on gold and Si/SiO$_2$ for immobilization of proteins and cells: SPR and QCM studies. *Biomacromolecules* **2007**, *8*, 3922–3929. [CrossRef]
32. Cai, L.L.; Liu, P.; Li, X.; Huang, X.; Ye, Y.Q.; Chen, F.Y.; Yuan, H.; Hu, F.Q.; Du, Y.Z. RGD peptide-mediated chitosan-based polymeric micelles targeting delivery for integrin-overexpressing tumor cells. *Int. J. Nanomed.* **2011**, *6*, 3499–3508.
33. Chen, H.; Zhang, Y.; Li, D.; Hu, X.; Wang, L.; McClung, W.G.; Brash, J.L. Surfaces having dual fibrinolytic and protein resistant properties by immobilization of lysine on polyurethane through a PEG spacer. *J. Biomed. Mater. Research. Part. A* **2009**, *90*, 940–946. [CrossRef] [PubMed]
34. Swar, S.; Zajícová, V.; Müllerová, J.; Šubrtová, P.; Horáková, J.; Dolenský, B.; Řezanka, M.; Stibor, I. Effective poly(ethylene glycol) methyl ether grafting technique onto Nylon 6 surface to achieve resistance against pathogenic bacteria Staphylococcus aureus and Pseudomonas aeruginosa. *J. Mater. Sci.* **2018**, *53*, 14104–14120. [CrossRef]
35. Jia, X.; Herrera-Alonso, M.; McCarthy, T.J. Nylon surface modification. Part 1. Targeting the amide groups for selective introduction of reactive functionalities. *Polymer* **2006**, *14*, 4916–4924. [CrossRef]
36. Liu, Z.; Qi, L.; An, X.; Liu, C.; Hu, Y. Surface Engineering of Thin Film Composite Polyamide Membranes with Silver Nanoparticles through Layer-by-Layer Interfacial Polymerization for Antibacterial Properties. *ACS Appl. Mater. Interfaces* **2017**, *9*, 40987–40997. [CrossRef] [PubMed]
37. Sileika, T.S.; Kim, H.D.; Maniak, P.; Messersmith, P.B. Antibacterial performance of polydopaminemodified polymer surfaces containing passive and active components. *ACS Appl. Mater. Interfaces* **2011**, *3*, 4602–4610. [CrossRef]
38. Arima, Y.; Iwata, H. Effect of wettability and surface functional groups on protein adsorption and celladhesion using well-defined mixed self-assembled monolayers. *Biomaterials* **2007**, *28*, 3074–3082. [CrossRef]
39. Vaillard, V.A.; Menegon, M.; Neuman, N.I.; Vaillard, S.E. mPEG–NHS carbonates: Effect of alkyl spacerson the reactivity: Kinetic and mechanistic insights. *Appl. Polym. Sci.* **2019**, *136*, 47028. [CrossRef]
40. Romero-Vargas Castrillón, S.; Lu, X.; Shaffer, D.L.; Elimelech, M. Amine enrichment and poly(ethylene glycol) (PEG) surface modification of thin-film composite forward osmosis membranes for organic fouling control. *J. Membr. Sci.* **2014**, *450*, 331–339. [CrossRef]
41. Hsieh, Y.-L.; Timm, D.A. Relationship of substratum wettability measurements and initial Staphylococcus aureau adhesion to films and fabrics. *J. Colloid Interface Sci.* **1988**, *123*, 275–286. [CrossRef]
42. Pei, J.; Hall, H.; Spencer, N.D. The role of plasma proteins in cell adhesion to PEG surface-density-gradient-modified titanium oxide. *Biomaterials* **2011**, *32*, 8968–8978. [CrossRef] [PubMed]

© 2020 by the authors. Licensee MDPI, Basel, Switzerland. This article is an open access article distributed under the terms and conditions of the Creative Commons Attribution (CC BY) license (http://creativecommons.org/licenses/by/4.0/).

Article

Enhanced Hydrophilic and Electrophilic Properties of Polyvinyl Chloride (PVC) Biofilm Carrier

Haifeng Cai, Yang Wang, Kai Wu and Weihong Guo *

Polymer Processing Laboratory, Key Laboratory for Preparation and Application of Ultrafine Materials of Ministry of Education, School of Materials Science and Engineering, East China University of Science and Technology, Shanghai 200237, China; 18721358176@163.com (H.C.); m18864832205@163.com (Y.W.); 13122320038@163.com (K.W.)
* Correspondence: guoweihong@ecust.edu.cn; Tel.: +86-21-64251844

Received: 12 May 2020; Accepted: 27 May 2020; Published: 29 May 2020

Abstract: Polyvinyl chloride (PVC) biofilm carrier is used as a carrier for bacterial adsorption in wastewater treatment. The hydrophilicity and electrophilicity of its surface play an important role in the adsorption of bacteria. The PVC biofilm carrier was prepared by extruder, and its surface properties were investigated. In order to improve the hydrophilicity and electrophilic properties of the PVC biofilm carrier, polyvinyl alcohol (PVA) and cationic polyacrylamide (cPAM) were incorporated into polyvinyl chloride (PVC) by blending. Besides, the surface area of the PVC biofilm carrier was increased by azodicarbonamide modified with 10% by weight of zinc oxide (mAC). The surface contact angle of PVC applied by PVA and cPAM at 5 wt %, 15 wt % was 81.6°, which was 18.0% lower than pure PVC. It shows the significant improvement of the hydrophilicity of PVC. The zeta potential of pure PVC was −9.59 mV, while the modified PVC was 14.6 mV, which proves that the surface charge of PVC changed from negative to positive. Positive charge is more conducive to the adsorption of bacteria. It is obvious from the scanning electron microscope (SEM) images that holes appeared on the surface of the PVC biofilm carrier after adding mAC, which indicates the increase of PVC surface area.

Keywords: polyvinyl alcohol; cationic polyacrylamide; polyvinyl chloride; azodicarbonamide

1. Introduction

With the continuous development of industry and the continuous improvement of people's living standards, more and more garbage has been produced, which has led to an increase in sewage discharge. The generation of various pollutants in the aquatic environment has become an issue of increasing global concern in the past few decades [1]. Wastewater treatment plants are designed to eliminate various chemical and microbial pollutants in wastewater [2]. Due to its low cost and high treatment efficiency, biological wastewater treatment processes remain the most widely used method for removing organic pollutants and nutrients [3–5]. Generally, biofilm-based wastewater treatment systems have several advantages: for example, their high active biomass concentration, short hydraulic residence time, low space requirements, and less sludge production. In particular, the microbial communities in biofilms are diverse, which allow degrading a wide range of organic pollutants [6]. Meanwhile, the attachment and formation of biofilms largely depend on the surface properties of the biofilm carrier, including the physical/chemical properties of the carrier surface, the charge properties of the carrier surface, and the surface roughness of the carrier and so on [7–12]. In recent studies, the hydrophilicity and electronegativity of the carrier surface play an important role in the formation and treatment efficiency of biofilms [13–15]. Therefore, the selection of the required carrier is considered to be the decisive factor affecting bacterial adhesion and biofilm formation [16,17].

The interaction between bacteria and the surface of the carrier is mainly influenced by interface interactions such as repulsive force/attraction and Van der Waals force [18]. It is widely believed that the cell wall surface of bacteria contains functional groups such as –OH, –COOH, and –CHO. Hydrogen bonds can be formed between the carrier surface and bacteria [18,19]. In earlier research, it was found that the adhesion of bacteria to hydrophobic surfaces is significantly reduced [20,21]. In addition, the microbial surface is negatively charged as phosphoric acid and carboxylic acid groups in the microbial cell membrane. Therefore, the electrophilic property of biofilm carriers has a great effect on the adhesion of microorganisms and the formation of biofilms [22–24].

An ideal biofilm carrier should have the following characteristics: low cost, excellent mechanical strength, low density, stability, large specific surface area, high bioaffinity, anti-biodegradability, and anti-aging [25–27]. Polyvinyl chloride (PVC) owns most of the advantages described above. PVC is an important thermoplastic that can be used in a wide range of applications such as pipes, profiles, bars, films, insulation materials, etc. Moreover, PVC has excellent properties such as non-flammability, corrosion resistance, insulation, and wear resistance. The most prominent advantage of PVC is its low price, higher tensile strength, and larger bending strength [28]. However, the bioaffinity of PVC is relatively low, and its hydrophilicity is relatively weak. At the same time, its surface is relatively smooth, and its specific surface area is relatively small. Moreover, the surface of pure PVC is negatively charged, which is the same as the surface of bacteria [6,29–31]. These disadvantages make it difficult for bacteria and microorganisms to attach to PVC biofilm carriers.

In this study, in order to solve the above-mentioned shortcomings of PVC, such as weak hydrophilicity, small surface area and negatively charged surface, a typical hydrophilic polymer PVA was incorporated into the PVC to improve its hydrophilicity. In order to change the surface chargeability of the PVC biofilm carrier, positively charged polymer cationic polyacrylamide (cPAM) was incorporated into the PVC. At the same time, a compound blowing agent azodicarbonamide (mAC) was added to increase the specific surface area and surface roughness of the PVC biofilm carrier.

2. Experimental

2.1. Materials

PVC (SG-5) was purchased from Yuyao Maiduo Plastic Chemical Co., Ltd. (Yuyao, China). PVA and Stearic acid were purchased from Shanghai Lingfeng Chemical Reagent Co., Ltd. (Shanghai, China). cPAM was purchased from Zhengzhou Jintai Environmental Protection Technology Co., Ltd. (Zhengzhou, China). Azodicarbonamide was purchased from Shanghai Tengzhun Biological Technology Co., Ltd. (Shanghai, China). Calcium zinc stabilizer was purchased from Guangdong Winner New Material Technology Co., Ltd. (Foshan, China). Dioctyl terephthalate (DOTP) was purchased from Jining Baichuan Chemical Co., Ltd. (Jining, China). Zinc oxide was purchased from Sinopharm Chemical Reagent Co., Ltd. (Shanghai, China). Antioxidant 1010 was purchased from Shanghai Xian Ding Biological Technology Co., Ltd. (Shanghai, China). The specific information of the materials is shown in Table 1.

2.2. Preparation of PVC Biofilm Carrier

PVC resin (100 phr) was mixed with 0–10 wt % of PVA and 0–15 wt % of cPAM (the weight ratios of PVC to PVA were 100/0, 100/5, 100/10 and the weight ratios of PVC to cPAM were 100/0, 100/3, 100/6, 100/9, 100/12, and 100/15) using calcium zinc stabilizer (8 phr) as a heat stabilizer, DOTP (60 phr) as a plasticizer, and mAC (1 phr) as a chemical blowing agents. Then, stearic acid (0.4 phr) and antioxidant 1010 (0.5 phr) were added, and the mixture was mixed thoroughly in a high-speed mixer (SHR-10A, 750 r/min). After the mixture was well mixed and the DOTP was fully absorbed by the mixture, the mixture appears loose. Then, the mixture is taken out and dried. Then, the mixture was extruded and winded in a twin-screw extruder (18 °C, 15 r/min). As a control, PVC without fillers was prepared following the same procedure [28,32]. The preparation of PVC biofilm carrier and the photo of sample are shown in Figures 1 and 2.

Table 1. The information of the materials. cPAM: cationic polyacrylamide, PVA: polyvinyl alcohol, PVC: Polyvinyl chloride.

Material	Sales/Manufacturer	Material Properties
PVC (SG-5)	Yuyao Maiduo Plastic Chemical Co., Ltd.	GB/T 5761-2006 Injection Grade
PVA	Shanghai Lingfeng Chemical Reagent Co., Ltd.	A.R, Average degree of polymerization:1750 ± 50
Stearic acid	Shanghai Lingfeng Chemical Reagent Co., Ltd.	A.R, MW:284.48
cPAM	Zhengzhou Jintai Environmental Protection Technology Co., Ltd.	A.R, MW:1.2×10^7
Azodicarbonamide	Shanghai Tengzhun Biological Technology Co., Ltd.	A.R, MW:116.08
Calcium zinc stabilizer	Guangdong Winner New Material Technology Co., Ltd.	WWP-F02 A.R
Dioctyl terephthalate	Jining Baichuan Chemical Co., Ltd.	A.R
Zinc oxide	Sinopharm Chemical Reagent Co., Ltd.	A.R, MW:81.39
Antioxidant 1010	Shanghai Xian Ding Biological Technology Co., Ltd.	C.P, MW:1177.63

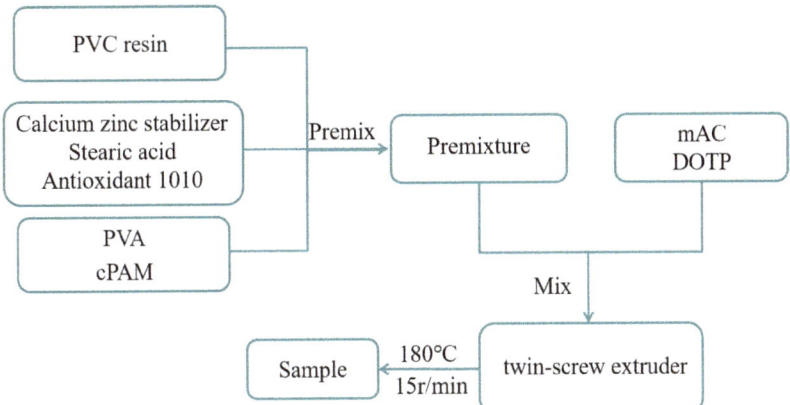

Figure 1. Preparation of PVC biofilm carrier.

Figure 2. The photo of sample (a) sample particles, (b) sample.

The formulation and abbreviation of the samples are listed in Table 2.

Table 2. The formulation and abbreviation of the samples.

Samples	PVC/g	PVA/g	mAC/g	cPAM/g
PVC	100	0	0	0
PVC/PVA-5	100	5	0	0
PVC/PVA-10	100	10	0	0
PVC/mAC-PVA	100	0	1	0
PVC/mAC-PVA-5	100	5	1	0
PVC/mAC-PVA-10	100	10	1	0
PVC/cPAM	100	5	1	0
PVC/cPAM-3	100	5	1	3
PVC/cPAM-6	100	5	1	6
PVC/cPAM-9	100	5	1	9
PVC/cPAM-12	100	5	1	12
PVC/cPAM-15	100	5	1	15

2.3. Surface Contact Angle

The surface contact angle of the resultant PVC biofilm carriers was measured by a Standard type contact angle meter (JC2000D2, Shanghai Zhongchen Digital Technology Equipment Co., Ltd., Shanghai, China). Distilled water was slowly dropped onto the surface of the specimens. Then, a photograph of the water droplets and the surface of the specimen were taken by the contact angle meter. At least five different locations were measured for each specimen. The conductivity of the used distilled water was 0.056 µS/cm, and the resistivity of the used distilled water was 18.2 MΩ·cm at 25 °C. The indoor temperature is 25 ± 0.5 °C, and the indoor humidity is 50% ± 1% during the test.

2.4. Fourier Transform Infrared Spectra

Fourier transform infrared (FTIR) spectra were measured on a NICOLET 6700 spectrometer (Thermo Scientific Co., Waltham, MA, USA) from 4000 to 400 cm^{-1} to study the interaction between PVA and PVC. The samples tested are in powder form, and potassium bromide pressed-disk technique was used.

2.5. Zeta Potential

The zeta potential of the specimens was measured by a laser particle size analyzer (LS230, Microtrac Inc., Clay, FL, USA). First, we dissolve the powder sample in distilled water and place it in the sample cell for measurement. The conductivity of the used distilled water was 0.056 µS/cm and the resistivity of the used distilled water was 18.2 MΩ·cm at 25 °C.

2.6. Field Emission Scanning Electron Microscope

The surface morphology of the specimens was observed by using an S-4800 field emission scanning electron microscope (Hitachi, SEM, Tokyo, Japan). Prior to SEM observation, all specimens were coated with a thin gold layer. The magnification power of SEM was 1.00k and 2.00k, and the voltage was 15.0 kv.

2.7. Mechanical Properties

The tensile properties were determined by using an MTSE44 universal testing machine (Jinan Yongce Industrial Equipment Co., Ltd., Jinan, China) in accordance with ISO 527 respectively. At least five independent measurements were conducted for each sample (75 mm × 4 mm × 2 mm for tensile test).

3. Results and Discussion

For PVC biofilm carriers prepared by different formulations, their surface contact angle, zeta potential, mechanical properties, and surface morphology were tested by using the corresponding equipment.

3.1. Hydrophilicity of PVC Biofilm Carrier

The hydrophilicity of the PVC biofilm carrier plays an important role in the adsorption of bacteria. Since the cell wall surface of bacteria contains functional groups such as –OH, –COOH, and –CHO, it prefers to be adsorbed on the surface of strong hydrophilic carriers [18,19]. The surface contact angle can be used intuitively to characterize the hydrophilicity of the carrier. In order to investigate the modification effect of polyvinyl alcohol on polyvinyl chloride, the surface hydrophilic properties of polyvinyl chloride were tested and characterized. The surface contact angle of unmodified PVC and modified PVC were measured by a standard-type contact angle meter. Figures 3 and 4 are the surface contact angle of the PVC biofilm carrier. It shows the change in the surface contact angle of the specimens after the addition of PVA (PVC/PVA-5, PVC/PVA-10) and mAC (PVC/mAC-PVA, PVC/mAC-PVA-5, PVC/mAC-PVA-10).

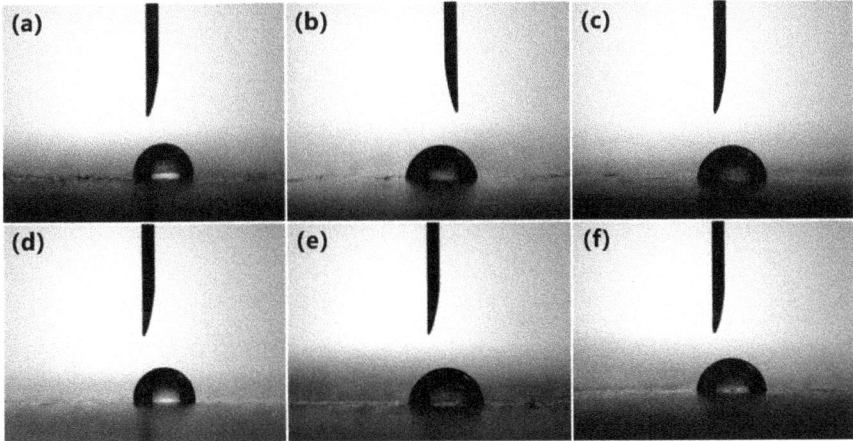

Figure 3. Surface contact angle images of the PVC biofilm carrier: (**a**) PVC, (**b**) PVC/PVA-5, (**c**) PVC/PVA-10, (**d**) PVC/mAC-PVA, (**e**) PVC/mAC-PVA-5, (**f**) PVC/mAC-PVA-10.

Compared with pure PVC, the contact angle of the PVC sample with 5 wt % of PVA has been significantly reduced, from 99.5° to 84.2°, which is reduced by 15.4%. The significant reduction of surface contact angle indicates that PVC was successfully modified by PVA. The specific reaction mechanism is shown in Figure 5. PVA is an aqueous polymer whose segment is rich in hydroxyl groups. When it is added to PVC, the hydroxyl groups on the segments partially interact with chlorine on the PVC to form hydrogen bonds. This makes PVA well integrated on PVC. At the same time, the hydroxyl group that exists on PVA alone improves the hydrophilicity of PVC. Hydroxyl is a hydrophilic group. The improvement of the hydrophilicity of PVC is due to the independent hydroxyl group on PVA. The more hydroxyl groups on the modified PVC, the better its hydrophilic property and the lower the surface contact angle. When the amount of PVA was increased to 10 wt %, the surface contact angle of the sample increased from 84.6° to 86.4°. This could be because the amount of PVA added is too large, causing the hydroxyl group between PVA to act to generate hydrogen bonds, which in turn reduces the binding of PVA to PVC and the number of PVA independent hydroxyl groups. Therefore, it is more appropriate to control the amount of PVA added to 5 wt %.

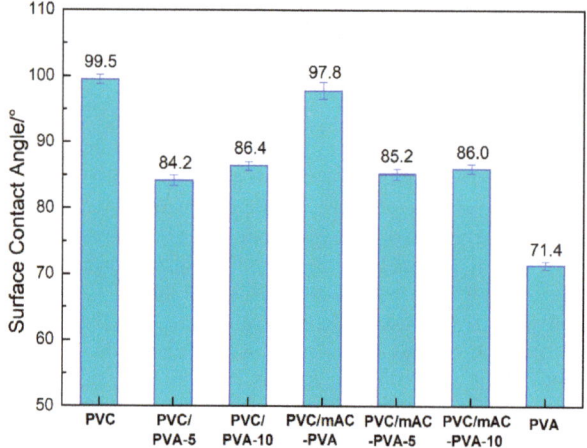

Figure 4. Surface contact angle of the PVC biofilm carrier.

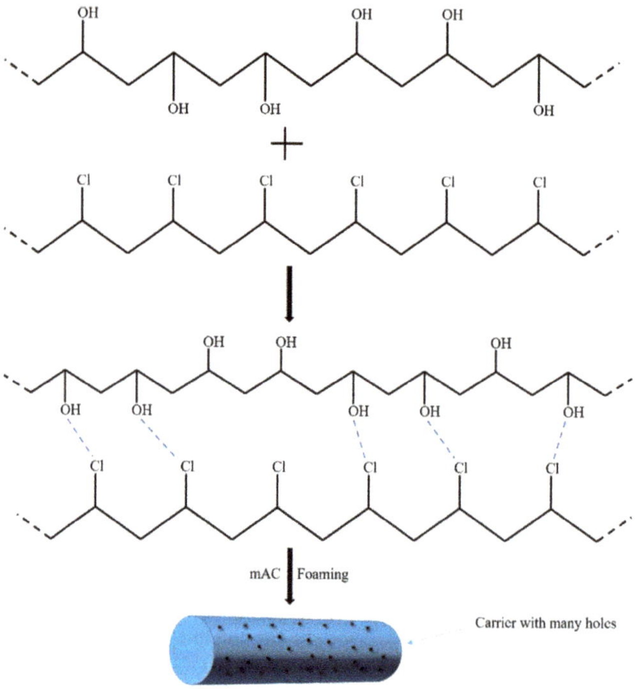

Figure 5. The reaction mechanism of PVC's modification.

Bacteria and microorganisms used to treat sewage are generally hydrophilic [18,20,21]. The hydrophilicity of PVC biofilm carrier modified with PVA was effectively improved. Therefore, bacteria and microorganisms can be more effectively attached to the modified PVC biofilm carrier. The increase in the amount of bacteria on the biofilm carrier can effectively improve the ability and efficiency of sewage treatment.

At the same time, as shown in Figure 4, after the addition of mAC, the change trend of the contact angle of the sample is consistent with the above description. This trend is expected. The role of the mAC is to increase the surface area of the PVC biofilm carrier by foaming and has no effect on the surface chemistry of the sample. Neither PVC nor PVA will react with mAC, so the chemical properties of the modified PVC will not be affected. This result shows that there is no side effect when adding PVA and mAC into PVC at the same time.

Figure 6and Figure 7 are the surface contact angle of the PVC biofilm carrier, which contains different amounts of cPAM. It shows the change in the surface contact angle of the sample after adding cPAM based on the addition of 5 wt % PVA and 1 wt % mAC. It can be seen that compared with pure PVC, after the amount of cPAM added reaches 12 wt %, the contact angle of PVC is further reduced to 80.8°. The data indicate that compared with the PVC biofilm carrier modified by PVA, the hydrophilic property of the PVC biofilm carrier added with a certain amount of cPAM and PVA is stronger. The reason for this phenomenon is that the amino group on the cPAM chain segment plays a role in enhancing the hydrophilicity of the PVC biofilm carrier. The mechanism between cPAM and PVC is similar to PVA and PVC. Although the polarity of the amino group is not as large as that of the hydroxyl group, after the amount reaches a certain amount, the amino group can also play a role in giving the PVC a certain hydrophilicity.

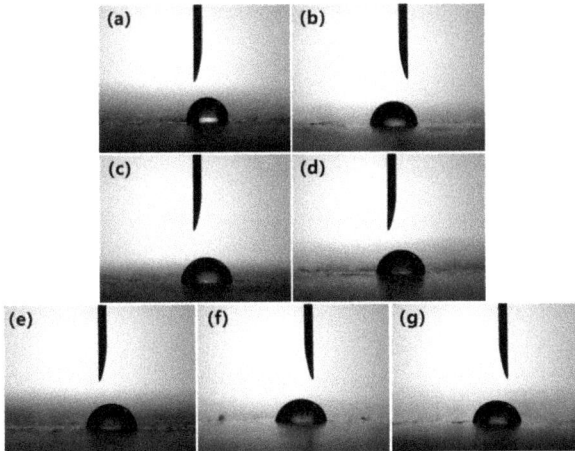

Figure 6. Surface contact angle images of the PVC modified by cPAM: (**a**) PVC, (**b**) PVC/cPAM, (**c**) PVC/cPAM-3, (**d**) PVC/cPAM-6, (**e**) PVC/cPAM-9, (**f**) PVC/cPAM-12, and (**g**) PVC/cPAM-15.

3.2. FTIR Spectra and Energy Dispersive Spectrum (EDS)—The Change of Hydroxyl Groups

In the above part, the change of the surface contact angle of the carrier indicates that the hydrophilicity of the carrier is enhanced after the addition of PVA. That is because the hydroxyl group that exists on PVA alone improves the hydrophilicity of PVC. The specific reaction mechanism is shown in Figure 5. In order to verify that the improvement of the hydrophilic property of the PVC biofilm carrier is indeed due to the reaction between PVA and PVC, the changes in surface functional groups and elements of the PVC before and after modification need to be tested and investigated. The FTIR and EDS spectra of modified PVC and unmodified PVC were tested and characterized. Figure 8 is the FTIR spectra of unmodified PVC (a) and PVC/PVA-5 (b). The purpose of testing the infrared spectrum of PVC is to investigate whether PVA is effective in modifying PVC. As shown in

Figure 8, the 2959 cm^{-1} absorption peak of PVC and the 2958 cm^{-1} absorption peak of PVC/PVA-5 are the stretching vibrations of CH2. The 1462 cm^{-1} absorption peak of PVC and the 1459 cm^{-1} absorption peak of PVC/PVA-5 are the bending vibration of CH$_2$. Meanwhile, the absorption peaks of 1265 and 1272 cm^{-1} are attributed to the wobbling vibration of the adjacent carbon atom of CH$_2$, which is connected to a chlorine atom. The 1724 cm^{-1} absorption peak of PVC/PVA-5 is the absorption peak of C=O remaining in PVA. Besides, in Figure 8, the absorption peak around 3400 cm^{-1} indicates the association absorption peak of O–H [28,33]. Compared with unmodified PVC, the absorption peak of O–H of PVC modified by PVA is obviously sharper and wider. It shows that a certain amount of O–H appears in the surface of the modified PVC. This result proves that the reaction mechanism between PVA and PVC that we mentioned earlier is reasonable and correct.

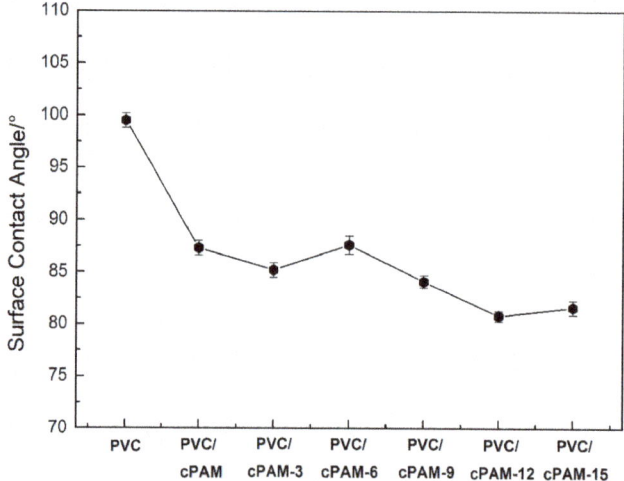

Figure 7. Surface contact angle of the PVC modified by cPAM.

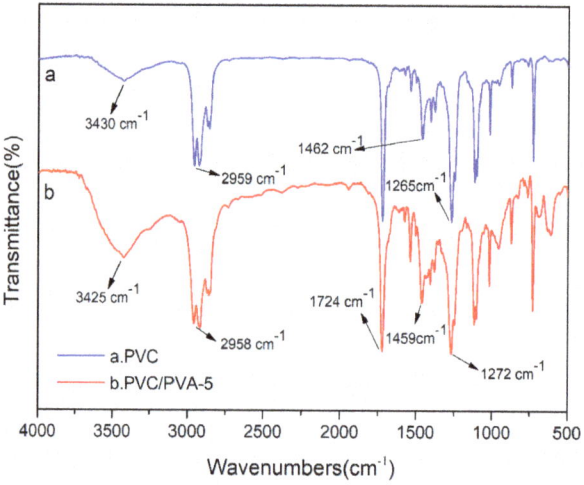

Figure 8. Fourier transform infrared (FTIR) spectra of (a) PVC and (b) PVC/PVA-5.

For further verification, EDS analysis was performed on the samples. The EDS results of PVC and PVC/PVA-5 are summarized in Figure 9 and Table 3.

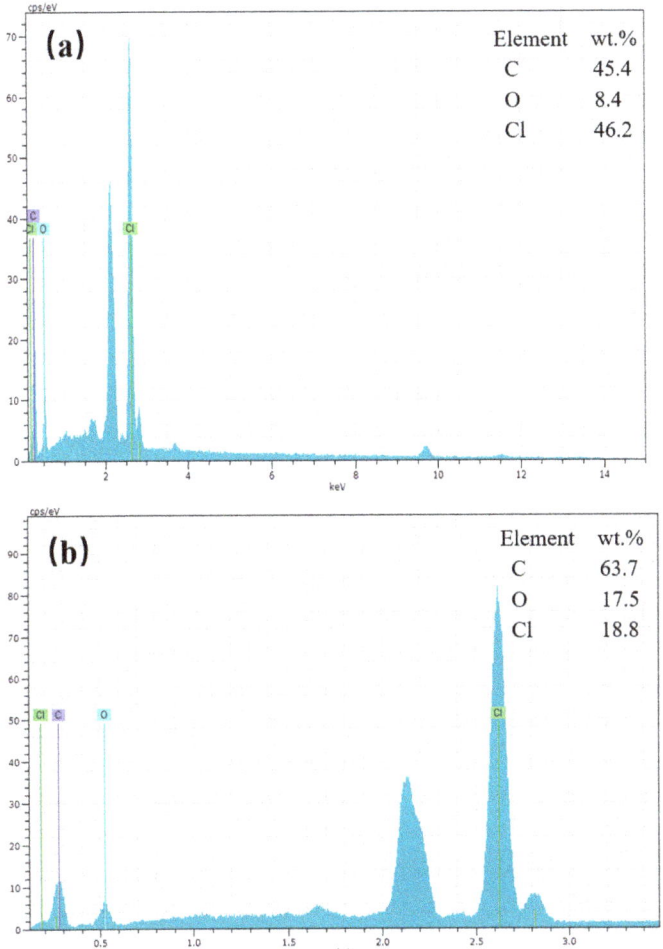

Figure 9. EDS spectra of (a) PVC and (b) PVC/PVA-5.

Table 3. The EDS results of PVC and PVC/PVA-5.

Sample	PVC		PVC/PVA-5	
	Weight (wt %)	Atom (at. %)	Weight (wt %)	Atom (at. %)
C	45.4 ± 2.4	67.4 ± 2.1	63.7 ± 4.2	76.6 ± 4.0
O	8.4 ± 0.5	9.3 ± 0.4	17.5 ± 1.4	15.8 ± 1.2
Cl	46.2 ± 0.6	23.3 ± 0.5	18.8 ± 0.3	7.6 ± 0.2

It can be seen from Table 3 that the element content of the unmodified PVC is C (45.4 wt %), O (8.4 wt %), and Cl (46.2 wt %), while the element content of the PVC modified by PVA is C (63.7 wt %), O (17.5 wt %), and Cl (18.8 wt %). Compared with unmodified PVC, the content of oxygen and carbon of the modified PVC increased significantly, while the content of chlorine decreased significantly.

This indicates that PVC successfully interacts with PVA, resulting in an increase in the content of oxygen and carbon element on the surface.

Moreover, in order to evaluate the miscibility of PVC and PVA in the present samples, the melting characteristics of modified PVC and unmodified PVC were tested and characterized. Figure 10 is the melting characteristics of unmodified PVC, PVA and PVC/PVA-5. It can be seen from Figure 10 that the melting points of unmodified PVC, PVA, and PVC/PVA-5 are 158.1, 192.7, and 170.6 °C. The melting point of PVC/PVA-5 is between unmodified PVC and PVA. Moreover, it can be clearly seen from the melting curve of PVC/PVA-5 in Figure 10 that only a relatively smooth melting peak appears in PVC/PVA-5, and no other peaks appear. This shows that PVC and PVA are well miscible together, and their miscibility is relatively good.

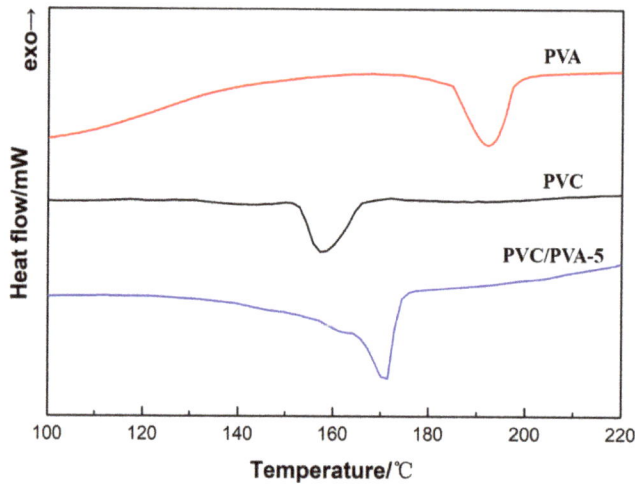

Figure 10. The melting characteristics of PVC, PVA, and PVC/PVA-5.

From the melting characteristics of PVC, PVA, and PVC/PVA-5 in Figure 10, it can be analyzed that the miscibility of PVC and PVA is relatively good. Combining the results of FTIR spectra and EDS spectra, it can be determined that PVC is successfully modified by PVA. The hydrophilic property of PVC is effectively enhanced.

3.3. Electrophilicity of PVC Biofilm Carrier

The microbial surface is negatively charged as phosphoric acid and carboxylic acid groups in the microbial cell membrane. Therefore, the electrophilic property of biofilm carriers has a great effect on the adhesion of microorganisms and the formation of biofilms [22–24]. In order to change the surface chargeability of the PVC biofilm carrier from negative to positive, positively charged polymer cationic polyacrylamide (cPAM) was incorporated into the PVC. The zeta potential can be used intuitively to characterize the electrophilicity of the carrier. In order to investigate whether the surface charge property of PVC have changed after the addition of cPAM, the charge properties of the surface of the PVC biofilm carrier were tested and characterized. The zeta potential of modified PVC and unmodified PVC were measured by a laser particle size analyzer.

Figure 11 is the zeta potential of the PVC biofilm carrier. It shows the change in zeta potential of the sample after adding different amounts of cPAM (0, 3, 6, 9, 12 and 15 wt %).

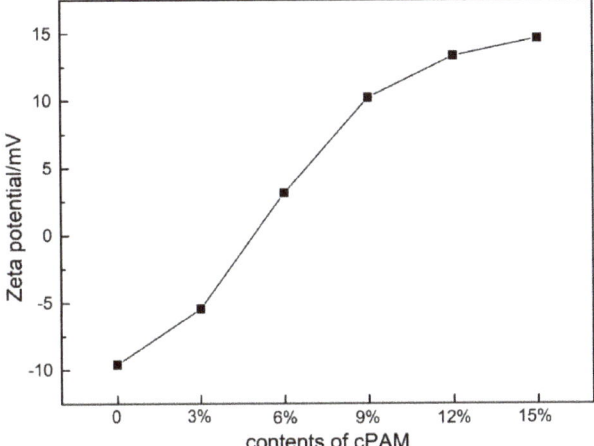

Figure 11. Zeta potential of the PVC biofilm carrier.

It can be seen that as the amount of cPAM added increases, the zeta potential of the sample also increases from −9.59 mV for pure PVC to 14.6 mV for 15 wt %. This shows that after the addition of cPAM, the surface chargeability of PVC gradually changes from negative to positive, and as the amount of cPAM increases, the zeta potential increases, indicating that the positive charge has also become stronger and stronger. Figure 12 shows the electrostatic interactions between unmodified PVC/modified PVC and bacteria. As shown in Figure 12, bacteria are generally negatively charged in water, and unmodified PVC is also negatively charged. There is a repulsive effect between the two, which is not conducive to the adhesion of bacteria. Meanwhile, the modified PVC has a positive charge, which will attract the bacteria and facilitate the adsorption of bacteria. The increase in the amount of bacteria on the biofilm carrier can effectively improve the ability and efficiency of sewage treatment [15,30,34].

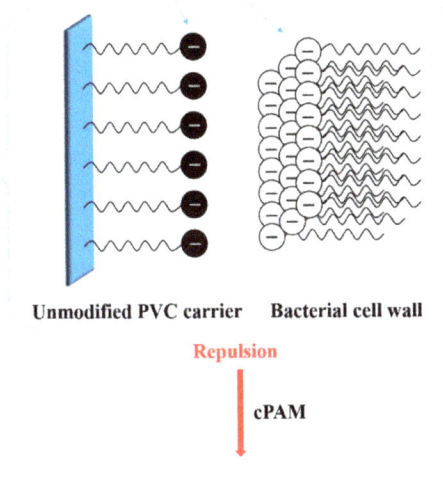

Figure 12. *Cont.*

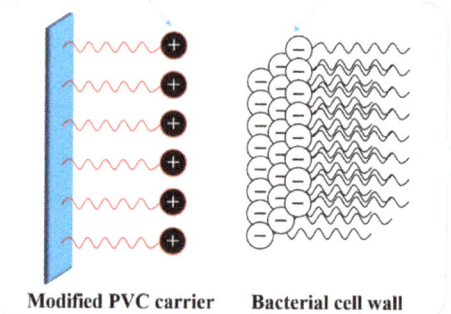

Figure 12. The electrostatic interactions between unmodified PVC/modified PVC and bacteria.

3.4. Surface Morphology of PVC Biofilm Carrier

In order to investigate the surface change of PVC after the addition of the compound chemical blowing agent mAC, the surface morphology of the PVC biofilm carrier before and after foaming was observed by S-4800 field emission scanning electron microscope. Figure 13 is the SEM images of the surface morphology of pure PVC (Figure 13a), 1 wt % mAC (Figure 13b), 5 wt % PVA (Figure 13c), and 1 wt % mAC with 5 wt % PVA (Figure 13d). The part circled by the red circle in Figure 13 is the hole. In a field emission scanning electron microscope, the magnification power of SEM was 2.00 k. Comparing Figure 13a,b, it can be seen that the surface of pure PVC is relatively smooth before the addition of mAC. After the addition of mAC, the surface of the PVC sample clearly shows holes, which indicates that mAC has successfully foamed. Thereby, the surface area of the sample was increased. Although the distribution of pores is relatively irregular, it can be seen from the figure that the size of the pores is more than the micron level. The general diameter of bacteria is about 1 μm, so these holes are completely satisfied for the adhesion and growth of bacteria on it. The principle of mAC is that at the processing temperature of PVC, zinc oxide will activate azodicarbonamide and reduce its decomposition temperature, so that its decomposition temperature is close to the processing temperature. Therefore, mAC will decompose during the processing of PVC. Its decomposition products are mainly harmless gases, such as nitrogen, which will escape and form pores on the surface of the material [35]. Comparing Figure 13c,d, after adding 5 wt % PVA and 1 wt % mAC, the formation of the pores on the surface of the sample is consistent with the one described above.

Figure 14 is the SEM surface morphology of the PVC biofilm carrier, which contains different amounts of cPAM based on the addition of 5 wt % PVA and 1 wt % mAC. It shows the SEM surface morphology of pure PVC (Figure 14a), 3 wt % cPAM (Figure 14b), 12 wt % cPAM (Figure 14c), and 15 wt % cPAM (Figure 14d). The part circled by the red circle in Figure 14 is the hole. At the same time, as shown in Figure 14, the individual holes in Figure 14a,c are enlarged, and the enlarged images of the holes are indicated by arrows. The SEM magnification of the enlarged image of the hole is 10.0 k. From these SEM images, we can find that the amount of cPAM added will not affect the foaming on the basis of adding the blowing agent mAC. The blowing agent can still foam normally to form pores on the surface of the sample, thereby increasing the surface area of the sample.

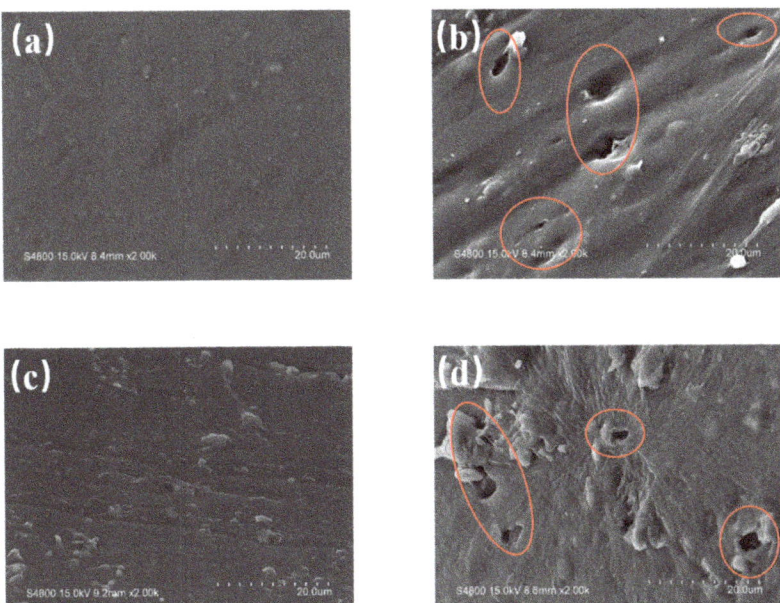

Figure 13. SEM images of surface morphology of the PVC biofilm carrier: (**a**) PVC, (**b**) PVC/mAC-PVA, (**c**) PVC/PVA-5, (**d**) PVC/mAC-PVA-5. The red circle is the hole.

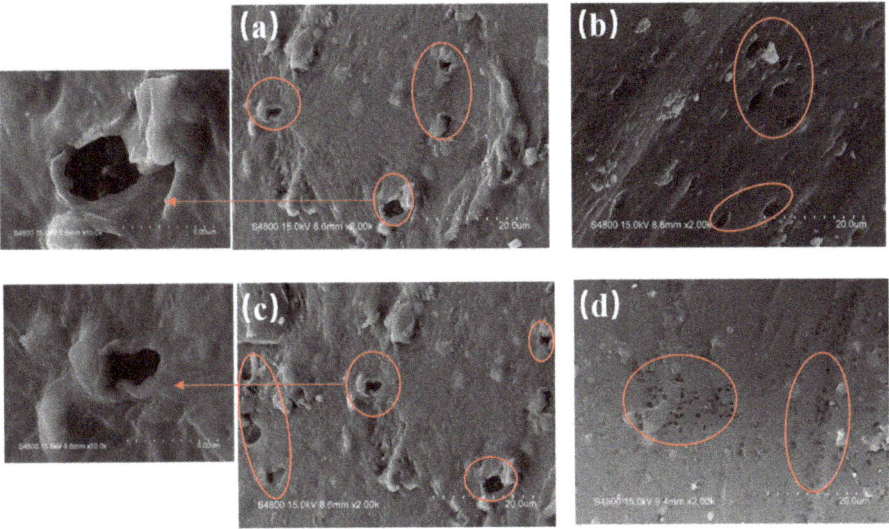

Figure 14. SEM images of surface morphology of the PVC biofilm carrier: (**a**) PVC/cPAM, (**b**) PVC/cPAM-3, (**c**) PVC/cPAM-12, and (**d**) PVC/cPAM-15. The red circle is the hole.

3.5. Mechanical Properties

Figures 15 and 16 is the tensile strength of the PVC biofilm carrier. It shows the tensile properties of each sample after adding PVA, mAC, and cPAM. It can be clearly seen from the figure that compared with pure PVC, after adding PVA, mAC, and cPAM, the tensile properties of the sample did not change much, the highest was 5.6%, and the lowest was 4.2%. This is because the added substances did not change the structure and segment of the PVC matrix. The main change was the surface properties of PVC. The most important effect was the combination of hydrogen bonding with PVC, and no other chemical reaction. The tensile properties of PVC are mainly imparted by its structure. Therefore, the tensile properties of the samples do not change greatly on the basis that the main structure and the segment of the PVC matrix are not greatly changed. This result illustrates that the amount of each filler added is appropriate [36–40].

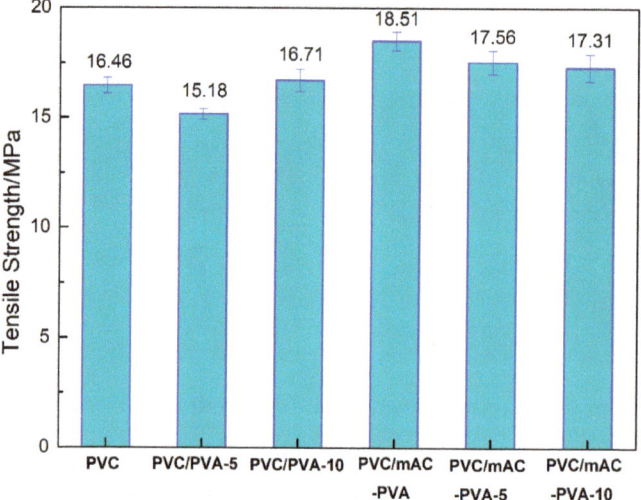

Figure 15. Tensile strength of the PVC biofilm carrier.

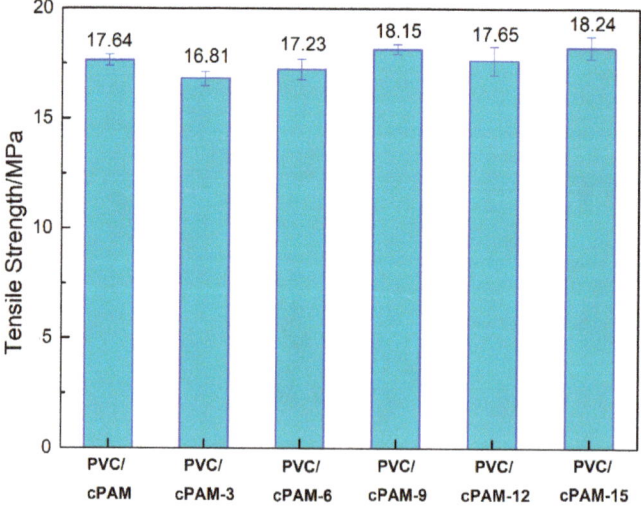

Figure 16. Tensile strength of the PVC modified by cPAM.

4. Conclusions

In this study, PVA and cPAM were incorporated into the PVC matrix. At the same time, mAC was also added. Then, the surface properties and zeta potential of the PVC biofilm carrier were investigated. The results showed that the surface contact angle of the PVC biofilm carrier was significantly lower than that of pure PVC after the addition of PVA and cPAM. This is the effect of the hydroxyl and amino groups contained in PVA and cPAM. In addition, due to the addition of cPAM, the zeta potential of the PVC biofilm carrier gradually increased. The surface chargeability of PVC gradually changes from negative to positive. When 5 wt % PVA and 15 wt % cPAM were added, the surface contact angle and zeta potential of PVC are 81.6° and 14.6 mV. In addition, the holes clearly appeared on the surface of PVC after 1 wt % mAC was added. This result shows that the hydrophilicity, electrophilicity, and specific surface area of the PVC biofilm carrier have been significantly improved. Therefore, it is beneficial to the adsorption of bacteria and microorganisms.

Author Contributions: Data curation, H.C.; Formal analysis, H.C.; Investigation, H.C., Y.W. and K.W.; Methodology, W.G.; Resources, W.G.; Software, Y.W. and K.W.; Writing—original draft, H.C.; Writing—review & editing, H.C. and W.G. All authors have read and agreed to the published version of the manuscript.

Funding: This research received no external funding.

Acknowledgments: The authors sincerely acknowledge the financial assistance of Tus-Water Group Co., LTD.

Conflicts of Interest: The authors declare no conflict of interest.

References

1. Luo, Y.; Guo, W.; Ngo, H.; Nghiem, L.; Hai, F.; Zhang, J.; Liang, S.; Wang, X. A review on the occurrence of micropollutants in the aquatic environment and their fate and removal during wastewater treatment. *Sci. Total Environ.* **2014**, *473*, 619–641. [CrossRef] [PubMed]
2. Tran, N.; Ngo, H.; Urase, T.; Gin, K. A critical review on characterization strategies of organic matter for wastewater and water treatment processes. *Bioresour. Technol.* **2015**, *193*, 523–533. [CrossRef] [PubMed]
3. Kikuchi, T.; Tanaka, S. Biological removal and recovery of toxic heavy metals in water environment. *Crit. Rev. Environ. Sci. Technol.* **2012**, *42*, 1007–1057. [CrossRef]
4. Zhang, Y.; Zhao, X.; Zhang, X.; Peng, S. A review of different drinking water treatments for natural organic matter removal. *Water Sci. Technol. Water Supply* **2015**, *15*, 442–455. [CrossRef]
5. Xu, H.; Yang, B.; Liu, Y.; Li, F.; Shen, C.; Ma, C. Recent advances in anaerobic biological processes for textile printing and dyeing wastewater treatment: A mini-review. *World J. Microbiol. Biotechnol.* **2018**, *34*, 165–169. [CrossRef]
6. Zhao, Y.; Liu, D.; Huang, W.; Yang, Y.; Ji, M.; Nghiem, L.; Trinh, Q.; Tran, N. Insights into biofilm carriers for biological wastewater treatment processes: Current state-of-the-art, challenges, and opportunities. *Bioresour. Technol.* **2019**, *288*, 1–15. [CrossRef]
7. Feng, L.; Chen, K.; Han, D.; Zhao, J.; Lu, Y.; Yang, G.; Mu, J.; Zhao, X. Comparison of nitrogen removal and microbial properties in solid-phase denitrification systems for water purification with various pretreated lignocellulosic carriers. *Bioresour. Technol.* **2016**, *224*, 236–245. [CrossRef]
8. Guo, W.; Ngo, H.; Dharmawan, F.; Palmer, C. Roles of polyurethane foam in aerobic moving and fixed bed bioreactors. *Bioresour. Technol.* **2010**, *101*, 1435–1439. [CrossRef]
9. Huang, T.; Xu, J.; Cai, D. Efficiency of active barriers attaching biofilm as sediment capping to eliminate the internal nitrogen in eutrophic lake and canal. *J. Environ. Sci.* **2011**, *23*, 38–43. [CrossRef]
10. Müller-Renno, C.; Buhl, S.; Davoudi, N.; Aurich, J.; Ripperger, S.; Ulber, R. Novel Materials for Biofilm Reactors and their Characterization. *Adv. Biochem. Eng. Biotechnol.* **2013**, *146*, 207–233. [CrossRef]
11. Tarjányi-Szikora, S.; Oláh, J.; Makó, M.; Palkó, G.; Barkács, K.; Záray, G. Comparison of different granular solids as biofilm carriers. *Microchem. J.* **2013**, *107*, 101–107. [CrossRef]
12. Ahmad, M.; Liu, S.; Mahmood, N.; Mahmood, A.; Ali, M.; Zheng, M.; Ni, J. Effects of porous carrier size on biofilm development, microbial distribution and nitrogen removal in microaerobic bioreactors. *Bioresour. Technol.* **2017**, *234*, 360–369. [CrossRef]

13. Chu, L.; Wang, J.; Quan, F.; Xing, X.; Tang, L.; Zhang, C. Modification of polyurethane foam carriers and application in a moving bed biofilm reactor. *Process. Biochem.* **2014**, *49*, 1979–1982. [CrossRef]
14. Deng, L.; Guo, W.; Ngo, H.; Zhang, X.; Wang, X.; Zhang, Q. New functional biocarriers for enhancing the performance of a hybrid moving bed biofilm reactor-membrane bioreactor system. *Bioresour. Technol.* **2016**, *208*, 87–93. [CrossRef]
15. Mao, Y.; Quan, X.; Zhao, H.; Zhang, Y.; Chen, S.; Liu, T.; Quan, W. Accelerated startup of moving bed biofilm process with novel electrophilic suspended biofilm carriers. *Chem. Eng. J.* **2017**, *315*, 364–372. [CrossRef]
16. Liu, Y.; Zhao, Q. Influence of surface energy of modified surfaces on bacterial adhesion. *Biophys. Chem.* **2005**, *117*, 39–45. [CrossRef] [PubMed]
17. Zhang, X.; Zhou, X.; Ni, H. Surface Modification of Basalt Fiber, with Organic/Inorganic Composites for Biofilm Carrier Used in Wastewater Treatment. *ACS Sustain. Chem. Eng.* **2018**, *6*, 2596–2602. [CrossRef]
18. Renner, L.; Weibel, D. Physicochemical regulation of biofilm formation. *Mrs Bull.* **2011**, *36*, 347–355. [CrossRef]
19. Kang, S.; Choi, H. Effect of surface hydrophobicity on the adhesion of S. cerevisiae onto modified surfaces by poly(styrene-ran-sulfonic acid) random copolymers. *Colloids Surf. B Biointerfaces* **2005**, *46*, 70–77. [CrossRef]
20. Feng, G.; Cheng, Y.; Wang, S.; Borca-Tasciuc, D.; Worobo, R.; Moraru, C. Bacterial attachment and biofilm formation on surfaces are reduced by small-diameter nanoscale pores: How small is small enough? *Npj Biofilms Microbiomes* **2015**, *1*, 15022. [CrossRef]
21. Yuan, Y.; Hays, M.; Hardwidge, P.; Kim, J. Surface characteristics influencing bacterial adhesion to polymeric substrates. *RSC Adv.* **2017**, *7*, 14254–14261. [CrossRef]
22. Terada, A.; Okuyama, K.; Nishikawa, M.; Tsuneda, S.; Hosomi, M. The effect of surface charge property on Escherichia coli initial adhesion and subsequent biofilm formation. *Biotechnol. Bioeng.* **2012**, *109*, 1745–1754. [CrossRef] [PubMed]
23. Chen, S.; Cheng, X.; Zhang, X.; Sun, D. Influence of surface modification of polyethylene biocarriers on biofilm properties and wastewater treatment efficiency in moving-bed biofilm reactors. *Water Sci. Technol.* **2012**, *65*, 1021–1026. [CrossRef] [PubMed]
24. Van Merode, A.; Van Der Mei, H.; Busscher, H.; Krom, B. Influence of Culture Heterogeneity in Cell Surface Charge on Adhesion and Biofilm Formation by Enterococcus faecalis. *J. Bacteriol.* **2006**, *188*, 2421–2426. [CrossRef]
25. Liu, Y.; Zhu, Y.; Jia, H.; Yong, X.; Zhang, L.; Zhou, J.; Cao, Z.; Kruse, A.; Wei, P. Effects of different biofilm carriers on biogas production during anaerobic digestion of corn straw. *Bioresour. Technol.* **2017**, *48*, 445–451. [CrossRef]
26. Xu, S.; Jiang, Q. Surface modification of carbon fiber support by ferrous oxalate for biofilm wastewater treatment system. *J. Clean. Prod.* **2018**, *194*, 416–424. [CrossRef]
27. Nguyen, V.; Karunakaran, E.; Collins, G.; Biggs, C. Physicochemical analysis of initial adhesion and biofilm formation of Methanosarcina barkeri on polymer support material. *Colloids Surf. B Biointerfaces* **2016**, *143*, 518–525. [CrossRef]
28. Lu, Y.; Wu, C.; Xu, S. Mechanical, thermal and flame retardant properties of magnesium hydroxide filled poly(vinyl chloride) composites: The effect of filler shape. *Compos. Part. A* **2018**, *113*, 1–11. [CrossRef]
29. Fulaz, S.; Vitale, S.; Quinn, L.; Casey, E. Nanoparticle–Biofilm Interactions: The Role of the EPS Matrix. *Trends. Microbiol.* **2019**, *27*, 915–926. [CrossRef]
30. Liu, T.; Jia, G.; Quan, X. Accelerated start-up and microbial community structures of simultaneous nitrification and denitrification by using novel suspended carriers. *J. Chem. Technol. Biotechnol.* **2018**, *93*, 577–584. [CrossRef]
31. Zhu, Y. Preparation and Characterization of a New Hydrophilic and Biocompatible Magnetic Polypropylene Carrier used in Wastewater Treatment. *Environ. Technol.* **2017**, *39*, 1–29. [CrossRef]
32. Lu, Y.; Jiang, N.; Li, X. Effect of inorganic–organic surface modification of calcium sulfate whiskers on mechanical and thermal properties of calcium sulfate whisker/poly(vinyl chloride) composites. *RSC Adv.* **2017**, *7*, 46486–46498. [CrossRef]
33. Yuan, W.; Cui, J.; Cai, Y.; Xu, S. A novel surface modification for calcium sulfate whisker used for reinforcement of poly(vinyl chloride). *J. Polym. Res.* **2015**, *22*, 173. [CrossRef]

34. Abbasnezhad, H.; Gray, M.; Foght, J. Two different mechanisms for adhesion of Gram-negative bacterium, Pseudomonas fluorescens LP6a, to an oil–water interface. *Colloids Surf. B Biointerfaces* **2008**, *62*, 36–41. [CrossRef]
35. Petchwattana, N.; Covavisaruch, S. Influences of particle sizes and contents of chemical blowing agents on foaming wood plastic composites prepared from poly(vinyl chloride) and rice hull. *Mater. Des.* **2011**, *32*, 2844–2850. [CrossRef]
36. Feng, Q.; Wang, Y.; Wang, T.; Zheng, H.; Chu, L.; Zhang, C.; Chen, H.; Kong, X.; Xing, X. Effects of packing rates of cubic-shaped polyurethane foam carriers on the microbial community and the removal of organics and nitrogen in moving bed biofilm reactors. *Bioresour. Technol.* **2012**, *117*, 201–207. [CrossRef] [PubMed]
37. Tran, N.; Gin, K. Occurrence and removal of pharmaceuticals, hormones, personal care products, and endocrine disrupters in a full-scale water reclamation plant. *Sci. Total Environ.* **2017**, *599–600*, 1503–1506. [CrossRef]
38. Sun, S.; Liu, J.; Zhang, M.; He, S. Simultaneous improving nitrogen removal and decreasing greenhouse gas emission with biofilm carriers addition in ecological floating bed. *Bioresour. Technol.* **2019**, *292*, 1–8. [CrossRef]
39. Zhong, H.; Wang, H.; Tian, Y.; Liu, X.; Yang, Y.; Zhu, L.; Yan, S.; Liu, G. Treatment of polluted surface water with nylon silk carrier-aerated biofilm reactor (CABR). *Bioresour. Technol.* **2019**, *289*, 1–9. [CrossRef]
40. Mulinari, J.; Andrade, C.; Brandao, H.; Silva, A.; Souza, S.; Souza, A. Enhanced textile wastewater treatment by a novel biofilm carrier with adsorbed nutrients. *Biocatal. Agric. Biotechnol.* **2020**, *24*, 1–8. [CrossRef]

© 2020 by the authors. Licensee MDPI, Basel, Switzerland. This article is an open access article distributed under the terms and conditions of the Creative Commons Attribution (CC BY) license (http://creativecommons.org/licenses/by/4.0/).

Article

Causes of the Gloss Transition Defect on High-Gloss Injection-Molded Surfaces

Jinsu Gim [1], Eunsu Han [2], Byungohk Rhee [2,*], Walter Friesenbichler [3] and Dieter P. Gruber [4]

1 Center for Coating Materials and Processing, Engineering Research Center, Seoul National University, 1, Gwanak-ro, Gwanak-gu, Seoul 08826, Korea; interactionjs@gmail.com
2 Department of Mechanical Engineering, Ajou University, 206, Worldcup-ro, Suwon 16499, Korea; leunsul@gmail.com
3 Department of Polymer Engineering and Science, Montanuniversität Leoben, A-8700 Leoben, Austria; walter.friesenbichler@unileoben.ac.at
4 Polymer Competence Center Leoben GmbH, A-8700 Leoben, Austria; dieter.gruber@pccl.at
* Correspondence: rhex@ajou.ac.kr; Tel.: +82-31-219-2347

Received: 24 August 2020; Accepted: 12 September 2020; Published: 15 September 2020

Abstract: The gloss transition defect of injection-molded surfaces should be mitigated because it creates a poor impression of product quality. Conventional approaches for the suppression of the gloss transition defect employ a trial-and-error approach and additional equipment. The causes of the generation of a low-gloss polymer surface and the surface change during the molding process have not been systematically analyzed. This article proposes the causes of the generation of a low-gloss polymer surface and the occurrence of gloss transition according to the molding condition. The changes in the polymer surface and gloss were analyzed using gloss and topography measurements. The shrinkage of the polymer surface generates a rough topography and low glossiness. Replication to the smooth mold surface compensates for the effect of surface shrinkage and increases the surface gloss. The surface stiffness and melt pressure influence the degree of mold surface replication. The flow front speed and mold temperature are the main factors influencing the surface gloss because they affect the development rate of the melt pressure and the recovery rate of the surface stiffness. Therefore, the mold design and process condition should be optimized to enhance the uniformity of the flow front speed and mold temperature.

Keywords: gloss transition defect; surface defect; surface gloss; shrinkage; mold surface replication; surface analysis; injection molding

1. Introduction

A gloss transition defect is a visually recognizable transition on the surfaces of injection-molded products. The gloss transition defect is indicated by the gloss transition line between relatively low- and high-gloss areas perpendicular to the flow direction, as shown in Figure 1. It creates a poor impression of product quality. A low-gloss area in a high-gloss surface appears dusty and cloudy. A high-gloss area on a low-gloss surface looks like the area is worn out. The defect is readily visible under bright light conditions such as sunlight. Large exterior parts such as automobile trims are susceptible to the gloss transition defect. Post-processes such as coating, painting, and vapor deposition could be affected by a gloss transition defect due to the difference in surface characteristics. Therefore, the gloss transition defect should be mitigated in the injection molding process.

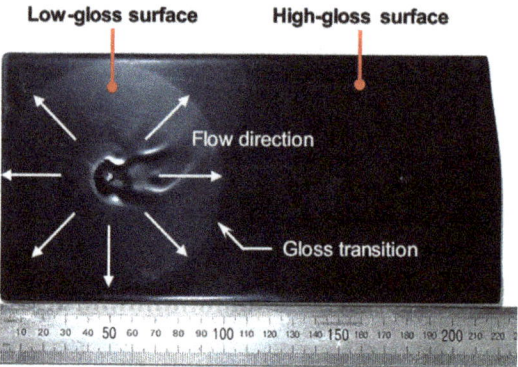

Figure 1. Example of the gloss defect on a high-gloss injection-molded surface.

Molding conditions enhancing the mold surface replication have been recommended to eliminate surface defects and increase surface gloss. High mold temperature has been suggested as the main parameter influencing the surface gloss because the formation of a frozen layer is affected by the temperature condition [1–11]. High cavity or packing pressure [1,2,4,8,10,11] and injection speed [1,2,10,11] increase the surface gloss but have a less significant effect compared to the mold temperature [2,10]. Additionally, the melt temperature [2,4,6,8], screw rotation speed [6], flow length [5,7], and melt viscosity [8] influence the surface gloss. The effect of these molding parameters was analyzed for overall surface gloss and not the gloss transition on a single surface.

Several methodologies have been proposed to mitigate the gloss transition defect. For a hot runner system applied to the molding of large-area products, a position-controlled valve pin was proposed to reduce the drastic fluctuations of the molding condition [12,13]. Owing to the lack of optimization methodologies for the position profile of valve pins, a trial-and-error approach needs to be applied. A high mold temperature enhancing the surface gloss increases the cooling and cycle times. Thus, rapid heat and cool molding (RHCM) and variothermal techniques have been suggested to avoid the increase in cooling and cycle times [14–17]. A mold surface coating with an insulation film was proposed to replicate the effect of the high mold temperature without the RHCM controller [18,19].

The surface defects mainly depend on the molding condition in the filling stage. Yoshii [20,21] and Tredoux [22,23] proposed that wavelike flow marks are induced by the go-over phenomena and nonuniform thermal contraction under a low injection speed in the filling stage. Yokoi [24] and other researchers [25,26] proposed that the tiger stripes observed in the alternating gloss-dull stripes are generated at high injection speeds, and they are attributed to the flow instability related to the Weissenberg number in the filling stage. For the gloss transition defect generated in the middle range of injection speeds, Isayev and Kim pointed out that stable surface gloss can be achieved by constantly maintained melt flow [27]. Jeon et al. [28], Yuan et al. [12], and Bott [29] proposed that the hesitation of the flow causes the gloss defect and referred to it as flow hesitation or the halo mark. Suhartono et al. [13] showed that the stress distribution of the numerical simulation result near the hot gates resembles the gloss transition and referred to it as a stress mark. Yuan et al. [30] observed that the sudden fluctuation of the melt pressure causes the gloss transition defect, and referred to it as a pressure transition mark. Prior studies on the gloss transition defect proposed that enhanced mold surface replication increases the surface gloss. However, existing studies have not investigated the generation of a low-gloss polymer surface in a highly polished mold, as shown in Figure 1, and the influence of the filling condition on the surface gloss.

In this article, the causes of the gloss difference and the molding conditions for the mitigation of the resultant gloss transition defect are proposed. The surface characteristics determining the gloss difference are analyzed using surface topography and gloss measurements. The generation of a low-gloss polymer surface and the change in the polymer surface during the filling stage are analyzed using a short-shot specimen. The effect of the main process parameters on the surface gloss is investigated by analyzing the design of experiment (DOE) and the proposed causes. Consequently, the molding conditions for mitigating the gloss transition defect are proposed.

2. Materials and Methods

2.1. Materials

2.1.1. Polymer Material

Poly (acrylonitrile-*co*-butadiene-*co*-styrene) (ABS) used herein is appropriate for the investigation of the gloss transition defect because its injection-molded surface is particularly sensitive to surface defects [1]. Furthermore, the black ABS HF380 (color code 9001) manufactured by LG Chem Ltd. (Seoul, Korea) was selected to recognize the gloss difference easily. The ABS was dried at 80 °C for 4 h to prevent bubble formation on the surface through the evaporation of moisture content.

2.1.2. Injection Molds

Two different injection molds were used in this study. The first mold (mold type A) was used for ASTM specimens, as shown in Figure 2a. The gloss transition from the flow front in the filling stage appears along the long flow length of the short-shot specimen. A cavity with a long flow length was selected in the mold for the investigation of the polymer surface in the filling stage. The dimensions of the cavity were 12.7 mm × 127 mm × 3.2 mm. The dimensions of the gate were 6.3 mm × 3.2 mm. The air vent with the same width of the cavity was located at the end of the filling position. The other cavities in the mold were blocked at the runner branches. The areal root-mean-square (RMS) roughness of the mirror-polished mold surface was under 5 nm, and it was measured using the GPI XP/D interferometer of Zygo Corporation (Middlefield, CT, USA).

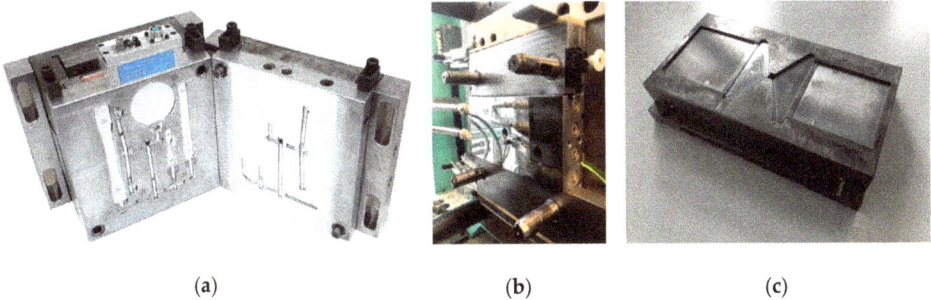

(a) (b) (c)

Figure 2. Injection molds: (**a**) mold type A, (**b**) fixed plate of the mold type B, and (**c**) moving plate of mold type B

The second mold (mold type B) was used for the ISO D2 specimen, as shown in Figure 2b,c. It was selected for the analysis of the effect of the molding condition on the surface gloss. The dimensions of the cavity were 60 mm × 60 mm × 2 mm. It had a fan gate for obtaining a uniform flow pattern, as shown in Figure 2c. The dimensions of the gate were 60 mm × 1.5 mm. The surface of the fixed mold was mirror-polished, as shown in Figure 2b. At the surface of the fixed mold, the cavity pressure was measured using the Type 6190CA pressure sensor and the Type 5018 charge amplifier of Kistler AG (Winterthur, Switzerland).

2.2. Experiment Conditions

2.2.1. Surface Changes in the Filling Stage

The change in the polymer surface in the filling stage needs to be investigated with respect to the gloss transition defect because a new polymer surface is generated, and it contacts the mold surface in the filling stage. The short-shot specimen represents the generation and change in the polymer surface during the filling stage. The high-speed electric injection-molding machine LGE150IIIDHS of LS Mtron Ltd. (Anyang, Korea) with a clamping force of 150 tons and the mold type A were used to mold the short-shot specimen. The flow front speed varies in the range of 10–100 mm/s for large area injection-molded parts such as TV back covers [31] and automotive bumpers. To investigate the change of the polymer surface in a wide range of filling conditions, the range of flow front speeds was set as 10–1000 mm/s. The jetting and burn marks were not generated in all experimental conditions because the mold type A had a large-area gate with the same thickness to the cavity and a wide air vent. The packing stage was not employed. Twenty cycles of the stabilization process were conducted before acquiring the specimen. The molding conditions are presented in Table 1.

Table 1. Injection-molding conditions for the short-shot specimen.

Process Parameter	Values
Coolant temperature (°C)	35
Barrel temperature (°C)	210
Flow front speed, FFS (mm/s)	17.8, 31.6, 56.2, 100, 178, 316, 562, 1000

2.2.2. Influence of Process Parameters on Surface Gloss

Three process parameters were selected based on the factors expected to influence the surface gloss. The flow front speed in the filling stage is expected to be the main factor influencing the gloss transition defect. The packing pressure in the packing stage is predicted to enhance the surface gloss because it pressurizes the polymer surface to the mold surface. The mold temperature directly determined by the coolant temperature suppresses the gloss transition defect. The values of the coolant temperature were selected in the range of molding conditions recommended by the material manufacturer. The values of the flow front speed reflect the range of flow front speed for conventional injection molding conditions of large-area parts [31]. The minimum value of the packing pressure maintained the maximum cavity pressure at the end of filling, and the maximum value of the packing pressure doubled the cavity pressure, as shown in Figure 3. Selected process parameters and levels are listed in Table 2. The DOE was a full factorial design. The data analysis software Minitab 16.2 analyzed the effect of the process parameters on the surface gloss. The electric injection-molding machine Allrounder 470 A 1000-400 of Arburg GmbH (Lossburg, Germany) with a clamping force of 100 tons and the mold type B were used. The barrel temperature was set to 215 °C. Twenty cycles of the stabilization process were conducted before each DOE condition. The cavity pressure measurement showed that the packing time of 5 s was longer than the gate solidification time.

Table 2. Process parameters for the design of experiment (DOE).

Process Parameter	Levels	Values
Flow front speed, FFS (mm/s)	5	25, 50, 100, 250, 375
Packing pressure (bar)	3	400, 500, 600
Coolant temperature (°C)	3	40, 55, 70

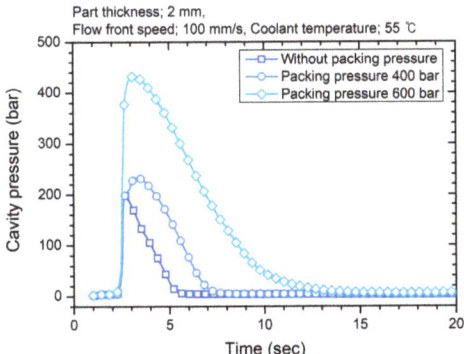

Figure 3. Cavity pressure under different packing pressure conditions.

2.3. Measurement and Analysis

2.3.1. Surface Gloss

The standard glossmeter GL0020 DuoGloss of TQC Sheen B.V. (Capelle aan den IJssel, Netherlands) was used to measure the surface gloss of the injection-molded surfaces. The geometry of the glossmeter complied with the standard test methods for surface gloss [32–34]. The surface gloss was quantified by the specular gloss value in gloss units (GU). The resolution of the glossmeter was 0.1 GU within the range of 0–100 GU. A measurement geometry of 20° was selected owing to the high glossiness of the specimen surface. The sizes of the measuring spot and detector aperture were 5 mm × 5 mm and 1.8° × 3.6°, respectively. A measuring pad with a black matt fabric material (0.0 GU for 20° geometry, 0.2 GU for 60° geometry) was used to eliminate the influence of surrounding light. The glossmeter was calibrated using the standard specimen included in the glossmeter set before the measurement of each specimen. For each specimen, a flat surface without a sink mark influencing the gloss measurement was measured.

2.3.2. Gloss Distribution

The intensity profile analysis (IPA) developed by the Polymer Competence Center Leoben (PCCL) was employed to measure the gloss distribution on the short-shot specimen [35,36]. The measurement of the gloss distribution using the standard glossmeter requires repetitive measurements at different surface positions due to the small measuring area. The IPA based on the evaluation of the modulation transfer function can measure the gloss distribution close to human vision using a single measurement with high precision [35,36]. According to the visual test method of gloss difference [37], the contrast of the reflected image reveals the quality of the surface reflection and gloss. Figure 4a shows the measurement setup. The measurement was conducted in a dark room to prevent the influence of external lights. The line chart was illuminated by the diffused light source on the rear side. The high-contrast image of the line chart was projected onto the specimen surface and reference mirror. The reflected images on the specimen and mirror were both captured using the digital single-lens reflex camera EOS 700D with the EF-S 18–55 mm lens of Canon Inc. (Tokyo, Japan). The aperture was closed to F/9 to ensure a deep focal depth. The reflected image of the reference mirror was employed to focus on the line chart and normalize the intensity of the captured image. The positional contrast was derived from the intensity profile, as shown in Figure 4b.

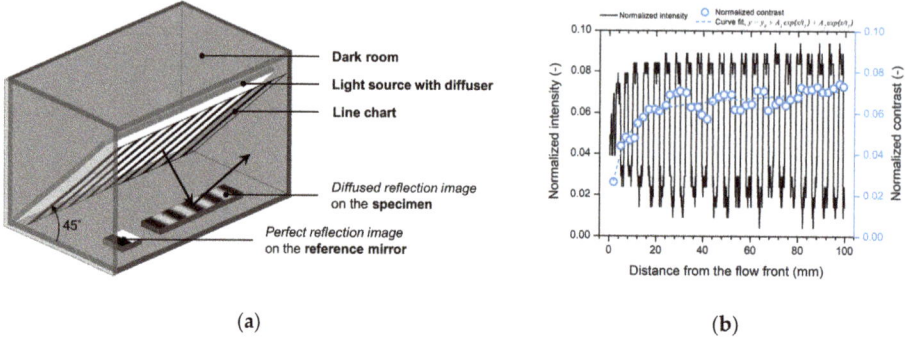

Figure 4. Gloss distribution measurement using the IPA technique [35,36]: (**a**) measurement setup and (**b**) intensity profile and contrast distribution.

2.3.3. Surface Topography

The interferometer DCM8 of Leica Microsystems GmbH (Wetzlar, Germany) was used to measure the surface characteristics of the high-gloss polymer surface. Interferometry is a suitable measurement technique for a high-gloss polymer surface because it does not damage the low-hardness polymer surface, and has sufficient resolution (0.1–1 nm) for the analysis of the glossy surface, which has a lower roughness than the wavelength of visible light (380–740 nm) [38]. The surface was evaluated using a Mirau 50× lens in the phase-shifting interferometry mode with a green light source. The vertical resolution was 0.1 nm, and the lateral resolution was approximately 0.3 μm. The measured area was 351 μm × 261 μm.

The topography parameters, RMS roughness (S_q), and lateral correlation length (L_c) were determined by the height–height correlation function (HHCF, C_z) of the measured surface, as shown in Figure 5. HHCF is defined as follows [39]:

$$C_z(\lambda_s) = [(z(x + \lambda_s) - z(x))^2] \tag{1}$$

where z is the surface height, x is the lateral position, and λ_s represents the spatial wavelength. The square bracket represents the average value of the term in the bracket for all lateral positions. The RMS roughness and lateral correlation length indicate the characteristic scales of the amplitude and wavelength of the surface fluctuation, respectively. The measured topography was analyzed using the open-source software for scanning probe microscopy, Gwyddion 2.52 [40].

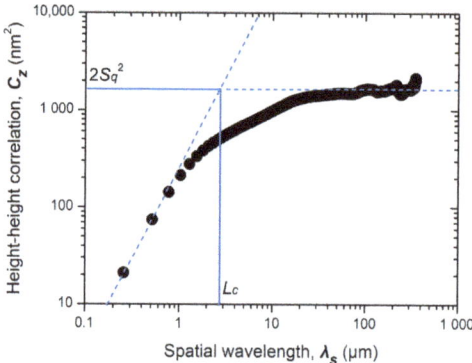

Figure 5. Determination of the root-mean-square (RMS) roughness (S_q) and lateral correlation length (L_c) using the height–height correlation function (HHCF).

3. Results and Discussion

3.1. Gloss Difference Induced by Surface Topography

The surface gloss and topography parameters show a similar tendency to the flow front speed. Figure 6a,b shows the measured surface gloss and the topography parameters of the specimen molded at various flow-front speeds, respectively. The surface gloss increased as the flow front speed increased. The RMS roughness and lateral correlation length decreased as the flow front speed increased. The surface gloss and topography parameters converged at a high flow-front speed. This similar trend for different molding conditions indicates a strong correlation between the surface topography and gloss.

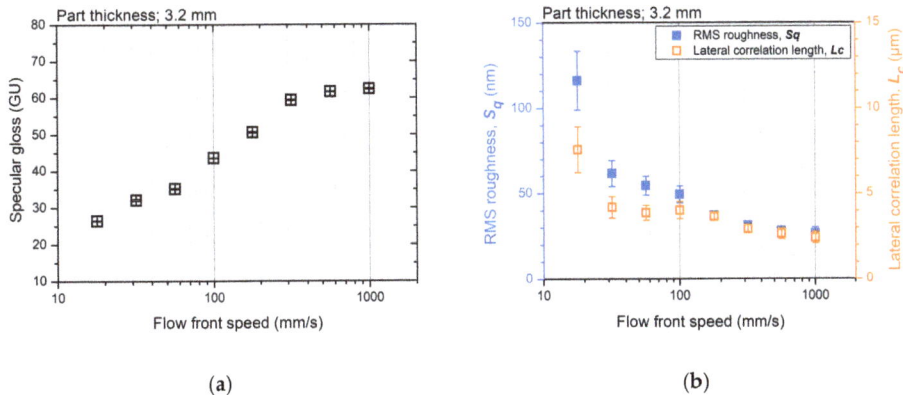

Figure 6. Surface measurement result: (**a**) surface gloss and (**b**) topography parameters.

The theory of light-scattering phenomena can explain the relationship between surface topography and gloss. This is because the gloss depends on the degree of the reflected light concentration at the specular angle. As shown in Figure 7, the relatively low-gloss surface molded by a low flow-front speed shows a rough topography. This rough surface scatters the reflected light over a broad angular range. The rough surface induces diffused reflection and appears less glossy. The surface gloss can be modeled by the Kirchhoff theory of light scattering according to the surface roughness [41]. According to the modified Kirchhoff theory suggested by Alexander-Katz and Barrera [42], the surface gloss can be modeled as a function of the RMS roughness and lateral correlation length.

Figure 8 shows a comparison of the measured and predicted surface gloss. The surface gloss was predicted using the Kirchhoff theory, measured topography parameters, the characteristics of the glossmeter [34], and the refractive index (1.515) of the ABS material [43]. The predicted surface gloss is consistent with the measured gloss. This indicates that the difference in the surface topography results in a gloss difference. The difference in surface gloss can represent the influence of the molding condition on the surface topography.

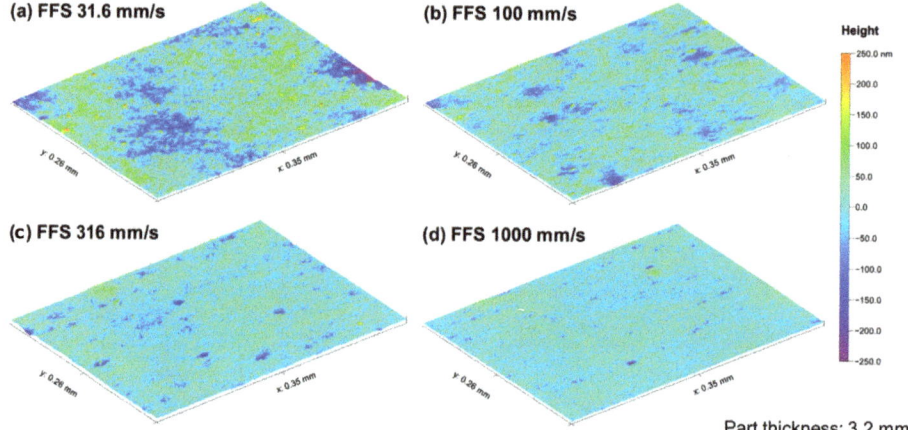

Figure 7. Surface topography molded by various flow-front speeds (FFS); (**a**) 31.6 mm/s, (**b**) 100 mm/s, (**c**) 316 mm/s, and (**d**) 1000 mm/s.

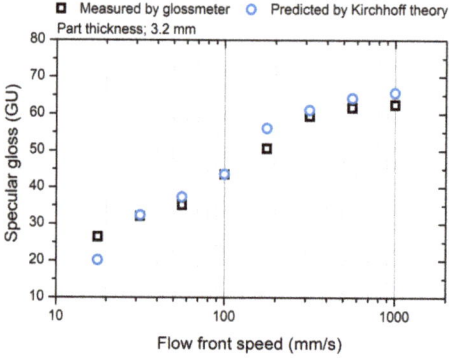

Figure 8. Comparison of the measured and predicted surface gloss.

As shown in Figure 6b, the surface roughness determining the gloss difference is in a higher range (>20 nm) than that of the mold surface (<5 nm) and varies with the molding condition. The larger roughness of the polymer surface than that of the mold surface results from surface shrinkage. The decrease in the surface roughness is related to the effect of the molding conditions on the mold surface replication, as Oliveira et al. proposed [8]. The generation of a rough polymer surface and the mold surface replication occur in the filling stage.

3.2. Surface Changes in the Filling Stage

The gloss distribution of the short-shot specimen was analyzed to investigate the change in the polymer surface during the molding process. Figure 9 shows the distributions of the surface gloss on the short-shot specimen along the distance from the flow front. The contrast in Figure 9 indicates the surface gloss. As the distance from the flow front increased, the surface gloss increased. In particular, the gloss increased rapidly at a distance less than 20 mm, and gradually at a distance greater than 20 mm. A higher flow-front speed resulted in a more rapid increase in the surface gloss near the flow front. A similar distribution was observed in all the specimens molded in a wide range of flow front speeds. The distribution of the surface gloss could be explained by two factors: the generation of the rough surface due to surface shrinkage and the mold surface replication due to the melt pressure.

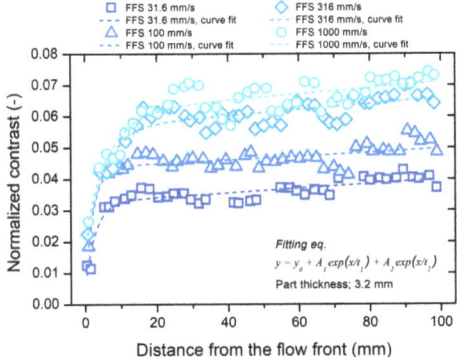

Figure 9. Contrast distribution of the short-shot specimen.

3.3. Rough Surface Generation

The rough polymer surface was generated after contact with the mold surface. During the filling stage, the fountain flow at the flow front generated a new polymer surface and transported the polymer surface to the mold surface, as shown in Figure 10. Then, the polymer surface touched the mold surface. The polymer surface was smooth owing to the elongational stress at the flow front. After contacting the mold surface, the polymer surface cooled down and shrunk.

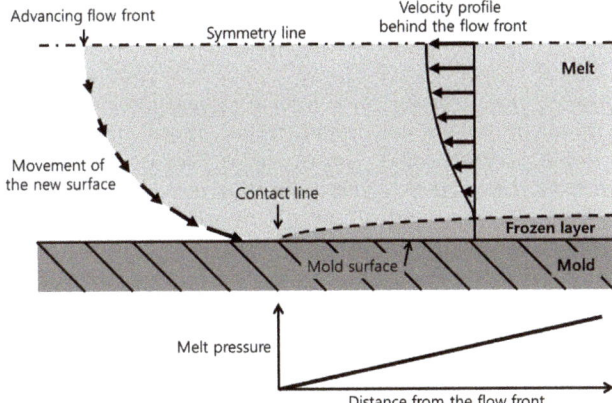

Figure 10. Fountain flow and melt pressure development at the flow front.

The shrinkage crumples the polymer surface, resulting in a rough surface. Inhomogeneous shrinkage due to subsurface morphology affects the surface topography [44]. ABS contains butadiene rubber particles in the poly(styrene-*co*-acrylonitrile) (SAN) matrix. The size of the rubber particles ranges from 0.5 to 5 μm [45,46]. The rubber particles make the subsurface morphology complex and induce rough topography. Rosato reported a similar generation of rough surfaces due to the two phases in ABS [2]. Therefore, surface shrinkage generates rough and low-gloss polymer surfaces in highly polished injection molds.

The generation of a rough surface by the surface shrinkage is maximized near the flow front of the short-shot specimen because the surface is not sufficiently pressurized by the melt pressure. If the filling stops before the cavity is fully filled, the generation of a new surface at the flow front stops. The polymer surface near the flow front freely shrinks because the distance near the flow front is too short to develop the melt pressure. Therefore, the surface at the flow front shows a rough topography.

Figure 11 shows the rough surface at the flow front. The surface gloss near the flow front is at a minimum owing to the maximized surface shrinkage, as shown in Figure 9.

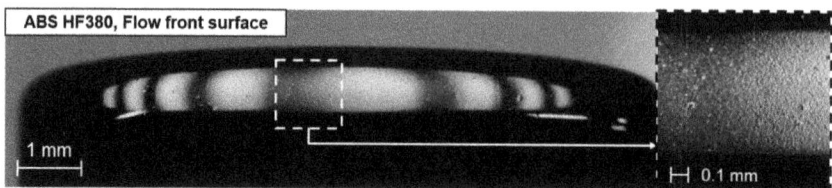

Figure 11. Rough surface at the flow front.

3.4. Mold Surface Replication

In Figure 9, the increasing gloss along the distance shows the effect of the melt pressure development on the surface topography. As the distance from the flow front increased, the melt pressure developed, as shown in Figure 10, and pushed the polymer surface to the smooth mold surface. The polymer surface contacting the mold surface shrunk simultaneously. A sufficient melt pressure enhanced the replication of the polymer surface to the smooth mold surface, and the surface shrinkage was compensated. The generation of a rough surface was due to surface shrinkage and mold surface replication due to the melt pressure occurring simultaneously. As the distance from the flow front increased, the generation of the rough surface was suppressed to a greater extent by a higher melt pressure. The surface gloss increased along the distance, as shown in Figure 9.

The rapid increase in the surface gloss near the flow front in Figure 9 was related to the stiffness recovery time of the polymer surface. The temperature of the polymer surface decreased from the melt temperature to the contact temperature immediately after contacting the mold surface. According to the heat conduction model between the melt and the mold suggested by Yoshii et al. [20], and Carslaw and Jaeger [47], the contact temperature was close to the mold temperature because the mold steel had a much higher thermal diffusivity than the polymer material. The subsurface temperature in the melt was higher than the contact temperature. As the contact time increased, the subsurface temperature converged to the contact temperature. Surface cooling recovered the surface stiffness. The recovery of the surface stiffness proceeded as the thickness of the frozen layer increased, as shown in Figure 10. The surface stiffness resisted the mold surface replication and the compensation of the shrinkage by the melt pressure. Until the surface stiffness recovered sufficiently to resist the melt pressure, the amount of mold surface replication increased rapidly. The surface gloss increased rapidly near the flow front, where the polymer surface was soft owing to the short cooling time.

The level of the surface gloss was determined in the region near the flow front. During the filling stage, the surface stiffness was recovered sufficiently far from the flow front. The melt pressure could not enhance the mold surface replication as much as near the flow front. The increasing rate of surface gloss reduced far from the flow front, as shown in Figure 9. This indicates that the difference in the surface gloss was determined near the flow front, and the filling stage dominantly affected the gloss transition defect. The effect of the molding conditions, including the filling condition, on the surface gloss was investigated via DOE analysis.

3.5. Influence of Molding Conditions on Surface Gloss

Figure 12 shows the surface gloss of the specimens molded under various injection-molding conditions. The correlation between the molding conditions and the surface gloss was analyzed using DOE. Figure 13 shows the factorial plots representing the effect of the molding conditions on the surface gloss. Table 3 shows the analysis of variance (ANOVA) result. Coolant temperature and flow front speed were considered significant factors influencing specular gloss because p-values of these factors were under 0.05. Packing pressure was not a significant factor.

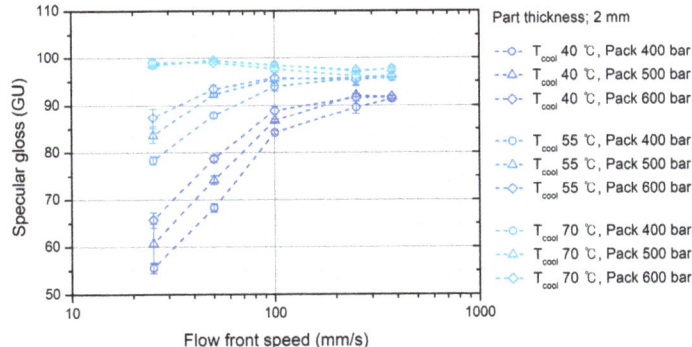

Figure 12. Surface gloss under various molding conditions.

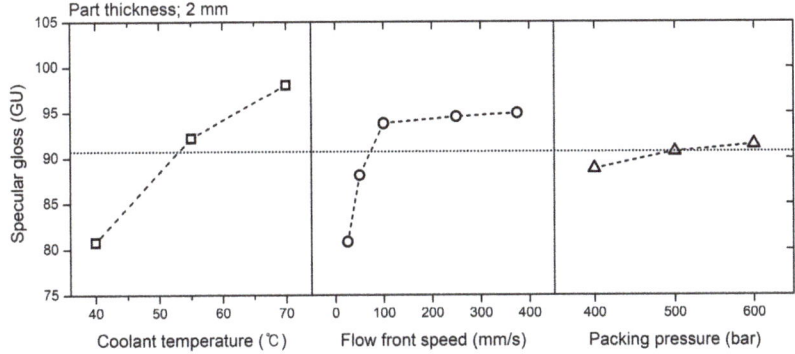

Figure 13. Effect of the molding conditions on the surface gloss.

Table 3. Analysis of variance (ANOVA) for specular gloss.

Design Parameters	Degree of Freedom	Sum of Squares	Mean Square	F-Ratio	p-Value
Coolant temperature	2	3552.56	1776.28	45.77	0.000
Flow front speed	4	2007.25	501.81	12.93	0.000
Packing pressure	3	204.81	68.27	1.76	0.167
Error	50	1940.26	38.81		
Total	59	7704.89			

3.5.1. Effect of Mold Temperature

The effect of the coolant temperature (T_{cool}) on the surface gloss represents the influence of the mold temperature because the coolant temperature dominantly determines the mold temperature. The mold temperature had the greatest influence on the surface gloss, as shown in Figure 13. An increase in the mold temperature enhanced the overall surface gloss. This result agrees with the previous research [1–11]. At a high mold temperature (T_{cool} of 70 °C), the surface shows the highest surface gloss value in Figure 12. The high mold temperature decreased the cooling rate of the polymer surface so that the temperature of the polymer surface was higher than that at a low mold temperature. The recovery rate of the surface stiffness and shrinking rate decreased. This indicates that the melt pressure pressurized the softer polymer surface over a longer time. The generation of the rough surface due to the shrinkage was suppressed, and the mold surface replication was enhanced at a high mold temperature.

An imbalance in the mold surface temperature can induce the gloss transition defect even if the mold surface was highly polished. Owing to the high thermal conductivity of the mold steel, the temperature did not change significantly over the mold surface in a single cycle. In the case of a mold with a poor thermal design, such as nonoptimized cooling channels with a high distance from each other and the cavity wall, the heat from the melt can be accumulated locally and the temperature deviation can be approximately 10 °C [48]. For example, the corner in the cavity is easily far from cooling channels and susceptible gloss defects [49]. It is desirable to maintain a uniform mold surface temperature to suppress the gloss transition defect. Optimization of the conventional cooling channel or conformal cooling channel is recommended in the mold design step.

The fluctuation of the surface gloss can be suppressed at high mold temperatures, as shown in Figure 12. The glass transition temperature of the ABS was 99 °C. As the mold temperature approaches the glass transition temperature of the polymer material, the recovery of the surface stiffness is considerably suppressed, and the polymer surface can replicate sufficiently to the mold surface even at a slow development of melt pressure. A high mold temperature is recommended when it is difficult to adjust the other molding conditions or to revise the mold design.

3.5.2. Effect of the Flow Front Speed

The flow front speed in the filling stage is a significant parameter on the surface gloss as much as the mold temperature. The mold temperature has the largest influence on the surface gloss as Figure 13 and as reported by prior research [1–11]. In comparison with the mold temperature inducing a gradual increase of the surface gloss, the flow front speed increased rapidly the surface gloss below 100 mm/s of the flow front speeds, as shown in Figure 13.

The flow front speed influenced the increasing rate of the melt pressure. The distance from the flow front and the flow front speed increased the melt pressure development at a specific location on the mold surface [50]. As the flow front speed increased, a higher melt pressure pressurized the soft polymer surface until the surface stiffness recovered sufficiently. The flow front speed mainly influenced the surface near the flow front during the filling stage, as shown in Figure 8.

The increment in the surface gloss due to the flow front speed converged at a high flow-front speed. For example, the surface gloss sharply increased at the coolant temperature of 40 °C and flow front speeds less than 100 mm/s, as shown in Figure 12. However, the surface gloss converged to approximately 90 GU at flow front speeds greater than 100 mm/s. This is due to the difference in the rates of surface cooling and pressure development. If the flow front speed is sufficiently high, the development of the melt pressure is faster than the recovery of the surface stiffness. A sufficiently high melt pressure maximizes the mold surface replication of the polymer surface. The fluctuation of the surface gloss due to the flow front speed stabilizes at high flow-front speeds. This indicates that the gloss transition defect can be suppressed by a high flow-front speed.

The fluctuation of the flow front speed can induce the gloss transition defect. The flow front speed may fluctuate in the filling stage and induce the gloss transition defect, as shown in Figure 14. The flow front speed increased to 250 mm/s at the middle of the surface and induced the strong gloss transition defect, as shown in Figure 14a. The fluctuation of the flow front speed in a high flow-front speed range decreased the degree of the gloss transition defect. The flow front speed was changed from 100 to 250 mm/s at the middle of the specimen in Figure 14c and made barely visible the gloss transition. This result represents that the gloss transition defect can be suppressed in a high flow-front speed range even if the flow front speed fluctuates in the filling stage.

The flow pattern according to the cavity geometry affects the distribution of the flow front speed. For example, the flow front speed gradually decreased in a radial flow pattern. In addition, the operation of the injection-molding machine and hot runner system influenced the fluctuation of the flow front speed. For example, the sequential operation of the valve gate in the hot runner system caused significant fluctuations in the flow front speed near the valve gate owing to the sudden release of a high pressure and melt compression in the manifold. Therefore, the mold design and the process parameters related to the flow front speed should be optimized to minimize the fluctuation of the flow front speed.

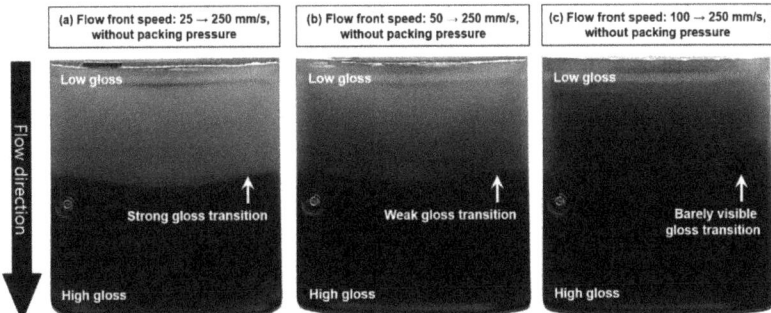

Figure 14. Effect of the flow front speed fluctuation on the gloss transition defect; (a) flow front speed fluctuation 25–250 mm/s, (b) flow front speed fluctuation 50–250 mm/s, and (c) flow front speed fluctuation 100–250 mm/s.

3.5.3. Effect of Packing Pressure

The packing pressure increased the surface gloss, as shown in Figure 12. The pressure that pushed the polymer surface to the mold surface enhanced the surface gloss. In the filling stage, the melt pressure pressurized the polymer surface. After the filling stage, the packing pressure additionally pressurized the polymer surface. The influence of the packing pressure on the surface gloss was not as high as that of the flow front speed and mold temperature, as shown in Figure 13 and this result is in agreement with the report proposed by Posciotti et al. [11]. The packing pressure could not eliminate the gloss transition defect already generated in the filling stage, as shown in Figure 15. This is because the packing pressure was applied to the already stiffened polymer surface. The surface stiffness was already recovered when the packing stage started. The effect of the packing pressure was similar to the slow increase of the surface gloss due to the melt pressure far from the flow front, as shown in Figure 9. This result indicates that the packing pressure cannot eliminate the gloss difference already created in the filling stage.

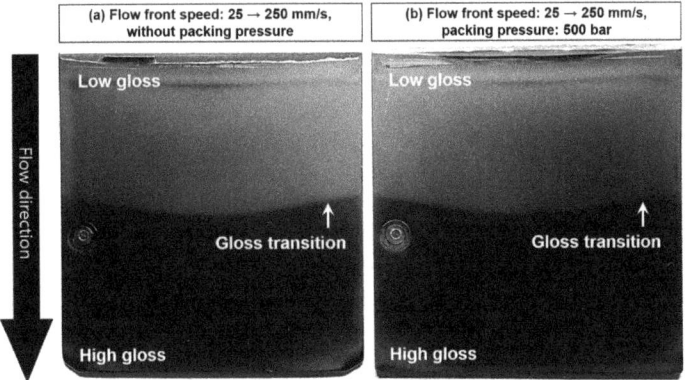

Figure 15. Effect of the packing pressure on the gloss transition defect caused by the flow fluctuation 25–250 mm/s; (a) without packing pressure, and (b) with 500 bar of packing pressure.

4. Conclusions

In this study, the causes of the gloss transition defect were investigated via surface topography and gloss measurements. The generation of a low-gloss surface and the change in the surface during the molding process were analyzed to determine the causes of the gloss transition defect. The effects of the molding conditions on the surface gloss were described. Consequently, the molding conditions suppressing the gloss transition defect were proposed.

A low-gloss polymer surface can be molded even if the mold surface is highly polished. This is due to the generation of a rough polymer surface due to surface shrinkage. For polymer materials with complex morphologies such as ABS containing rubber particles, inhomogeneous shrinkage due to the complex morphology made the polymer surface rough. The rough surface induced diffused reflection, resulting in low glossiness. The mold surface replication compensated for the effect of the surface shrinkage and increased the surface gloss. The melt pressure and surface stiffness influenced the degree of mold surface replication. The melt pressure pushing the polymer surface to the mold surface enhanced the replication and generated high surface gloss. The surface stiffness resisted the melt pressure and mold surface replication. Therefore, the surface shrinkage and the difference in the mold surface replication caused the gloss transition defect.

The filling condition mainly determines the surface gloss. The mold temperature and flow front speed were the parameters having the greatest influence on the surface gloss. The mold temperature influenced the recovery rate of the surface stiffness. At a high mold temperature, the polymer surface was soft for a longer time. The softer polymer surface could better replicate the mold surface. The flow front speed affected the development rate of the melt pressure, pushing the polymer surface to the mold surface. At a high flow front speed, rapidly increasing melt pressure pressurized the polymer surface and replicated the mold surface better. Therefore, the fluctuation of the flow front speed and the nonuniformity of the mold temperature in the filling stage generated the gloss transition defect.

The nonuniformity of the mold temperature and the fluctuation of the flow front speed should be minimized in the filling stage to suppress the gloss transition defect. The thermal design of the mold, such as the cooling channels, should be optimized to minimize the temperature deviation of the mold surface. It is desirable to optimize the operation of the injection unit of the molding machine and sequential valve gates to reduce the fluctuation of the flow front speed.

Author Contributions: Conceptualization, J.G.; methodology, J.G. and D.P.G.; investigation, J.G. and E.H.; data curation, J.G. and E.H.; writing—draft preparation, J.G.; writing—review and editing, B.R., W.F., and D.P.G.; visualization, J.G.; supervision, B.R. and W.F.; funding acquisition, B.R. and W.F. All the authors have read and agreed to the published version of the manuscript. All authors have read and agreed to the published version of the manuscript.

Funding: This research was partially funded by the Erasmus+ international mobility grant (KA107) of the European Commission. The publication was also prepared with the support of the National Research Foundation of Korea (NRF) grant funded by the Korea government (MSIT) (No. NRF-2018R1A5A1024127).

Acknowledgments: The authors thank LG Chem Ltd., Republic of Korea for providing the materials and Thomas Ules of Polymer Competence Center Leoben GmbH, Austria, for supporting the topography measurement.

Conflicts of Interest: The authors declare no conflict of interest.

References

1. Beaumont, J.P. *Runner and Gating Design Handbook: Tools for Successful Injection Molding*, 2nd ed.; Hanser Gardner Publications: Cincinnati, OH, USA, 2008.
2. Rosato, D.V. *Injection Molding Handbook: The Complete Molding Operation: Technology, Performance, Economics*, 2nd ed.; Capman & Hall: New York, NY, USA, 1994.
3. Wang, G.; Zhao, G.; Wang, W. Effect of cavity surface temperature on reinforced plastic part surface appearance in rapid heat cycle moulding. *Mater. Des.* **2013**, *44*, 509–520. [CrossRef]
4. Berger, G.R.; Gruber, D.P.; Friesenbichler, W.; Teichert, C.; Burgsteiner, M. Replication of stochastic and geometric micro structures – aspects of visual appearance. *Int. Polym. Process.* **2011**, *26*, 131–322. [CrossRef]

5. Dawkins, E.; Engelmann, P.; Horton, K. Color and gloss- the connection to process conditions. *J. Injection Molding Technol.* **1998**, *2*, 1–7.
6. Dawkins, E.; Horton, K.; Engelmann, P.; Monfore, M. The effects of injection molding parameters on color and gloss. In *Coloring Technology for Plastics*; Harris, R.M., Ed.; William Andrew: Norwich, NY, USA, 1999; pp. 149–155.
7. Koppi, K.; Ceraso, J.M.; Cleven, J.A.; Salamon, B.A. Gloss modeling of injection molded rubber-modified styrenic polymers. In Proceedings of the Society of Plastics Engineers' Annual Technical Conference (SPE ANTEC), San Francisco, CA, USA, 5–9 May 2002.
8. Oliveira, M.J.; Brito, A.M.; Costa, M.C.; Costa, M.F. Gloss and surface topography of ABS: A study on the influence of the injection molding parameters. *Polym. Eng. Sci.* **2006**, *46*, 1394–1401. [CrossRef]
9. Wang, G.; Zhao, G.; Li, H.; Guan, Y. Research on a new variotherm injection molding technology and its application on the molding of a large LCD panel. *Polym.-Plast. Technol. Eng.* **2009**, *48*, 671–681. [CrossRef]
10. Theilade, U.A.; Hansen, H.N. Surface microstructure replication in injection molding. *Int. J. Adv. Manuf. Technol.* **2007**, *33*, 157–166. [CrossRef]
11. Pisciotti, F.; Boldizar, A.; Rigdahl, M.; Ariño, I. Effect of injection-molding conditions on the gloss and color of pigmented polypropylene. *Polym. Eng. Sci.* **2005**, *45*, 1557–1567. [CrossRef]
12. Yuan, Z.; Astbury, D.; Costa, F.S.; Ward, D.; Prey, M. Simulation and validation of mold filling with velocity controlled valve gates. In Proceedings of the Society of Plastics Engineers' Annual Technical Conference (SPE ANTEC), Orlando, FL, USA, 23–25 March 2015.
13. Suhartono, E.; Chiu, H.-S.; Hsu, C.-C.; Wang, C.-C.; Wang, C.-W.; Pavan, N. Predict and solve stress mark on product's cosmetic surface using controlled sequential valve gating simulation. In Proceedings of the Society of Plastics Engineers' Annual Technical Conference (SPE ANTEC), Anaheim, CA, USA, 8–10 May 2017.
14. Li, X.-P.; Zhao, G.-Q.; Guan, Y.-J.; Ma, M.-X. Optimal design of heating channels for rapid heating cycle injection mold based on response surface and genetic algorithm. *Mater. Design.* **2009**, *30*, 4317–4323. [CrossRef]
15. Li, J.; Li, T.; Jia, Y.; Jiang, S.; Turng, L.-S. Modeling and characterization of crystallization during rapid heat cycle molding. *Polym. Test.* **2018**, *71*, 182–191. [CrossRef]
16. Berger, G.R.; Friesenbichler, W.; Gruber, D.P.; Pacher, G.A.; Macher, J. Rapid heat cycle molding, surface topography and visual appearance of injection molded parts. Research work in Leoben in the last 10 years. In Proceedings of the Society of Plastics Engineers' Annual Technical Conference (SPE ANTEC), Las Vegas, NV, USA, 28–30 April 2014.
17. Chen, S.-C.; Chang, J.-A.; Cin, J.-C. Dynamic mold surface temperature control for molding high gloss, painting free part surface achieving cycle time and cost reduction. In Proceedings of the Society of Plastics Engineers' Annual Technical Conference (SPE ANTEC), Chicago, IL, USA, 22–26 June 2009.
18. Lee, J.; Turng, L.-S. Improving surface quality of microcellular injection molded parts through mold surface temperature manipulation with thin film insulation. *Polym. Eng. Sci.* **2010**, *50*, 1281–1289. [CrossRef]
19. Lee, J.; Turng, L.-S.; Dougherty, E.; Gorton, P. A novel method for improving the surface quality of microcellular injection molded parts. *Polymer* **2011**, *52*, 1436–1446. [CrossRef]
20. Yoshii, M.; Kuramoto, H.; Kawana, T.; Kato, K. The observation and origin of micro flow marks in the precision injection molding of polycarbonate. *Polym. Eng. Sci.* **1996**, *36*, 819–826. [CrossRef]
21. Yoshii, M.; Kuramoto, H.; Kato, K. Experimental study of transcription of smooth surfaces in injection molding. *Polym. Eng. Sci.* **1993**, *33*, 1251–1260. [CrossRef]
22. Tredoux, L.; Satoh, I.; Kurosaki, Y. Investigation of wave-like flow marks in injection molding: Flow visualization and micro-geometry. *Polym. Eng. Sci.* **1999**, *39*, 2233–2241. [CrossRef]
23. Tredoux, L.; Satoh, I.; Kurosaki, Y. Investigation of wavelike flow marks in injection molding: A new hypothesis for the generation mechanism. *Polym. Eng. Sci.* **2000**, *40*, 2161–2174. [CrossRef]
24. Yokoi, H.; Mashda, N.; Mitsuhata, H. Visualization analysis of flow front behavior during filling process of injection mold cavity by two-axis tracking system. *J. Mater. Process. Technol.* **2002**, *130–131*, 328–333. [CrossRef]
25. Bogaerds, A.C.B.; Hulsen, M.A.; Peters, G.W.M.; Baaijens, F.P.T. Stability analysis of injection molding flows. *J. Rheol.* **2004**, *48*, 765–785. [CrossRef]
26. Baltussen, M.G.H.M.; Hulsen, M.A.; Peters, G.W.M. Numerical simulation of the fountain flow instability in injection molding. *J. Non Newton. Fluid Mech.* **2010**, *165*, 631–640. [CrossRef]

27. Isayev, A.I.; Kim, N.A. Co-injection molding of polymers. In *Injection Molding Technology and Fundamentals*; Kamal, M.R., Isayev, A.I., Liu, S.-J., White, J.L., Eds.; Hanser Publications: Cincinnati, OH, USA, 2009.
28. Jeon, J.; Kim, H.; Rhee, B.; Chio, J.; Park, E.; Jung, K. A study on the halo surface defects of injection molded products. In Proceedings of the Society of Plastics Engineers' Annual Technical Conference (SPE ANTEC), Anaheim, CA, USA, 8–10 May 2017.
29. Bott, J. Do you still get stubborn surface defects, even with sequential valve gating? *Plast. Technol.* **2012**, *58*, 26–29.
30. Yuan, Z.; Ward, D.; Prey, M. Application and simulation of velocity-controlled valve gates. In Proceedings of the Autodesk University, Las Vegas, NV, USA, 2–4 December 2014.
31. Gim, J.; Kim, B.; Rhee, B.; Choi, J.; An, S.; Jung, K. Valve gate open lag time in conventional hot runner system. In Proceedings of the Society of Plastics Engineers' Annual Technical Conference (SPE ANTEC), Orlando, FL, USA, 7–9 May 2018.
32. ASTM D523-08. *Standard Test Method for Specular Gloss*; ASTM International: West Conshohocken, PA, USA, 2008.
33. ASTM D2457-03. *Standard Test Method for Specular Gloss of Plastic Films and Solid Plastics*; ASTM International: West Conshohocken, PA, USA, 2003.
34. ISO 2013:2014(E). *Paints and Varnishes—Determination of Gloss Value at 20 Degrees, 60 Degrees and 85 Degrees*; International Organization for Standardization: Geneva, Switzerland, 2014.
35. Gruber, D.P.; Buder-Stroisznigg, M.; Wallner, G.; Strauss, B.; Jandel, L.; Lang, R.W. A novel methodology for the evaluation of distinctness of image of glossy surfaces. *Prog. Org. Coat.* **2008**, *63*, 377–381. [CrossRef]
36. Gruber, D.P.; Wallner, G.; Buder-Stroisznigg, M. Method for Analysing the Surface Properties of a Material. PCT-Patent WO2006135948, 28 December 2006.
37. ASTM D4449-90. *Standard Test Method for Visual Evaluation of Gloss Differences between Surfaces of Similar Appearance*; ASTM International: West Conshohocken, PA, USA, 2004.
38. Stout, K.J.; Blunt, L. *Three Dimensional Surface Topography*, 2nd ed.; Penton Press: London, UK, 2000; pp. 19–94.
39. Ariño, I.; Kleist, U.; Barros, G.G.; Johansson, P.-Å.; Rigdahl, M. Surface texture characterization of injection-molded pigmented plastics. *Polym. Eng. Sci.* **2004**, *44*, 1615–1626. [CrossRef]
40. Nečas, D.; Klapetek, P. Gwyddion: An open-source software for SPM data analysis. *Cent. Eur. J. Phys.* **2012**, *10*, 181–188. [CrossRef]
41. Ogilvy, J.A. *Theory of Wave Scattering from Random Rough Surfaces*; IOP Publishing: London, UK, 1991.
42. Alexander-Katz, R.; Barrera, R.G. Surface correlation effects on gloss. *J. Polym. Sci.* **1998**, *36*, 1321–1334. [CrossRef]
43. Ariño, I.; Kleist, U.; Mattsson, L.; Rigdahl, M. On the relation between surface texture and gloss of injection-molded pigmented plastics. *Polym. Eng. Sci.* **2005**, *45*, 1343–1356. [CrossRef]
44. Lednický, F.; Pelzbauer, Z. Gloss as an inner morphology characteristic of ABS polymers. *Angew. Makromol. Chem.* **1986**, *141*, 151–160. [CrossRef]
45. Fink, J.K. *Handbook of Engineering and Specialty Thermoplastics*; Scrivener Publishing: Salem, MA, USA, 2010; Volume 1, pp. 211–268.
46. Wypych, G. *Handbook of Polymers*; ChemTec Publishing: Toronto, ON, Canada, 2012; pp. 3–10.
47. Carslaw, H.S.; Jaeger, J.C. *Conduction of Heat in Solids*, 2nd ed.; Oxford University Press: Oxford, UK, 1959; pp. 50–91.
48. Shayfull, Z.; Sharif, S.; Zain, A.M.; Ghazili, M.F.; Saad, R.M. Potential of conformal cooling channels in rapid heat cycle molding: A review. *Adv. Polym. Technol.* **2013**, *33*, 21381. [CrossRef]
49. Goodship, V. *Troubleshooting Injection Moulding*; Rapra Technology Ltd.: Shrewsbury, UK, 2004.
50. Morrison, F.A. *Understanding Rheology*; Oxford University Press: Oxford, UK, 2001; pp. 59–98.

© 2020 by the authors. Licensee MDPI, Basel, Switzerland. This article is an open access article distributed under the terms and conditions of the Creative Commons Attribution (CC BY) license (http://creativecommons.org/licenses/by/4.0/).

Article

Design of Experiment Approach to Optimize Hydrophobic Fabric Treatments

Iva Rezić [1,*] and Ana Kiš [2]

[1] Department of Applied Chemistry, Faculty of Textile Technology, University of Zagreb, 10000 Zagreb, Croatia
[2] Textile Company, Čateks, d.d. Ul. Zrinsko Frankopanska 25, 40000 Čakovec, Croatia; a.kis@cateks.hr
* Correspondence: iva_rezic@net.hr

Received: 29 August 2020; Accepted: 16 September 2020; Published: 18 September 2020

Abstract: Polymer materials can be functionalized with different surface treatments. By applying nanoparticles in coating, excellent antimicrobial properties are achieved. In addition, antimicrobial properties are enhanced by hydrophobic surface modification. Therefore, the goal of this work was to modify the process parameters to achieve excellent hydrophobicity of polymer surfaces. For this purpose, a Design of Experiment (DoE) statistical methodology was used to model and optimize the process through six processing parameters. In order to obtain the optimum and to study the interaction between parameters, response surface methodology coupled with a center composite design was applied. The ANNOVA test was significant for all variables. The results of the influence of process parameters showed that, by increasing the pressure, concentration of hydrophobic compounds and dye concentration, water vapor permeability was enhanced, while by decreasing weight, its efficiency was enhanced. Moreover, the increase in the temperature enhanced water vapor permeability but decreased the resistance to water wetting. An optimal process with ecologically favorable 6C fluorocarbon (68.802 g/L) surpassed all preliminary test results for 21.15%. The optimal process contained the following parameters: 154.3 °C, 1.05 bar, 56.07 g/L dye, 220 g/m^2 fabric. Therefore, it is shown that DoE is an excellent tool for optimization of the parameters used in polymer surface functionalization.

Keywords: polymer surface modification; hydrophobic properties; optimization; mathematical modeling; hydrophobicity

1. Introduction

Hydrophobic finishing of textile materials is one of the most important processes in the textile industry. Excellent finished compounds repel water, oil and dry dirt which is extremely important in clothing, sports, military, medical and technical textiles [1]. Polymer materials can be functionalized to become active against resistant microorganisms with different surface treatments. For example, by applying nanoparticles inside the surface coating, excellent antimicrobial properties are achieved. In addition, antimicrobial properties are enhanced by further surface modification that enables excellent hydrophobic properties.

Historical usage of water repellant materials covers a broad area of wax and resin compounds that can be easily washed out [2]. Today perfluoroalkyl compounds (PFAS's) are widely used in almost all staining repellant finishes since those are the only chemicals capable of simultaneously repelling water, oil, dirt and all other staining compounds [3]. In addition, by enhancing the repellant properties, other fabric properties show much better performances, including increased resistance to acids, bases and other chemicals, more rapid drying, and better durable press properties [4].

Fluorocarbons are synthetically produced organic chemicals in which all hydrogen atoms are replaced by fluorine containing a perfluoroalkyl residue (Figure 1) [1]. Those chemicals exhibit a significant reduction in surface tension (due to their incompatibility with water and oil), and outstanding stability (both chemical and thermal) [3].

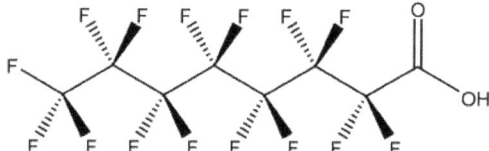

Figure 1. Perfluorooctane alcohol: $F_3C-CF_2-CF_2-CF_2-CF_2-CF_2-CF_2-CF_2-CF_2-CH_2-CH_2-OH$ [3].

The mechanism of repellency includes reducing the free energy at the material surface [5]. Therefore, if the adhesive forces between material and drops of liquid on the material are greater than the internal cohesive interactions within the liquid, the drops will spread. In contrast, in cases when adhesive interactions between the material and the liquid are smaller than the internal cohesive interactions within the liquid, the liquid drops will not spread [6].

There are many parameters which influence the surface tension and thereby repellency, and one of them is the length of the chain [3]. The literature data show that the critical surface tension decreases rapidly as the chain length increases from one to eight, and after that decreases [7]. However, fluorocarbons with 8 C atoms ("C8") have a negative impact on the environment and human health so, from September 2020, the Croatian industry will replace all ecological non favorable substances with C6 fluorocarbons which are allowed by strict regulations. This puts extra requirements on industrial managers and process engineers that need to meet demands of rigorous ecological standards, as well as customer requirements for super hydrophobic garments.

Different treatments are developed using cost effective nanotechnology-based repellents, but new formulae need improvement since current products have a negative impact on comfort in hot and humid environments [8]. Therefore, in order to enhance the repellency and other protective properties of garments, many efforts are being conducted in different processing areas: fabric weave technology [9], coating technology, surface nanocomposite modifications with resins and nanoparticles [10] and double-sided knitting consisting of two different kinds of fibers [11]. Moreover, achieving waterproof breathable fabrics [12], super-hydrophobic and multi-responsive fabric composite with excellent electro-photo-thermal effects [13] with applications not only to textiles but also in composite insulators [14] and producing intelligent electronic clothing systems [15], is under investigation. Recent developments also cover the area of medical materials with antibacterial capacity [16,17], as well as methods for the incorporation of membranes, coatings, fabrics, lining material and other vital parameters for designing breathable garments [18].

Many parameters influence the wear comfort of materials i.e., porosity and strength of the micro-porous layer on textile material, such as: crystallinity, temperature and rate of stretching, pressure, temperature and duration of the heat treatment, the choice of reagents and others [1]. In order to find optimal conditions for the multi-parameter process, Design of Experiment (DoE) statistical methodology is often used [19–23]. For example, Moussa et al. used a factorial experimental design for optimization of the waterproof breathable property of samples [11] and optimized vital parameters. Coronado et al. used the mixture design of experiments to assess the environmental impact of clay-based structural ceramics containing hazardous metals As, Ba, Cd, Cr, Cu, Mo, Ni, Pb, and Zn [19], and Waha et al. optimized the process of biodegradation by the Taguchi design of experiment [20].

Long chain fluoro-chemicals with at least eight perfluorinated carbon atoms are synthetic compounds that have been used since the 1940s in a wide variety of consumer and industrial products such are firefighting foams, surfaces, and food contact paper. Due to their widespread usage, today they can be detected in the environment, as well as in humans. Unfortunately, recent evidence shows that continued exposure to above specific levels of certain long chain fluorochemicals may lead to adverse health effects [24]. Recent investigations have proved that PFAS's are associated with reproductive toxicity, reduced growth metrics in newborns and elevated cholesterol levels in humans. Moreover, PFAS are highly persistent and resistant to degradation and are therefore a serious global concern.

According to some predictions, by 2050, more people could die from the infections caused by antibiotic-resistant bacteria than from cancer. Only in Europe 25,000 deaths per year and costs over EUR 1.5 billion are associated with resistant microorganisms. Especially dangerous infections are bacteria such as *Staphylococcus aureus*, *Methicillin-resistant Staphylococcus aureus* (MRSA), known as "super-bacteria" or "golden staphylococci", which is increasingly difficult to cope with due to its' resistance to a wide range of antibiotic-based penicillin drugs (β-lactam antibiotics such as methicillin, dicloxacillin, nafcylin and oxacillin). *Staphylococcus aureus* is a member of the *Staphylococcaceae* family of gram-positive bacteria of spherical forms and is one of the most significant pathogens in the world. It is an infection with a frequency ranging from 20 to 50 cases per 100,000 inhabitants per year, with 10% to 30% of infections ending with a deadly outcome. This number is greater than the sum of the deaths caused by AIDS, tuberculosis and viral hepatitis combined.

Therefore, the aim of this research was to develop a hydrophobic polymer surface that can be further used as potential antimicrobial material. The idea was to achieve excellent results of hydrophobic properties through the optimization of textile functionalization process parameters. Optimization was performed through six process parameters using Design of Experiment. The goal of optimization was to find optimal conditions of the water vapor permeability and the resistance to water wetting. The analysis of water vapor permeability is related to the level of comfort at low physical activity but does not give the information about condensation on a textile surface.

2. Materials and Methods

2.1. Samples

In this research three different samples of textile material intended to be used as military garments were investigated. Samples were made of cotton and polyamide yarns with weights of 190 g/m^2, 220 g/m^2 and 240 g/m^2. The mechanical properties of samples are presented in Table 1.

Table 1. Mechanical properties of investigated samples.

Weight, g/m^2	Fabric Composition	Density, Thread per 1 cm		Yarn Count, Tex *		Construction of Fabric
		Warp	Weft	Warp	Weft	
190	50% cordura (PA 6.6. Type 420 HT dull)/50% cotton	35.8	19.5	14 × 2 tex	14 × 2 tex	Ripstop (square 7 ± 1 mm × 7 ± 1 mm)
220		35.8	20.5	14 × 2 tex	38.4 × 1 tex	
240		35.8	20.5	15 × 2 tex	50 × 1 tex	

* *Tex* is a direct measure of linear density and represents grams per kilometre of the yarn.

2.2. Reagents and Chemicals

The hydrophobic compounds tested were a fluorocarbon agent based on C8 fluorochemicals, Sevophob HFK–F, (fluorocarbon resin which is used for permanent water, oil and dirt-repellent finishing, producer: Textil Color, Sevelen, Switzerland), and a fluorocarbon agent based on C6 fluorochemicals, Tubiguard SCS-F, (low viscous liquid dispersion, producer: CHT) in concentrations from 35 to 70 g/L. The dye used during the textile finishing treatment was Bezathren Navy Blue GN vat dye (producer: CHT, Montlingen, Switzerland) since it offers an outstanding light, wet and chlorine fastness level like no other dye class on cellulose fibers. It was applied in the form of a micro-disperse powder which is easily dispersed and used in textile modification.

2.3. Textile Finishing Treatment

The textile finishing treatment used for the modification of the material into a hydrophobic textile was pad impregnation of the dyeing of fabric and for the application of finishing chemicals. It had several steps: (1) pre-treatment for removal of waste, oil, dry matter and other impurities; (2) dyeing;

and (3) finishing with two different types of fluorocarbon agents to achieve water and oil repellency. The second and the third steps (dyeing and finishing experiments) were performed on a laboratory scale using laboratory machinery for padding in which the fabric passes into a solution of dye and chemicals under a submerged roller, and then it goes out of the bath is squeezed to remove excess solution by pressure occurring between two cylinders. The objective of this process is to mechanically impregnate the fabric with the solution or dispersion of chemicals.

2.4. Determination of Water Vapor Permeability

Moisture transport through textiles is the factor which influences thermological and physiological comfort of the material. The moisture is transferred through a material in the form of vapors or liquids.

Water vapor permeability, (WVP) is expressed as the time rate of water vapor transmission through a unit area induced by unit vapor pressure difference between two specific surfaces, under specified temperature and humidity conditions [25]. It is calculated as:

$$WVP = \frac{G}{t} \times A \times \Delta p \tag{1}$$

where Δp, (Pa) is the difference of partial pressure between two sides of material, G, (g)—weight change, t, (h)—time during which G occurred, and A, (m^2)—test area.

In the ASTM E96 standard procedure the test cup is filled with the distilled water and the circular sample is tightly covered onto the cup. The cup prepared for testing is placed in a controlled environment at an ambient temperature of 23.0 °C, with a relative humidity of 100% inside the cup and 50% outside the cup. Due to the forces of the differences in concentration and pressure, vapor diffusion occurs through a textile from the cup in this environment [25].

After this treatment, the resistance of the hydrophobic modification was tested in a 5-cycle process washing according to the norm EN ISO 6330:2012 Textiles—Domestic washing and drying procedures for textile testing, testing procedures at 60 °C. In this work, the WVP was tested according to the ASTM E96 standard procedure in which the vapor passes from inside of the cup to the outside of the cup (Figure 2).

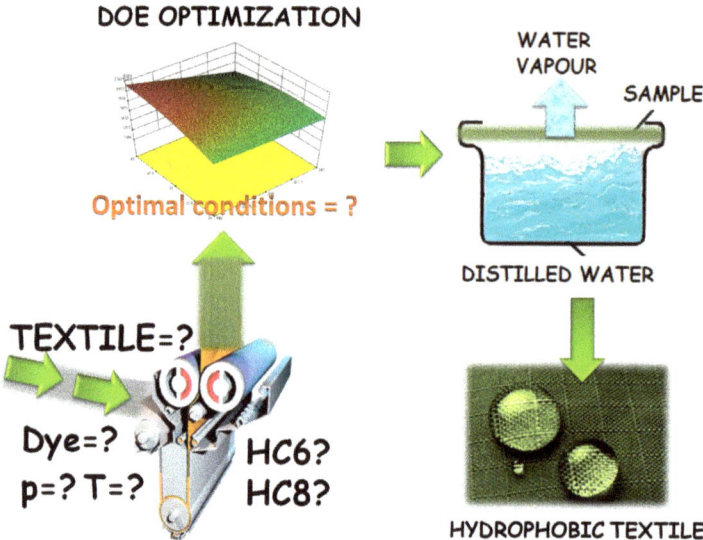

Figure 2. Schematic overview of process for optimizing the highest water vapor permeability by the ASTM E96 method.

The drawback of this methodology is the fact that the permeability that is measured by this method depends not only on properties of a material, but also on the thickness of the air layer near the surface of the textile material. Therefore, the resistance to surface wetting was measured in another set of experiments (Section 2.5).

2.5. Resistance to Surface Wetting

A resistance to surface wetting test was performed according to the ISO 4920:2012 standard procedure [26]. This method specifies spray test conditions for determining the resistance of fabric to surface wetting by water in the following manner: a specified volume of distilled water is sprayed onto a test specimen that has been mounted on a ring and placed at an angle of 45°. It is placed in a way that the centre of the specimen is at a specified distance below the spray nozzle. The spray rating is determined by comparing the appearance of the specimen with descriptive standards.

2.6. Mathematical and Statistical Procedures

Sections 2.1–2.5. were statistically analyzed using Design Expert Stat Ease software 9.1 (Minneapolis, MN, USA). A central composite design was chosen for the modeling of six process parameters with replicates, so the total number of experiments was 42. The optimal conditions for six independent variables, i.e., fabric weight, dye concentration, concentration of hydrophobic compounds Sevophob HFK–F and Tubiquard SCS–F, temperature, pressure and the type of the hydrophobic compound, were obtained using algorithms with the Design Expert State Ease version 9.1 software (Minneapolis, MN, USA). The critical surface tension of hydrophobic compounds decreases rapidly as the chain length increases from 1 to 8, and after that decreases [7]. In majority of industrial processes fluorocarbons with 8 C atoms were frequently used due to their extraordinary hydrophobic properties. However, due to their extremely negative impact on the environment and human health, their usage is prohibited so the industry needs to find a suitable replacement. Therefore, in this work the efficiency of different treatments using C 6 and C 8 fluorocarbons was investigated and optimized through six different industrial process parameters.

This statistical program DoE searches for an optimal combination of factor levels that simultaneously satisfies the requirements placed on each of the responses and factors.

3. Results

This section is be divided by subheadings. It should provide a concise and precise description of the experimental results, their interpretation as well as the experimental conclusions that can be drawn. Sevophob HFK–F and Tubiquard SCS–F are fluoro-based finishing compounds that were investigated in this research as efficient hydrophobic reagents that enable durable and high-performing finishing of military garments. Sevophob HFK-F is much a more efficient reagent, but contains eight C atoms in its structure, while Tubiquard SCS-F has only six C atoms.

In order to achieve equally satisfying results with an ecologically acceptable reagent, multi-parameter optimization using Design of Experiment was performed. Therefore, in this work six process parameters were investigated, namely: the weight of the military fabric (ranging from 190 to 240 g/m^2), concentration of the dye (10 to 60 g/L), concentration of hydrophobic compounds (35 to 70 g/L), type of hydrophobic compounds (ranging from six to eight carbon atoms), temperature (from 150 to 170 °C) and pressure (from 1.0 to 2.0 bar). In total, six input process variables were varied for optimization purposes in 42 preliminary experiments, with the goal to optimize the hydrophobic properties of the garments, while preserving the wear comfort of the materials.

3.1. Optimization by Design of Experiments

A traditional optimization protocol monitoring one parameter at the time cannot provide information on interactions between the process parameters and their outcome for the industrial process. Design of experiment (DoE) statistical methodology offers the answer to this problem and the possibility to study the effects of process variables and their responses within the minimal number of preliminary experiments [27].

Using DoE based on the response surface methodology (RSM) within the State Ease software, the aggregate mix proportions were derived, and the total number of experiments was drastically reduced. In order to examine whether there is a relationship between the selected parameters and the response variables investigated, the collected data needed to be analyzed in a statistical manner using regression according to Equation (2):

$$x_i = (X_i - X_i^x)/\Delta X_i \qquad (2)$$

where x_i is the coded value of the independent variable i, X_i the natural value of the variable i, X_i^x the natural value of the variable i in the central point and ΔX_i is the value of the step change.

The response y (estimation of the coefficients of a quadratic model) is represented by Equation (4), within the central composite design:

$$y = \beta_0 + \sum_{i=1}^{k} \beta_i x_i + \sum_{i=1}^{k} \beta_{ii} x_i^2 + \sum_{i<j} \beta_{ii} x_i x_j + \varepsilon \qquad (3)$$

where, y is the measured response, β_0 is the intercept term, β_i, β_{ii} and β_{ij} are the measures of the effects of variables x_i, $x_i x_j$ and x_i^2, respectively. The central composite design was selected since this is one of the most efficient and popular classes of second order designs recommended in the literature [19]. A very important optimization step is the selection of parameters and their ranges to be studied, which was done based on previous analysis. The factors were studied in several levels, from low to high, as is presented in Table 2.

As can be seen from the Table 2, three experiments (runs 3 and 39, 4 and 28, 11 and 12) were repeated to check the reproducibility and to estimate an experimental error. All three responses gave a reproducible result where the deviation of each run was found within an experimental error (i.e., ±0.075% for water vapor permeability, and ±0.053 for resistance to water wetting). The worst result in the preliminary tests was recorded in the eighth experiment (with 1196 g/m² for water vapor permeability), for the hydrophobic compound with six C atoms.

The center composite design matrix of 42 experiments covering full design of three level factors was used to build a quadratic model of the experimental data of the observed responses. Standard error of design is shown in Figure 3. It reports the standard error of the predicted mean in a way in which larger standard error means less reliable the estimates.

Table 2. Experimental planning in Design of Experiment, central composite design, six parameters, 42 preliminary experiments, six input parameters and two output responses.

	PARAMETERS						RESPONSES	
Random Run	Fabric Weight g/m²	Dye Concentration g/L	Concentration of Hydrophobic Compound FC, g/L	Temperature °C	Pressure Bar	Type of Hydrophobic Compound	WVP, g/m², 24 h	Resistance to Water Wetting
1	240	60	70	170	1.66	Tubiguard SCS-F	1658	70
2	190	60	70	150	1	Tubiguard SCS-F	2090	80
3	220	40	35	161	1.85	Sevophob HFK-F	1646	90
4	190	59.2	47.6	159.8	1.39	Sevophob HFK-F	1824	90
5	240	10	62.1	166.5	1.43	Tubiguard SCS-F	1738	100
6	220	60	70	164	1.3	Sevophob HFK-F	1517	90
7	220	29	70	153	1.77	Tubiguard SCS-F	1963	60
8	220	40	41.8	160.7	1.2	Tubiguard SCS-F	1196	70
9	240	16.5	38.2	170	1.38	Tubiguard SCS-F	1617	80
10	190	22.8	35	150	2	Sevophob HFK-F	2000	95
11	190	13.8	60.2	157.4	1.66	Tubiguard SCS-F	1466	80
12	190	13.8	60.2	157.4	1.66	Tubiguard SCS-F	1390	80
13	220	52.5	35	170	1	Sevophob HFK-F	1765	95
14	240	35	50.9	159.9	2	Tubiguard SCS-F	1463	70
15	240	30.0	67.9	170	2	Tubiguard SCS-F	1522	80
16	240	10	70	164.6	2	Sevophob HFK-F	1696	100
17	220	49.8	61.3	150.3	1.05	Sevophob HFK-F	1926	100
18	190	12.8	70	150	1	Sevophob HFK-F	1770	95
19	190	13.5	70	170	1	Tubiguard SCS-F	1825	80
20	220	60	51.7	150	2	Tubiguard SCS-F	2007	80
21	220	10	54.5	158	1.25	Sevophob HFK-F	1247	100
22	240	35	68.0	166.0	1	Tubiguard SCS-F	1720	100
23	240	60	35	150	1.59	Tubiguard SCS-F	1560	70
24	220	10	35	150	2	Tubiguard SCS-F	1296	60
25	220	60	42.7	170	2	Sevophob HFK-F	1362	90
26	240	60	70	150.5	2	Sevophob HFK-F	1558	90
27	240	26.9	54.3	150	1.52	Sevophob HFK-F	1653	90
28	190	59.2	47.6	159.8	1.39	Sevophob HFK-F	1647	95
29	240	10	70	150	1	Tubiguard SCS-F	1585	50
30	240	56.3	61.1	155.9	1.45	Tubiguard SCS-F	1796	85
31	240	27.0	54.3	150	1.52	Sevophob HFK-F	1995	95
32	240	10	35	157.5	1	Sevophob HFK-F	1834	100
33	220	60	35	150	1	Sevophob HFK-F	1720	95
34	190	60	35	169.5	2	Tubiguard SCS-F	1973	90
35	190	10	35	170	1.6	Sevophob HFK-F	1892	100
36	240	60	47.3	166	1	Tubiguard SCS-F	1560	90
37	220	10	53.7	170	2	Tubiguard SCS-F	1457	80
38	220	31.8	70	159	1.01	Tubiguard SCS-F	1783	80
39	220	40	35	161	1.85	Sevophob HFK-F	1521	95
40	190	29.5	40.3	150	1	Tubiguard SCS-F	1596	70
41	220	40	41.8	160.7	1.2	Tubiguard SCS-F	1723	70
42	190	37.5	63.7	170	1.95	Sevophob HFK-F	1711	95

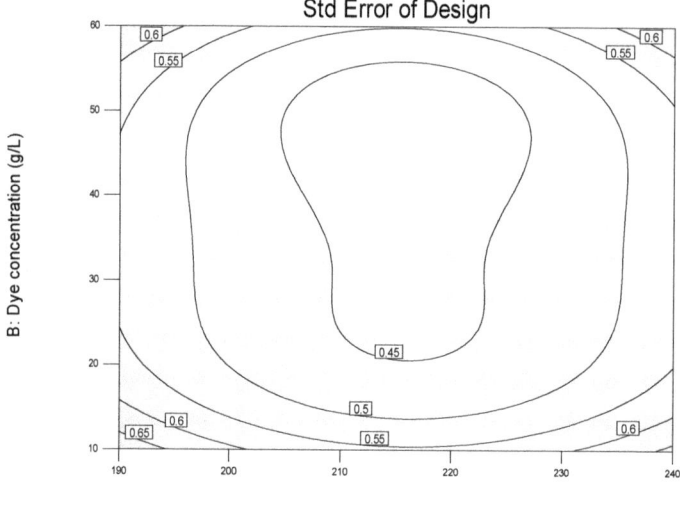

Figure 3. Standard error of design.

3.2. ANNOVA Report

The analysis of variance (ANOVA) was carried out to establish the significance of different parameters for the quadratic model. The quadratic model for the water vapor permeability in terms of coded factors is presented in Equation (4) as:

$$\begin{aligned}\text{Water Repellency} &= 87.41 - 1.19 \times A + 0.12 \times B + 0.49 \times C + 4.02 \times D \\ &\quad - 0.20 \times E - 9.87 \times F + 0.15 \times AB + 0.33 \times AC + 3.81 \\ &\quad \times AD - 3.96 \times AE - 2.23 \times AF - 1.64 \times BC - 6.21 \times BD \\ &\quad - 0.012 \times BE + 3.56 \times BF - 0.34 \times CD - 1.21 \times CE \\ &\quad + 0.048 \times CF - 0.91 \times DE + 2.89 \times DF + 0.49 \times EF \\ &\quad + 3.07 \times B^2 - 4.85 \times C^2 - 1.14 \times D^2 + 0.64 \times E^2\end{aligned} \quad (4)$$

where coded parameters are: A Fabric weight, B Dye concentration, C Hydrofobe concentration. D Temperature and E the Pressure. The significance of parameters is confirmed if their p-values are below value of 0.05 (which is the significance limit).

In this model, all parameters were significant having p values of 0.0020, 0.0109, 0.0005, 0.0220, 0.0010 and 0.0001 for fabric weight, dye concentration, hydrophobic compound concentration, temperature, and pressure, respectively. The model F-value of 4.22 implies that the model is significant according to a 95% level of confidence.

Values of "$p > F$" less than 0.0500 indicate that the model terms are significant and in this work the p-value was 0.0022. The lack of fit was calculated from the experimental error (pure error) and residuals. The lack of fit is the ratio between the residuals and pure error. "Lack of fit F-value" of 10.13 with its p-value of 0.0096 implies the lack of fit was significant. Therefore, the suggested model for water vapor permeability presented in Equation (4) is valid for the present study.

3.3. Significant Graphics

Figure 4 compares experimental values with the predicted model value obtained from Equation (4). The value of the correlation coefficient, R^2 was found to be high (i.e., close to unity) which confirms the accuracy of the model [20,21].

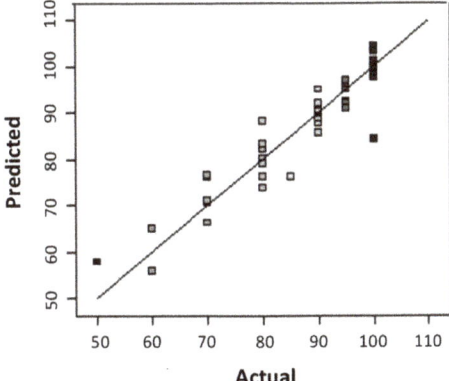

Figure 4. Plot of actual value and the value predicted from the Equation (4).

The Durbin–Watson statistic presented in the Figure 5 shows that no correlation can be observed between the experimental and calculated values so there is no evidence of correlation in the residuals' series, and therefore there is no accumulation of experimental error [19].

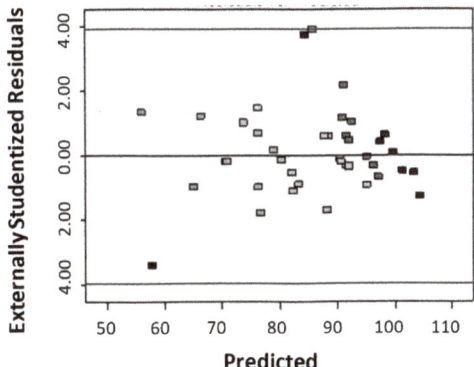

Figure 5. Residual correlation: q(experimental)–q(calculated).

3.4. Selectivity

The model from Equation (4) is presented graphically in Figure 6, and the study of the effects of process variables on the response water vapor permeability is presented in Figure 7.

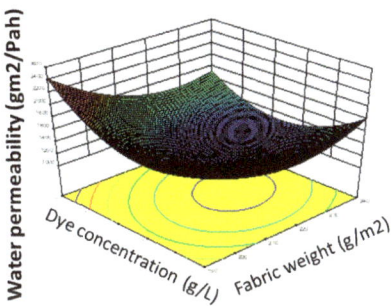

Figure 6. Graphical presentation of the model equation, effect of fabric weight and the dye concentration on the response parameter water vapor permeability.

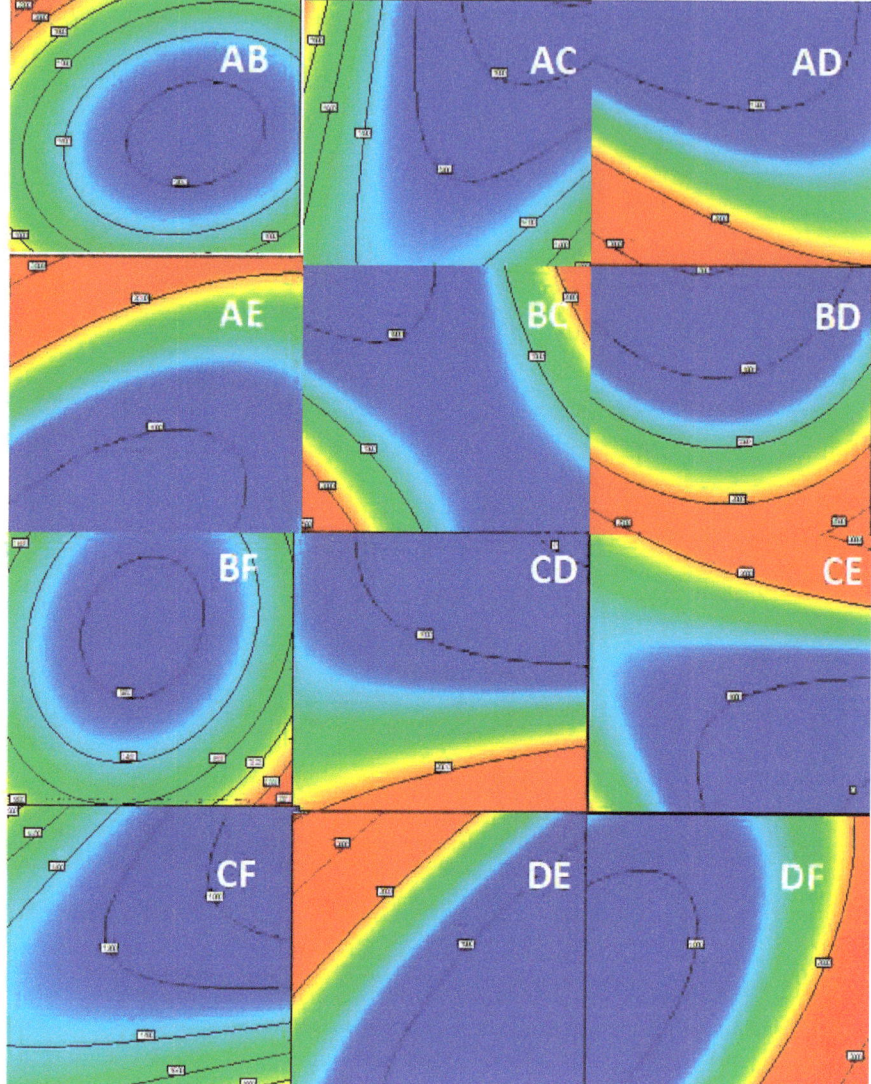

Figure 7. Effects of industrial process parameters on response water vapor permeability: (**A**—fabric weight, **B**—dye concentration, **C**—hydrophobic concentration, **D**—temperature, **E**—pressure and **F**—the type of the hydrophobic compound). Interactions shown are: (**AB,AC,AD,AE,BC,BD,BF,CD,CE,CF,DE,DF**).

As can be seen in Figure 7, six parameters mutually influence the response water vapor permeability, but also show interactions and synergistic effects on the output variable.

3.5. Optimization Using Response Surface Methodology

Resistance to water wetting occurred both when using Sevophob HFK–F and Tubiquard SCS–F through lowering the surface energy of the fabric, so that water did not wet out the garment. Although similar finishes could be achieved with some other types of finishers (such are waxes, oils and silicones), those other reagents are not repellent to oil, sand or lotion compounds.

For this reason, the fluorocarbons such are Sevophob HFK–F and Tubiquard SCS–F, were used as the most effective compounds for repelling both oil and water, which is a very important parameter in finishing military garments. However, Sevophob HFK-F contains eight C atoms so the optimal process in which Turbiquard can be used, but with equal efficiency needed to be developed.

Zahid et al. reported that among materials with extra developed hydrophobic properties, textile materials are come in contact with the human skin most frequently. Therefore, the authors have emphasized that textile treatments for water or oil repellency should be non-toxic, biocompatible, and comply with stringent health standards [28]. Moreover, due to the large quantities of water, chemicals and reagents used worldwide in the textile industry, treatments should be scalable, sustainable, and eco-friendly. Due to this awareness, new eco-friendly processes are being developed and adopted.

Moreover, the review article of Zahid et al. reported that although fluorinated polymers with C8 chemistry are the best performing materials to render textiles water or oil repellent, they pose substantial health and environmental problems and are being banned. Therefore C8-free vehicles for non-wettable treatment formulations are probably the only ones that can offer important commercialization prospects. In addition, their review article indicated promising future strategies and new materials that can transform the process for non-wettable textiles into an all-sustainable technology [28].

In this work, the model Equation (4) was used to find the manner in which the parameters need to be varied in order to achieve the optimal solution for obtaining hydrophobic properties with less harmful reagents that contain six C atoms. The corresponding optimized surface response of the quadratic model is shown in Figure 8.

The surface contour plots of parameter interactions between industrial process variables are elliptical, and the central point is the point in which the slope of the contour in all directions is equal to zero. The minimal predicted response yield is indicated by the surface confined in the smallest curve of the contour diagram, so the optimal maximized solutions are outside this area, as is shown in Figure 8.

The analysis of the results presented in Figure 8 shows the significant influence of the process parameters on the results. As can be seen from this Figure, by increasing the concentration of hydrophobic compounds, concentration of the dye and pressure, leads to the enhancement of water vapor permeability. In contrast, by decreasing the garment weight, its efficiency was enhanced. Moreover, the increase in temperature enhanced water vapor permeability but decreased the resistance to water wetting. Those results can be explained by the fact that, at high concentration of reagents and rigorous process variables, optimal coating finishing will occur.

Statistical analysis of this model results in the optimal solution for maximized water vapor permeability. Therefore, the additional experiments were carried out at optimized conditions and the results of the responses predicted by DoE were compared to experimental results as is shown in Table 3. Table 3 presents verification experiments of the model at three optimal solutions for three different kinds of textile fabric (weights 190, 220 and 240 g/m^2, respectively).

Table 3. Verification experiments at optimum conditions of process parameters (A—fabric weight, B—dye concentration, C—hydrophobic concentration, D—temperature, E—pressure and F—the type of the hydrophobic compound) obtained by Design of Experiment for maximized water vapor permeability and maximized resistance to water wetting.

Nr.	A	B	C	D	E	F	Resistance to Water Wetting, % PREDICTED	Resistance to Water Wetting, % EXPERIMENT	Water Vapor Permeability PREDICTED g/m^2	Water Vapor Permeability EXPERIMENT g/m^2
1.	190	16.931	41.066	168.903	1.046	Sevophob	102	100	2670	2609
2.	220	56.067	68.802	154.276	1.048	Tubiquard	100	100	2176	2620
3.	240	11.106	38.815	152.460	1.908	Sevophob	88	100	2355	2530

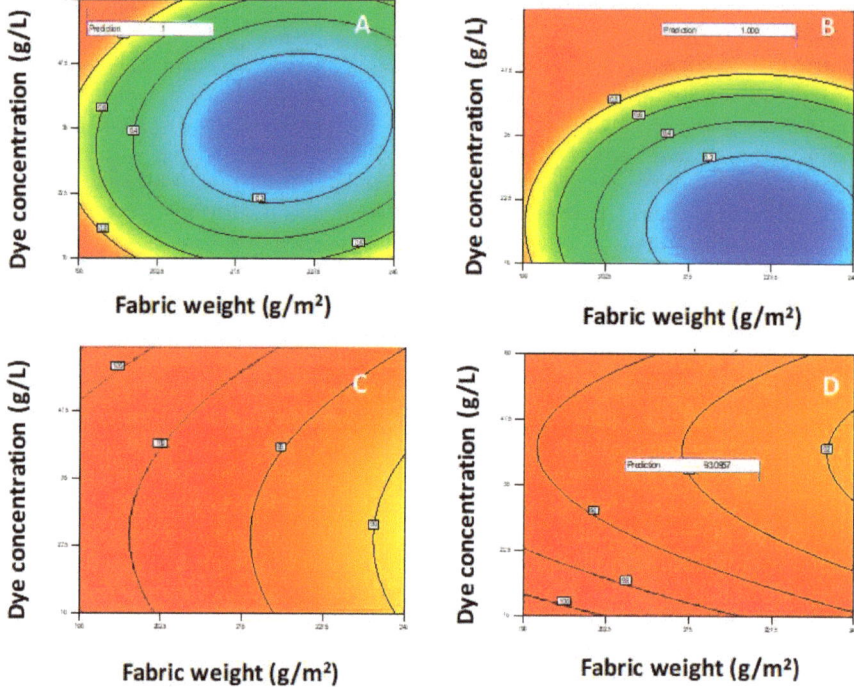

Figure 8. (**A–D**): Surface contour plots for desirability, water vapor permeability, and resistance to water wetting at optimal values of initial fabric weight (g/m^2) and dye concentration (g/L) for different values of other parameters: (**A**) C = 51.386, D = 159.109, E = 1.504, F = Sevophob HFK-F; (**B**) C = 61.437, D = 156.768, E = 1.228, F = Tubiquard SCS-F; (**C**) C = 51.756, D = 150.309, E = 1.995, F = Sevophob HFK—F; (**D**) C = 38.136, D = 159.208, E = 1.852, F = Sevophob HFK—F, where the other parameters are C—hydrophobic concentration, D—temperature, E—pressure and F—the type of the hydrophobic compound.

As can be seen from this table, two optimal solutions were obtained with eight C atoms, but another two were obtained with six C atoms, and this result surpassed all preliminary tests for both six C atoms and eight C atoms by 21.15%. However, when this optimal result of 2620 g/m^2 is compared with the statistical mean value obtained only for six C atoms (1726.86 g/m^2), the global optimum was 51.72% better. This means that the industrial process can be performed under the presented variable parameters achieving satisfying results for both ecologically and economically.

Excellent resistance to water wetting remained even after five washing cycles, which is an excellent result. Without optimization, the resistance to water wetting effects would drop to 50–70%, and our optimized process resulted with a modification that preserved resistance to water wetting (at 90–100%).

Figure 9 shows the desirability for each factor and all responses individually for the selected optimal solution. As can be seen from the Table 3, for the optimal result, there was an error of ±1.96% for resistance to water wetting and error of ±2.28% for water vapor permeability with a 95% confidence level (i.e., 94 ± 0.52%).

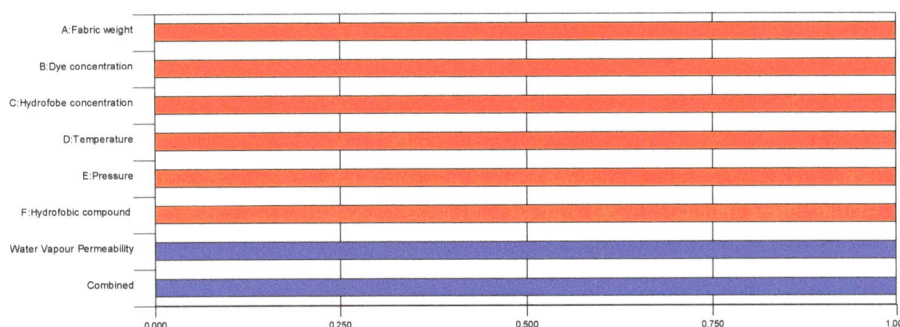

Figure 9. Desirability for each factor and each response individually for the optimal solution.

The verified optimized solution of the model responses of water vapor permeability and resistance to water wetting are shown in Figure 10.

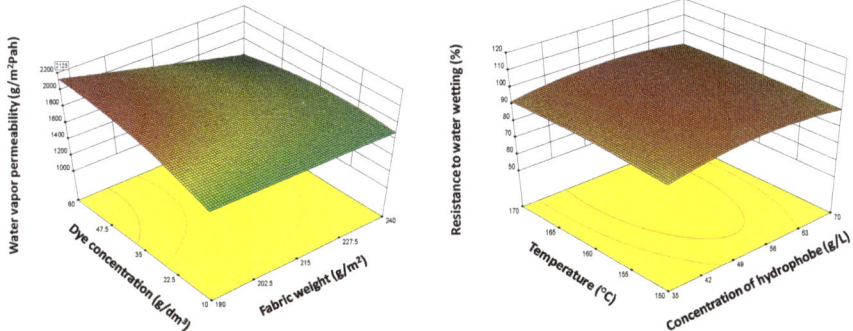

Figure 10. Response surface of the Central composite design, optimized response variables for: water vapor permeability and resistance to water wetting.

Based on the optimization model shown in Figure 10, the optimum of responses (water vapor permeability and resistance to water wetting) could be predicted and verified by comparing the predicted result with the new experimental output.

An optimal process with an ecologically favorable fluorocarbon with six C atoms (68.802 g/L, Tubiquard) surpassed all test results by 21.15%. It was found at 154.3 °C, 1.05 bar, 56.07 g/L dye, 220 g/m² fabric, which surpassed the results of statistical mean for six C atoms by 51.72%. Moreover, the best solution obtained with hydrophobic compound of eight C atoms was 2609 g/m². It was found at 152.5 °C, 1.9 bar, 38.815 g/L of Sevophob HFK-F, 11.1 g/L dye on 240 g/m² fabric, and surpassed the statistical mean of all test results with eight C atoms (1619.2 g/m²) by 61.12%. From this it can be concluded that Design of Experiment is an excellent tool for optimization of multi-parameter complex systems because with high numbers of factors only a fraction of the experiments need to be completed to efficiently estimate the main effects and parameter interactions [27].

4. Conclusions

The presented work shows that Design of Experiment is a convivial tool for optimization of modification of the surface since the influence of the various process parameters can be easily investigated. In this work, fabric weight, dye concentration, hydrophobic reagent concentration, temperature, pressure, and the type of the hydrophobic compound, have been investigated to determine the optimal conditions. Validity of the regression equation has been controlled by a statistical approach, and the results have revealed following conclusions:

- Sevophob HFK–F and Tubiquard SCS–F are efficient hydrophobic reagents, with Sevophob HFK–F being much more efficient but ecologically unacceptable.
- Central composite design of experiments was used based on 42 preliminary experiments in order to investigate six process parameters within laboratory scale optimization, to find an alternative solution using Tubiquard SCS-F with six C atoms.
- An Ecologically favorable optimum using a fluorocarbon with six C atoms (68.802 g/L, Tubiquard) surpassed all the test results by 21.15%. It was found at 154.3 °C, 1.05 bar, 56.07 g/L dye, 220 g/m^2 fabric.

From this it can be concluded that efficient water repellency can be achieved using six C fluorocarbons while ecologically harmful and forbidden hydrophobic compounds with eight C atoms can be successfully replaced by reagents acceptable for the industry purposes.

Author Contributions: Conceptualization, I.R.; methodology, I.R. and A.K.; software, I.R.; validation, I.R.; formal analysis A.K.; investigation, I.R.; resources, I.R.; data curation, I.R. and A.K.; writing—original draft preparation, I.R.; writing—review and editing, I.R. and A.K.; visualization, I.R.; supervision, I.R.; project administration, I.R.; funding acquisition, I.R. All authors have read and agreed to the published version of the manuscript.

Funding: This research was funded by the Croatian Science Foundation grant number HRZZ-IP-2019-04-1381. Any opinions, findings and conclusions or recommendations expressed in this material are those of the authors and do not necessarily reflect the views of Croatian Science Foundation.

Conflicts of Interest: The authors declare no conflict of interest.

Abbreviations

ASTM	American Society for Testing and Materials
EN ISO	European Norm
ISO	International Organization for Standardization
HT	high tension
PA	polyamide

References

1. Sayed, U.; Dabhi, P. Finishing of textiles with fluorocarbons. In *The Textile Institute Book Series Waterproof and Water Repellent Textiles and Clothing, Chapter 6*; Williams, J., Ed.; Mahendra Publications; Woodhead Publishing: Cambridge, UK; Sawston, UK, 2014; pp. 139–153.
2. Rezić, I.; Krstić, D.; Bokić, L. Ultrasonic extraction of resins from an historical textile. *Ultrason. Sonochem.* **2008**, *15*, 21–24. [CrossRef] [PubMed]
3. Audenaert, F.; Lens, H.; Rolly, D.; Vander Elst, P. Fluoro-chemical textile repellents—Synthesis, and applications: A 3M perspective. *J. Text. Inst.* **1999**, *90*, 76–94. [CrossRef]
4. Schindler, W.; Hauser, P. *Chemical Finishing of Textiles*; Woodhead Publishing Ltd.: Boca Raton, FL, USA; Cambridge, UK, 2004; pp. 74–91.
5. Kissa, E.; Lewin, M.; Sello, S. *Handbook of Fiber Science and Technology, Chemical Processing of Fibers and Fabrics, Part B Functional Finishes*; Marcel Dekker: New York, NY, USA, 1984; Volume 2, p. 143.
6. Holmes, D.; Horrocks, A.; Anand, S. (Eds.) *Handbook of Technical Textiles*; Woodhead Publishing Ltd.: Cambridge, UK, 2000; p. 461.
7. Lu, D.; Sha, S.; Luo, J.; Huang, Z.; Zhang Jackie, X. Treatment train approaches for the remediation of per- and polyfluoroalkyl substances PFAS: A. critical review. *J. Hazard. Mater.* **2020**, *386*, 121963. [CrossRef] [PubMed]
8. Gibson, P. Water-repellent Treatment on Military Uniform Fabrics: Physiological and Comfort. *J. Ind. Text.* **2008**, *38*, 43–54. [CrossRef]
9. Murray, C.C.; Vatankhah, H.; McDonough, C.A.; Nickerson, A.; Hedtke, T.T.; Higgins, C.P.; Bellona, C.L. Removal of per- and polyfluoroalkyl substances using super-fine powder activated carbon and ceramic membrane filtration. *J. Hazard. Mater.* **2019**, *366*, 160–168. [CrossRef]

10. Zahid, M.; Heredia-Guerrero, J.A.; Athanassiou, A.; Bayer, I.S. Robust water repellent treatment for woven cotton fabrics with eco-friendly polymers. *Chem. Eng. J.* **2017**, *319*, 321–332. [CrossRef]
11. Moussa, A.; Marzoug, I.B.; Bouchereb, H. Development and optimisation of waterproof breathable double-sided knitting using a factorial experimental design. *J. Ind. Text.* **2015**, *45*, 437–466. [CrossRef]
12. Ravindra, D.; Kale, D.; Vade, A.; Potdar, T. Optimization study for Waterproof breathable Polyester fabric. *Int. J. Innov. Res. Technol.* **2016**, *3*, 16–24.
13. Huo, L.; Huang, X.; Li, J.; Guo, Z.; Gao, Q.; Hu, M.; Xue, H.; Gao, J. Superhydrophobic and multi-responsive fabric composite with excellent electro-photo-thermal effect and electromagnetic interference shielding performance. *Chem. Eng. J.* **2019**, 123537, in press, corrected proof. [CrossRef]
14. Peng, W.; Qin, H.; Zhao, M.; Zhao, X.; Guo, Z. Creation of a multifunctional superhydrophobic coating for composite insulators. *Chem. Eng. J.* **2018**, *352*, 774–781. [CrossRef]
15. Wang, J.; He, J.; Ma, L.; Zhang, Y.; Shen, L.; Xiong, S.; Li, K.; Qu, M. Multifunctional conductive cellulose fabric with flexibility, superamphiphobicity and flame-retardancy for all-weather wearable smart electronic textiles and high-temperature warning device. *Chem. Eng. J.* **2020**, *390*, 124508. [CrossRef]
16. Nabil, B.; Christine, C.; Julien, V.; Abdelkrim, A. Polyfunctional cotton fabrics with catalytic activity and antibacterial capacity. *Chem. Eng. J.* **2018**, *351*, 328–339. [CrossRef]
17. Rezić, I. Determination of engineered nanoparticles on textiles and in textile wastewaters. *TrAC Trends Anal. Chem.* **2011**, *30*, 1159–1167. [CrossRef]
18. Mukhopadhyay, A.; Midha, V.K. A Review on Designing the Waterproof Breathable Fabrics Part I: Fundamental Principles and Designing Aspects of Breathable Fabrics. *J. Ind. Text.* **2008**, *37*, 225–262. [CrossRef]
19. Coronado, M.; Segadães, A.M.; Andrés, A. Using mixture design of experiments to assess the environmental impact of clay-based structural ceramics containing foundry wastes. *J. Hazard. Mater.* **2015**, *299*, 529–539. [CrossRef] [PubMed]
20. Wahla, A.Q.; Iqbal, S.; Anwar, S.; Firdous, S.; Mueller, J.A. Optimizing the metribuzin degrading potential of a novel bacterial consortium based on Taguchi design of experiment. *J. Hazard. Mater.* **2019**, *366*, 1–9. [CrossRef]
21. Rezić, I. Prediction of the surface tension of surfactants mixtures for detergent formulation using Design Expert software. *Mon. Chem.* **2011**, *142*, 1219–1225. [CrossRef]
22. Rezić, T.; Rezić, I.; Blaženović, I.; Šantek, B. Optimization of corrosion process of stainless steel coating during cleaning in steel brewery tanks. *Mater. Corros.* **2013**, *64*, 321–327. [CrossRef]
23. Andlar, M.; Rezić, I.; Oros, D.; Kracher, D.; Ludwig, R.; Rezić, T.; Šantek, B. Optimization of enzymatic sugar beet hydrolysis in a horizontal rotating tubular bioreactor. *J. Chem. Technol. Biotechnol.* **2017**, *92*, 623–632. [CrossRef]
24. United States Environmental Protection Agency. *EPA's Per- and Polyfluoroalkyl Substances PFAS Action Plan*; EPA 823R18004; United States Environmental Protection Agency: Washington, DC, USA, 2019. Available online: https://www.epa.gov/pfas (accessed on 2 May 2020).
25. Arabuli, S.; Vlasenko, V.; Havelka, A.; Kus, Z. Analysis of Modern Methods for Measuring Vapor Permeability Properties of Textiles. In Proceedings of the 7th International Conference—TEXSCI, Liberec, Czech Republic, 6–8 September 2010.
26. *ISO 4920:2012 Textile Fabrics—Determination of Resistance to Surface Wetting Spray Test*; International Organization for Standardization: Geneva, Switzerland, 2012.
27. Rezić, I. Optimization of ultrasonic extraction of 23 elements from cotton. *Ultrason. Sonochem.* **2009**, *16*, 63–69. [CrossRef]
28. Zahid, M.; Mazzon, G.; Athanassiou, A.; Bayer, I.S. Environmentally benign non-wettable textile treatments: A review of recent state-of-the-art. *Adv. Colloid Interface Sci.* **2019**, *270*, 216–250. [CrossRef] [PubMed]

© 2020 by the authors. Licensee MDPI, Basel, Switzerland. This article is an open access article distributed under the terms and conditions of the Creative Commons Attribution (CC BY) license (http://creativecommons.org/licenses/by/4.0/).

Article

Improvement of the Space Charge Suppression and Hydrophobicity Property of Cellulose Insulation Pressboard by Surface Sputtering a ZnO/PTFE Functional Film

Yanqing Li [1], Jian Hao [1,*], Jinfeng Zhang [2], Wei Hou [1], Cong Liu [1] and Ruijin Liao [1]

[1] State Key Laboratory of Power Transmission Equipment & System Security and New Technology, Chongqing University, Chongqing 400044, China; starwade@163.com (Y.L.); cquhouwei@163.com (W.H.); cqu_lc@163.com (C.L.); rjliao@cqu.edu.cn (R.L.)
[2] Key Laboratory of Engineering Dielectrics and Its Application, Ministry of Education, Harbin University of Science and Technology, Harbin 150080, China; zhangjinfeng_phd16@hrbust.edu.cn
* Correspondence: cquhaojian@126.com; Tel.: +86-182-2301-0926

Received: 1 September 2019; Accepted: 30 September 2019; Published: 3 October 2019

Abstract: Oil-impregnated cellulose insulation polymer (oil-paper/pressboard insulation) has been widely used in power transformers. Establishing effective ways of improving the physical and chemical properties of the cellulose insulation polymer is currently a popular research topic. In order to improve the charge injection inhibition and hydrophobic properties of the cellulose insulation polymer used in power transformers, nano-structure zinc oxide (ZnO) and polytetrafluoroethylene (PTFE) films were fabricated on a cellulose insulation pressboard surface via reactive radio frequency (RF) magnetron sputtering. Before the fabrication of their composite film, Accelrys Materials Studio (MS) software was applied to simulate the interaction between the nanoparticles and cellulose molecules to determine the depositing sequence. Simulation results show that the ZnO nanoparticle has a better adhesion strength with cellulose molecules than the PTFE nanoparticle, so ZnO film should be sputtered at first to fabricate the ZnO/PTFE composite film for better film quality. The sputtered, thin films were characterized by X-ray photoelectron spectroscopy (XPS), scanning electron microscopy (SEM), and X-ray diffraction (XRD). The space charge injection behavior and the hydrophobicity performance of the untreated pressboard; and the cellulose insulation pressboard with sputtered nano-structure ZnO, PTFE, and the ZnO/PTFE functional films were compared with each other. X-ray photoelectron spectroscopy results showed that ZnO, PTFE, and ZnO/PTFE functional films were all successfully fabricated on the cellulose insulation pressboard surface. Scanning electron microscopy and XRD results present the nano-structure of the sputtered ZnO, PTFE, and ZnO/PTFE functional films and their amorphous states, respectively. The ZnO/PTFE composite functional film shows an apparent space charge suppression effect and hydrophobicity. The amount of the accumulated space charge in the pressboard sputtered ZnO/PTFE composite functional film decreased by about 40% compared with that in untreated cellulose insulation pressboard, and the water contact angle (WCA) increased from 0° to 116°.

Keywords: nano-structure functional film; magnetron sputtering; cellulose insulation polymer; space charge; hydrophobicity; zinc oxide; polytetrafluoroethylene

---

1. Introduction

High voltage direct current (HVDC) power systems have been utilized in long-distance power energy transmission. The most important apparatus in HVDC systems is the HVDC converter transformer. The converter transformer mainly consists of a steel core, and winding and insulating materials. The insulation materials are mainly the combination of oil and cellulose insulation polymers (cellulose insulation paper/pressboard). The oil-paper/pressboard insulation in the valve winding and outlet bushing of the HVDC converter transformer, simultaneously experience AC, DC, and transient impulse voltages under operation [1]. Under the voltage of DC component, space charge accumulation within the solid insulation material has been regarded as a major issue affecting the safe and reliable operation of the converter transformer [1]. The formation of space charge results in a distortion of the electric field distribution and may lead to local electric field enhancement, thus further inducing aging, partial discharge, and even breakdown, which could ultimately bring about insulation failure [2–4]. Therefore, effective methods to suppress space charge accumulation in the oil-paper insulation have always been the focus of research in recent years.

Moreover, moisture is regarded as "the first enemy" after temperature [5,6], which can reduce the thermal life and electrical breakdown strength of the oil-paper insulation. Water is produced during the decomposition of the cellulose. The water produced undergoes a migration from cellulose to oil and vice versa. The hygroscopicity of the cellulose and the water solubility in oil determines the equilibrium of water migration. Water can accelerate the decomposition of the cellulose, which means the more water content in the insulation paper, the faster the thermal aging process. Water content in oil can also dramtically lower the dielectric strength of oil when exceeding the saturation limit. In addition, the appearance of water at the interface of cellulose and oil may lead to partial discharges on the surface [7]. Field experience shows that the moisture content of the oil-impregnated insulation paper in a transformer is usually below 0.5% in the initial stage of operation; however, it may increase to 2%–4% at the terminal stage of its lifetime [6]. Therefore, improving the hydrophobicity of insulating paper is conducive to a partial suppression effect on water migration, thus ensuring reliable insulation performance.

As mentioned above, research on finding effective ways to improve the space charge accumulation phenomenon and hydrophobicity of the pressboard is of great significance. The introduction of nanomaterials is considered to be a popular way of modifying materials. The nanocomposite was first reported by Lewis in 1994 [8], and since then, nano-modification has become a popular way to improve the performance of insulating materials especially polymers. It was demonstrated by some researches that the bulk doping of different nano-particles could improve some properties of dielectric materials. For the cellulose insulation paper, the electrical properties of nano-Al_2O_3 and nano-SiO_2 doped paper are better than those of conventional paper. In addition, there is less space charge accumulation in the bulk of the modified paper [9,10]. Researches on nano-doping in other insulating materials, such as low-density polyethylene (LDPE) and epoxy resin, also showed a dramatic effect of doping nanoparticles on the suppression in space charge accumulation [11–14]. Nevertheless, the positive effect of nano-doping on the materials is limited by the aggregation of nanoparticles, due to their high surface energy. Besides bulk doping, another way to utilize the nano-effect is surface treatment. The surface modification of materials has gained enormous importance due to the ability to controllably change physical and chemical properties of solid surfaces without affecting the bulk properties [15]. Milliere et al. managed to mitigate the charge injection from an electrode into LDPE by magnetron sputtering a polymer composite layer containing silver nanoparticles [16]. However, rarely does research go into achieving the space charge suppression of the insulation pressboard by surface treatment. From the above, it is worth investigating the fabrication of a special functional nano-structure film on the surface of the cellulose pressboard, which could inhibit space charge accumulation and simultaneously make the cellulose polymers hydrophobic.

Zinc oxide is a wide band gap (3.4 eV) transparent oxide semiconductor material, with the advantages of high electron mobility and visible light transparency [17]. In recent years, ZnO has been attracting more and more attention. Nano-ZnO is believed to be beneficial to charge transport [18]. Moreover, fluorocarbon polymer-like films have been utilized in low friction coatings, excellent dielectric films, and optical coatings [19,20]. Polytetrafluoroethylene also has excellent insulation and hydrophobicity performance [19,20]. It can be imagined that if the coated functional film can have both the advantages of ZnO and PTFE somehow, it will play a role in improving the performance of the cellulose insulation pressboard.

In this paper, first, the ZnO, PTFE, and ZnO/PTFE functional films were each deposited on the surface of the cellulose insulation pressboard by radio frequency (RF) magnetron sputtering. Then, the physical and chemical characteristics of the as-prepared functional films were analyzed. Finally, we investigated the influence of the deposited functional films on the space charge behavior and the hydrophobicity of the cellulose insulation pressboard.

2. Experiments

2.1. Material Studio Simulation

The optimal depositing sequence should be determined before the fabrication of the composite film on the insulation pressboard surface. We believe that a better film adherence contributes to better film quality. From this perspective, we decided to calculate the interactions bewteen different nanoparticles and cellulose molecules with the help of Accelrys Materials Studio (MS) software (BIOVIA, San Diego, CA, USA). All simulations were carried out in the Forcite and Amorphous Cell modules included in the MS software. The amorphous region of the insulation pressboard was built following a method proposed by Theodorou et al [21]. Previous simulation results show that the length of cellulose chains has little influence on the molecular conformation and physico-chemical properties [22]. Therefore, in this simulation, the cellulose chains with 3, 4 and 5 degrees of polymerization (DP) were used to establish amorphous region models of the pressboard—each being comprised of two cellulose chains, which thereby took the interactions between cellulose chains into consideration. In addition, the radius of the ZnO and PTFE nanoparticle was set at 5 Å according to reference [23]. To make the built model reasonable, geometry and energy optimization was required [24]. The treatment process contained structural refinement, volume relaxation, and annealing. In the structural refinement and volume relaxation process, the default Smart algorithm was applied, which meant a rough optimization by the 'Steepest descent' method followed by a further optimization in conjugate gradient method with 10,000 steps. Then, in the annealing treatment, the temperature started from 300 to 650 K and then dropped in the increment of 43.75 K until the initial state 300 K was reached under a canonical ensemble (i.e., NVT ensemble, in which the values of particle number N, volume P, and Temperature T are fixed). The annealing time was set at 100 ps and energy minization was carried out for every annealing step. After the above treatment, the optimized models used for the next simulation in molecular dynamics were built. Figures 1 and 2 show the initial and optimized models in a and b, respectively.

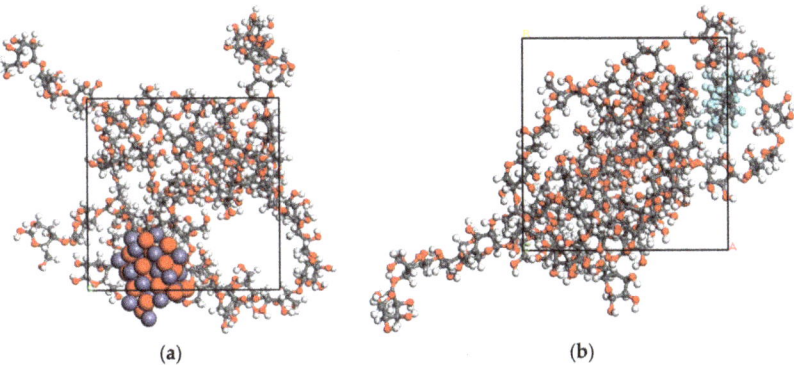

Figure 1. The initial models of (**a**) a ZnO nanoparticle and cellulose, and (**b**) a PTFE nanoparticle and cellulose.

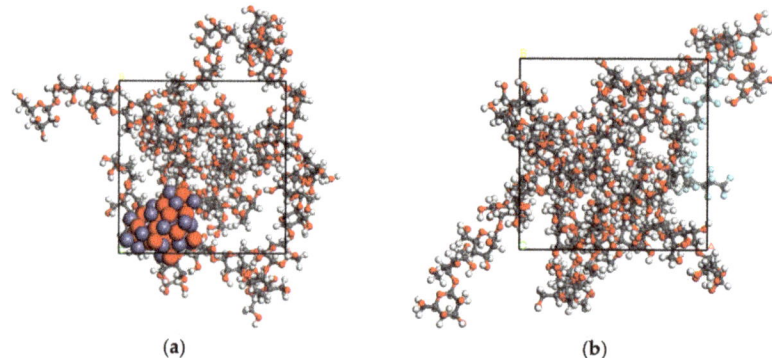

Figure 2. The optimized models of (**a**) a ZnO nanoparticle and cellulose, and (**b**) a PTFE nanoparticle and cellulose.

In the next stage, a molecular dynamics (MD) simulation was performed in the NVT ensemble. The nose temperature control method was used and the pressure was set to standard atmospheric pressure by using the Berendsen control method [25]. The force field adopted was a COMPASS II force field, which was a high-quality molecular force field that integrated organic and inorganic molecule parameters into the same force field. The time of MD simulation process was set at 500 ps, and the dynamics information of each atom in the system was collected once every 1000 fs.

2.2. Sample Preparation

The cellulose insulation pressboard (thickness 0.5 mm) was provided by the NARI Borui transformer factory, Chongqing, China. The insulation pressboard was cut into the size of 15 cm × 10 cm and then used in the RF magnetron sputtering experiment. The JPGF-480 RF magnetron sputtering device (Beijing Instrument Factory, Beijing, China) at 13.56 mHZ was used in this experiment. The cellulose insulation pressboard surface was initially deposited by RF magnetron sputtering of Zn target and PTFE target separately. The Zn and PTFE targets were provided by Zhongnuo XinCai Company, Beijing, China. The diameter was 61.5 mm, the thickness was 6 mm, and the purity was 99.999% for both targets. For the Zn target, sputtering was conducted in argon (Chongqing Hong Hao Gas Co., Ltd., Chongqing, China) plasma under a working pressure of 1.5 Pa and a constant sputtering power at 100 W with a fixed target-substrate distance of 10 cm. Oxygen (Chongqing Hong Hao Gas Co., Ltd., Chongqing, China) was used as the reactive gas at a flow of 20 sccm. The deposition time was 10 min for a better film performance according to multiple trials and to reference [26]. For the PTFE target, the cellulose insulation pressboard surface was sputtered for 20 min (1.5 Pa, 100 W), without reactive gas. To deposit the composited functional film, the cellulose pressboard surface was sputtered by Zn target for 10 min, and then sputtered by PTFE target for 20 min. The deposition parameter was the same as above. Figure 3 is the schematic diagram for the sample sputtering, and abbreviations for each sample are listed in Table 1.

Table 1. Sample composition.

Sample	Abbreviation
Untreated pressboard	UP
Pressboard sputtered Zn for 10 min (reactive O_2)	Z10
Pressboard sputtered PTFE for 20 min	P20
Pressboard sputtered Zn for 10 min (reactive O_2) and PTFE for 20 min	Z10+P20

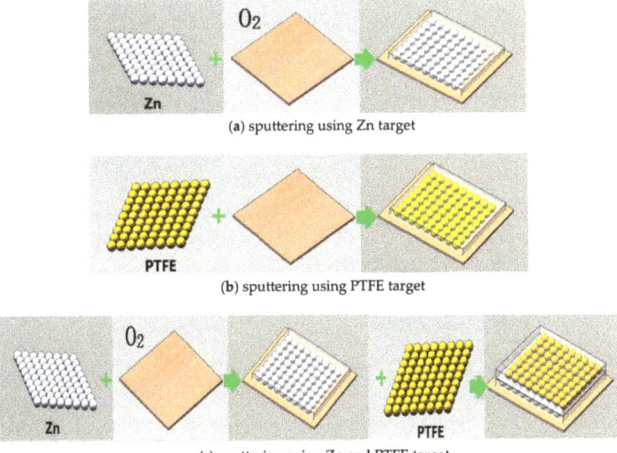

Figure 3. Radio frequency (RF) magnetron sputtering functional film on the surface of cellulose pressboard.

2.3. Characterization Method

Characterization of chemical composition was performed by XPS (Thermo Escalab 250 Xi, Waltham, MA, USA) with Al Kα X-rays source. The surface topographies and morphologies of cellulose polymer and sputtered films were analyzed through field emission scanning electron microscope (FE-SEM; JEOL JSM-7800F, Tokyo, Japan). In addition, XRD (PANalytical Empyrea, Almelo, the Netherlands) was performed to analyze the crystalline structure of the deposited film.

Besides the characterization on the structural property, the space charge suppression effect and the contact angle with water were also studied. Researchers have used the pulsed electro-acoustic (PEA) method to measure the space charge in solid dielectrics. The principle of the PEA method can be seen in many studies [4,27]. In brief, it consists of detecting the acoustic waves generated by internal charges under the Coulomb force of a pulsed electric field. The waves are detected by an external piezoelectric transducer, which converts the acoustic signal into an electrical signal. Then, the internal charge density is deduced by signal processing and mathematic treatment. The PEA principle is schematically represented in Figure 4 [27], where q(t) is the electric charge distributed in the sample, P(t) is the acoustic pressure wave as a function of time, the shape of P(t) is the same as the pulse electric field, and Vs(t) is the transducer output as a voltage signal. The PEA system (Shanghai Jiaotong University, Shanghai, China) has a pulse voltage of 600 V and a width of 5 ns. The bottom electrode is made of a 10 mm thick aluminium plate, and the top electrode is the semiconducting polymer film.

Before the space charge measurement, the sputtered pressboard was cut to 30 mm × 30 mm and dried in the vacuum chamber under 50 Pa/90 °C for 24 h. The moisture content of the samples was less than 1%, and technical function of the samples fulfilled the standard IEC 60641-3-1:2008. Meanwhile, new transformer mineral oil was treated by oil filter to remove gas, moisture, and impurities, and also dried under 50 Pa/105 °C for 24h. Moisture content of the oil was less than 10 ppm and the electrical property of the oil fulfilled the standard IEC 60296-2012. The insulation pressboard samples and oil were sealed into bottles and set under 50 Pa/40 for 48 hours to make the insulation pressboard immersed sufficiently. In this test, the applied direct current (DC) electric field was 15kV/mm, and the voltage-on time was 30 min.

Finally, the static water contact angle (WCA) was measured with a Kyowa contact angle meter (Kyowa Electronic Instruments Co., Ltd., Tokyo, Japan) by dispensing 8 μL distilled deionized water droplets. Three different spots for one sample were measured each time, and the average value was regarded as the contact angle and used for subsequent analysis.

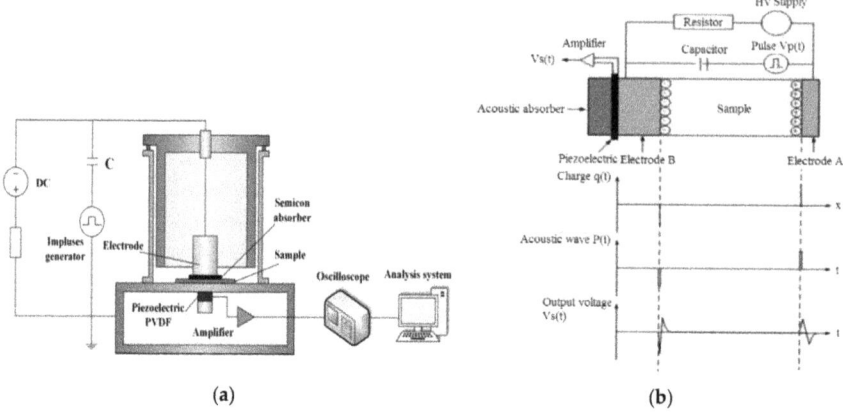

Figure 4. Schematic principle PEA method [27]: (**a**) experiment setup; (**b**) pulsed electro-acoustic (PEA) principle.

3. Results and Discussion

3.1. Molecular Dynamics Simulation Results

Figure 5 demonstrates some of the molecular dynamics simulation results by giving the detail of cell volume in ZnO-cellulose and PTFE-cellulose system. The grey part shows the occupied volume and the blue one stands for free volume. The occupied volumes for the ZnO-cellulose and PTFE-cellulose systems were 8409.06 Å^3 and 8421.44 Å^3, while the free volumes were 1620.01 Å^3 and 2440.65 Å^3 respectively. The introduction of fractional free volume (FFV) is to measure the void between the molecules inside the material which gives the molecular chains spaces to move. We can calculate the FFV based on the Equation (2) when the temperature is below glass transition temperature. The FFV was 16.15% for ZnO-cellulose system and 22.5% for PTFE-cellulose system, which means there are more free spaces for the motion of molecules in PTFE-cellulose system. This could be attributed to the ZnO nanoparticle's stronger adherence with cellulose molecules than the PTFE nanoparticle.

$$FFV = \frac{Volume_{free}}{Volume_{free} + Volume_{occupied}} \quad (1)$$

The interaction between nanoparticles and cellulose molecules could be quantified by the energy calculated during the molecular dynamics simulation. In the ZnO-cellulose model, we calculated the total energy of the system, which was −20854.7 kcal/mol. Then, the energies of ZnO and cellulose were calculated respectively, by getting rid of the other molecules. The total energy of cellulose molecules was calculated to be 1995.8 kcal/mol, and the figure for ZnO nanoparticle was −22506.4 kcal/mol. The energy value does not have a practical meaning because the standard zero value was set randomly during the simulation. However, the difference between the sum of energy of these two materials and the total energy of the system does mean a lot. Based on the Equation (2), we can calculate the interaction energy of ZnO nanoparticle and cellulose molecules. The result turns out to be 344.1 kcal/mol. For another PTFE-cellulose model, the same calculation method was applied and the interaction energy between PTFE nanoparticle and cellulose molecules is 77.6 kcal/mol. We can see that the interaction energy of ZnO and cellulose is higher than PTFE and cellulose. As a result, the sputtered ZnO film will have a better adhesion strength than PTFE film. This indicates that sputtering ZnO film first is the optimal choice. Moreover, the physici-chemical properties of PTFE are so stable as to enjure the acid, alkali, and some other extreme environments. In other words, PTFE film may act as a protective layer on the surface. The sputtering sequence is also reasonable in this form.

$$\Delta E = (E_1 + E_2) - E_{total} \qquad (2)$$

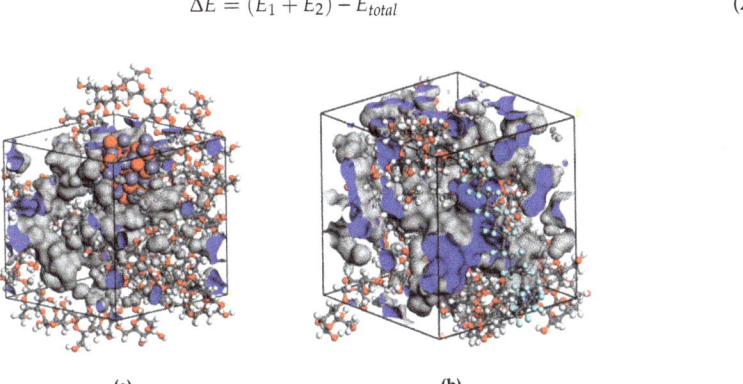

(a)   (b)

Figure 5. RF magnetron sputtering functional film on the surface of the cellulose pressboard.

3.2. Chemical Composition Analysis

Figure 6 presents a comparison of the XPS survey spectra of the untreated pressboard (UP) and the sputtered pressboard. Cellulose insulation pressboard consists of linear, polymeric chains of cyclic, β-d-glucopyranose units, which are composed of C, H, and O elements [28]. There was only the C_{1s} peak, the O_{1s}, peak, and the auger peak of C and O in UP, as labeled in Figure 6. After sputtering, new peaks of Zn atoms and F atoms appeared for sample Z10 and sample P20, respectively. As for sample Z10+P20, only peaks of F atoms were observed in the survey XPS spectra. The peak height of F_{1s} was 40,123, whereas that of Zn_{2p} was 2053. The reason is that the film containing zinc oxide is that is was beneath the fluorocarbon film. Therefore, the peaks of Zn_{2p} atoms are so weak as to be invisible in Figure 6. The above results indicate that the zinc oxide and fluorocarbon film has been fabricated on the pressboard surface.

The existence of chemical bonds sputtered on the cellulose polymer surface is determined by high-resolution XPS spectra. Figure 7a,b show the narrow scan spectra results of UP and Z10+P20, respectively. In Figure 7a, the peak at 284.6 eV corresponds to carbon–carbon (C–C) bond or carbon–hydrogen (C–H) bond; the peak at 286.4 eV is due to bonding of carbon to a single non-carbonyl oxygen (C–O); and the peak at 287.9 eV represents bonding of one carbonyl oxygen to a carbon atom (C=O). These three peaks are consistent with the results for cellulose [28]. After the functional film deposition, high-resolution scans revealed C_{1s} spectra containing peaks at 289.08 eV, 291.28 eV, and 293.68 eV, corresponding to C–F, C–F_2, and C–F_3 bonds, respectively. The O_{1s} peak in Figure 7c confirms the C–O bond at 533.08 eV on the surface of UP. After sputtering, the O_{1s} peak position shifted to lower binding energy at 531.58 eV. The binding energy of ZnO in O_{1s} spectra stands at 531.1 eV. This results of peak shifting are consistent with oxygen binding mainly to zinc. In addition, the O_{1s}'s peak height decreased sharply after sputtering due to the formation of film on the cellulose surface; thus, little information from cellulose was detected, and the zinc content was not very abundant because of the short sputtering time (10 min). In Figure 7d, two peaks appear at 1021.8 eV for $Zn_{2p3/2}$ and 1044.8 eV for $Zn_{2p1/2}$, confirming the formation of Zn–O bond. In addition, there is no peak located at around 88 eV corresponding to a Zn–Zn bond. The results demonstrate that the zinc was completely oxidized during the reactive RF magnetic sputtering treatment using O_2. Figure 7e shows the F_{1s} spectral result. There is also a new peak located at 688.88 eV corresponding to a C–F bond after sputtering. From the above, it could be concluded that the ZnO/PTFE film was successfully fabricated on the cellulose insulation pressboard surface.

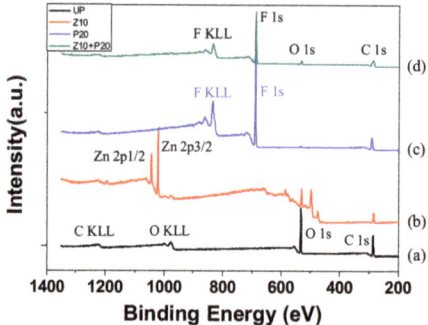

Figure 6. The X-ray photoelectron spectrpmetry (XPS) spectra of (**a**) UP, (**b**) Z10, (**c**) P20, and (**d**) Z10+P20.

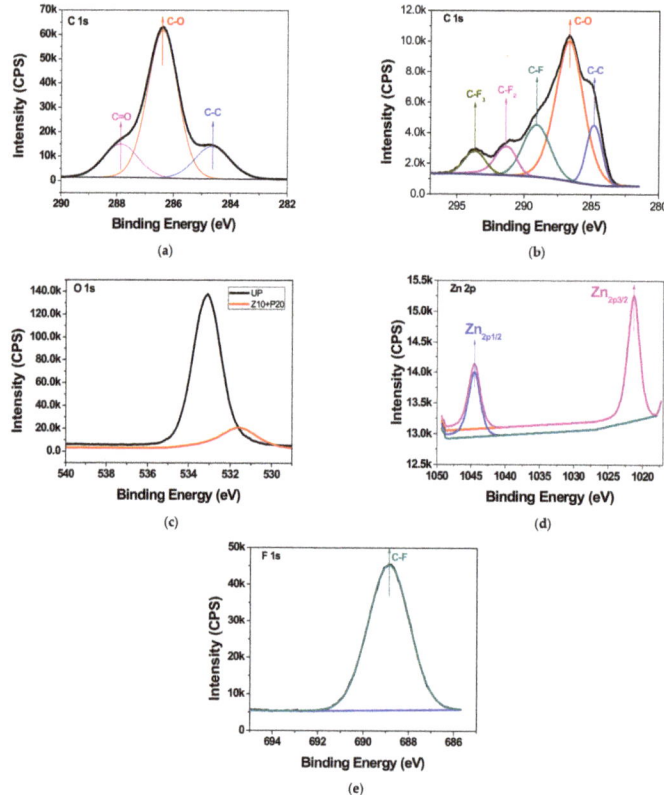

Figure 7. XPS resolution spectra of (**a**) C_{1s} for UP, (**b**) C_{1s} for Z10+P20, (**c**) O_{1s} for UP and Z10+P20, (**d**) Zn_{2p} for Z10+P20, and (**e**) F_{1s} for Z10+P20.

3.3. Surface Morphology

Scanning electron microscopy images of the untreated pressboard and the sputtered pressboard surface are illustrated in Figure 8. It can be observed that the fibers of UP (Figure 8a) intersecting each other and the surface were relatively rough. There were some cracks where the fibers intersect. The characteristic structure of the untreated pressboard was highly porous and laminar. The surface of sputtered pressboard (Z10+P20) was smoother and denser, as shown in Figure 8b. To better illustrate the surface morphology of sample Z10+P20, it was necessary to study the morphology of samples Z10 and P20. As presented in Figure 8c,d, the film was comprised of nano-particles with dozens of nanometers in diameter. To be specific, ZnO nanoparticles were small and well isolated from each other. In addition, the distribution of ZnO nano-particles was more uniform and denser than that of PTFE. For P20, the average size of PTFE nano-particles was larger than ZnO nano-particles, and the PTFE nano-particles tended to be in alignment. Scanning electron microscopy images for Z10+P20 at 10,000× magnification (Figure 8e) indicat that an obvious change occurs on the surface. The porosity decreases and individual fibers become harder to identify. With even higher magnification, at 60,000×, there is a dense and uniform distribution of tightly arranged particles deposited on the surface (Figure 8f). It can be seen that the sputtered film is comprised of nano-particles ranging from 40 to 70 nm in diameter. Some nano-particles are clearly seen, whereas some nano-particles are agglomerated. It could be inferred that the newly sputtered particles continually attach and aggregat to the previously deposited particles. The introduction of the nano-particles growing on the fibers could affect the material properties.

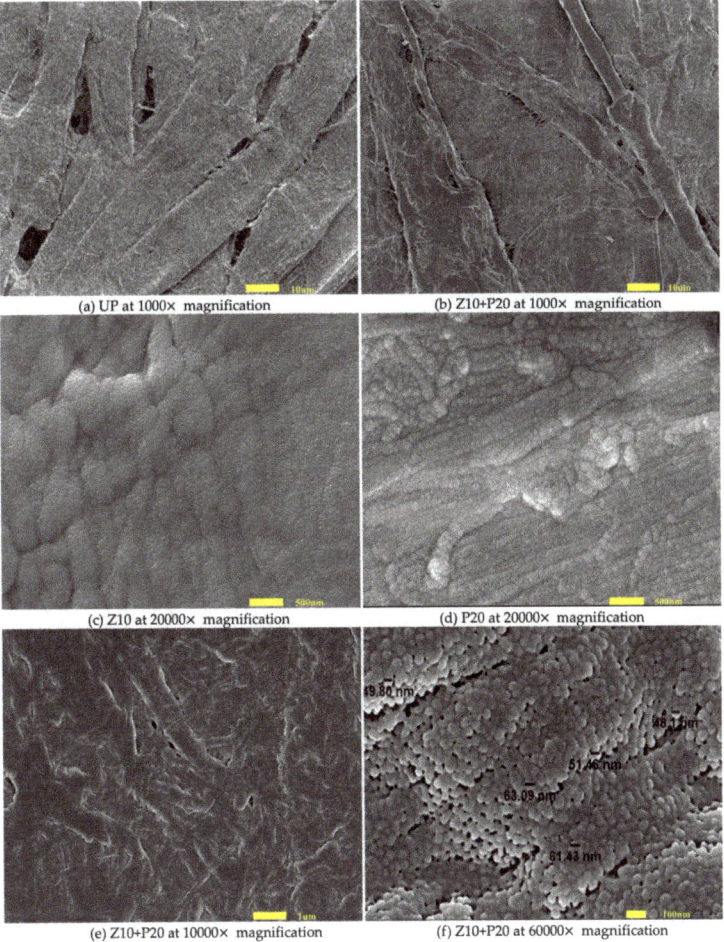

Figure 8. SEM images of (**a**) UP, (**b**) Z10+P20, (**c**) Z10, (**d**) Z10+P20, (**e**) Z10+P20, and (**f**) Z10+P20.

3.4. Crystalline Structure

X-ray diffraction analysis is very helpful for investigating the crystal structure of the material. Considering that the film sputtered on the surface, the small-angle X-ray diffraction (SAXD) method was used for precise characterization. The gracing angle was set at 1.5°, and the scanning step was 0.02°. Results for the samples—UP, Zn10, P20, and Zn10+P20—are given in Figure 9. The obvious diffraction peaks at $2\theta = 14.93°$, $2\theta = 22.60°$ and $2\theta = 34.85°$ are the characteristic phase (101), (002), and (040) diffraction peaks of cellulose I, respectively [29]. There are sharp peaks and some dispersive diffraction peaks, which means that the cellulose had a mixed structure of a crystallized and amorphous phase. However, there was no obvious difference between the untreated pressboard and the sputtered pressboard in this experiment. It could be inferred that the film is still on the growth stage when the sputtering time is short, thus ZnO did not show any specific crystal structure. Furthermore, the PTFE film also existed in an amorphous form on the surface of insulation pressboard.

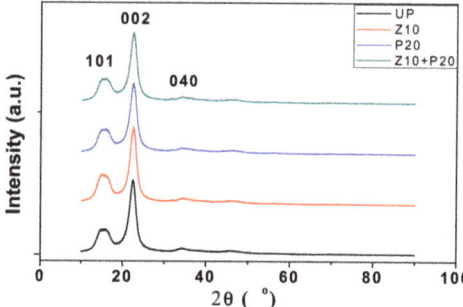

Figure 9. The results of small-angle X-ray diffraction.

3.5. Space Charge Suppression Effect

The threshold electric field for space charge in oil-impregnated insulation paper is usually about 10–12 kV/mm [30]. When the applied electric field is higher than 10–12 kV/mm, charge injection from the metal electrode to the pressboard through the contact is inevitable due to the Schottky injection mechanism and the tunneling effect [2]. Injected charges undergo several processes, such as migration, trapping, detrapping, and neutralization. Charges will keep migrating until these four processes reach an equilibrium. Charge accumulation occurs when the injection process prevails over the charge migrating process due to the high electric field. As mentioned in the introduction, nano-ZnO is believed to speed up the charge migration rate, which could reduce the space charge accumulation. Therefore, in this paper, ZnO nanoparticles are mainly used for the inhibition of space charge accumulation. To confirm the restraining function of the sputtered ZnO film on charge accumulation, the space charge distribution of the untreated and sputtered pressboard under DC electric field 15 kV/mm for 30 min was measured.

In Figure 10a, the anode charge density peak decreases from 11.6 C/m^3 at 5 s to 8.3 C/m^3 at 1800 s, whereas the cathode charge density peak decreases from −4.7 C/m^3 at 5 s to −3.8 C/m^3 at 1800 s. The charge density on the electrode decreases with the voltage applied time, indicating that space charge moves away from the electrodes into the inner part of the sample. It can be seen in Figure 10b that the anode charge density peak decreases from 12.6 C/m^3 at 5 s to 9.2 C/m^3 at 1800 s. It is clear that the space charge accumulated within sample Z10 bulk is obviously less than that in UP. Around the middle position within the sample bulk, the space charge density is around zero value. That is to say, no obvious charge accumulates here. The sample P20 was also tested. As shown in Figure 10c, a similar situation of charge injection occurs, and the charge injection from both anode and cathode takes place, though the trapped charge amount is less compared with UP (Figure 10a). This means the nano-structure PTFE film also seems to have a positive impact on charge accumulation suppression, albeit to a smaller degree. Figure 10d shows the result of sample Z10+P20. It can be seen from the charge distribution that the inhibition effect of space charge is almost the same as the result shown in Figure 10b for Z10. From the comparison we can see that the PTFE film does not weaken the effect of ZnO nano-particles on inhibiting the space charge injection.

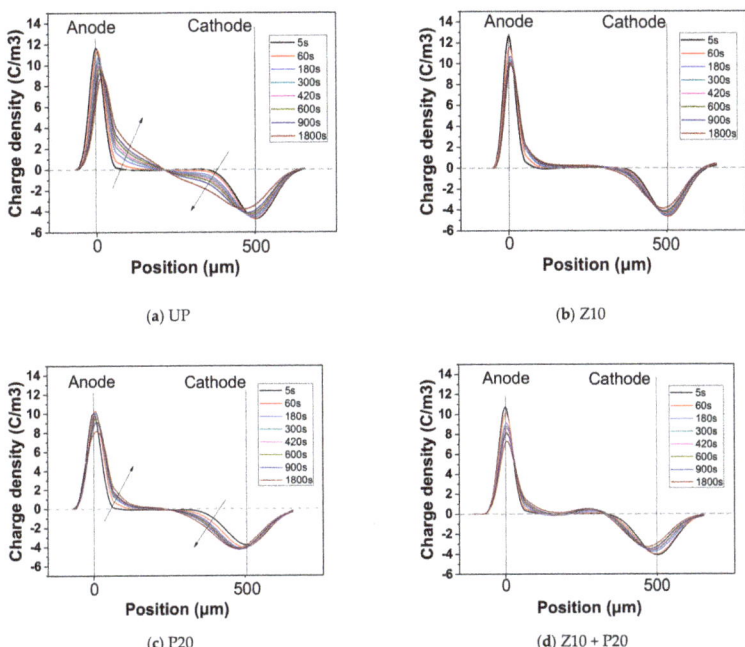

Figure 10. The space charge behavior of (**a**) UP, (**b**) Z10, (**c**) P20, and (**d**) SP under DC 15kV/mm.

The total absolute amount of space charge during the voltage-on period can be calculated using Equation (3), where L is the thickness of the sample, $\rho(x, t)$ is the charge density at position x, t is the DC voltage applied time, and S is the area of the electrode.

$$Q(t) = \int_0^L |\rho|x,t||Sdx \qquad (3)$$

The number of charges trapped in the bulk of sample UP, Zn10, P20, and Zn10+P20, during the DC electric field 15 kV/mm applied for 0–30 min are shown in Figure 11. The charge amount rises as the voltage-on time increases. It can be clearly seen that the increasing rate of charge amount and the final charge amount of the untreated pressboard are higher than other sputtered pressboards. Specifically, PTFE film made no difference in the initial stage of charge injection, but the charge amount at quasi-equilibrium state is lower than the UP. As for ZnO film, the lower increasing rate of charge amount confirmed that it clearly plays a significant role in charge injection inhibition. Moreover, the composited functional film sample ZnO/PTFE has the best result among all the samples, with both the lowest increasing rate of charge amount and the lowest total charge amount at equilibrium. Compared with untreated cellulose insulation pressboard, the amount of the trapped charges in the pressboard sputtered ZnO/PTFE composited functional film decreases by about 40%. From this perspective, the pressboard sputtered composited functional film ZnO/PTFE presents considerable performance for the space charge inhibition.

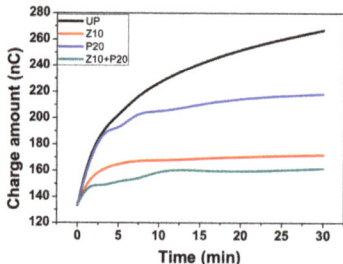

Figure 11. Absolute charge amount in the treated and untreated pressboard.

To study the impact of space charge accumulation on the electric field distortion, the electric field strength was calculated by Equation (4), where $\rho(x)$ is the charge density, ε_0 is the vacuum permittivity, ε_r is the relative permittivity of the sample, and L is the thickness of the sample.

$$E(x) = \int_0^x \frac{\rho(x)}{\varepsilon_0 \varepsilon_r} dx \quad 0 \leq x \leq L \tag{4}$$

The calculated electric field distribution within each sample is shown in Figure 12. For UP, the injected homo-charge reduces the electric field in the region between electrodes and sample. However, with the charges moving deeper into the bulk, the electric field gradually increases along the abscissa axis and peaks in the middle region of the bulk. Although the applied electric field is 15 kV/mm, the actual maximum electric field reaches 20 kV/mm, which means that the field enhancement, due to the presence of space charge, is around 33%. As for the Z10 sample (Figure 12b), the result shows that there is no obvious electric field distortion due to little space charge accumulation in the bulk, with the highest electric field strength standing at roughly 16 kV/mm. For P20, a weak mitigation effect results in the same obvious electric field distortion as in UP, but the peak value is lower than that in UP. For Z10+P20, the space charge suppression effect is better than Z10, since the actual electric field in the bulk is even lower than the applied electric stress.

3.6. Hydrophobicity Analysis

The insulation paper is developed from natural fiber. The structural characteristics of the fiber determine that it absorbs water easily. However, the hygroscopicity of the insulation paper is a very bad feature when used for insulation in a transformer. The moisture content would lead to the decomposition of cellulose and partial discharge. Therefore, if the insulation paper has better hydrophobicity, the negative effect of moisture could be suppressed. Figure 13 shows the contact angle results of water droplets dripping on the surface of untreated and sputtered pressboard.

For UP, the water contact angle was about 0°, indicating that water droplets were absorbed immediately by the pressboard due to the existence of hydroxyl groups in cellulose. For Z10, the water contact angle was high at first but decreased dramatically in few minutes. The surface of the pressboard was coated and the cellulose molecule was surrounded by ZnO nanoparticles, so the hydrophilic effect of the hydroxyl group was weakened. However, as it could be seen in Figure 8b, there were also some defects and cracks in the sputtered pressboard, though less than for UP. Therefore, the water would penetrate and then be absorbed by the cellulose molecules in few minutes. As reported in reference [31], PTFE film has many C–F groups, which show dramatic hydrophobicity. When the pressboard ws sputtered by the PTFE target, the newly formed C-F groups on the surface made the pressboard hydrophobic to some extent, so we can see the contact angle of P20 is 116.6° at the initial stage, as shown in Figure 13. Nevertheless, it could be seen that the contact angle decreased gradually with time, which meant the hydrophobicity of the PTFE film above the cellulose substrate could not endure a quite long time. This phenomenon also resulted from the property of the cellulose substrate, where the effect of hydroxyl group could just be weakened but not eliminated. Thus, the water droplets were absorbed into the substrate eventually. Figure 13 also showed the comparative results of sample Z10+P20 and the P20 sample. The starting value of sample Z10+P20 was almost the same as sample P20. In the beginning, the contact angle of Z10+P20 decreased faster than P20, but it was reversed in the later stage. Overall, the hydrophobicity of P20 and Z10+P20 were almost the same. It could be inferred that the deposited Z10+P20 composited functional film has a lower moisture absorption rate according to reference [32], which is good for insulation pressboard used in power transformers.

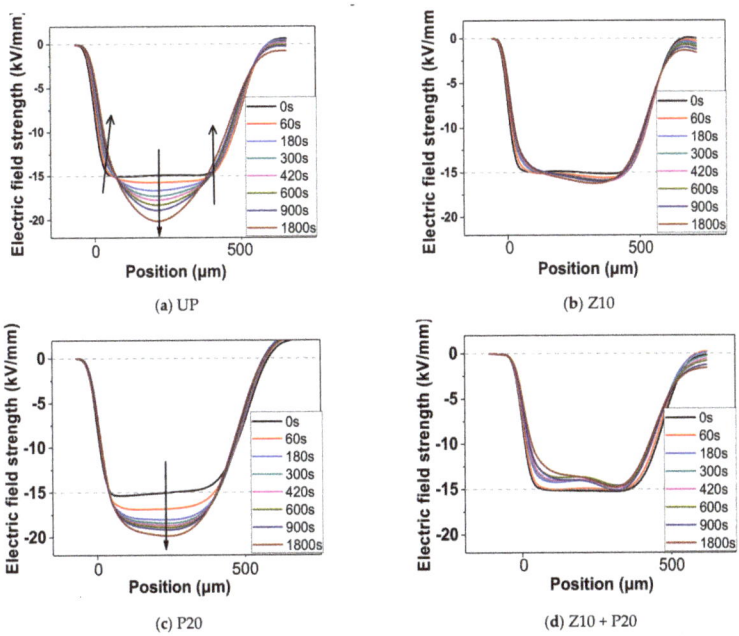

Figure 12. Electric field distribution for different samples: (a) UP, (b) Z10, (c) P20, and (d) Z10+P20.

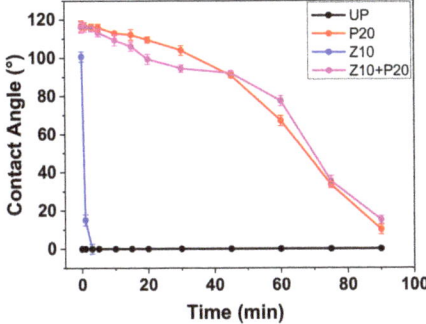

Figure 13. The contact angle of the untreated and sputtered pressboard.

3.7. Tensile Strength and Electric Breakdown Values

Tensile strength and electric breakdown values are two basic and critical parameters for the insulation pressboard, so we compared these two parameters of the untreated and treated pressboard, and the results were shown in Table 2. The tensile strength of the composite pressboard was directly tested after sputtering according to ISO 1924-2:2008. The samples used for DC and AC breakdown test were prepared the same way as the samples for PEA test, and the test was conducted by plate-plate electrodes according to IEC 60243-1:2013 standard. Every test was performed eight times and the mean value was calculated for Table 2. It is shown that there was no big difference between the untreated and sputtered composite pressboard. The sputtered pressboard also accompolished the standard IEC 60641-3-1:2008.

Table 2. Tensile strength and electric breakdown test results.

Sample	Tensile Strength (Mpa)	Standard Value (Mpa)	AC Breakdown (kV/mm)	DC Breakdown (kV/mm)	Standard Value (kV/mm)
UP	49.91	>40	55.28	136.37	>40
Z10+P20	50.48	>40	56.26	144.27	>40

4. Conclusions

The nano-structures ZnO, PTFE, and their composite films are designed to improve the insulation performance of the cellulose pressboard. Results from Materials Studio show that the ZnO nanoparticle has a better adhesion strength with cellulose molecules than the PTFE nanoparticle. Hence, ZnO film should be sputtered first to fabricate the ZnO/PTFE composite film.

Materials Studio simulation results show that ZnO nanoparticle has a better adhesion strength with cellulose molecules than PTFE nanoparticle. The ZnO film should be sputtered at first to fabricate the ZnO/PTFE composite film for better film quality.

X-ray photoelectron spectroscopy analysis shows that the ZnO/PTFE composited functional film was successfully fabricated on the cellulose insulation polymer surface by reactive RF magnetron sputtering. The SEM results show that there were dense and uniform nano-particles ranging from 40 to 70 nm in diameter deposited on the surface of cellulose. The XRD results show that the nano-structure of ZnO and PTFE was of an amorphous form.

The sputtered nano-structure ZnO film on the cellulose polymer surface has an obvious space charge suppression effect, and the sputtered nano-structure PTFE film could turn the cellulose surface from hydrophilicity into hydrophobicity. The nano-structure ZnO/PTFE composited functional film could integrate the advantages of nano-structure ZnO and PTFE film at the same time. The amount of the accumulated space charge in the pressboard sputtered ZnO/PTFE composite functional film decreases by about 40% compared with that in untreated cellulose insulation pressboard, and the water contact angle increases from 0° to 116°.

The tensile strength and electrical breakdown test results showed that the basic properties of the sputtered composite pressboard were not weakened, and achieved the standard ISO 5269-2:2004.

It is concluded that ratio frequency magnetron sputtering is an effective way to enhance the performance of the insulation pressboard in some aspects, without weakening the basic properties. This work provides a method for the development of high-performance cellulose insulation polymer used in HVDC equipment.

Author Contributions: J.H. and R.L. designed the experiments and contributed to the paper's supervision; Y.L. performed the reactive RF magnetron sputtering and hydrophobicity experiment and wrote the paper; J.Z. conducted the space charge measurement. W.H. did the molecular dynamics simulation part. J.H. and C.L. contributed to SEM, XPS, and XRD characterization.

Acknowledgments: This research was financially supported by the National Natural Science Foundation of China (51707022), the China Postdoctoral Science Foundation (2017M612910), Chongqing Special Funding Project for Post-Doctoral (Xm2017040), and Funds for Innovative Research Groups of China (51321063).

Conflicts of Interest: The authors declare no conflict of interest.

References

1. CIGRE Joint Working Group A2/B4.28. *HVDC Converter Transformers Guide Lines for Conducting Design Reviews for HVDC Converter Transformers*; CIGRE: Paris, France, 2010.

2. Dissado, L.A.; Fothergill, J.C. *Electrical Degradation and Breakdown in Polymers.*; Peter Peregrinus: London, UK, 1992; Volume 620.
3. Wei, Y.H.; Mu, H.B.; Deng, J.B. Effect of space charge on breakdown characteristics of aged oil-paper insulation under DC voltage. *IEEE Trans. Dielectr. Electr. Insul.* **2016**, *23*, 3143–3150. [CrossRef]
4. Hao, M.; Zhou, Y.; Chen, G.; Wilson, G.; Jarman, P. Space charge behavior in oil gap and impregnated pressboard combined system under HVDC stresses. *IEEE Trans. Dielectr. Electr. Insul.* **2016**, *23*, 848–858. [CrossRef]
5. Hao, J.; Chen, G.; Liao, R.J. Influence of moisture and temperature on space charge dynamics in multilayer oil-paper insulation. *IEEE Trans. Dielectr. Electr. Insul.* **2012**, *19*, 1456–1464. [CrossRef]
6. Emsley, A.M.; Stevens, G.C. Review of chemical indicators of degradation of cellulosic electrical paper insulation in oil-filled transformers. *IEE Proc. Sci. Meas. Technol.* **1994**, *141*, 324–334. [CrossRef]
7. Sikorski, W.; Walczak, K.; Przybylek, P. Moisture Migration in an Oil-Paper Insulation System in Relation to Online Partial Discharge Monitoring of Power Transformers. *Energies* **2016**, *9*, 1082. [CrossRef]
8. Lewis, T.J. Nanometric dielectrics. *IEEE Trans. Dielectr. Electr. Insul.* **1994**, *1*, 812–825. [CrossRef]
9. Liao, R.J.; Lv, C.; Wu, W.Q.; Liang, N.C.; Yang, L.J. Insulating properties of insulation paper modified by nano-Al_2O_3 for power transformer. *J. Electr. Power Sci. Technol.* **2014**, *29*, 3–7.
10. Yan, S.; Liao, R.J.; Yang, L.J.; Zhao, X.T.; Yuan, Y. Influence of nano-Al_2O_3 on electrical properties of insulation paper under thermal aging. In Proceedings of the 2016 IEEE International Conference on High Voltage Engineering and Application (ICHVE), Chengdu, China, 19–22 September 2016; pp. 1–4.
11. Wang, S.J.; Zha, J.W.; Li, W.K.; Dang, Z.M. Distinctive electrical properties in sandwich-structured Al_2O_3/low density polyethylene nanocomposites. *Appl. Phys. Lett.* **2016**, *108*, 031605. [CrossRef]
12. Fleming, R.J.; Ammala, A.; Casey, P.S. Conductivity and space charge in LDPE containing nano- and micro-sized ZnO particles. *IEEE Trans. Dielectr. Electr. Insul.* **2008**, *15*, 118–126. [CrossRef]
13. Katayama, J.; Ohki, Y.; Fuse, N.; Kozako, M.; Tanaka, T. Effects of nano-filler materials on the dielectric properties of epoxy nanocomposites. *IEEE Trans. Dielectr. Electr. Insul.* **2013**, *20*, 157–165. [CrossRef]
14. Andritsch, T.; Kochetov, R.; Morshuis, P.H.F.; Smit, J.J. Dielectric properties and space charge behaviour of MgO epoxy nanocomposites. In Proceedings of the 2010 10th IEEE International Conference on Solid Dielectrics, Potsdam, Germany, 4–9 July 2010; pp. 1–4.
15. Kylián, O.; Choukourov, A.; Biederman, H. Nanostructured plasma polymers. *Thin Solid Films* **2013**, *548*, 1–17. [CrossRef]
16. Milliere, L.; Makasheva, K.; Laurent, C. Silver nanoparticles as a key feature of a plasma polymer composite layer in mitigation of charge injection into polyethylene under dc stress. *J. Phys. D Appl. Phys.* **2016**, *49*, 15304. [CrossRef]
17. Takahashi, K.; Yoshikawa, A.; Sandhu, A. *Wide Bandgap Semiconductors, Fundamental Properties and Modern Photonic and Electronic Devices*; Springer: New York, NY, USA, 2007.
18. Ellmer, K.; Klein, A.; Rech, B. *Transparent Conductive Zinc Oxide*; Springer: Berlin/Heidelberg, Germany, 2008.
19. Yong, J.; Fang, Y.; Chen, F. Femtosecond laser ablated durable superhydrophobic PTFE films with penetrating microholes for oil/water separation: Separating oil from water and corrosive solutions. *Appl. Surf. Sci.* **2016**, *389*, 1148–1155. [CrossRef]
20. Toosi, S.F.; Moradi, S.; Kamal, S.; Hatzikiriakos, S.G. Superhydrophobic laser ablated PTFE substrates. *Appl. Surf. Sci.* **2015**, *349*, 715–723. [CrossRef]
21. Theodorou, D.N.; Suter, U.W. Detailed molecular structure of a vinyl polymer glass. *Macromolecules* **1985**, *18*, 1467–1478. [CrossRef]
22. Mazeau, K.; Heux, L. Molecular Dynamics Simulations of Bulk Native Crystalline and Amorphous Structures of Cellulose. *J. Phys. Chem. B* **2008**, *107*, 2394–2403. [CrossRef]
23. Tang, C.; Zhang, S.; Wang, X.; Hao, J. Enhanced mechanical properties and thermal stability of cellulose insulation paper achieved by doping with melamine-grafted nano-SiO_2. *Cellulose* **2018**, *25*, 3619–3633. [CrossRef]
24. Tang, C.; Zhang, S.; Xie, J.; Lv, C. Molecular simulation and experimental analysis of Al_2O_3-nanoparticle-modified insulation paper cellulose. *IEEE Trans. Dielectr. Electr. Insul.* **2017**, *24*, 1018–1026. [CrossRef]
25. Berendsen, H.J.C.; Postma, J.P.M.; Van Gunsteren, W.F.; DiNola, A.; Haak, J.R. Molecular dynamics with coupling to an external bath. *J. Chem. Phys.* **1984**, *81*, 3684. [CrossRef]

26. Pan, C.; Zhao, Z.; Wang, C. Effects of sputtering time on the properties of ZnO thin films prepared by magnetron sputtering. In Proceedings of the 2015 IEEE International Vacuum Electronics Conference (IVEC), Beijing, China, 27–29 April 2015.
27. Xu, Z.Q. Space Charge Measurement and Analysis in Low Density Polyethylene Film. Ph.D. Thesis, University of Southampton, Southampton, UK, 2009.
28. Yang, L.J.; Liao, R.J.; Sun, C.X.; Zhu, M.Z. Influence of vegetable oil on the thermal aging of transformer paper and its mechanism. *IEEE Trans. Dielectr. Electr. Insul.* **2012**, *18*, 2059. [CrossRef]
29. Liu, J.F.; Zhang, Y.Y.; Xu, J.J.; Rui, J.; Zhang, G.J.; Liu, L.W. Quantitative relationship between aging condition of transformer oil-paper insulation and large time constant of extend debye model. *Electr. Power Autom. Equip.* **2017**, *37*, 197–202.
30. Hao, M.; Zhou, Y.; Chen, G. Space charge behavior in thick oil-impregnated pressboard under HVDC stresses. *IEEE Trans. Dielectr. Electr. Insul.* **2015**, *22*, 72–80. [CrossRef]
31. Belgacem, M.N.; Salon-Brochier, M.C.; Krouit, M. Recent advances in surface chemical modification of cellulose fibres. *J. Adhesion Sci. Technol.* **2011**, *25*, 661–684. [CrossRef]
32. Hao, J.; Liu, C.; Li, Y.; Liao, R.; Liao, Q.; Tang, C. Preparation nano-structure polytetrafluoroethylene (PTFE) functional film on the cellulose insulation polymer and its effect on the breakdown voltage and hydrophobicity properties. *Materials* **2018**, *11*, 851. [CrossRef] [PubMed]

© 2019 by the authors. Licensee MDPI, Basel, Switzerland. This article is an open access article distributed under the terms and conditions of the Creative Commons Attribution (CC BY) license (http://creativecommons.org/licenses/by/4.0/).

Article

Microstructural and Tribological Properties of a Dopamine Hydrochloride and Graphene Oxide Coating Applied to Multifilament Surgical Sutures

Gangqiang Zhang [1,2,*], Jiewen Hu [1], Tianhui Ren [3] and Ping Zhu [1,*]

1. College of Textile & Clothing, Institute of Functional Textiles and Advanced Materials, State Key Laboratory of Bio-Fibers and Eco-Textiles, Collaborative Innovation Center of Marine Biomass Fibers Materials and Textiles of Shandong Province, Qingdao University, Qingdao 266071, China; Hujiesen@163.com
2. Shandong Jiejing Group, Rizhao 276800, China
3. School of Chemistry and Chemical Engineering, Key Laboratory for Thin Film and Microfabrication of the Ministry of Education, Shanghai Jiao Tong University, Shanghai 200240, China; thren@sjtu.edu.cn
* Correspondence: gqzhang@qdu.edu.cn (G.Z.); pzhu99@qdu.edu.cn (P.Z.)

Received: 23 June 2020; Accepted: 21 July 2020; Published: 22 July 2020

Abstract: With the development of fine surgery and desire for low-injury methods, the frictional properties of surgical sutures are one of the crucial factors that can cause damage to tissue, especially for some fragile and sensitive human tissues such as the eyeball. In this study, dopamine hydrochloride and graphene oxide were used as external application agents to prepare a biological coating for the surface of multifilament surgical sutures. The effects of this biocoating on the surface morphology, chemical properties, mechanical properties, and tribological properties of surgical sutures were studied. The friction force and the coefficient of friction of surgical sutures penetrating through a skin substitute were evaluated using a penetration friction apparatus and a linear elastic model. The tribological mechanism of the coating on the multifilament surgical sutures was investigated according to the results of the tribological test. The results showed that there were uniform dopamine and graphene oxide films on the surface of the surgical sutures, and that the fracture strength and yield stress of the coated sutures both increased. The surface wettability of the surgical sutures was improved after the coating treatment. The friction force and the coefficient of friction of the multifilament surgical sutures with the dopamine hydrochloride and graphene oxide coating changed little compared to those of the untreated multifilament surgical sutures.

Keywords: surface coating; dopamine hydrochloride; graphene oxide; surgical suture; friction

1. Introduction

Surgical sutures are a fundamental material in surgical operation, which directly affect the results of suturing [1]. With the development of delicate surgery and desire for low-injury methods, the frictional properties of surgical sutures are one of the crucial factors that can cause damage to tissue. Multifilament surgical sutures with excellent mechanical properties and significant flexibility and pliability are crucial for suturing [2,3]. The twisted structure and the surface roughness of surgical multifilament sutures increase penetration and frictional resistance [4]. Generally, the high frictional behavior of surgical sutures is related to tissue inflammation and increases the recovery time of scars, which results in a second trauma for patients [5].

Coating is a rough surface treatment method for multifilament surgical sutures that fills the interstices between the twisted fibers and reduces the frictional resistance [6]. Various coating materials have been used to improve the frictional properties of surgical sutures; for instance, antibiotic ointment has been used to coat prophylactic surgical sutures, which decreased the coefficient of friction of the sutures when passed through tissue [6]. Antibacterial materials have been shown to reduce the maximum friction force of braided silk interacting with a skin substitute. Dopamine hydrochloride and cardiomyopathy chitosan coatings have been used to treat multifilament surgical sutures, which barely changed the coefficient of friction of the surgical sutures when sliding through a skin substitute [7].

Surgical sutures are a kind of implant material. The coating materials for sutures should be biocompatible and should barely react with tissue. Graphene oxide (GO) is widely used in the biomaterial coating field due to its superior biocompatibility and mechanical strength [8]. GO coatings have been applied to implant materials to increase their frictional performance, such as magnesium and titanium alloys [9,10]. The application of a GO coating on fibers and fabrics has also been investigated. Cai et al. [11] applied a GO coating to cotton fabric by thermal reduction under the protection of nitrogen. Chen et al. [12] grafted a GO coating onto poly(p-phenylene benzobisoxazole) (PBO) fiber by a silane coupling agent, which improved the surface roughness and wettability of the grafted fiber. Hu et al. [13] used GO, chitosan, and polyvinyl alcohol as the functional finishing agents to carry out hydrogen bond layer-by-layer self-assembly to modify the surface of cotton fabric, and the results showed that this process can form a film on the fabric's surface. Dopamine hydrochloride (DA) is a biomaterial [14], and researchers have found that DA can be deposited onto the surfaces of various materials in a buffer solution to form a versatile platform for secondary reactions, which improves the cohesiveness and functionalization of a material [15–17].

The objective of this research was to prepare a DA and GO composite coating for the surface of multifilament surgical sutures and to investigate the influence of said coating on the frictional properties of the surgical sutures when penetrated through a skin substitute. The coating was characterized by mass change, a static contact angle, tensile strength, bending yield strength, and surface morphology. The impact of the coating treatment on the frictional properties of surgical sutures was investigated. The friction force of the surgical sutures was tested by using a penetration friction apparatus (PFA) [18,19], and the coefficient of friction was calculated by the elastic model and finite element simulation [20].

2. Materials and Methods

2.1. Materials

Polyglycolic acid (PGA) multifilament surgical sutures and straight stainless-steel tapered needles were purchased from Weigao Medical Instruments Co. Ltd. Sil8800 (Red, 80IRHD) artificial skin from Superior Seals has a similar toughness and constitutive function to human skin [21–23]. The chemical agents were of analytical grade and obtained from Aladdin Chemistry (Shanghai, China), including dopamine hydrochloride (DA), tetrahydrofuran (THF), potassium permanganate ($KMnO_4$), sulfuric acid (H_2SO_4), sodium hydroxide (NaOH), peroxide (H_2O_2), phosphoric acid (H_3PO_4), chlorhydric acid (HCl), and absolute ethanol.

2.2. Synthesis and Characterization of Graphene Oxide

According to the improved Hummers method [24], GO was prepared with flake graphite as raw material, concentrated sulfuric acid as an expanding agent, and potassium permanganate as an oxidant. The preparation process was as follows: H_2SO_4/H_3PO_4 was mixed in a flask at the ratio of 9:1 (180:20 mL), graphite powder (1.5 g) was added into the continuously stirred mixture, and $KMnO_4$ (9.0 g) was then slowly added. The whole mixing process was carried out in an ice bath under continuous stirring. Then, the flask was put into an oil bath and the temperature was slowly raised to 50 °C, which was then maintained for 12 h with magnetic stirring. After the reaction, the temperature of the mixture was reduced to room temperature, and 500 mL of ice was slowly added. Then, 30% H_2O_2 (3 mL) and 37% HCl (200 mL) were added to the lower slurry mixture, respectively. The mixture was repeatedly stirred and washed with deionized water until the pH value of the upper clear liquid was constant at 7. The microstructure, morphology, and chemical compositions were characterized by an

SEM-450 (FEI Company, Hillsborough, OR, USA), AFM (Bruker, Germany), and FT-IR (Perkin Elmer, Waltham, MA, USA), respectively.

2.3. Coating of the Multifilament Surgical Sutures

2.3.1. The Coating Treatment

Four steps were followed to treat the surgical sutures with the coating, namely, pre-treatment (boiling), etching, DA coating, and GO coating, as shown in Figure 1. First, the sutures' surface coating was removed by boiling the THF solution. Second, the clear surface was etched by NaOH. Third, the etched surgical suture samples were immersed in DA Tris buffer and continuously stirred for 12 h at room temperature. Finally, the surgical sutures were coated with the GO slurry by the dip-coating method.

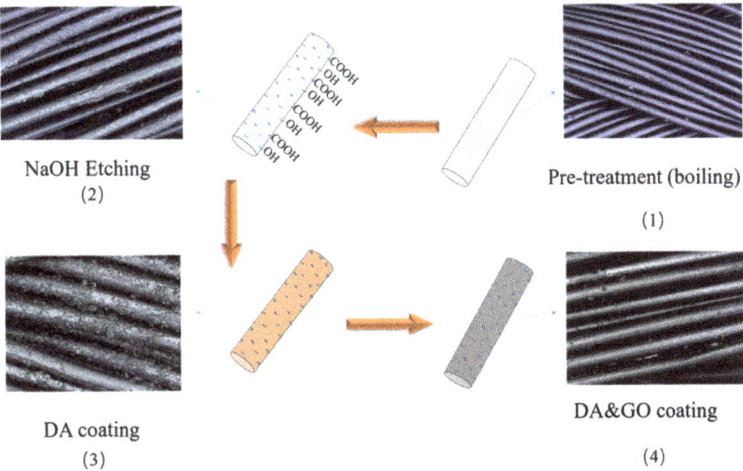

Figure 1. Schematic illustration of the coating process of polyglycolic acid (PGA) multifilament surgical suture. (**1**) Pre-treatment (boiling), (**2**) NaOH etching, (**3**) dopamine hydrochloride (DA) coating, (**4**) DA and graphene oxide (GO) coating.

2.3.2. Surface Analysis

A laser scanning microscope (VK 9700, KEYENCE, Osaka, Japan) was used to evaluate the topography and three-dimensional (3D) profiles of the sutures. FT-IR (Perkin Elmer, USA) was used to characterize the chemical composite of the surface of the surgical sutures.

2.3.3. Mechanical Properties

Weight Change

The weight change (w) of the treated surgical sutures was calculated by Equation (1):

$$w = \frac{w_2 - w_1}{w_1} \times 100\% \tag{1}$$

where w_1 is the weight of the untreated sutures and w_2 is the weight of the treated sutures.

Contact Angle

Water contact angle measurements (Data Physics OCA20, Germany) were used to characterize the hydrophilicity of the surface of the sutures by the sessile drop method at room temperature [25]. A camera recorded images of 3 µL of water dropped onto tight sutures. The contact angle was measured by the OCA20 software. The five measurements were repeated.

Tensile Strength Test

A tensile tester (Zwick/Roell 500N, Germany) was used to measure the strength of the sutures. The samples were cut into lengths of 25 cm with a moving velocity of 10 mm/min until the sutures broke. The tensile strength and the elongation at the break of the sutures were recorded by a computer [26].

Bending Stiffness

Bending stiffness is an important parameter that influences the friction properties of surgical sutures [27]. In this study, the cantilever method was used to estimate the bending stiffness of the surgical sutures. The load was applied to 1 cm parts of the sutures, with a total length of 5 cm. Then, 15 s later, the distance at which the end of the surgical sutures falls under the load and is placed on the horizontal plane was measured [28]. Next, 1 cm-long sutures were suspended in the air, and the bending stiffness was computed according to Equation (2).

$$B = \frac{F * l^3}{3f} \quad (2)$$

where B is the bending stiffness, l is 1 cm in this study, f is the deflection of the sutures' loading end, and F is the dead load.

2.4. Tribological Measurement

The friction and wear properties of the surgical sutures using artificial skin were tested by a PFA [20], as shown in Figure 2. The specific operation was as follows: The sample fixture was assembled on a Zwick/Roell tensile strength tester (500 N, Germany). When the surgical sutures penetrated through the friction tester, the friction force of the surgical sutures passing through the tissues or organs was evaluated. In the friction measurement, the artificial skin was fixed in the gripper. Then, one end of a surgical suture was left free, and the other penetrated the skin using the surgical suture needle and was fixed onto the force measuring sensor. The experiment was repeated three times for each experimental parameter, and the average value of the experimental results was taken. Table 1 shows the experimental parameter conditions.

Table 1. Experimental parameters.

Test-Related Factors	Instructions
Equipment	Penetration friction apparatus
The diameter of the space of gripper	25 ± 0.5 mm
Puncture angle	90°
Puncture velocity of the needle	60 mm/min
Puncture distance of the needle	10 mm
Penetration velocity of the suture	100 mm/min
Penetration distance of the suture	100 mm

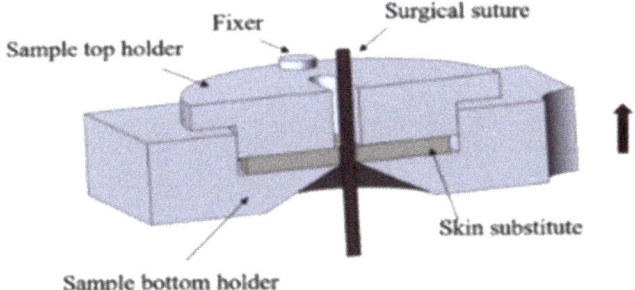

a. skin substitute gripper

b. penetration friction apparatus

Figure 2. Penetration friction apparatus (PFA): (**a**) Skin substitute gripper; (**b**) penetration friction apparatus.

3. Results and Discussion

3.1. Characterization of Graphene Oxide

The chemical components of GO were evaluated by FT-IR. From Figure 3a, in the spectrum, the O–H groups of GO at 3419 cm^{-1} can be observed. The stretching vibrations of the C=O and C=C of GO were 1734 and 1627 cm^{-1}, respectively. The C–O vibrations in C–OH and the C–O–C vibrations in epoxy were observed at 1384 and 1051 cm^{-1}, respectively. The interlayer spacing of graphite and GO was assessed by XRD according to Bragg's law, as shown in Equation (3).

$$n\lambda = 2d \sin\theta \tag{3}$$

From Figure 3b, it can be seen that the reflection of GO is a single peak at $2\theta = 10.3°$, illustrating that the layer spacing is larger. Due to the oxide groups in GO, the water molecules were trapped between the graphene oxide sheets [29,30]. No obvious peak was found in the profile of GO, indicating that the graphite was successfully oxidized to GO. From Figure 3c, it can be seen that the thickness of the GO sheet was 0.93 nm and that the stacking of the GO sheet was 2.1 nm, which is in accordance with the values for the single-layer GO sheet [31].

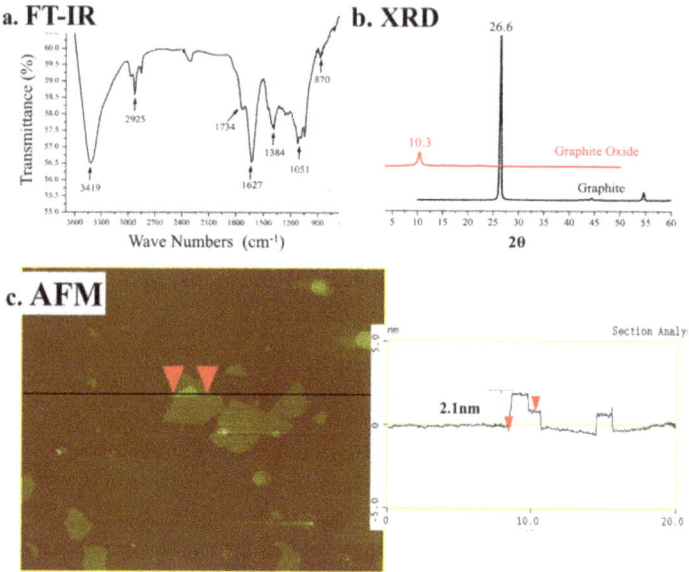

Figure 3. Characterization of GO: (**a**) FT-IR, (**b**) XRD, and (**c**) AFM tapping mode image and height profile.

3.2. Surface Chemical Composite and Morphology

The surface chemical composites of the PGA surgical sutures with different treatments were confirmed by FT-IR. From Figure 4, the stretching vibration of –OH with a carboxyl group (–C=O–OH) can be found in the FT-IR spectrum at 3515 cm^{-1}. The bending vibration peaks of C=O in carboxyl (–C=O–OH) were at 1080 and 1414 cm^{-1}, and the stretching vibration peaks of C=O in carboxyl (–C=O–OH) was observed at 1739 cm^{-1}. The characteristic absorption peak on the surface of the surgical sutures etched by NaOH was stronger. Therefore, it can be concluded that more carboxyl groups were produced after NaOH treatment. It can also be seen from the figure that the stretching vibration absorption peak of –NH/–OH appeared between 3600 and 3100 cm^{-1} for the PGA multifilament surgical sutures after the DA and GO coating treatment. The bending characteristic absorption peak of –NH was 1578 cm^{-1}, and the stretching characteristic absorption peaks of C–O were 1151 and 1080 cm^{-1} [32–34]. This confirms that the DA and GO coating adhered to the surface of the surgical sutures.

Figure 5 shows the surface morphology of the surgical sutures with and without a coating treatment. After the THF treatment, the coating material was removed from the surface of the commercialized surgical sutures. After NaOH etching, the small cracks and dents on the surface of the surgical sutures and fibers became thinner. The defects and specific surface area of the surface of the fiber became larger, providing more adhesion points for the subsequent coating.

After the DA coating, more DA coating particles accumulated on the surface of the sutures. The small DA particles in the buffer solution deposited onto the surface of the fiber to form a uniform film [14], so the adhesion to the surface of the sutures was enhanced. The DA coating provided good reaction conditions for the subsequent coating treatment as a secondary reaction platform [15–17]. After the GO coating, a uniform GO film formed on the surface of the suture, which became smoother, and the gap between the fibers was filled.

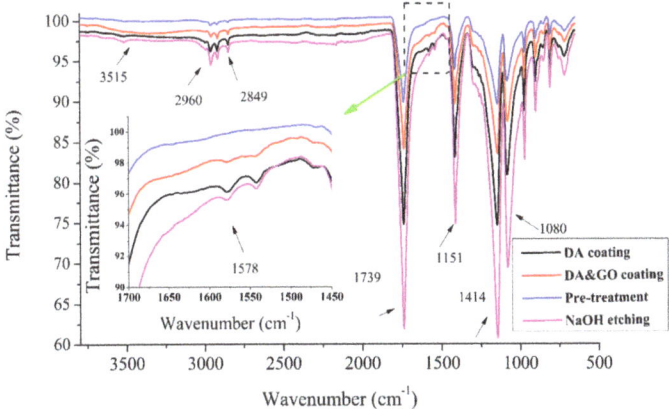

Figure 4. FT-IR of the polyglycolic acid (PGA) multifilament surgical sutures with and without different treatments (i.e., NaOH treatment and DA/GO coatings). The plot inside is a magnification of the infrared spectrum.

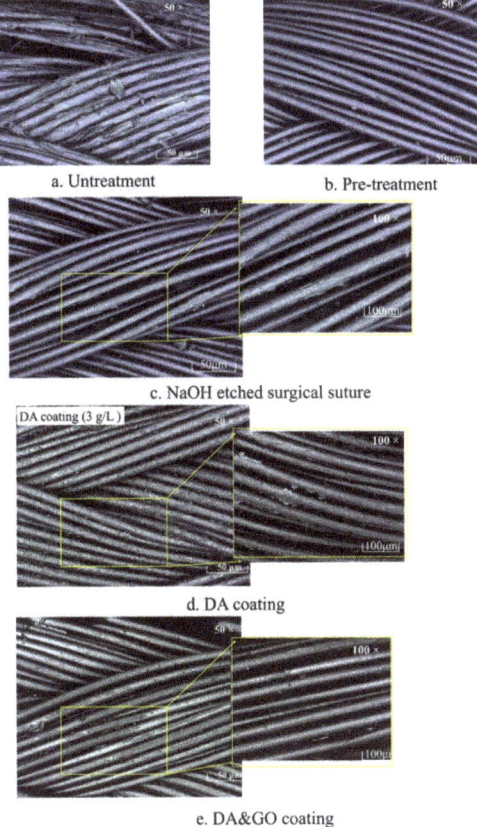

Figure 5. Morphology of the surface of the surgical sutures with different treatments. (**a**) Untreated suture; (**b**) THF treatment; (**c**) NaOH-etched suture; (**d**) DA coating; (**e**) DA and GO coating.

3.3. Mechanical Properties

Tensile fracture strength is an important parameter of surgical sutures. If the tensile breaking strength is too low, the sutures are easily pulled and slipped when they pass through tissues, leading to knots. Too low a tensile fracture strength shortens the absorption time of sutures after an operation, resulting in unsatisfactory suturing and increasing the risk of operation failure. Figure 6a shows the tensile strength and elongation at break of the sutures left untreated and those with NaOH etching and a DA and GO coating. The error bars in the graphs indicate the standard deviation of each measurement. It can be seen that when the sutures were treated with NaOH, the tensile strength of the PGA multifilament surgical sutures decreased from 58.9 to 50.0 N and the elongation at break from 30.4% to 24.2%. This may be because the loss of surgical sutures with the NaOH etching destroyed the structure of the fiber and created defects. It can also be seen from the figure that the increasing range of the tensile strength and elongation at break of the surgical sutures after the DA and GO coating treatment is limited. This is because the adsorption of the coating on the surgical suture is limited, and the adsorption capacity of the coating material cannot continue to increase during the coating treatment.

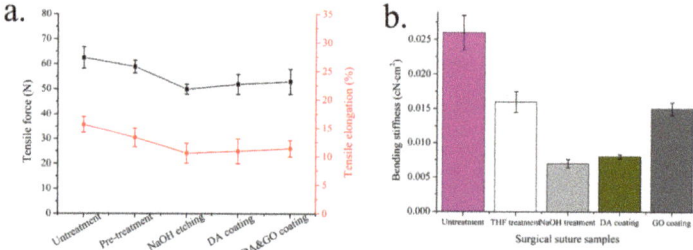

Figure 6. Tensile force, tensile elongation, and bending stiffness of the PGA multifilament surgical sutures. (**a**) Tensile force and elongation; (**b**) bending stiffness.

The bending yield property of surgical sutures is an important parameter that affects the friction between surgical sutures and tissues. From Figure 6a, it can be seen that the bending yield strength of the commercial surgical sutures without any treatment was 0.026 cN·cm^2, which reduced to 0.016 cN·cm^2 after removing the protective layer by THF treatment, and 0.007 cN·cm^2 after NaOH treatment. After the DA coating, the suture bending yield strength increased from 0.007 to 0.008 cN·cm^2. Then, with GO coating treatment, the bending yield strength of the suture increased to 0.015 cN·cm^2. These results are similar to previous research results [7,28].

The surfaces of the surgical sutures currently on the market are covered by a coating. If the coating is removed, the fibers of multifilament surgical sutures with a multifilament braided structure have no adhesion for the coating, the cohesion between the fibers is reduced, and the bending yield strength is reduced more. NaOH treatment further reduced the strength and diameter of the fibers, and, at the same time, the cohesion between the fibers was further decreased and the surgical sutures were able to bend more easily. However, after the DA coating treatment, the bond strength of the fiber was weak, and so an increase in the bending yield strength was not obvious. For the surgical sutures treated with GO, because GO can form a film on the surface of surgical sutures with DA particles, it filled the gaps between the fibers of the surgical sutures, which improved the cohesion between the fibers, increased the overall hardness of the surgical sutures, and improved the bending yield strength.

Mass change rate is an important parameter to evaluate the effects of treatment on surgical sutures. Figure 7a shows the mass change rate of the surgical sutures with and without a coating treatment. It can be seen that after NaOH etching, the mass loss rate of the PGA multifilament surgical sutures was 16.5%. When the surgical sutures were coated with DA, the mass increase rate was 3.92%. After three GO coating treatments, the mass increase rate of the PGA multifilament sutures was 0.75%.

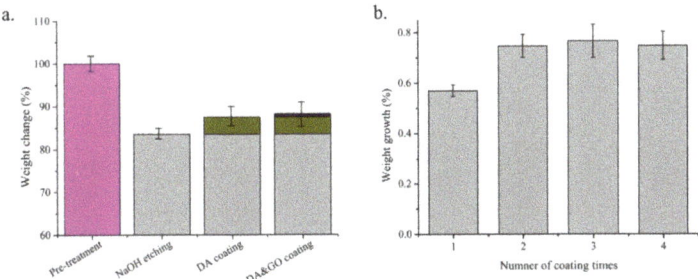

Figure 7. Weight change of the surgical sutures. (**a**) Weight change of the multifilament surgical sutures treated with different treatments; (**b**) weight change of the multifilament surgical sutures with a GO coating.

Figure 7b shows the effect of the GO coating treatment on the mass change rate of the surgical sutures with and without a coating. It can be seen that with the increase in the number of coating treatments, the weight of the PGA multifilament surgical suture increases. After three coating treatments, the increase in the mass change rate of the PGA multifilament sutures tended to stabilize. In this experiment, the PGA sutures were treated with three GO coatings.

3.4. Wettability Properties

NaOH etching and coating of multifilament surgical sutures influence the wettability properties of the surface. Figure 8 shows the static water contact angle of the surgical sutures with and without a coating treatment. After NaOH etching, the static contact angle of the surface of the sutures decreased to 58°. This is due to the fact that the hydrolysis reaction on the surface of the sutures increased the number of –OH and –COOH groups, which improved their wettability. After the DA and GO coating treatments, the static water contact angle of the surgical sutures increased to 91° and 72°, respectively.

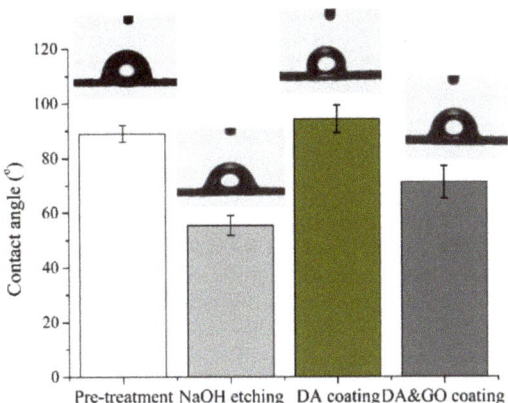

Figure 8. Water contact angle of the multifilament surgical sutures with different treatments.

3.5. Tribological Properties

3.5.1. Frictional Properties

Figure 9a shows the impact of the coating treatment on the friction force of the multifilament surgical sutures. The friction force of the sutures with THF, DA, and GO coating treatments was 0.72, 0.75, and 0.779, respectively. The friction force differed little between those sutures that did or did not receive a coating treatment.

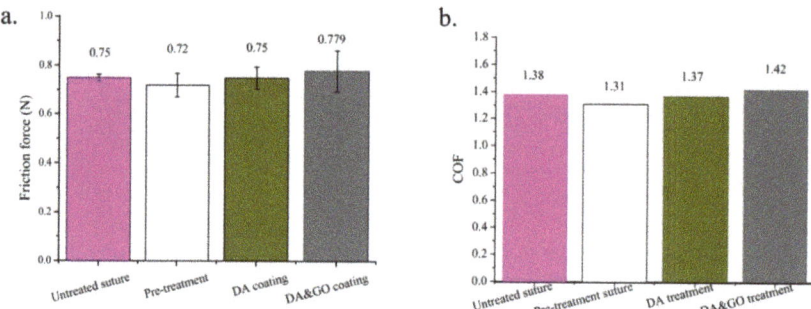

Figure 9. Friction properties of the PGA multifilament surgical sutures with DA and GO coatings. (**a**) The friction force; (**b**) the coefficient of friction.

The coefficient of friction (COF) is a curricular impact factor that assesses tribological properties, which are defined as the friction force and the normal force. The friction force was tested by PFA, while the normal force of the surgical sutures penetrating through the artificial skin was predicted by a simplified linear elastic model, which is based on the Hertz contact model [18,35].

$$F_N = 2EhL\frac{r_{suture}}{a} \int_0^a \sqrt{\frac{a^2 - x^2}{R^2 - x^2}}\, dx \tag{4}$$

According to the classical frictional equation (Equation (5)):

$$\mu = F_f / F_N \tag{5}$$

The thickness of the etching and coating was much less than the diameter of the surgical sutures in this contact condition. From Figure 9b, it can be seen that the COF of the surgical sutures without treatment, with DA, and with GO was 1.38, 1.37, and 1.42, respectively.

3.5.2. Wear Properties

The wear morphologies of the surgical sutures and skin substitutes are shown in Figure 10. It can be seen that there is little wear debris present on the commercialized and coated surface of the sutures, as shown in Figure 10a,b. There is little difference between the coated surgical sutures and the commercialized surgical sutures. In order to further study the wear performance of the surgical sutures with the DA and GO coating, we observed the three-dimensional wear morphology of the surface of the artificial skin. The wear crack in the skin substitute was similar to that of the commercialized and coated sutures, as shown in Figure 10c,d.

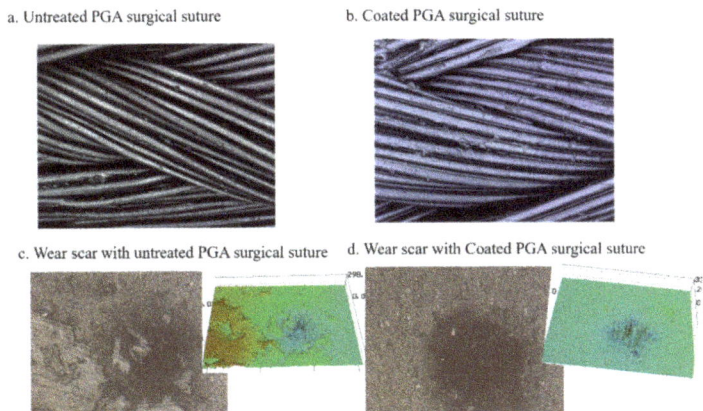

Figure 10. Wear morphology of the surgical sutures and the skin substitute. (**a**) Untreated surgical suture; (**b**) coated surgical suture; (**c**) wear scar with untreated surgical sutures; (**d**) wear scar in skin substitute with coated surgical sutures.

3.5.3. Mechanism

In this study, the surgical sutures with different treatments—including commercial suture and sutures with either a DA coating or a DA and GO coating—presented similar friction and wear properties. This may be explained by the fact that there was little change in the high stiffness and roughness of sutures after the coating treatment. Moreover, the experimental condition in this paper was dry friction.

In the contact model, the friction force was composited with adhesion and the deformation component [36].

$$F_f = F_{f,adh} + F_{f,def} \tag{6}$$

The adhesion component was the main component of the friction force due to the surgical sutures and the skin substitute being viscoelastic materials. The adhesion component was determined by the product of the shear strength τ and the real contact area A_r between them [37]:

$$F_{f,adh} = \tau A_r \tag{7}$$

The shear strength of the surgical sutures and the skin substitute was identified. The contact area was the most effective factor in the frictional properties of the surgical sutures. To better reveal the influence of the GO coating on the tribological mechanism of the skin substitute, the artificial skin was penetrated with a surgical suture, as shown in Figure 11a,b. It can be seen that when the surgical suture was inserted into the planar crack in the skin substitute, the surgical suture deformed under the pressure. Furthermore, the cohesion force of the twisted fibers was low. The resilience force of the skin substitute acted symmetrically on the two surfaces of the surgical suture, as shown in Figure 11c,d. Hence, the contact area was similar between surgical sutures with and without a coating, which presented similar frictional properties.

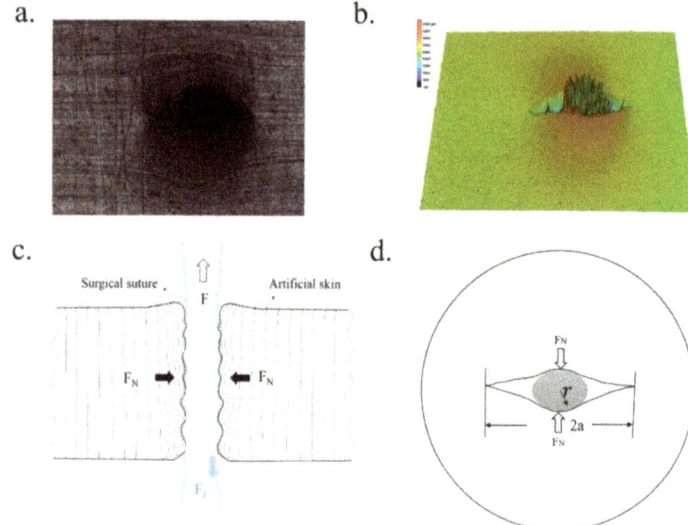

Figure 11. (**a**) Image of a surgical suture penetrating the skin substitute; (**b**) three-dimensional (3D) image of the surgical suture inserted into the skin substitute; (**c**) sketch of the surgical suture penetrating through the skin substitute; (**d**) force analysis of the surgical suture.

4. Conclusions

In this study, a DA and GO coating was applied to the surface of multifilament sutures to form a uniform film. The COF of the PGA multifilament sutures with and without a DA and GO coating was 1.37 and 1.42, respectively, which were in the same range. According to the contact mode and the way in which the surgical sutures are inserted, the contact surface was the main factor affecting the friction properties of the surgical sutures penetrating through the skin substitute. Due to the twist structure of surgical sutures, the contact area was similar when the surgical sutures were inserted into the skin substitute; hence, the coating barely affected the frictional properties of the surgical sutures.

Author Contributions: All authors listed have made a substantial, direct and intellectual contribution to the work, and approved it for publication. All authors have read and agreed to the published version of the manuscript.

Funding: This research received no external funding.

Conflicts of Interest: The authors declare no conflict of interest.

References

1. Misra, S.; Ramesh, K.; Okamura, A. Modeling of tool-tissue interactions for computer-based surgical simulation: A literature review. *Presence* **2008**, *17*, 463–491. [CrossRef] [PubMed]
2. Gilbert, J.L. Wound closure biomaterials and devices. *Shock* **1999**, *11*, 226. [CrossRef]
3. Moy, R.L.; Waldman, B.; Hein, D.W. A review of sutures and suturing techniques. *J. Dermatol. Surg. Oncol.* **1992**, *18*, 785–795. [CrossRef] [PubMed]
4. Rao, Y.; Farris, R.J. A modeling and experimental study of the influence of twist on the mechanical properties of high-performance fiber yarns. *J. Appl. Polym. Sci.* **2000**, *77*, 1938–1949. [CrossRef]
5. Apt, L.; Henrick, A. "Tissue-drag" with polyglycolic acid (dexon) and polyglactin 910 (vicryl) sutures in strabismus surgery. *J. Pediatr. Ophthalmol.* **1976**, *13*, 360–364. [CrossRef]
6. Ajmeri, J.R.; Ajmeri, C.J. Surgical sutures: The largest textile implant material. In *Medical Textiles and Biomaterials for Healthcare*; Woodhead Publishing: Gujarat, India, 2006; pp. 432–440.

7. Zhang, G.; Zheng, G.; Ren, T.; Zeng, X.; Der Heide, E.V. Dopamine hydrochloride and carboxymethyl chitosan coatings for multifilament surgical suture and their influence on friction during sliding contact with skin substitute. *Friction* **2020**, *8*, 58–69. [CrossRef]
8. Li, M.; Liu, Q.; Jia, Z.; Xu, X.; Cheng, Y.; Zheng, Y.; Xi, T.; Wei, S. Graphene oxide/hydroxyapatite composite coatings fabricated by electrophoretic nanotechnology for biological applications. *Carbon* **2014**, *67*, 185–197. [CrossRef]
9. Tong, L.; Zhang, J.; Xu, C.; Wang, X.; Song, S.; Jiang, Z.H.; Kamado, S.; Cheng, L.; Zhang, H. Enhanced corrosion and wear resistances by graphene oxide coating on the surface of mg-zn-ca alloy. *Carbon* **2016**, *109*, 340–351. [CrossRef]
10. Li, P.; Zhou, H.; Cheng, X. Nano/micro tribological behaviors of a self-assembled graphene oxide nanolayer on ti/titanium alloy substrates. *Appl. Surf. Sci.* **2013**, *285*, 937–944. [CrossRef]
11. Cai, G.; Xu, Z.; Yang, M.; Tang, B.; Wang, X. Functionalization of cotton fabrics through thermal reduction of graphene oxide. *Appl. Surf. Sci.* **2017**, *393*, 441–448. [CrossRef]
12. Chen, L.; Wei, F.; Liu, L.; Cheng, W.; Hu, Z.; Wu, G.; Du, Y.; Zhang, C.; Huang, Y. Grafting of silane and graphene oxide onto pbo fibers: Multifunctional interphase for fiber/polymer matrix composites with simultaneously improved interfacial and atomic oxygen resistant properties. *Compos. Sci. Technol.* **2015**, *106*, 32–38. [CrossRef]
13. Hu, X.; Tian, M.; Qu, L.; Zhu, S.; Han, G. Multifunctional cotton fabrics with graphene/polyurethane coatings with far-infrared emission, electrical conductivity, and ultraviolet-blocking properties. *Carbon* **2015**, *95*, 625–633. [CrossRef]
14. Yan, J.; Yang, L.; Lin, M.F.; Ma, J.; Lu, X.; Lee, P.S. Polydopamine spheres as active templates for convenient synthesis of various nanostructures. *Small* **2013**, *9*, 596–603. [CrossRef] [PubMed]
15. Ku, S.H.; Ryu, J.; Hong, S.K.; Lee, H.; Park, C.B. General functionalization route for cell adhesion on non-wetting surfaces. *Biomaterials* **2010**, *31*, 2535–2541. [CrossRef]
16. Barras, A.; Lyskawa, J.; Szunerits, S.; Woisel, P.; Boukherroub, R. Direct functionalization of nanodiamond particles using dopamine derivatives. *Langmuir* **2011**, *27*, 12451–12457. [CrossRef]
17. Lee, H.; Dellatore, S.M.; Miller, W.M.; Messersmith, P.B. Mussel-inspired surface chemistry for multifunctional coatings. *Science* **2007**, *318*, 426–430. [CrossRef]
18. Zhang, G.; Ren, T.; Zhang, S.; Zeng, X.; van der Heide, E. Study on the tribological behavior of surgical suture interacting with a skin substitute by using a penetration friction apparatus. *Colloids Surf. B Biointerfaces* **2018**, *162*, 228–235. [CrossRef]
19. Zhang, G.; Zeng, X.; Su, Y.; Borras, F.X.; de Rooij, M.B.; Ren, T.; van der Heide, E. Influence of suture size on the frictional performance of surgical suture evaluated by a penetration friction measurement approach. *J. Mech. Behav. Biomed. Mater.* **2018**, *80*, 171–179. [CrossRef]
20. Zhang, G.; Ren, T.; Lette, W.; Zeng, X.; van der Heide, E. Development of a penetration friction apparatus (pfa) to measure the frictional performance of surgical suture. *J. Mech. Behav. Biomed. Mater.* **2017**, *74*, 392–399. [CrossRef]
21. Shergold, O.A.; Fleck, N.A. Experimental investigation into the deep penetration of soft solids by sharp and blunt punches, with application to the piercing of skin. *J. Biomech. Eng.* **2005**, *127*, 838–848. [CrossRef]
22. Azar, T.; Hayward, V. Estimation of the Fracture Toughness of Soft Tissue from Needle Insertion. In *International Symposium on Biomedical Simulation*; Springer: Berlin/Heidelberg, Germany, 2008; pp. 166–175.
23. Shergold, O.A.; Fleck, N.A.; Radford, D. The uniaxial stress versus strain response of pig skin and silicone rubber at low and high strain rates. *Int. J. Impact Eng.* **2006**, *32*, 1384–1402. [CrossRef]
24. Marcano, D.C.; Kosynkin, D.V.; Berlin, J.M.; Sinitskii, A.; Sun, Z.; Slesarev, A.; Alemany, L.B.; Lu, W.; Tour, J.M. Improved synthesis of graphene oxide. *ACS Nano* **2010**, *4*, 4806–4814. [CrossRef] [PubMed]
25. Peng, F.; Olson, J.R.; Shaw, M.T.; Wei, M. Influence of pretreatment on the surface characteristics of plla fibers and subsequent hydroxyapatite coating. *J. Biomed. Mater. Res. B Appl. Biomater.* **2009**, *88*, 220–229. [CrossRef]
26. De Simone, S.; Gallo, A.; Paladini, F.; Sannino, A.; Pollini, M. Development of silver nano-coatings on silk sutures as a novel approach against surgical infections. *J. Mater. Sci. Mater. Med.* **2014**, *25*, 2205–2214. [CrossRef] [PubMed]
27. Chen, X.; Hou, D.; Tang, X.; Wang, L. Quantitative physical and handling characteristics of novel antibacterial braided silk suture materials. *J. Mech. Behav. Biomed. Mater.* **2015**, *50*, 160–170. [CrossRef]

28. Rodeheaver, G.T. Knotting and handling characteristics of coated synthetic absorbable sutures. *J. Surg. Res.* **1983**, *35*, 525–530. [CrossRef]
29. Singh, V.K.; Elomaa, O.; Johansson, L.-S.; Hannula, S.-P.; Koskinen, J. Lubricating properties of silica/graphene oxide composite powders. *Carbon* **2014**, *79*, 227–235. [CrossRef]
30. Some, S.; Kim, Y.; Yoon, Y.; Yoo, H.; Lee, S.; Park, Y.; Lee, H. High-quality reduced graphene oxide by a dual-function chemical reduction and healing process. *Sci. Rep.* **2013**, *3*, 1929. [CrossRef] [PubMed]
31. Stankovich, S.; Dikin, D.A.; Dommett, G.H.; Kohlhaas, K.M.; Zimney, E.J.; Stach, E.A.; Piner, R.D.; Nguyen, S.T.; Ruoff, R.S. Graphene-based composite materials. *Nature* **2006**, *442*, 282–286. [CrossRef]
32. Brugnerotto, J.; Lizardi, J.; Goycoolea, F.M.; Argüelles-Monal, W.; Desbrières, J.; Rinaudo, M. An infrared investigation in relation with chitin and chitosan characterization. *Polymer* **2001**, *42*, 3569–3580. [CrossRef]
33. Chen, X.-G.; Park, H.-J. Chemical characteristics of o-carboxymethyl chitosans related to the preparation conditions. *Carbohydr. Polym.* **2003**, *53*, 355–359. [CrossRef]
34. Zheng, M.; Han, B.; Yang, Y.; Liu, W. Synthesis, characterization and biological safety of o-carboxymethyl chitosan used to treat sarcoma 180 tumor. *Carbohydr. Polym.* **2011**, *86*, 231–238. [CrossRef]
35. Zhang, G.; Ren, T.; Zeng, X.; Van Der Heide, E. Influence of surgical suture properties on the tribological interactions with artificial skin by a capstan experiment approach. *Friction* **2017**, *5*, 87–98. [CrossRef]
36. Roselman, I.; Tabor, D. The friction of carbon fibres. *J. Phys. D Appl. Phys.* **1976**, *9*, 2517. [CrossRef]
37. Roselman, I.C.; Tabor, D. The friction and wear of individual carbon fibres. *J. Phys. D Appl. Phys.* **1977**, *10*, 1181. [CrossRef]

© 2020 by the authors. Licensee MDPI, Basel, Switzerland. This article is an open access article distributed under the terms and conditions of the Creative Commons Attribution (CC BY) license (http://creativecommons.org/licenses/by/4.0/).

Article

Protection of Poly(Vinyl Chloride) Films against Photodegradation Using Various Valsartan Tin Complexes

Alaa Mohammed [1], Gamal A. El-Hiti [2,*], Emad Yousif [1,*], Ahmed A. Ahmed [3], Dina S. Ahmed [4] and Mohammad Hayal Alotaibi [5,*]

1. Department of Chemistry, College of Science, Al-Nahrain University, Baghdad 64021, Iraq; alaaalqaycy7@gmail.com
2. Cornea Research Chair, Department of Optometry, College of Applied Medical Sciences, King Saud University, P.O. Box 10219, Riyadh 11433, Saudi Arabia
3. Polymer Research Unit, College of Science, Al-Mustansiriyah University, Baghdad 10052, Iraq; drahmed625@gmail.com
4. Department of Medical Instrumentation Engineering, Al-Mansour University College, Baghdad 64021, Iraq; dinasaadi86@gmail.com
5. National Center for Petrochemicals Technology, King Abdulaziz City for Science and Technology, P.O. Box 6086, Riyadh 11442, Saudi Arabia
* Correspondence: gelhiti@ksu.edu.sa (G.A.E.-H.); emad_yousif@hotmail.com (E.Y.); mhhalotaibi@kacst.edu.sa (M.H.A.); Tel.: +966-11469-3778 (G.A.E.-H.); Fax: +966-11469-3536 (G.A.E.-H.)

Received: 7 April 2020; Accepted: 20 April 2020; Published: 21 April 2020

Abstract: Poly(vinyl chloride) is a common plastic that is widely used in many industrial applications. Poly(vinyl chloride) is mixed with additives to improve its mechanical and physical properties and to enable its use in harsh environments. Herein, to protect poly(vinyl chloride) films against photoirradiation with ultraviolet light, a number of tin complexes containing valsartan were synthesized and their chemical structures were established. Fourier-transform infrared spectroscopy, weight loss, and molecular weight determination showed that the non-desirable changes were lower in the films containing the tin complexes than for the blank polymeric films. Analysis of the surface morphology of the irradiated polymeric materials showed that the films containing additives were less rough than the irradiated blank film. The tin complexes protected the poly(vinyl chloride) films against irradiation, where the complexes with high aromaticity were particularly effective. The additives act as primary and secondary stabilizers that absorb the incident radiation and slowly remit it to the polymeric chain as heat energy over time at a harmless level.

Keywords: tin compounds; valsartan; poly(vinyl chloride); additives; average molecular weight; weight loss; functional group index

1. Introduction

Plastics are extensively used as replacements for metals, glass, and wood in many modern applications [1]. Plastics have unique performance and superior properties compared with other materials [2]. The properties of plastic such as the toughness, rigidity, density, color, and transparency can be controlled during the manufacture process. Moreover, plastics can be cheaply produced and last for a long time. The most common plastics are polyethylene, polyethylene terephthalate, polypropylene, polystyrene, and poly(vinyl chloride) (PVC) [3]. PVC has a high chlorine content (ca. 57% by weight) and is thus non-combustible [1]. Therefore, PVC can be used in furniture; construction; and many construction applications such as upholstery, pipes, windows shutters, roofing foils, flooring, and fire retardants [4]. In addition, PVC has good mechanical and chemical properties, is easy to produce in large quantitates, and resists ecological strain cracking [5].

The performance of PVC can be enhanced by incorporating additives to enable its use in outdoor applications. PVC can be mixed with plasticisers to enable the production of flexible polymeric materials for certain applications [4,5]. PVC undergoes gradual degradation in harsh environments, such as under exposure to heat and direct ultraviolet (UV) light for a long period [6]. The degradation of PVC leads to a reduction in its mechanical integrity, change in color, and formation of micro-cracks within the surface. Therefore, PVC cannot be used on its own and must be combined with stabilizers to enhance its photostability [7]. The additives should be easy and cheap to produce and well incorporated within the PVC polymeric chains. Further, the additives should not alter the color of PVC and should be non-volatile, chemically stable, non-toxic, and should not pollute the surrounding environment. The most common PVC industrial additives act as smoke suppressors, flame retardants, thermal and impact modifiers, heat stabilizers, UV stabilizers, screeners, absorbers, and free radical scavengers [8–10]. PVC stabilizers are classified into various types [8], that is, primary stabilizers that deactivate the allylic chlorides that are generated in the photodegradation of the polymeric chains and secondary additives that act as scavengers of chloride radicals and hydrogen chloride [11]. Many non-toxic organic materials have been used as PVC additives [12].

bis(2-Ethylhexyl)phthalate (Figure 1) can be obtained from phthalic acid and has been used as a PVC plasticizer in the past. It is non-volatile, oily, has a low production cost, and is compatible with PVC [8]. However, the safety and health hazards associated with phthalates hinder their use in medicinal products (e.g., blood bags). Tetrachlorobiphenyl (Figure 1) has been used in the past as a PVC flame retardant and stabilizer [13]. However, chlorinated aromatics were banned owing to their carcinogenic and environmentally hazardous nature [13]. Stabilizers containing metals are common. For example, stabilizers containing a mixture of barium and zinc are considered non-hazardous under normal use. However, a co-stabilizer is required along with the barium and zinc mixture, for example [14–16]. Alternative stabilizers have been used to replace those that pose a risk to the environment and humans. For example, *tris*(di-*tert*-butylphenyl)phosphite (Figure 1) has been used as a PVC additive and antioxidant on the commercial scale [17].

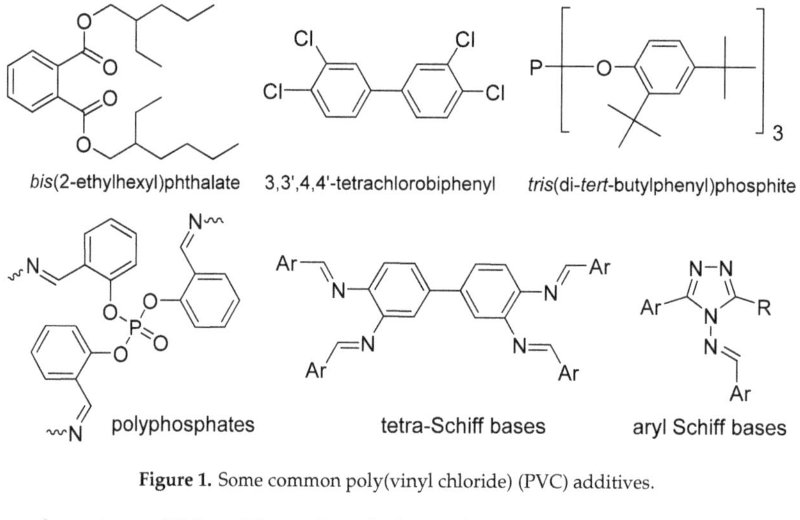

Figure 1. Some common poly(vinyl chloride) (PVC) additives.

Recently, various additives (Figure 1), including polyphosphates [18–20], Schiff bases [21–25], aromatic compounds [26,27], and organic–metal complexes [28–31], were investigated for use as PVC stabilizers. Some success has been achieved; however, research into the design of new and efficient PVC stabilizers for use on the commercial scale is ongoing. In the current work, we report the synthesis of a number of new tin complexes containing valsartan. The substituents on the synthesized tin complexes are varied to include aliphatic (butyl groups) or aromatic substituents (phenyl groups).

Valsartan is commercially available, non-toxic, highly aromatic, and contains a high level of oxygen and nitrogen (heteroatoms). Therefore, it was expected that the synthesized tin complexes, and in particular those with a high degree of aromaticity, would act as efficient PVC stabilizers.

2. Materials and Methods

2.1. General

Chemicals, reagents, and solvents were obtained from Merck (Gillingham, UK). PVC ($\overline{M}_V$ = ca. 171,000) was supplied by Petkim Petrokimya (Istanbul, Turkey). An AA-6880 Shimadzu atomic absorption flame spectrophotometer (Shimadzu, Tokyo, Japan) was used to measure the tin content of the complexes. The FTIR spectra were recorded on an FTIR 8300 Shimadzu spectrophotometer (Shimadzu, Tokyo, Japan). ^{1}H NMR (500 MHz) spectra were recorded on a Varian INOVA spectrometer (Palo Alto, CA, USA). ^{119}Sn NMR (107 MHz) spectra were recorded on a Bruker DRX spectrophotometer (Bruker, Zürich, Switzerland). An accelerated weather-meter QUV tester obtained from Q-Panel Company (Homestead, FL, USA) was used to irradiate the PVC samples with UV light (λ_{max} = 365 nm) at 25 °C. An Ostwald U-Tube Viscometer was used to measure the viscosity of PVC. The surface morphology of the PVC films was inspected using a Veeco system (Plainview, NY, USA) and a TESCAN MIRA3 LMU instrument (Kohoutovice, Czech Republic) at an accelerating voltage of 10 kV.

2.2. Synthesis of Tin Complexes 1 and 2

A mixture of valsartan (0.87 g; 2.0 mmol) and triphenyltin or tributyltin chloride (2.0 mmol) in boiling methanol (MeOH; 30 mL) was stirred for 6 h. The solid obtained after cooling the mixture to 25 °C was filtered, washed with MeOH, and dried to give 1 or 2 (Figure 2).

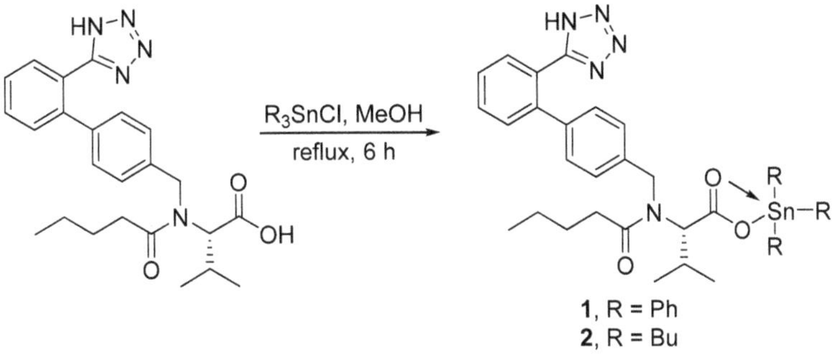

Figure 2. Synthesis of tin complexes 1 and 2.

2.3. Synthesis of Tin Complexes 3 and 4

A mixture of valsartan (0.87 g; 2.0 mmol) and diphenyltin or dibutyltin chloride (1.0 mmol) in boiling MeOH (30 mL) was stirred for 8 h. The solid obtained was filtered, washed with MeOH, and dried to give 3 or 4 (Figure 3).

2.4. Preparation of PVC Films

A mixture of PVC (5.0 g) and the tin complex (25 mg) in tetrahydrofuran (THF; 100 mL) was stirred at 25 °C for 2 h. The mixture was transferred onto glass plates with a thickness of 40 µm. The films produced were left to dry under vacuum for 18 h.

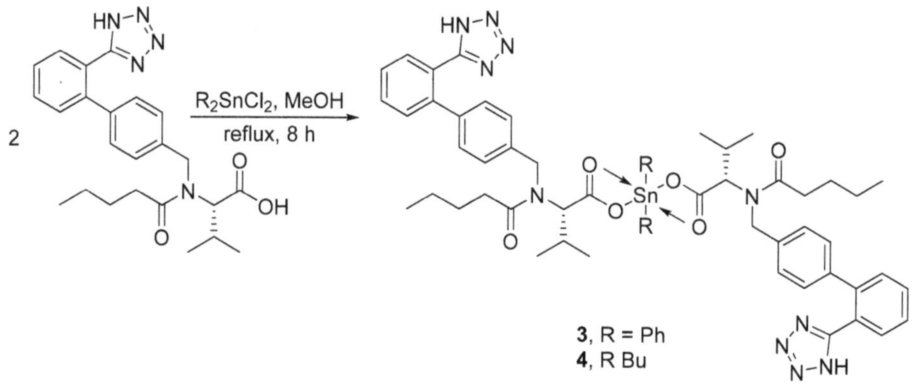

Figure 3. Synthesis of tin complexes 3 and 4.

2.5. Assessment of PVC Photodegradation Using FTIR Spectrophotometry

Photodegradation of the PVC films was investigated using FTIR spectrophotometry. The changes in the intensities of the absorption peak of the carbonyl (C=O; 1722 cm^{-1}) and polyene (C=C; 1602 cm^{-1}) groups were monitored. These peaks arise owing to the formation of small fragments containing C=O and C=C groups generated during PVC photooxidation. The change in the intensity of these peaks was monitored relative to the intensity of a standard peak (the C–H bond of the CH$_2$ groups; 1328 cm^{-1}). Equation (1) was used to calculate the functional group (C=O or C=C) index (I_s) from the absorbance of the functional group (A_s) and that for the standard peak (A_r) [32].

$$I_s = A_s/A_r \tag{1}$$

2.6. Assessment of PVC Photodegradation Using Weight Loss

The weight loss of the PVC films due to photodegradation was calculated from the weight of PVC before (W_0) and after irradiation (W_t) using Equation (2) [30].

$$\text{Weight loss } (\%) = [(W_0 - W_t)/W_0] \times 100 \tag{2}$$

2.7. Assessment of PVC Photodegradation Using Average Molecular Weight ($\overline{M}_V$)

The $\overline{M}_V$ of PVC after irradiation was calculated from the intrinsic viscosity [η] of the solution of the polymeric materials using Equation (3), known as the Mark–Houwink equation [33].

$$[\eta] = 1.63 \times 10^{-2} M_v^{0.766} \tag{3}$$

3. Results and Discussion

3.1. Synthesis of Tin Complexes 1–4

The reaction of valsartan and the appropriate tributyltin or triphenyltin chloride (in a 1:1 ratio) in boiling MeOH for 6 h gave **1** or **2** (Figure 2) in 82% and 79% yield, respectively (Table 1). Similarly, reaction of excess valsartan (two mole equivalents) and the appropriate dibutyltin or diphenyltin chloride for 8 h in refluxing MeOH gave **3** and **4** (Figure 3) in 75% and 88% yield, respectively (Table 1). The purity and elemental composition of the synthesized tin complexes **1–4** were confirmed by elemental analysis. The color, melting point, yield, and elemental analysis data for the tin complexes **1–4** are summarized in Table 1.

Table 1. Color, melting point, yield, and elemental analysis of **1–4**.

Complex	Color	Melting Point (°C)	Yield (%)	Found (Calculated) (%)			
				C	H	N	Sn
1	white	257–259	82	64.29 (64.30)	5.50 (5.52)	8.92 (8.93)	15.10 (15.13)
2	white	112–114	79	59.65 (59.68)	7.62 (7.65)	9.64 (9.67)	16.34 (16.38)
3	off-white	103–105	75	63.07 (63.11)	5.79 (5.83)	12.23 (12.27)	10.37 (10.40)
4	off-white	76–78	88	61.01 (61.04)	6.74 (6.77)	12.68 (12.71)	10.73 (10.77)

The FTIR spectra of **1–4** (Figures S1–S4) show absorption bands corresponding to the symmetrical and asymmetrical vibrations of the carbonyl group in the regions of 146–1477 and 1732–1735 cm^{-1}, respectively (Table 2) [34]. For valsartan, these absorption bands appeared at lower wavenumbers (1442 and 1670 cm^{-1}, respectively), which clearly indicated the formation of a bond between the tin atom and oxygen of the carboxylate group [35]. Indeed, the absorption bands that appeared at 443–455 cm^{-1} are attributed to the Sn–O bonds. In addition, the Sn–C absorption bands appeared in the 559–563 cm^{-1} region. The difference [Δv (asym – sym)] between the symmetric (sym) and asymmetric (asym) vibrational frequencies for the carbonyl group was in the range of 258–270 cm^{-1}. The value of Δv indicates that valsartan acts as an asymmetric bi-dentate ligand [36].

Table 2. Select FTIR spectral data for **1–4**.

Complex	Wavenumber (cm^{-1})				
	C=O sym	C=O asym	Δv (asym – sym)	Sn–C	Sn–O
1	1477	1735	258	559	447
2	1462	1732	270	563	455
3	1473	1735	262	559	443
4	1477	1735	258	559	447

The ^{1}H NMR spectra of **1–4** (Figures S5–S8) show the absence of the proton resonance at 12.63 ppm for the carboxylic group of valsartan [37]. Clearly, the tin complexes were produced as a result of replacement of the carboxylic group proton with a tin atom. The ^{1}H NMR spectra of **1–4** showed the presence of protons of both valsartan and the substituents (phenyl and butyl moieties) at the expected chemical shifts (Table 3).

Table 3. ^{1}H and ^{119}Sn NMR spectral data for **1–4**.

Complex	NMR (DMSO-d_6; δ in ppm and J in Hz)	
	^{1}H (500 MHz)	^{119}Sn (107 MHz)
1	7.73–7.06 (m, 23H, Ar), 6.95 (s, exch., 1H, NH), 4.62 (s, 2H, CH$_2$), 4.47 (d, J = 7.2 Hz, 1H, CH), 2.22 (br, 1H, CH), 2.14 (t, J = 7.4 Hz, 2H, CH$_2$), 1.40–1.29 (m, 4H, CH$_2$CH$_2$), 0.94 (d, J = 7.2 Hz, 6H, 2 Me), 0.78 (t, J = 7.4 Hz, 3H, Me)	−137.8
2	7.66–7.55 (m, 6H, Ar), 7.20 (d, J = 8.1 Hz, 1H, Ar), 7.08 (d, J = 8.1 Hz, 1H, Ar), 6.95 (s, exch., 1H, NH), 4.63 (s, 2H, CH$_2$), 4.51 (d, J = 7.2 Hz, 1H, CH), 2.21 (br, 1H, CH), 2.08 (m, 2H, CH$_2$), 1.60–1.30 (m, 22H, 11 CH$_2$), 1.13–0.85 (m, 18H, 6 Me)	−131.2
3	7.70–7.53 (m, 8H, Ar), 7.37–7.00 (m, 18H, Ar), 6.97 (s, exch., 2H, 2 NH), 4.66 (s, 4H, 2 CH$_2$), 4.49 (d, J = 7.1 Hz, 2H, 2 CH), 2.22 (br, 2H, 2 CH), 2.09 (t, J = 7.5 Hz, 4H, 2 CH$_2$), 1.38–1.25 (m, 8H, 2 CH$_2$CH$_2$), 0.92 (d, J = 7.1 Hz, 12H, 4 Me), 0.80 (t, J = 7.5 Hz, 6H, 2 Me)	−406.1
4	7.68–7.62 (m, 12H, Ar), 7.20 (d, J = 8.2 Hz, 2H, Ar), 7.08 (d, J = 8.2 Hz, 2H, Ar), 6.97 (s, exch., 2H, 2 NH), 4.63 (s, 4H, 2 CH$_2$), 4.46 (d, J = 7.1 Hz, 2H, 2 CH), 2.20 (br, 2H, 2 CH), 2.09 (m, 4H, 2 CH$_2$), 1.65–1.29 (m, 20H, 10 CH$_2$), 1.16–0.87 (m, 24H, 8 Me)	−218.1

The ^{119}Sn NMR spectra of **1–4** (Figures S9–S12) showed characteristic signals at −137.8 and −131.2 ppm owing to the tin atom in **1** and **2**, respectively. These chemical shifts (Table 3) revealed the formation of five-coordinated complexes [38–40]. For **3** and **4**, the signals corresponding the Sn atom appeared at −406.1 and −218.1 ppm, respectively. The chemical shifts for these signals indicated the formation of six coordinated complexes. Clearly, the geometry of the complexes affects the chemical shifts as a result of the shielding effect of the tin atom and substituents [38–40].

3.2. Assessment of Photodegradation of PVC Using Energy Dispersive X-ray (EDX) Mapping

The elemental composition of **1–4** was analyzed using energy dispersive X-ray (EDX). EDX confirmed the elements within complexes **1–4** (Figures S13–S18) [41]. PVC was mixed with complexes **1–4** (0.5 wt.%) and thin (40 μm) films were made. The films were irradiated with UV light and EDX was used to determine the elemental composition of the polymer blends. The EDX mapping images revealed that the tin complexes were well-distributed throughout the films [42]. For the unmodified (blank) PVC, the percentage of chlorine in the films was reduced from 64.8% before irradiation to 55.8% after irradiation (300 h). These results indicate significant dehydrochlorination, where hydrogen chloride was eliminated from the blank PVC as a result of photodegradation. After irradiation, the reduction in the chlorine content of the PVC films containing complexes **1–4** was lower compared with that of the blank PVC film. The chlorine content was highest (56.4%) in the case of the irradiated PVC/**1** blend. Complex **1**, which is highly aromatic (three phenyl, two aryl, and one tetrazole moieties) was the most efficient additive for stabilizing the polymeric materials. It has been reported that additives containing aromatic moieties are more efficient PVC photostabilizers compared with the corresponding ones containing aliphatic residues [31]. Complex **1** absorbs UV irradiation directly and releases the adsorbed energy over a long period of time at a rate that is not harmful to the PVC chains.

3.3. Assessment of Photodegradation of PVC Using FTIR Spectrophotometry

PVC undergoes photooxidative degradation upon irradiation in the presence of an oxygen source [43,44]. This process leads to the formation of small polymeric fragments containing carbonyl (C=O; carboxyl and ketone) and polyene (C=C; carbon–carbon double bond residues) groups [43,44]. Such functional groups can be detected using FTIR spectroscopy. In addition, the intensity of the FTIR signals can be monitored during the photooxidation of PVC and compared with the intensity of the signals of the C–H bond of the CH_2 moieties (1328 cm^{-1}) within the polymeric chains. The absorption of the C–H bond is not altered during the irradiation process. Figure 4 shows that the intensity of the signals of both the C=O (1722 cm^{-1}) and C=C (1602 cm^{-1}) groups was significantly higher for the irradiated blank PVC film. For the PVC film containing complex **1**, it was clear that the intensity of the peaks of both functional groups was significantly lower than that of the corresponding peaks that appeared for the irradiated PVC (blank) film.

Equation (1) was used to calculate the functional group indices ($I_{C=O}$ and $I_{C=C}$) for the blank PVC and those containing complexes **1–4** at 50 h intervals for an irradiation period of up to 300 h. The $I_{C=O}$ and $I_{C=O}$ values for the PVC films were plotted against the irradiation time at 50 h intervals (Figures 5 and 6). Both $I_{C=O}$ and $I_{C=O}$ changed significantly for the PVC film in the absence of any additives compared with the PVC films containing **1–4** as additives. Clearly, complexes **1–4** stabilized PVC substantially. For example, the $I_{C=O}$ after 300 h of irradiation was 0.26, 0.13, 0.18, 0.16, and 0.20 for the PVC, PVC/**1**, PVC/**2**, PVC/**3**, and PVC/**4** films, respectively. Similarly, the $I_{C=C}$ for blank PVC was 0.27 after irradiation (300 h) compared with 0.12, 0.16, 0.14, and 0.19 for the PVC/**1**, PVC/**2**, PVC/**3**, and PVC/**4** films, respectively. Clearly, the minimum change in the indices of the C=O and C=C groups was achieved when complex **1** (highly aromatic) was used. The efficiency of complexes **1–4** for photostabilizing PVC against irradiation followed the order: **1** (triphenyltin) > **3** (diphenyltin) > **2** (tributyltin) > **4** (dibutyltin).

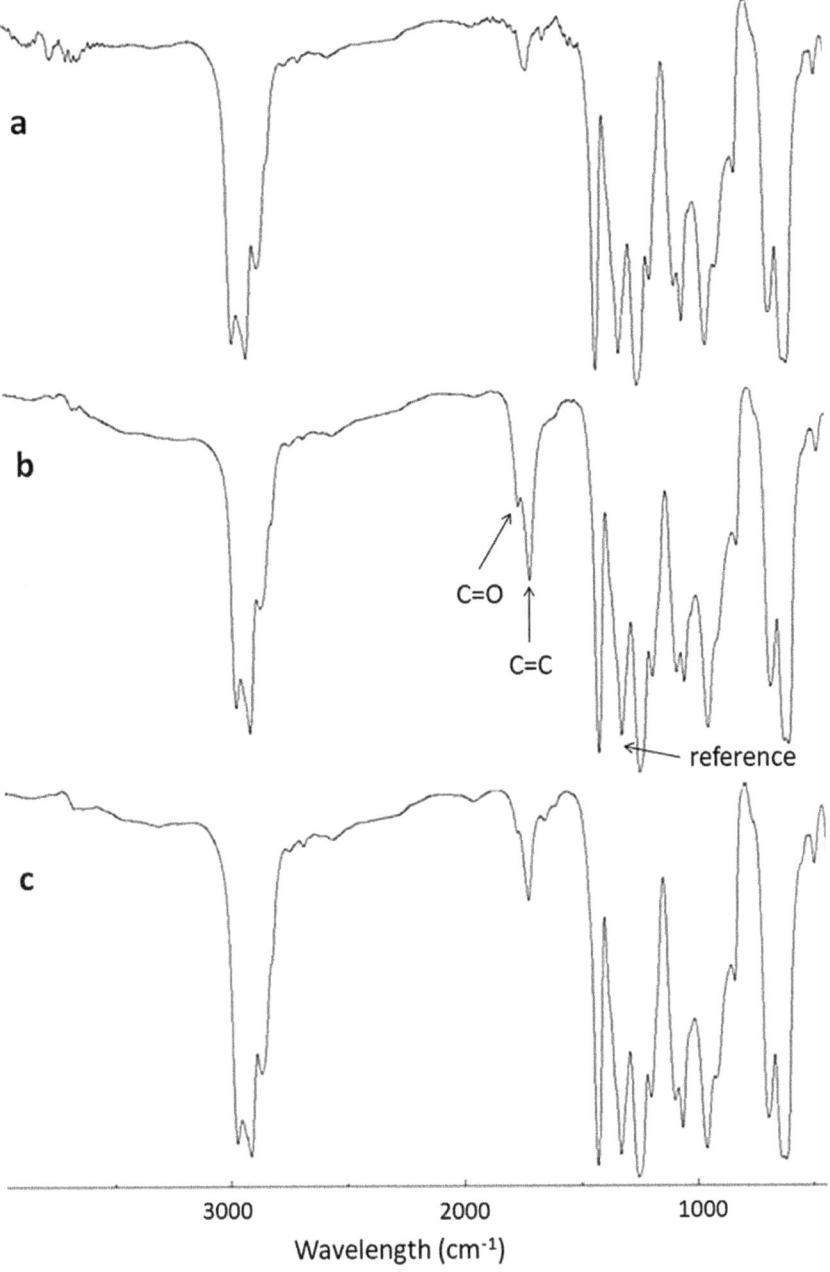

Figure 4. FTIR spectra of (**a**) poly(vinyl chloride) (PVC) film before irradiation, (**b**) PVC film after irradiation (300 h), and (**c**) PVC + **1** blend after irradiation (300 h).

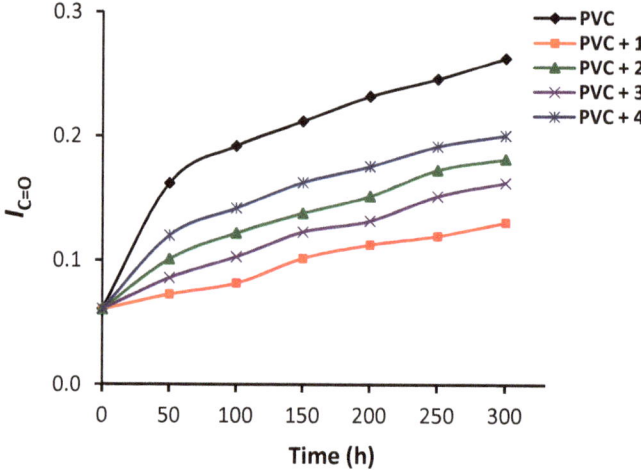

Figure 5. Changes in the $I_{C=O}$ index for PVC films.

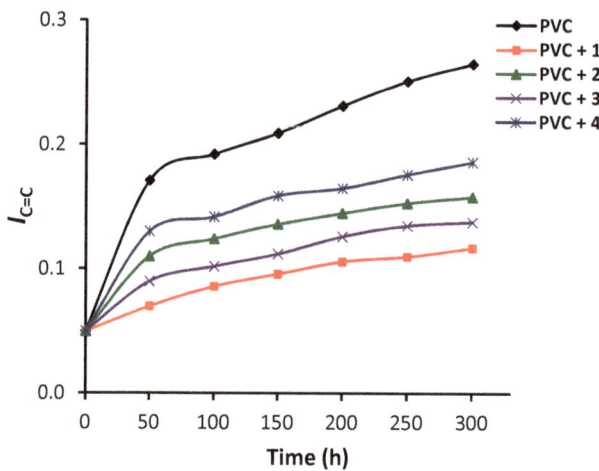

Figure 6. Changes in the $I_{C=C}$ index for PVC films.

3.4. Assessment of Photodegradation of PVC Using Weight Loss

The photooxidation of PVC causes cross-linking of the polymeric chains owing to the production of free radical moieties. As a result, hydrogen chloride (dehydrochlorination) and volatile small organic residues are eliminated, accompanied by PVC discoloration. Such processes lead to weight loss at a variety of relatively high temperatures [45–47]. To determine the efficiency of complexes **1–4** as stabilizers, the PVC films were irradiated and the weight loss was calculated at 50 h intervals during irradiation using Equation (2). The results obtained are presented in Figure 7. The PVC weight loss was sharp at the beginning of irradiation (50 h) and increased gradually and reached a maximum after 300 h of continuous irradiation. The PVC weight loss was highest (3.5%) for unmodified PVC. In the presence of complexes **1–4**, the PVC weight loss ranged from 1.7%–2.6% after 300 h of continuous irradiation. The PVC weight loss was lowest (1.7%) when complex **1** was used as the additive. The weight loss percentage for the PVC/**2**, PVC/**3**, and PVC/**4** blends was 2.3%, 1.9%, and 2.6%, respectively. Clearly, the complexes, and in particular the highly aromatic additives (complex **1** and **3**), enhanced the photostability of the PVC films significantly.

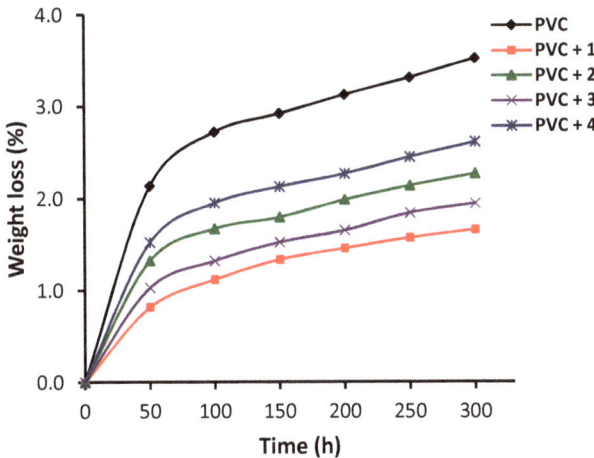

Figure 7. Changes in weight loss of PVC films.

3.5. Assessment of PVC Photodegradation Using Viscosity Average Molecular Weight ($\overline{M}_V$)

Photodegradation of PVC leads to a decrease in its molecular weight as a result of chain scission [48]. The viscosity of PVC in solution is used as a measure of the $\overline{M}_V$. The PVC films irradiated for different periods were dissolved in THF and their viscosity was measured using a viscometer [33]. The $\overline{M}_V$ for each film at different irradiation times (from 50 to 300 h) was calculated using Equation (3). The results are presented in Figure 8.

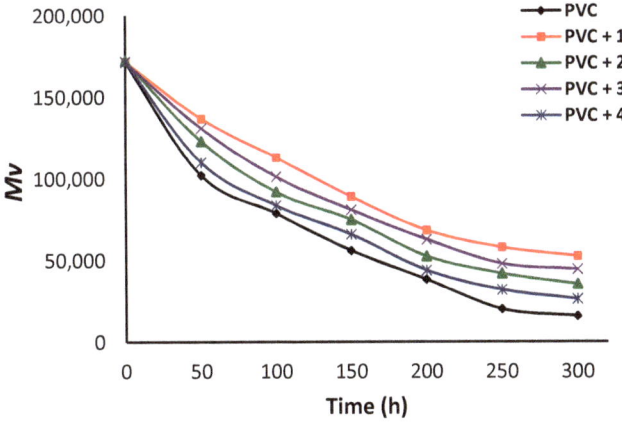

Figure 8. Changes in the $\overline{M}_V$ for PVC films.

A clear decrease in the $\overline{M}_V$ was observed during the irradiation process and was more pronounced for the blank PVC film. For example, the $\overline{M}_V$ for blank PVC was approximately 171,300 at the start of the irradiation process and declined to 78,800 after 100 h, to 38,300 after 200 h, and to only 15,700 at the end of the irradiation process (300 h). At the end of the irradiation process, the $\overline{M}_V$ for the PVC/**1**, PVC/**2**, PVC/**3**, and PVC/**4** blends was 52,600, 35,400, 44,600, and 26,300, respectively. The $\overline{M}_V$ for the blank PVC film decreased by more 90% after the irradiation process, compared with 70% when complex **1** was used as the additive. Clearly, the tin complexes, and in particular **1**, stabilized PVC against irradiation to a significant degree.

3.6. Surface Analysis of PVC

Scanning electron microscopy (SEM) can be used to provide less distorted, clear, and high-resolution images of the particles in materials, and is a useful technique for investigating the variation of the surface, particle size and shape, homogeneity, and cross sections within the blends [49–52]. The surface morphology of the PVC films was examined by field-emission scanning electron microscopy (FESEM). Figure 9 shows the FESEM images of the surface of the PVC film before and after irradiation. It has been reported that non-irradiated polymers have a smooth surface and a high level of homogeneity [19]. Indeed, the surface of the blank PVC film before irradiation was smoother, flatter, less lumpy, and more homogeneous compared with that obtained after irradiation. The roughness and cracks formed on the PVC surface after irradiation are the result of bond breaking within the polymeric chains and elimination of hydrogen chloride [53]. Figure 10 shows the FESEM images of the particles within the PVC films containing complexes **1–4** after irradiation, at two magnification powers. The surface of the PVC/**1** blend was more or less smooth and clean, similar to that for the non-irradiated PVC film. Clearly, complexes **1–4** were very effective in stabilizing PVC against irradiation. The other complexes provided some degree of stabilization to the PVC films.

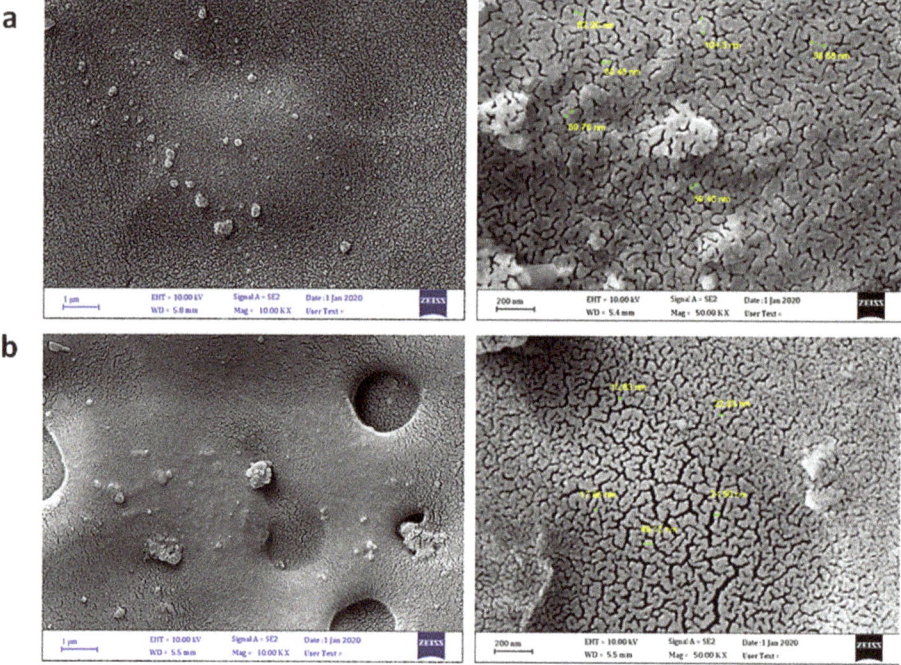

Figure 9. Field-emission scanning electron microscopy (FESEM) images of (**a**) PVC film before irradiation and (**b**) PVC film after irradiation.

The surface morphology of the PVC film was also observed using atomic force microscopy (AFM). AFM provides useful information about the features and roughness of the PVC surface. Irradiation of PVC for a long duration leads to bond breaking to produce a rough and broken surface [54,55]. Figure 11 shows the topographic AFM images (two- and three-dimensional) of the surface of the PVC films after 300 h of irradiation. The addition of complexes **1–4** clearly improved the photostability of the PVC films. The surface of the irradiated PVC blends containing complexes **1–4** was less rough than that of the blank PVC film. The roughness factor (R_q) for the PVC, PVC/**1**, PVC/**2**, PVC/**3**, and PVC/**4** films was 452, 61, 95, 80, and 112, respectively. Clearly, the use of complex **1** improved the roughness of the PVC film 7.4-fold. This improvement in R_q is better than that reported with many PVC additives (Table 4). Clearly, complexes **1–4** are better PVC photostabilizers compared with triazole-3-thiol Schiff bases [21], biphenyl tetraamine Schiff bases [22], naproxen tin complexes [30], melamine Schiff bases [24], 2-(4-isobutylphenyl)propanoate tin complexes [56], and furosemide tin complexes [28].

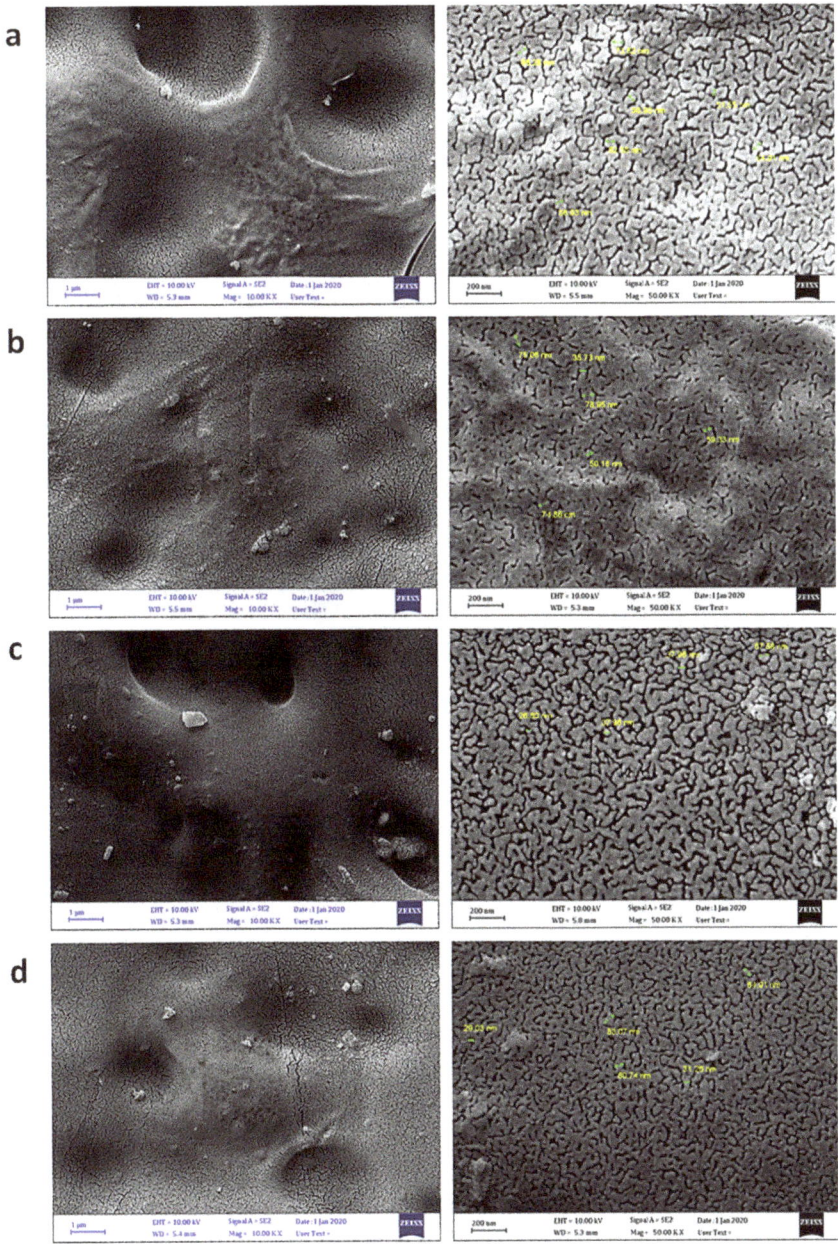

Figure 10. FESEM images of (**a**) PVC + **1**, (**b**) PVC + **2**, (**c**) PVC + **3**, and (**d**) PVC + **4** films after irradiation.

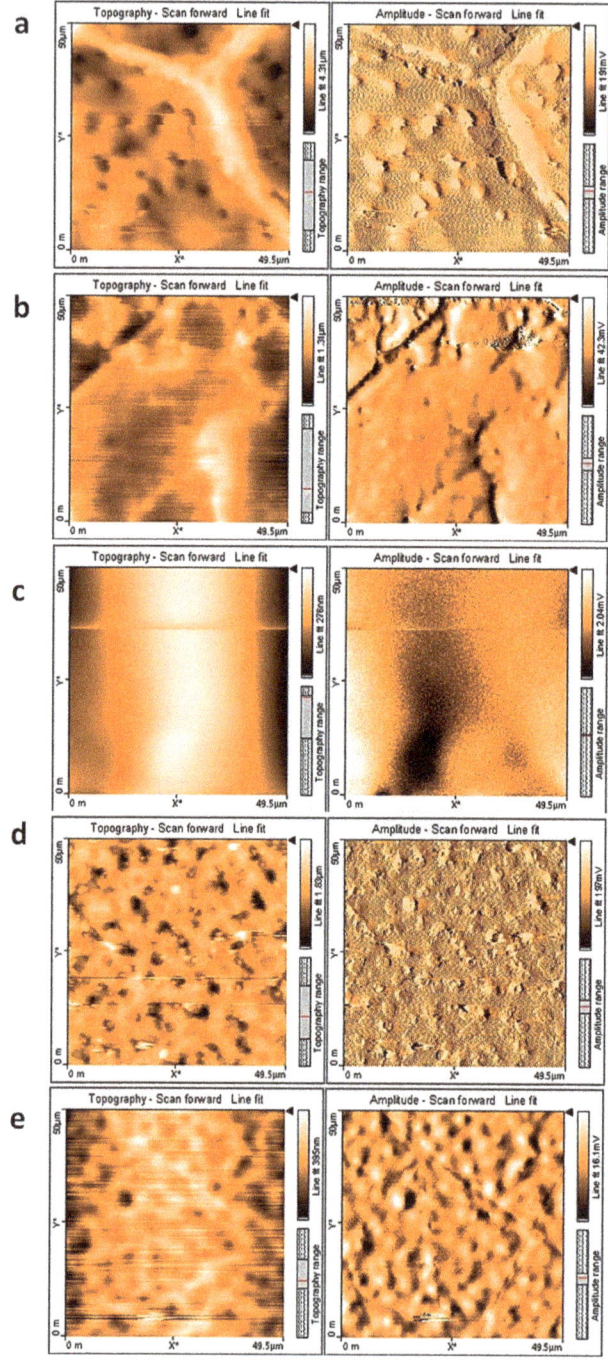

Figure 11. Atomic force microscopy (AFM) images of (**a**) PVC, (**b**) PVC + **1**, (**c**) PVC + **2**, (**d**) PVC + **3**, and (**e**) PVC + **4** films.

Table 4. Effect of additives on roughness factor (R_q; %) for poly(vinyl chloride) (PVC).

PVC Additive	Improvement in R_q (Fold)	Reference
Valsartan tin complex	7.4	[current work]
Furosemide tin complex	6.6	[28]
Ciprofloxacin tin complex	16.6	[29]
Naproxen tin complex	5.2	[30]
Telmisartan tin complex	9.4	[31]
2-(4-Isobutylphenyl)propanoate tin complex	6.2	[56]
Triazole-3-thiol Schiff base	3.3	[21]
Biphenyl tetraamine Schiff base	3.6	[22]
Melamine Schiff base	6.0	[24]
Polyphosphate	16.7	[18]

3.7. PVC Photostabilization Mechanisms

The use of tin complexes **1–4** as additives significantly reduced photodegradation of the PVC films. The effectiveness of the synthesized tin complexes as PVC photostabilizers followed the order: **1 > 3 > 2 > 4**. Complexes **1** and **3** are highly aromatic (phenyl moieties), while complexes **2** and **4** contain butyl substituents (aliphatic moieties). In addition, complexes **1–4** contain two aryl and one tetrazole ring within their skeletons. Therefore, these complexes are able to absorb UV light directly. After irradiation, the complexes re-emitted the absorbed radiation as a heat energy over a period of time at a rate that is not harmful to the PVC polymeric chains (Figure 12). The high-energy state of the tetrazole, aryl, or phenyl ring within complexes **1–4** can be stabilized by the resonance of the aromatic moieties [26].

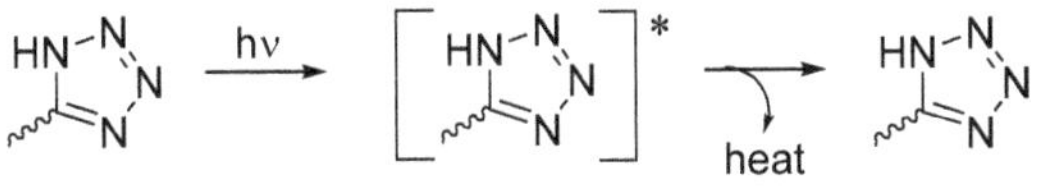

Figure 12. Function of tetrazole unit as a UV absorber.

Complexes **1–4** contain the tin atom, which acts as an acidic center. Tin abstracts the chloride ion from hydrogen chloride that is eliminated from the PVC chains upon irradiation to produce a stable substituted tin chloride (Figure 13; for triphenyltin complex **1**). Therefore, the tin complexes, and **1** in particular, are secondary PVC photostabilizers and act as hydrogen chloride scavengers [7,28].

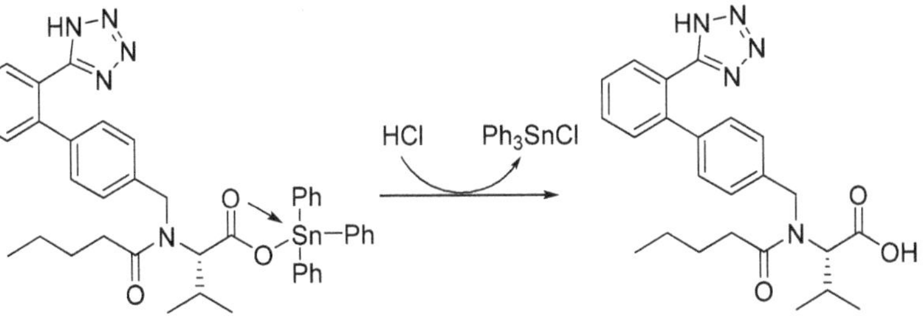

Figure 13. Function of complex **1** as a hydrogen chloride scavenger.

PVC undergoes photooxidation in the presence of oxygenated species such as hydrogen peroxides (PO_2H) [55,56]. The tin complexes can induce the decomposition of hydroperoxides by displacing the acidic tin atom within the additive (Figure 14; for triphenyltin complex **1**). This process inhibits PVC photodegradation significantly.

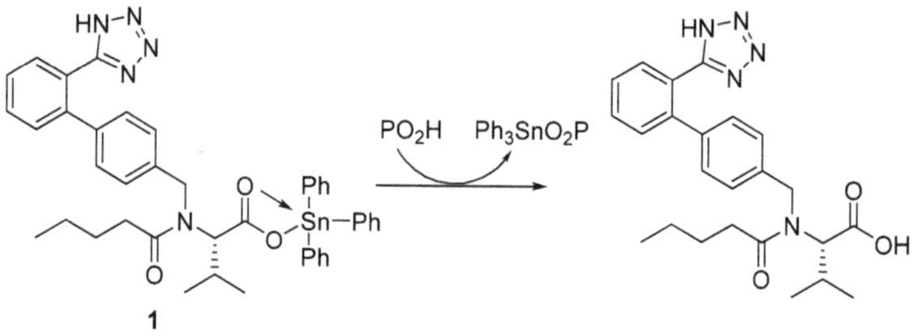

Figure 14. Function of complex **1** to induce hydroperoxide decomposition.

Peroxide radicals (POO$^\bullet$) have a negative impact on the PVC films and lead to the formation of various photooxidative products. The synthesized tin complexes can act as radical scavengers. The intermediate containing both the peroxide radical (chromophore) and the aryl moieties within the additives (Figure 15; for triphenyltin complex **1**) is highly stable via resonance [57]. Therefore, complexes **1–4** inhibit PVC photooxidation and provide a degree of stabilization against irradiation.

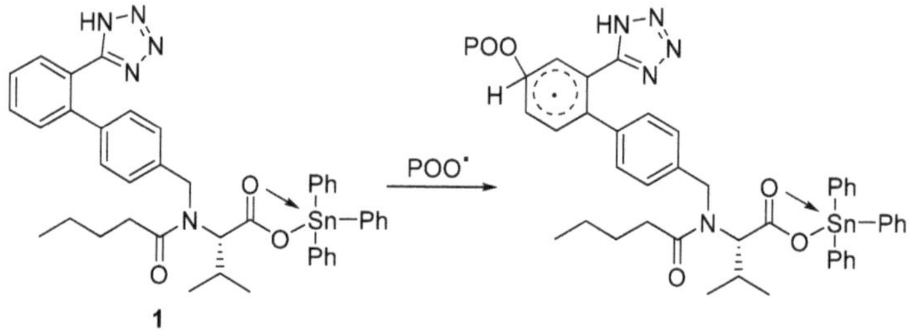

Figure 15. Function of complex **1** as a radical scavenger.

The polarity of the C–Cl bonds within the PVC chains and that of the nitrogen atoms in the tetrazole ring and the oxygen of carboxylate and amide groups might facilitate attractive interactions between PVC and the additives (Figure 16; for triphenyltin complex **1**). This attraction may augment energy transfer from the polymeric chains to the additives before the energy from photoirradiation can be dissipated [28]. However, this hypothesis does not take into account the complications due to steric hindrance within macromolecules.

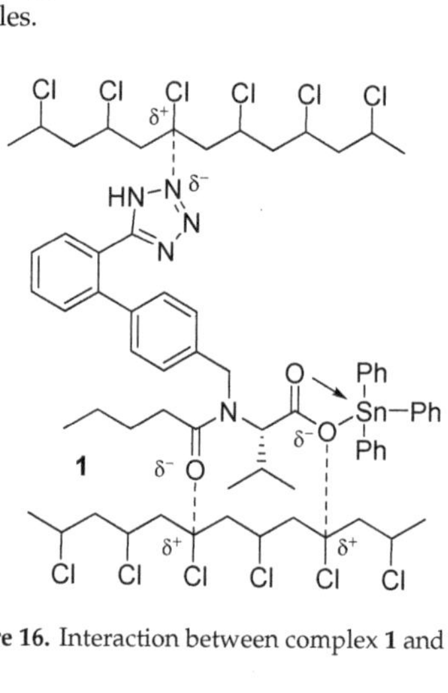

Figure 16. Interaction between complex **1** and PVC.

4. Conclusions

A number of tin complexes containing valsartan were synthesized in high yields using simple procedures. The synthesized tin complexes varied in the number and type of substituents. The chemical structures and elemental compositions of the tin complexes were established using various analytical tools. The effectiveness of the new tin complexes as photostabilizers for poly(vinyl chloride) films was tested after irradiation with ultraviolet light for up to 300 h. The undesirable changes in the carbonyl and polyene functional group indices, weight, and molecular weight of the polymeric films were significantly reduced in the presence of the tin complexes. The surface of the polymeric blends containing the tin complexes was smoother, flatter, and less rough than that of the blank film. Clearly, the synthesized tin complexes, and in particular those with a high level of aromaticity (phenyl derivatives), act as photostabilizers to protect the films against photodegradation. The main function of the tin complexes is to directly absorb ultraviolet irradiation, with subsequent slow release of the energy over time in a manner that does not degrade the polymeric chains and free radicals and hydrogen chloride scavengers. Valsartan tin complexes could be used as poly(vinyl chloride) photostabilizers in an industrial scale, but the production cost needs to be assessed.

Supplementary Materials: The following are available online at http://www.mdpi.com/2073-4360/12/4/969/s1, Figure S1: FTIR spectrum of **1**; Figure S2: FTIR spectrum of **2**; Figure S3: FTIR spectrum of **3**; Figure S4: FTIR spectrum of **4**; Figure S5: ^{1}H NMR spectrum of **1**; Figure S6: ^{1}H NMR spectrum of **2**; Figure S7: ^{1}H NMR spectrum of **3**; Figure S8: ^{1}H NMR spectrum of **4**; Figure S9: ^{119}Sn NMR spectrum of **1**; Figure S10: ^{119}Sn NMR spectrum of **2**; Figure S11: ^{119}Sn NMR spectrum of **3**; Figure S12: ^{119}Sn NMR spectrum of **4**; Figure S13: EDX graphs of blank PVC before irradiation; Figure S14: EDX graphs of PVC film after irradiation; Figure S15: EDX graphs of PVC film containing **1**; Figure S16: EDX graphs of PVC film containing **2**; Figure S17: EDX graphs of PVC film containing **3**; Figure S18: EDX graphs of PVC film containing **4**.

Author Contributions: Conceptualization and experimental design, G.A.E.-H., E.Y., A.A.A., D.S.A., and M.H.A.; experimental work and data analysis, A.M.; writing, G.A.E.-H., E.Y., and D.S.A. All authors discussed the results and improved the final text of the paper. All authors have read and agreed to the published version of the manuscript.

Funding: The authors are grateful to the Deanship of Scientific Research, King Saud University for funding through Vice Deanship of Scientific Research Chairs.

Acknowledgments: The authors thank Al-Nahrain University for technical support.

Conflicts of Interest: The authors declare no conflicts of interest.

References

1. Andrady, A.L.; Neal, M.A. Applications and societal benefits of plastics. *Phil. Trans. R. Soc. B* **2009**, *364*, 1977–1984. [CrossRef] [PubMed]
2. Crawford, C.B.; Quinn, B. *Microplastic Pollutants*, 1st ed.; Elsevier Science: Cambridge, UK, 2017.
3. Feldman, D. Polymer history. *Des. Monomers Polym.* **2008**, *11*, 1–15. [CrossRef]
4. Burgess, R.H. *Manufacture and Processing of PVC*; Elsevier Applied Science Publishers LTD: Cambridge, UK, 2005.
5. Titow, W.V. *PVC Plastics Properties, Processing, and Applications*; Elsevier Applied Science Publishers LTD: Cambridge, UK, 1990.
6. Yu, J.; Sun, L.; Ma, C.; Qiao, Y.; Yao, H. Thermal degradation of PVC: A review. *Waste Manag.* **2016**, *48*, 300–314. [CrossRef] [PubMed]
7. Folarin, O.M.; Sadiku, E.R. Thermal stabilizers for poly(vinyl chloride): A review. *Int. J. Phys. Sci.* **2011**, *6*, 4323–4330. [CrossRef]
8. Cadogan, D.F.; Howick, C.J. Plasticizers. In *Ullmann's Encyclopedia of Industrial Chemistry*; Wiley-VCH: Weinheim, Germany, 2000.
9. Gao, A.X.; Bolt, J.D.; Feng, A.A. Role of titanium dioxide pigments in outdoor weathering of rigid PVC. *Plast. Rubber Compos.* **2008**, *37*, 397–402. [CrossRef]

10. Chai, R.D.; Zhang, J. Synergistic effect of hindered amine light stabilizers/ultraviolet absorbers on the polyvinyl chloride/powder nitrile rubber blends during photodegradation. *Polym. Eng. Sci.* **2013**, *53*, 1760–1769. [CrossRef]
11. Karayıldırım, T.; Yanık, J.; Yüksel, M.; Sağlam, M.; Haussmann, M. Degradation of PVC containing mixtures in the presence of HCl fixators. *J. Polym. Environ.* **2005**, *13*, 365–379. [CrossRef]
12. Mohamed, N.A.; Yassin, A.A.; Khalil, K.D.; Sabaa, M.W. Organic thermal stabilizers for rigid poly (vinyl chloride) I. Barbituric and thiobarbituric acids. *Polym. Degrad. Stab.* **2000**, *70*, 5–10. [CrossRef]
13. Porta, M.; Zumeta, E. Implementing the Stockholm treaty on persistent organic pollutants. *Occup. Environ. Med.* **2002**, *59*, 651–652. [CrossRef] [PubMed]
14. Grossman, R.F. Mixed metal vinyl stabilizer synergism. II: Reactions with zinc replacing cadmium. *J. Vinyl Addit. Technol.* **1990**, *12*, 142–145. [CrossRef]
15. Li, D.; Xie, L.; Fu, M.; Zhang, J.; Indrawirawan, S.; Zhang, Y.; Tang, S. Synergistic effects of lanthanum-pentaerythritol alkoxide with zinc stearates and with beta-diketone on the thermal stability of poly(vinyl chloride). *Polym. Degd. Stab.* **2015**, *114*, 52–59. [CrossRef]
16. Fu, M.; Li, D.; Liu, H.; Ai, H.; Zhang, Y.; Zhang, L. Synergistic effects of zinc-mannitol alkoxide with calcium/zinc stearates and with β-diketone on thermal stability of rigid poly(vinyl chloride). *J. Polym. Res.* **2016**, *23*, 13. [CrossRef]
17. Wolf, R.; Kaul, B.L. Plastics, Additives. In *Ullmann's Encyclopedia of Industrial Chemistry*; Wiley-VCH: Weinheim, Germany, 2000. [CrossRef]
18. Ahmed, D.S.; El-Hiti, G.A.; Yousif, E.; Hameed, A.S. Polyphosphates as inhibitors for poly(vinyl chloride) photodegradation. *Molecules* **2017**, *22*, 1849. [CrossRef] [PubMed]
19. Alotaibi, M.H.; El-Hiti, G.A.; Yousif, E.; Ahmed, D.S.; Hashim, H.; Hameed, A.S.; Ahmed, A. Evaluation of the use of polyphosphates as photostabilizers and in the formation of ball-like polystyrene materials. *J. Polym. Res.* **2019**, *26*, 161. [CrossRef]
20. El-Hiti, G.A.; Ahmed, D.S.; Yousif, E.; Alotaibi, M.H.; Star, H.A.; Ahmed, A.A. Influence of polyphosphates on the physicochemical properties of poly(vinyl chloride) after irradiation with ultraviolet light. *Polymers* **2020**, *12*, 193. [CrossRef] [PubMed]
21. Yousif, E.; Hasan, A.; El-Hiti, G.A. Spectroscopic, physical and topography of photochemical process of PVC films in the presence of Schiff base metal complexes. *Polymers* **2016**, *8*, 204. [CrossRef] [PubMed]
22. Ahmed, D.S.; El-Hiti, G.A.; Hameed, A.S.; Yousif, E.; Ahmed, A. New tetra-Schiff bases as efficient photostabilizers for poly(vinyl chloride). *Molecules* **2017**, *22*, 1506. [CrossRef]
23. Shaalan, N.; Laftah, N.; El-Hiti, G.A.; Alotaibi, M.H.; Muslih, R.; Ahmed, D.S.; Yousif, E. Poly(vinyl chloride) photostabilization in the presence of Schiff bases containing a thiadiazole moiety. *Molecules* **2018**, *23*, 913. [CrossRef]
24. El-Hiti, G.A.; Alotaibi, M.H.; Ahmed, A.A.; Hamad, B.A.; Ahmed, D.S.; Ahmed, A.; Hashim, H.; Yousif, E. The morphology and performance of poly(vinyl chloride) containing melamine Schiff bases against ultraviolet light. *Molecules* **2019**, *24*, 803. [CrossRef]
25. Ahmed, A.A.; Ahmed, D.S.; El-Hiti, G.A.; Alotaibi, M.H.; Hashim, H.; Yousif, E. SEM morphological analysis of irradiated polystyrene film doped by a Schiff base containing a 1,2,4-triazole ring system. *Appl. Petrochem. Res.* **2019**, *9*, 169–177. [CrossRef]
26. Balakit, A.A.; Ahmed, A.; El-Hiti, G.A.; Smith, K.; Yousif, E. Synthesis of new thiophene derivatives and their use as photostabilizers for rigid poly(vinyl chloride). *Int. J. Polym. Sci.* **2015**, *2015*, 510390. [CrossRef]
27. Sabaa, M.W.; Oraby, E.H.; Abdel Naby, A.S.; Mohammed, R.R. Anthraquinone derivatives as organic stabilizers for rigid poly(vinyl chloride) against photo-degradation. *Eur. Polym. J.* **2005**, *41*, 2530–2543. [CrossRef]
28. Ali, M.M.; El-Hiti, G.A.; Yousif, E. Photostabilizing efficiency of poly(vinyl chloride) in the presence of organotin(IV) complexes as photostabilizers. *Molecules* **2016**, *21*, 1151. [CrossRef] [PubMed]
29. Ghazi, D.; El-Hiti, G.A.; Yousif, E.; Ahmed, D.S.; Alotaibi, M.H. The effect of ultraviolet irradiation on the physicochemical properties of poly(vinyl chloride) films containing organotin (IV) complexes as photostabilizers. *Molecules* **2018**, *23*, 254. [CrossRef] [PubMed]
30. Hadi, A.G.; Yousif, E.; El-Hiti, G.A.; Ahmed, D.S.; Jawad, K.; Alotaibi, M.H.; Hashim, H. Long-term effect of ultraviolet irradiation on poly(vinyl chloride) films containing naproxen diorganotin (IV) complexes. *Molecules* **2019**, *24*, 2396. [CrossRef] [PubMed]

31. Hadi, A.G.; Jawad, K.; El-Hiti, G.A.; Alotaibi, M.H.; Ahmed, A.A.; Ahmed, D.S.; Yousif, E. Photostabilization of poly(vinyl chloride) by organotin (IV) compounds against photodegradation. *Molecules* **2019**, *24*, 3557. [CrossRef]
32. Gaumet, S.; Gardette, J.-L. Photo-oxidation of poly(vinyl chloride): Part 2—A comparative study of the carbonylated products in photo-chemical and thermal oxidations. *Polym. Degrad. Stab.* **1991**, *33*, 17–34. [CrossRef]
33. Mark, J.E. *Physical Properties of Polymers Handbook*; Springer: New York, NY, USA, 2007.
34. Sharma, Y.R. *Elementary Organic Spectroscopy: Principles and Chemical Applications*; S. Chad & Company Ltd.: New Delhi, India, 2008.
35. Ibrahim, M.; Nada, A.; Kamal, D.E. Density functional theory and FTIR spectroscopic study of carboxyl group. *Indian J. Pure Appl. Phys.* **2005**, *43*, 911–917.
36. Alcock, N.W.; Culver, J.; Roe, S.M. Secondary bonding. Part 15. Influence of lone pairs on coordination: Comparison of diphenyl-tin (IV) and –tellurium (IV) carboxylates and dithiocarbamates. *J. Chem. Soc. Dalton Trans.* **1992**, 1477–1484. [CrossRef]
37. Mohammed, A.; Yousif, E.; El-Hiti, G.A. Synthesis and use of valsartan metal complexes as media for carbon dioxide storage. *Materials* **2020**, *13*, 1183. [CrossRef]
38. Pejchal, V.; Holeček, J.; Nádvorník, M.; Lyčka, A. ^{13}C and ^{119}Sn NMR Spectra of some mono-*n*-butyltin (IV) compounds. *Collect. Czech. Chem. Commun.* **1995**, *60*, 1492–1501. [CrossRef]
39. Shahid, K.; Ali, S.; Shahzadi, S.; Badshah, A.; Khan, K.M.; Maharvi, G.M. Organotin (IV) complexes of aniline derivatives. I. Synthesis, spectral and antibacterial studies of di- and triorganotin (IV) derivatives of 4-bromomaleanilic acid. *Synth. React. Inorg. Met Org. Chem.* **2003**, *33*, 1221–1235. [CrossRef]
40. Rehman, W.; Baloch, M.K.; Badshah, A.; Ali, S. Synthesis and characterization of biologically potent di-organotin (IV) complexes of mono-methyl glutarate. *J. Chin. Chem. Soc.* **2005**, *52*, 231–236. [CrossRef]
41. Blanco Jerez, L.M.; Rangel Oyervides, L.D.; Gómez, A.; Jiménez-Pérez, V.M.; Muñoz-Flores, B.M. Electrochemical metallization with Sn of (*E*)-4-((4-nitrobenzylidene) amino) phenol in non-aqueous media: Characterization and biological activity of the organotin compound. *Int. J. Electrochem. Sci.* **2016**, *11*, 45–53.
42. Farjamia, M.; Vatanpourb, V.; Moghadassi, A. Fabrication of a new emulsion polyvinyl chloride (EPVC) nanocomposite ultrafiltration membrane modified by *para*-hydroxybenzoate alumoxane (PHBA) additive to improve permeability and antifouling performance. *Chem. Eng. Res. Design* **2020**, *153*, 8–20. [CrossRef]
43. Gardette, J.L.; Gaumet, S.; Lemaire, J. Photooxidation of poly(vinyl chloride). 1. A re-examination of the mechanism. *Macromolecules* **1989**, *22*, 2576–2581. [CrossRef]
44. Bacaloglu, R.; Fisch, M. Degradation and stabilization of poly (vinyl chloride). V. Reaction mechanism of poly (vinyl chloride) degradation. *Polym. Degrad. Stab.* **1995**, *47*, 33–57. [CrossRef]
45. Jafari, A.J.; Donaldson, J.D. Determination of HCl and VOC emission from thermal degradation of PVC in the absence and presence of copper, copper (II) oxide and copper (II) chloride. *E-J. Chem.* **2009**, *6*, 685–692. [CrossRef]
46. Jiménez, A.; López, J.; Vilaplana, H.; Dussel, H.-J. Thermal degradation of plastisols. Effect of some additives on the evolution of gaseous products. *J. Anal. Appl. Pyrol.* **1997**, *40–41*, 201–215. [CrossRef]
47. Blazsó, M.; Jakab, E. Effect of metals, metal oxides, and carboxylates on the thermal decomposition processes of poly(vinyl chloride). *J. Anal. Appl. Pyrol.* **1999**, *49*, 125–143. [CrossRef]
48. Allcock, H.; Lampe, F.; Mark, J.E. *Contemporary Polymer Chemistry*, 3rd ed.; Pearson Prentice-Hall: Upper Saddle River, NJ, USA, 2003.
49. Alotaibi, M.H.; El-Hiti, G.A.; Hashim, H.; Hameed, A.S.; Ahmed, D.S.; Yousif, E. SEM analysis of the tunable honeycomb structure of irradiated poly(vinyl chloride) films doped with polyphosphate. *Heliyon* **2018**, *4*, e01013. [CrossRef] [PubMed]
50. Hashim, H.; El-Hiti, G.A.; Alotaibi, M.H.; Ahmed, D.S.; Yousif, E. Fabrication of ordered honeycomb porous poly(vinyl chloride) thin film doped with a Schiff base and nickel (II) chloride. *Heliyon* **2018**, *4*, e00743. [CrossRef] [PubMed]
51. Mehmood, N.; Andreasson, E.; Kao-Walter, S. SEM observations of a metal foil laminated with a polymer film. *Procedia Mater. Sci.* **2014**, *3*, 1435–1440. [CrossRef]
52. Nikafshar, S.; Zabihi, O.; Ahmadi, M.; Mirmohseni, A.; Taseidifar, M.; Naebe, M. The effects of UV light on the chemical and mechanical properties of a transparent epoxy-diamine system in the presence of an organic UV absorber. *Materials* **2017**, *10*, 180. [CrossRef] [PubMed]

53. Shi, W.; Zhang, J.; Shi, X.-M.; Jiang, G.-D. Different photodegradation processes of PVC with different average degrees of polymerization. *J. Appl. Polym. Sci.* **2008**, *107*, 528–540. [CrossRef]
54. Zheng, X.-G.; Tang, L.-H.; Zhang, N.; Gao, Q.-H.; Zhang, C.-F.; Zhu, Z.-B. Dehydrochlorination of PVC materials at high temperature. *Energy Fuels* **2003**, *17*, 896–900. [CrossRef]
55. Pospíšil, J.; Klemchuk, P.P. *Oxidation Inhibition in Organic Materials*; CRC Press: Boca Raton, FL, USA, 1989.
56. Mohammed, R.; El-Hiti, G.A.; Ahmed, A.; Yousif, E. Poly(vinyl chloride) doped by 2-(4-isobutylphenyl) propanoate metal complexes: Enhanced resistance to UV irradiation. *Arab. J. Sci. Eng.* **2017**, *42*, 4307–4315. [CrossRef]
57. Sabaa, M.W.; Oraby, E.H.; Abdul Naby, A.S.; Mohamed, R.R. *N*-Phenyl-3-substituted-5-pyrazolone derivatives as organic stabilizer for rigid PVC against photodegradation. *J. Appl. Polym. Sci.* **2005**, *101*, 1543–1555. [CrossRef]

© 2020 by the authors. Licensee MDPI, Basel, Switzerland. This article is an open access article distributed under the terms and conditions of the Creative Commons Attribution (CC BY) license (http://creativecommons.org/licenses/by/4.0/).

Article

Study of Modified Area of Polymer Samples Exposed to a He Atmospheric Pressure Plasma Jet Using Different Treatment Conditions

Thalita M. C. Nishime [1,2,*], **Robert Wagner** [1] **and Konstantin G. Kostov** [2]

1. Leibniz Institute for Plasma Science and Technology (INP), D-17489 Greifswald (MV), Germany; robert.wagner@inp-greifswald.de
2. Faculty of Engineering (FEG)—São Paulo State University (UNESP), Guaratinguetá (SP) 12516-410, Brazil; konstantin.kostov@unesp.br
* Correspondence: thalita.nishime@inp-greifswald.de

Received: 3 April 2020; Accepted: 23 April 2020; Published: 1 May 2020

Abstract: In the last decade atmospheric pressure plasma jets (APPJs) have been routinely employed for surface processing of polymers due to their capability of generating very reactive chemistry at near-ambient temperature conditions. Usually, the plasma jet modification effect spans over a limited area (typically a few cm^2), therefore, for industrial applications, where treatment of large and irregular surfaces is needed, jet and/or sample manipulations are required. More specifically, for treating hollow objects, like pipes and containers, the plasma jet must be introduced inside of them. In this case, a normal jet incidence to treated surface is difficult if not impossible to maintain. In this paper, a plasma jet produced at the end of a long flexible plastic tube was used to treat polyethylene terephthalate (PET) samples with different incidence angles and using different process parameters. Decreasing the angle formed between the plasma plume and the substrate leads to increase in the modified area as detected by surface wettability analysis. The same trend was confirmed by the distribution of reactive oxygen species (ROS), expanding on starch-iodine-agar plates, where a greater area was covered when the APPJ was tilted. Additionally, UV-VUV irradiation profiles obtained from the plasma jet spreading on the surface confirms such behavior.

Keywords: PET; polymer; plasma jet; tilted application; ROS distribution; UV; VUV

1. Introduction

Atmospheric pressure plasma jets (APPJs) have drawn much attention mainly due to their successful application for treatment of materials [1–3] as well as biological targets in so-called plasma medicine [4,5]. APPJs are capable of generating chemically rich plasma plumes in open air where ions, electrons, photons, and reactive species are transported to the target allowing the treatment of large and irregular surfaces [6]. They have a great flexibility concerning their construction designs and sizes allowing the generation from micro-scaled jets [7] to plasma plumes up to several centimeters in length [8]. Over the years, many different APPJ configurations have been reported in the literature, such as the ones summarized by Lu et al. [9] where different electrode geometries and excitation sources were highlighted. Depending on the electrode geometry, feeding gas, and power supply, APPJs can be used for a diversity of applications.

Regarding polymer treatment, active species generated in the APPJs are able to interact with only the material's surface while the bulk is kept unchanged. This characteristic is very important in applications that require biocompatibility and interphase processing, such as painting, dying, and composite manufacturing. During plasma treatment of polymeric surfaces, a series of interactions take place. The first is surface functionalization, where new functional groups are introduced onto the surface due to reactions with reactive species (like oxygen moieties) [10]. The second is surface etching or ablation, in which atoms and molecules of the polymeric chain are abstracted from the surface resulting in the formation of volatile compounds or short polymer chain fragments (low molecular weight oxidized materials) that can be easily washed away [1]. The third is cross-linking at the surface, where two or more parallel polymeric chains are bonded together. This can occur due to either ion bombardment or exposition to UV radiation [1,11]. Thus, after plasma treatment, the surface is modified and can exhibit new properties, such as improved adhesion [12] and wettability [2,13–15].

Cold APPJs are commonly applied for the treatment of heat sensitive surfaces, such as thermoplastic polymers [1,16]. Polyethylene terephthalate (PET) is one of the most common thermoplastic polymers used in the food packaging industry [17]. However, it exhibits a chemically inert surface with a low surface energy which leads to poor adhesion properties requiring pre-processing measures. Thus, plasma treatment can help altering the surface characteristics of PET by improving its wettability [12,18,19]. APPJs have been reported as a good alternative surface modification technique where pronounced water contact angle (WCA) reductions of polymers were observed [18,19]. This wettability enhancement was mostly associated with surface oxidation and the formation of polar groups observed by XPS and FTIR analysis [16,18,19]. Onyshchenko and coauthors [19] observed significant changes in oxygen content on a treated PET surface within a circular area with 10 mm radius centered at the plasma plume application point. Vesel and coauthors [18] preformed a detailed study of effective modification zone on PET samples using WCA mappings for different treatment times and distances. They observed that shorter treatment distances allowed the achievement of larger modified areas with lower WCA values while the same surface modifications could be obtained for greater distances, but much longer treatment times. The authors correlated the wettability modifications with the surface functionalization caused by long-living oxygen species.

However, plasma jet treatments usually provide a punctual and localized modification, which is interesting for some applications, but disadvantageous for the treatment of large surfaces and irregular objects. One alternative is the construction of an array or a matrix of plasma jets, where tens of APPJs can be operated in parallel, thus allowing the coverage of larger areas [20,21]. However, it has been reported that the simultaneous operation of many plasma jets can be challenging due to the strong electric repulsion between the neighboring plasma plumes resulting in plasma plume misalignment and, consequently, a non-uniform treatment [21]. Another alternative was reported by Mui et al. [22], where a plasma jet terminating with a horn-shaped nozzle was used for the treatment of PET samples. The 60-mm-diameter conical nozzle allowed a homogeneous treatment over the entire PET sample that lies inside the horn. The uniform modification of PET substrates placed vertically inside the reactor was also demonstrated. Although the obtained WCA reduction extends over a relatively large region, the treatment using such a device relies on the possibility of fitting the object inside the horn nozzle which is not always feasible. For instance, treatment of irregular structures with hollow cavities represents a great challenge. Therefore, a flexible plasma jet that can be used to reach regions of difficult access in the substrate while keeping a small treatment distance would be very important for the modification process. In previous papers we have reported the development and biomedical applications of a bendable plasma jet produced at the end of a flexible plastic tube [23–27]. By any means, the use of flexible devices does not ensure the possibility of keeping the conventional perpendicular arrangement (between the plasma plume and target) as it is commonly used for investigations. Thus, the study of surface modification using different incidence angles of a plasma plume is necessary.

Summarizing the literature data about plasma surface modification of polymers, it can be concluded that the treated area and its radial distribution depend upon many parameters, such as the jet design and power, the distance to the target, the gas flow rate, the treatment time, the plume incidence angle and the target condition (floating or grounded). Most previous works usually investigate the effect of some, but not all, of these parameters together (like plume incidence angle [28–30], distance to target [18,19], gas flow rate [31], and treatment time [18,19]) on the target modification process, but very few studies, if none, consider the effect of all parameter. More specifically the effect of target condition being floating or grounded has been frequently overlooked.

In the present work, the modification of PET samples using a malleable plasma jet was carried out. Due to its flexibility, the treatment was performed using different incidence angles between the plasma plume and the PET surface. A similar method to the one used by Vesel and coauthors [18] was adopted here for the WCA measurements. The surface modifications induced by a plasma jet launched parallel to the polymeric surface like in [31] were also investigated. For all studied configurations, the spreading of reactive oxygen species (ROS) on starch-iodine-agar plates and the irradiation profiles of UV and VUV from the plasma plume were correlated with the areas of changed wettability.

2. Materials and Methods

2.1. Plasma Jet

The plasma source used in this work was already described in previous studies [23–27]. It consists of a dielectric barrier discharge (DBD) reactor connected to a long and flexible plastic tube containing a floating metal electrode inside it. The wire upstream end pops up a few mm into the DBD reactor, while the downstream wire tip terminates a few mm before the plastic tube end. When a noble gas is fed into the primary DBD reactor and high voltage (HV) is applied, plasma is ignited, and the wire tip acquires the plasma potential. The gas is flushed through the polyurethane tube (12 Fr/Ch, Kangaroo^{TM} Nasogastric Feeding Tube, CardinalHealth, OH, USA) and due to the enhanced electric field at the downstream metal wire tip, a secondary (remote) plasma plume can be ejected from the plastic tube end. A schematic layout of the plasma jet device is shown in Figure 1a.

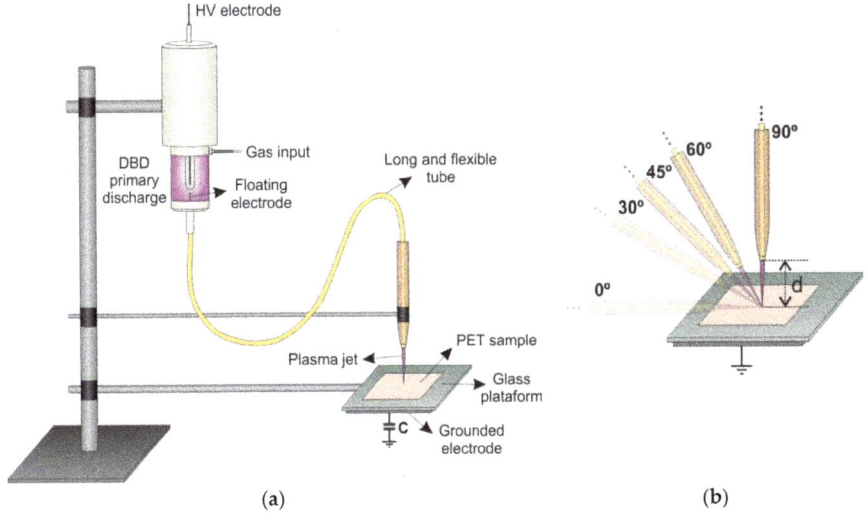

Figure 1. (**a**) Schematic setup of the plasma jet device used in this study and (**b**) arrangement of the plasma plume with different incidence angles. The distance between the tube tip and the substrate is assigned as d.

This plasma source was fed with helium with different gas flow rates up to 3.0 slm. The HV electrode was powered by a Minipuls4 AC power supply (GBS Elektronik GmbH, Radeberg, Germany) keeping the same parameters used for the previous biological applications [23,24] (voltage amplitude of 14 kV and frequency of 32 kHz). The voltage signal was amplitude modulated using the burst mode of operation with a duty cycle of 22%, which helps cool down the discharge region, avoiding damage to the plastic tube and treated targets. With a He flow rate of 2.0 slm and the given electrical parameters, the plasma plume has a length of approximately 15 mm. A serial capacitor (1.0 nF) was used for measurement of transferred charge to the grounded substrate.

2.2. Sample Preparation and Treatment

Polyethylene terephthalate (PET) samples with a thickness of 0.5 mm obtained from commercial plastic bottles were cut in squares (50 × 50 mm). All samples were cleaned in three steps using an ultrasound cleaner device. First, for 10 min using distilled water and detergent. Second, to remove organic impurities from the surface, they were rinsed with isopropyl alcohol (99.9% purity) for 20 min. Third, the samples were washed for 20 min with only distilled water to remove possible contaminant residues. All samples were dried in a controlled environment at room temperature. After cleaning, the PET samples were glued on glass microscope slides of 1-mm thickness using double-sided tape to ensure a completely flat surface for treatment.

The static PET samples were punctually treated for 60 s of exposure time, with a gas flow rate of 2.0 slm and with different plasma plume incidence angles of 90° (a normal incidence), 45°, and 0°. A schematic drawing of the treatment is presented in Figure 1b. The samples were placed underneath the plasma jet at distances (d) of 5 mm, 10 mm, and 15 mm with different target conditions. In the floating condition, no grounded metal electrode was placed underneath the 5.85-mm-thick glass platform, while for the grounded condition the grounded metal electrode was used. For the sample parallel treatment (0°), the plasma jet was placed 1 mm away from the PET surface and the gas flow rate was varied between 1.0 slm and 3.0 slm.

2.3. Water Contact Angle (WCA) Analysis

All treated samples were analyzed by water contact angle (WCA) measurements using the sessile drop method. The WCA measurements were performed in static mode using deionized water with droplets of 0.3 µL. The entire polymeric surface was covered by evenly distributed drops with 3 mm distance within one row and a space of 6 mm in between columns. A similar approach was adopted by Vesel et al. [18]. This method provides a 2D mapping of the treated surface wettability allowing the identification of the actual modification area caused by the plasma jet.

2.4. Starch-Iodine-Agar Plates for Reactive Oxygen Species (ROS) Detection

Starch-iodine-agar plates are commonly used as an indicative method for ROS distribution detection from APPJs where the color of reached areas changes to purple. In the presence of water, potassium iodide (KI) reacts with ROS generated in the discharge resulting in formation of iodine (I_2). This last combines with iodine ions (I^-) formed from the dissociation of potassium iodide producing triiodide ions (I_3^-) [32,33]. Triiodide ions can combine with the starch forming amylose complexes that result in changes of color (purple) enabling the local evaluation of the treated area.

To prepare the plates, 20 g of starch and 25 g of agar-agar were dissolved in 1 L of warm demineralized water. This solution was heated until boiling point and then cooled down to around 80 °C. Only after cooling, 20 g of potassium iodide was added. For experiment execution, Petri dishes were filled with the prepared solution until the starch-iodine-agar layer was around 4 mm thick, which corresponds to approximately 7 mL for 55 mm Petri dishes and around 12 mL for 90 mm ones.

The starch-iodine-agar plates were placed underneath the plasma plume with different distances between the tube tip and the agar surface (1 mm, 2 mm, 3 mm, 5 mm, 10 mm, and 15 mm), for different exposure times and different incidence angles (0°, 30°, 45°, 60°, 90°). In the parallel configuration (0°), the plates were used to evaluate the effect of gas flow variation on the ROS spreading.

After exposition to plasma, pictures of each plate were taken individually under the same illumination, distance, and exposure time conditions. Three different regions could be visualized on the agar surface, however, only the outermost reached zone was considered here. The evaluation was carried out using the free software ImageJ (U. S. National Institutes of Health, Bethesda, MD, USA). All pictures were converted to 32-bit grey scale images and the outermost spreading regions were detected by selecting 94% of the darker pixels within the Petri dish area. Each parameter variation was performed in triplicate. Due to the time delay between plasma treatment and photo shooting, the purple shade of ROS distribution zones cannot be visualized in the pictures presented here.

2.5. Ultraviolet (UV) and Vacuum Ultraviolet (VUV) Irradiation Profile Pictures

For a qualitative visualization of UV and VUV effective zones during plasma treatment, phosphor coated plates were used. Based on the fluorescence behavior for specific wavelengths, particular coatings were chosen to detect each kind of radiation. For UV detection a quartz plate with a $BaMgAl_{10}O_{17}:Eu^{2+}$ coating without binder was employed. The combination of this fluorescent dye and the quartz plate has a spectral sensitivity between 200 nm and 370 nm which allows the irradiation from OH and N_2 to pass through the coating filter with blue luminescence [34]. The VUV detection device consists of a deposited layer of Zn_2SiO_4: Mn also without binder on top of an MgF_2 optical window (to allow the VUV radiation to be transmitted). Thus, the spectral sensitivity of this device is between 175 nm and 225 nm, which allows the detection of the atomic nitrogen line at 174 nm with green luminescence [35].

For posterior analyses, pictures of the luminescent regions caused by the plasma jet positioning were taken with same illumination, distance, and exposure time conditions. Here, the plasma jet was placed in same positions as for the PET treatment.

3. Results and Discussion

3.1. Plasma Jet: Electrical Characterization

The long tube plasma jet used in this work has already been reported in previous studies [23–27]. In [27], where the same electrical parameters were applied, details of the discharge current and applied voltage waveforms are given as well as specific aspects from the voltage modulation using burst mode with a duty cycle of 22%. Figure 2a presents an optical emission spectrum of the He plasma plume produced at the end of the flexible plastic tube when placed at 5 mm from the substrate with a gas flow rate of 2.0 slm. The spectrum was acquired close to the sample and exhibits a predominance of excited nitrogen emission lines (N_2- second positive system and N_2^+- first negative system). Ionized nitrogen species are mostly produced by Penning ionization process via He metastables. The presence of excited atomic oxygen and OH confirms the generation of ROS in the plasma plume, which are commonly associated with polymer surface modification [18,19]. Some week emission lines of He can also be observed in Figure 2a.

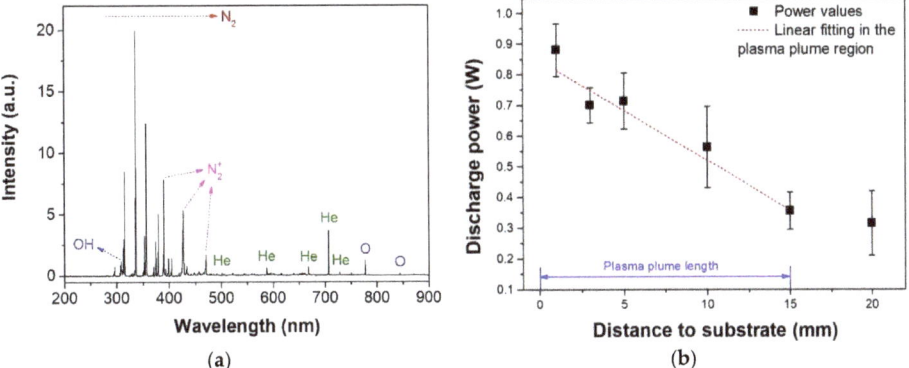

Figure 2. Discharge characterization. (**a**) Emission spectrum of the plasma plume close to the substrate at a distance of 5 mm from the nozzle and (**b**) plasma plume power as a function of the distance between the tube tip and the grounded substrate.

As already described previously [27], the discharge power from one burst can be obtained from the Q-V Lissajous figure and, for the presented plasma jet, the total power measured corresponds to the power of both discharge regions: primary DBD and secondary plasma plume. For treatments, only the discharge power correspondent to the plasma plume plays a role. In [27], the method used for calculating the discharge power values for each discharge region is explained. Figure 2b presents the plasma plume power for different distances between the tube tip and the PET sample placed on top of a grounded platform. The plasma jet was operated here with a gas flow rate of 2.0 slm and, therefore, had a length of around 15 mm, as indicated in Figure 2b. It can be observed that this plasma jet device works with low power (below 1.0 W) which contributes with avoiding heating damages to the sample. Figure 2b also shows that higher power values were measured for smaller distances. For the region where the plasma plume touches the PET sample (until 15 mm), a linear dependence of the power with distance was obtained, after which no further reduction was detected. This result indicates that the use of a grounded platform influences the plasma jet increasing the power and leading to a more intense discharge.

3.2. Normal (90°) Incidence: Distance and Substrate Regime Effect

Among the processes that promote surface modification in polymers, the attachment of new radicals generated in the discharge also play a role [36]. Such chemical changes and roughness modifications alter the material characteristics and lead to changes in the polymer wettability. Reactive oxygen species (ROS) generated in the plasma plume, e.g., O, O_3, and OH, are able to diffuse radially where long-lived species, like ozone, can reach longer distances and interact with the polymer surface. Thus, starch-iodine-agar plates were used to identify the maximum spreading area of ROS on the surface due to plasma treatment. Here, the plates were placed at 5 mm from the tube tip (forming 90° with the surface). Figure 3 shows the area variation of spreading ROS when the plasma exposure time is increased (up to 120 s). It can be observed that the affected area on the starch-iodine-agar plates increases with the treatment time tending to achieve saturation. In order to visualize the modified area on PET samples after plasma treatment, an exposure time of only 60 s was chosen. Thus, the treated areas could be easily identified by WCA measurements on 50 × 50 mm polymer samples.

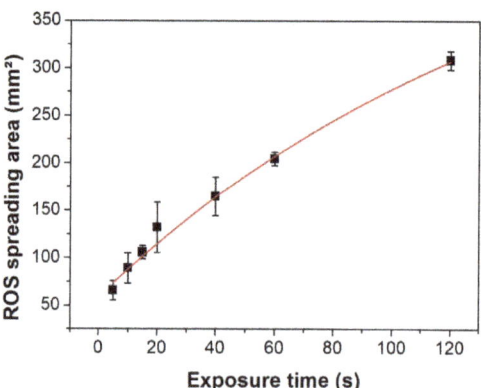

Figure 3. Area covered by ROS on starch-iodine-agar plates for different treatment times and a gas flow rate of 2.0 slm.

In similar conditions, PET samples (50 × 50 mm) were arranged underneath the plasma jet (at the perpendicular position). The polymer was treated for 60 s with a He gas flow rate of 2.0 slm. Figure 4a exhibits the 2D wettability profiles of PET samples measured soon after treatment for different distances: 5 mm, 10 mm, and 15 mm, where the green color corresponds to the untreated regions with WCA of around 90°. In addition, the treatments were performed in two different substrate conditions: floating (right column of Figure 4a) and grounded (left column of Figure 4a). It can be seen that in all cases the modified areas are approximately round, where the major WCA reduction is centered just below the plastic tube exit and from there the WCA values increase radially outwards. Moreover, the treated areas were much larger than the tube diameter (2.5 mm), where a 12-times greater diameter was obtained for the grounded jet at 5 mm (Figure 4a, left column, top map). On the other hand, for the greatest tested distance (15 mm) in Figure 4a for the floating target (right column bottom map), only a six-time greater diameter zone was achieved. The most pronounced WCA drop was observed for the grounded substrate condition, in which case the minimal WCA within the treated area reached about 33° after 60 s of treatment. Besides the lowest contact angle, the grounded jet condition also leads to a more homogeneous WCA distribution.

The area values for all WCA maps from Figure 4a are summarized in Figure 4b. The WCA reduction areas outlines were determined as the regions with around 10% of value reduction (~80°). It is evident in Figure 4b that the use of grounded target condition leads to formation of larger modified areas, where a difference of approximately 30% was obtained when comparing grounded and floating substrate conditions for 5 mm and 10 mm distances. This difference increases to around 50% when comparing both conditions at 15 mm. Therefore, the best results obtained correspond to the grounded target condition for all tested distances. This improvement in achieved area for shorter distances and grounded target are related to the larger treatment dose caused by the increase in power when the discharge is disturbed by the grounded electrode, shown in Figure 2b. Moreover, with shorter distances the plasma species can reach the surface faster and, in case of long-lived species, act for a longer time. The same effect of lesser degree of surface modification induced by the plasma jet at longer distances was observed in a study performed by Vesel and coauthors [18], where PET was treated using seven different distances between 2 mm and 40 mm. Therefore, the data presented in Figure 4 is in good agreement with the WCA behavior observed by these authors.

ROS are generated in the post-discharge region of APPJs due to air mixing into the plasma plume and they are pointed out as the major agent responsible for PET surface modification [18]. Figure 5a presents pictures of the zones affected by ROS spreading on starch-iodine-agar plates for different distances and grounded platform condition, as performed for the polymer treatment in Figure 4a, left column. In comparison, images of the irradiation profiles of UV (blue) and VUV (green) are also displayed in Figure 5a at the same scale (20 × 20 mm). Differently from the trend observed for the PET treatment, the ROS seem to spread further away from the plasma plume, especially for longer distances. It is important to notice that properties of the two surfaces are very distinct once the starch-iodine-agar plates feature a soft and much more humid surface than the PET polymer, which is solid and dry. Therefore, at shorter distances the He flow deforms significantly the surface creating an indentation (see Figure 5a top line) which, in turn, can perturb the admixture of ambient air into the plume as well as the radial spreading of ROS by the buoyancy effect [37]. In the case of bacteria disinfection on agar plates, Goree and coauthors [38] observed that shorter treatment distances led to a non-homogeneous ring-shaped zone, indicating a non-uniform radial spreading of ROS. Another factor that should be considered for the starch-iodine-agar plates experiments conducted at shorter distances is a discharge intensification with a corresponding plume temperature rise caused by the agar humidity. Furthermore, high temperature and humidity presence are known for hindering the production of ozone, which contributes to the smaller areas observed in the starch-iodine-agar plates test [39]. Thus, due to these important differences between two target surface characteristics, it is expected that the starch-iodine-agar plates experiment cannot be directly compared to the polymer modification at short tested distances once the discharge is intensified and the surface is deformed at this condition. However, for long treatment distances one can expect at least some similarities concerning the size of modified area on both different targets. For instance, for 15 mm treatment, the starch-iodine agar plate and the PET samples exhibit modification areas with similar sizes. Such differences among surfaces is also expected for the treatment of, for example, samples with complex geometries when compared to the flat thin PET substrate, where, for the last, larger modification areas are expected once the plasma species can easily spread on the surface reaching further. Additionally, the treatment of irregular surfaces using a grounded condition might be difficult depending on the sample geometry and its treatment can lead to inhomogeneous modification caused by surface irregularities. However, the use of flat PET samples can be used as a model for surface modification investigations using different processing parameters.

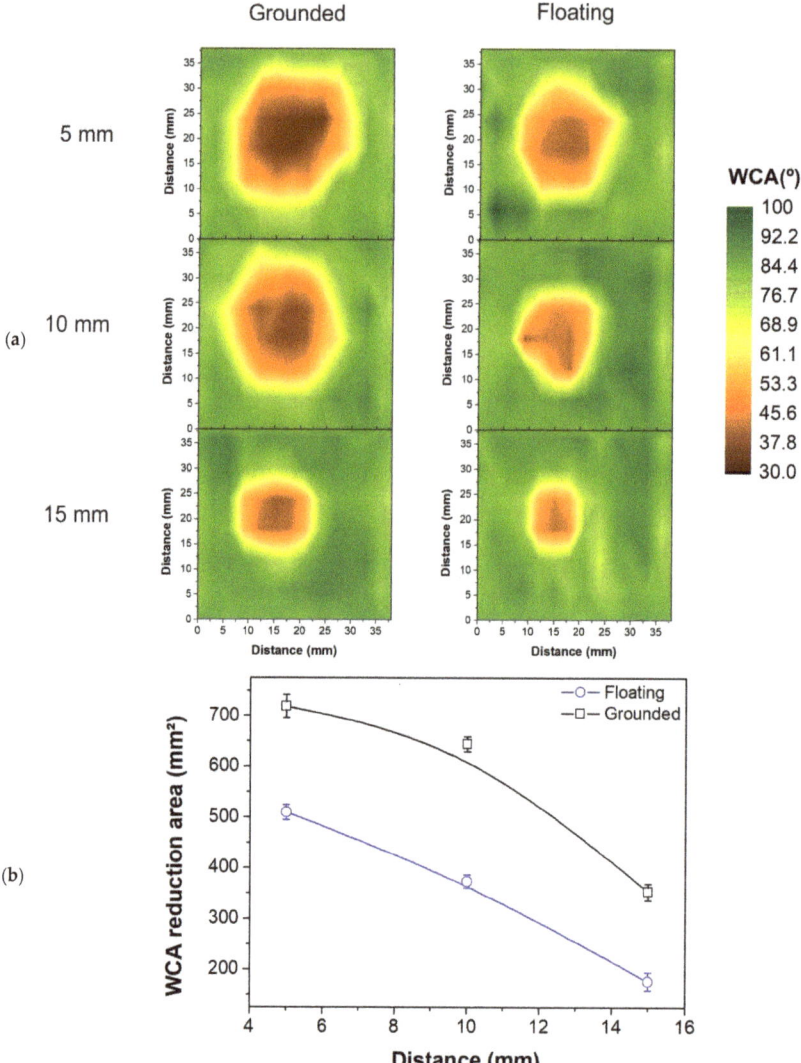

Figure 4. WCA profile of PET samples for a perpendicular (90°) treatment using different distances between the tube tip and the polymer. The treatment was performed under two different substrate conditions: grounded (left column) and floating (right column). The areas of WCA reduction from (**a**) are summarized in (**b**).

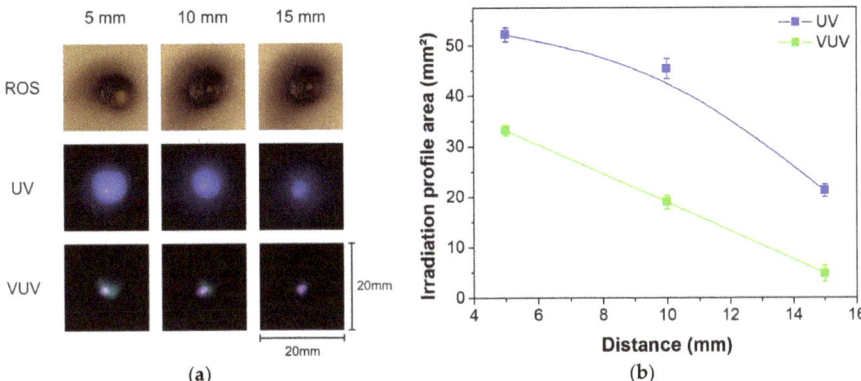

Figure 5. (a) Pictures of the affected zones on the starch plates in same scale as UV and VUV irradiation profiles for different distances between the tube tip and the surface and (b) calculated areas from the UV and VUV profiles exhibited in (a).

It is well known that UV and VUV radiation generated in plasma jets can cause crosslinking on polymeric surfaces [40]. In He plasma jets, the emissions in UV and VUV come from Penning ionization processes involving admixed air molecules in the plasma plume. As observed in [41] for treatment of PET, the amount of UV radiation produced by a plasma jet is not enough to produce large modifications. However, UV and VUV photons have sufficient energy to break weak bonds at the polymer's surfaces, thus benefiting oxidation and cross-linking processes [42]. Comparing the sizes of the profiles from Figure 5a, it is clear, that substantial amount of the analyzed plasma agents do not reach as far radial distances as observed for the wettability distribution of treated PET in Figure 4. However, as was already concluded by several authors [18,19,42] the simultaneous action of these active plasma species (even in a small amount) can lead to a strong synergetic effect greatly enhancing the modification potential of the plasma jet. The calculated areas for irradiation profiles of UV and VUV of Figure 5a are presented in Figure 5b. The characteristic size and area of both, the UV and VUV irradiation profiles, tends to diminish when the distance to the plasma plume is increased, as observed for the PET modification in Figure 4b. Additionally, the detected VUV emission for the longest distance of 15 mm is negligible which suggests that, in case of the applied He plasma jet, VUV emission does not play a crucial role.

3.3. Incidence Angle Variation: Effective Treatment Area Modification

In most applications plasma jets are normally kept in a perpendicular position to the target, as for the polymer treatments shown in Figure 4. However, in some cases, like the treatment of irregular surfaces, a manipulation of the plume may be required to reach regions of difficult access. Thus, the remote plasma jet system presented in this study is a good alternative for those cases, allowing an easy handling of the plasma plume enabling the device to be comfortably and safely bent. This manipulation flexibility is also extremely important when the plasma jet is applied in the biomedical field, for instance for treatment inside a patient's mouth or internal organs. Therefore, the treatment of PET samples and Petri dishes with starch-iodine-agar carried out in this work can be considered as a model study of effective treatment zone variation as a function of the plume incidence angle. To do so, the plasma jet generated at the end of the plastic long tube was directed to the surface using different tilting angles, such as the normal 90° already presented, 45°, and parallel to the surface (0°). Figure 6 shows the 2D WCA mappings for PET samples treated with the plasma jet placed at three different incidence angles for the two different target conditions: grounded platform (WCA left column) and floating condition (WCA right column). The 90° color maps (top line) are the same presented in the Figure 4a for 5 mm distance. For the 45° case, the tube end was positioned 5 mm above the sample center and the plasma plume expands rightward (as demonstrated in the schematic setup of Figure 1b). On the other hand, for parallel plasma plume application (0°) and distances higher than 3 mm, apparently the plasma does not affect the polymeric surface [31]. Thus, in this case, the distance to the PET surface was set to 1 mm, and the tube end was positioned at sample central line close to the sample's left side as marked in black in the contour maps (in Figure 6 the plasma jet expands rightward). For all three incidence angles, the He flow rate was set to 2.0 slm and the PET samples were exposed to plasma for 60 s.

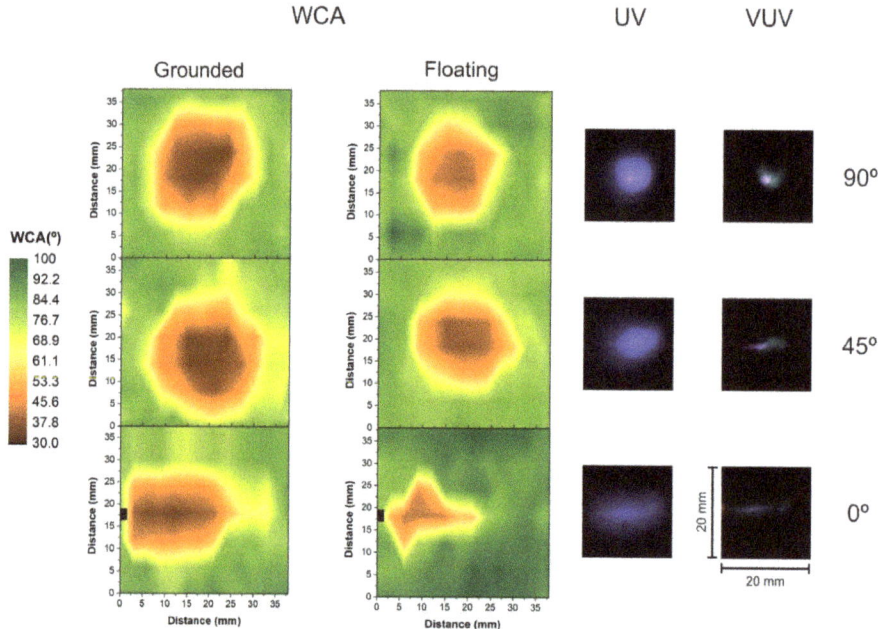

Figure 6. WCA color maps of treated PET for 60 s and three different plasma plume tilting angles, 90°, 45° and 0°, in grounded (left column) and floating (right column) condition. The plasma plume was directed to the right-hand side of the contour maps and the tube tip position is marked in black for the 0° cases. UV (blue) and VUV (green) emission profiles are also displayed for each condition in the same scale as the WCA maps.

Comparing the WCA reduction zones for grounding and floating conditions of Figure 6, it is clear that the modified areas for the three different incidence angles treated with grounding condition (WCA left column) were larger than for the floating one (WCA right column). The obtained difference for the normal incidence angle (90°) and 45° was around 30% while, for the parallel treatment (0°), the implementation of a grounded electrode underneath the dielectric platform led to an increase of approximately 60% of the PET modified region. Similarly to the observations in Section 3.2, again the grounding target case led, for all cases of the Figure 6, to the formation of larger and more homogeneous modification regions. For the plasma jet leaning at 45°, a difference in the treated zone shape can be noticed for the two target conditions, where a more elliptically-shaped WCA reduction zone was obtained for the floating platform and a rounder treated region was measured for the grounded case. The shape difference and larger uniform region obtained for the grounding condition were even more pronounced for the parallel (0°) treatment, where a much smaller and completely inhomogeneous modification area was measured for the floating case presented in the right column of WCA in Figure 6.

Tilting the plasma jet for polymer treatment can clearly increase the treated area. The treatments with inclination of 45° (grounded and floating target conditions) presented modification areas around 20% bigger than the ones performed perpendicularly (90°). Another aspect that should be considered is the minimum contact angle value that was achieved. For the grounded condition, the minimum WCA value obtained was around 33°, but no substantial difference in the area of the minimum contact angle was noticeable here between 90° and 45°, while for the floating condition, the treatment led to minimum contact angle value of around 40°, where the 45°-leaning treatment parameter presented a clearly larger zone of the same degree of surface modification. In both cases (grounding and floating) a saturation of the minimum contact angle seems to have been reached. A similar effect was detected by Vesel and coauthors [18] for the increase of treatment time where a wider modification region was obtained with same saturated minimum WCA for longer exposure times. Here, similar effect could be obtained by simply tilting the plasma jet and keeping a short treatment time of 60 s. Comparing the total modified areas on PET samples for the different angles, an increase of around 20% was obtained for the 45°-tilted plasma jet treatments for grounded and floating conditions when compared to the traditional 90° treatment configuration. Damany and coauthors [29] studied the role of the incidence angle of an argon plasma jet for the desorption of organic molecules. The obtained results regarding the affected area are in accordance with the ones presented here. Comparing the tested angles of 90° and 45° to the dielectric surface, the 45° condition was the treatment that led to the wider deposition area of bibenzyl. The authors also observed that tilting the plasma jet and varying the gas flow rate strongly modifies the spatial extension of the developed discharge at the dielectric surface [29].

Further tilting the plasma jet can also be interesting for some applications. Placing a several-cm-long plasma plume parallel to the polymer increases the interaction area [31] and it can help in situations in which the surface needs to be scanned for uniform treatments. For the parallel treatment (0°) presented in Figure 6, the treated area is similar to the one of other two incidence angles (20% smaller than the one obtained for the treatment at 45° but comparable to the 90° case). Additionally, the obtained minimum contact angle is as low as the other tested angles confirming the efficiency of the parallel jet application for surface modification. Differently from the grounded case, the parallel treatment with a floating target condition led to a completely inhomogeneous modification area around 40% smaller than the treatment at 90°. Therefore, the elongated shape of the homogeneous WCA reduction obtained in the grounded condition could be useful for providing a more uniform treatment by one-dimensional scanning of the polymeric surface.

Pictures of irradiation profiles of UV (blue) and VUV (green) are presented next to the WCA color mappings in Figure 6. The scale of all images is the same for better comparison. For the 45° and 0° configurations, the plasma jet tube is leaning to the left-hand side as for the contact angle maps in the left corner in the Figure 6 (also as schematized by Figure 1b). Basically, the UV profiles of Figure 6 present similar shapes to the WCA measured on the treated PET, however they span over smaller areas (the UV distribution being always wider than the VUV one). For the plume incidence angle of 45° the UV distribution exhibits an elliptically-shaped profile that is similar, but with smaller size in comparison to the polymer surface modification. The area of this elliptical profile is around 27% bigger than the profile obtained for the 90° configuration, similar to the increase observed for the WCA measurements. However, the VUV irradiation profiles for 90° and 45° exhibited very peculiar shapes that can be hardly correlated with the WCA patterns. Therefore, the VUV radiation produced by the He plasma jet for distance of 5 mm can be considered as a minor agent in the PET surface modification. However, for the 1 mm distance used for the 0° condition, the VUV emission profile can also be correlated to the modified zone. Thus, the VUV radiation produced by the He plasma jet seems to contribute for PET surface modification only at extremely short treatment distances.

Figure 7 exhibits the area on the starch-iodine-agar plates over which the ROS spread for different incidence angles and distances to the surface. Once again, similar behavior like the one in Figure 5 (ROS spreading area increases when the plasma jet is placed further away from the surface) could be observed. From the data presented in Figure 7, it is possible to notice a tendency of rapid area increase when the angle between the plasma jet and the surface is decreased. This trend is in accordance with the area raise obtained for PET treatment and showed by the WCA reduction. The increase in affected area for decreasing incidence angle was observed for all tested distances between 1 mm and 5 mm. For the distances of 2 mm and 3 mm, the affected areas for 0° were very small so that they were not considered. Above 3mm distance, no color modification was observed on the starch-iodine-agar plates for the parallel jet configuration. It is relevant to point out the distinct zone shapes produced on the starch-iodine-agar plates for each jet incidence angle. The pictures of the plates exposed to the parallel plasma jet at 1 mm distance and different incidence angles are displayed in Figure 8 (flexible tube leaning to the left-hand-side). It can be observed that, for the normal jet incidence (90° position), a roughly round figure was obtained. For tilted plume application the WCA patterns gradually evolve into oval-shaped profiles with the rounder part close to the inclination side (side A in Figure 8), that is, close to the lower side of the plastic tube tip. Along the direction of the plasma plume expansion (pointed out as side B in Figure 8), a wider and more diffuse ROS spreading, indicated by the lightest color modification, could be detected. Another interesting observation for the 90° and 60° treatment positions is the formation of a round region (around 4.5-mm-diameter) just where the plasma plume touches the agar, in which no color change was detected. The formation of these ROS-depleted regions at short distance and close to normal incidence can be explained by a strong backward He flow produced by gas reflection from the surface, which causes a blocking effect to the surrounding air molecules. They are hindered from diffusing into the effluent, leading to a drastic reduction of ROS formation. Starting from 45° jet inclination this effect is strongly reduced and probably an efficient gas mixing is obtained in the plasma effluent. The shape of ROS affected area evolves from an oval to elliptical pattern for the 30° treatment. In this case a darker spreading zone, when compared to the larger angles, suggests that more ROS are produced and they are able to propagate further. Finally, the parallel position of the plasma jet (0°) led to formation of a very elongated ROS pattern in a shape similar to the plasma plume. The shapes of spreading reactive oxygen species can be correlated to the ones obtained in the WCA analysis and to the UV irradiation profiles for 0°, 45°, and 90°.

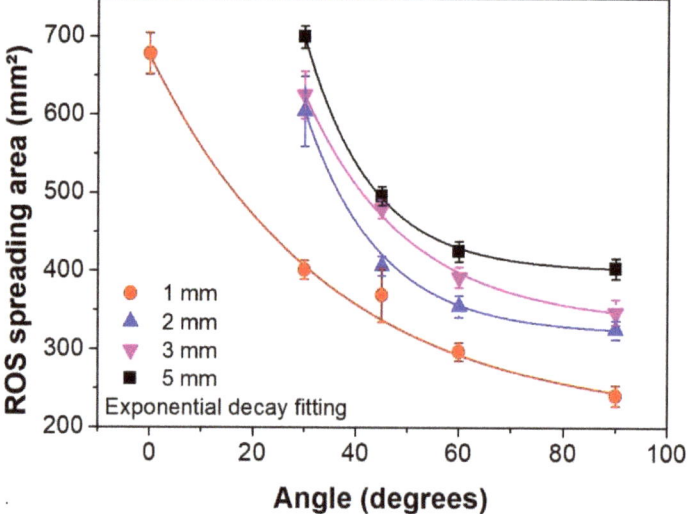

Figure 7. ROS spreading area detected on starch-iodine-agar plates exposed to a plasma jet positioned with different distances (1 mm, 2 mm, 3 mm, and 5 mm) from the agar surface and with different incidence angles (0°, 30°, 45°, 60°, and 90°).

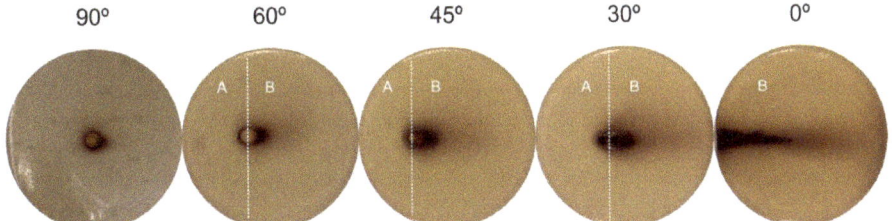

Figure 8. Visualization of the regions on starch-iodine-agar plates (90-mm-diameter) affected by ROS spreading due to plasma jet exposure (60 s) at 1 mm distance with different incidence angles. Here, the plastic tube end is leaning to the left-hand side of the Petri dishes as schematized in Figure 1b. "A" stands for the side of tube inclination and "B" for the plasma plume propagation side. The pictures shown here correspond to the red curve of Figure 7.

3.4. Parallel Treatment (0°): Gas Flow Variation

The easy handling characteristic of this plasma jet device can be especially interesting when it comes to surface modification and deposition once these applications require short distances in order to obtain good results. A good example of treatment of irregular objects is the modification of the inner surface of plastic tubes for surface activation studied by Prysiazhnyi et al. [3]. However, as mentioned above, for surface scanning purposes, applying the plasma plume in parallel can be advantageous and then, in this case, adjustment of the applied parameters is necessary. In this configuration (0° between the plume and the surface), regulating the gas flow rate can lead to a more homogeneous modification and also to larger treatment areas. In [31] the authors investigated the surface modification of polystyrene (PS) by side-on plasma jet under the floating target condition and found that the gas flow rate was an important parameter. The data in Figures 4 and 6 show that the grounded target condition always led to better surface modification (wider treated area and lower WCA). Moreover, the comparison between floating and grounded conditions for the parallel treatment (0°) showed a considerable difference of 60% larger and a more homogeneous surface modification area for the grounded case (Figure 6). Thus, all further treatments using parallel plume application were made in this condition (grounded).

The 2D WCA mappings for the parallel (0°) treatment of PET samples are presented in Figure 9. Each letter in this figure corresponds to a different He gas flow rate from 1.0 slm (a) to 3.0 slm (e) in ascending order. To compare with the contour maps, pictures of the UV and VUV irradiation profiles are shown at the same scale. In all pictures exhibited in Figure 9, the parallel plasma plume is directed to the right, where the position of the plastic tube end is marked by a black rectangle in the WCA maps. Olabanji and Bradley [31] reported the presence of untreated zones close to the plasma jet nozzle for high gas flow rates when treating PS in the floating condition. Differently from what they observed, in all cases the area of the PET samples modified by the plasma plume and grounded condition has an elongated (roughly elliptical) shape that starts at the plastic tube end and extends rightward. This difference might lie on the different substrate regimes studied in each case (floating target in [31] and grounded substrate here). Observing Figure 9 it is evident that the longitudinal extension of the surface modification varies with the gas flow rate, while its transversal size stays more or less unchanged. The white dotted line on top of the WCA contour maps shows the limit of the treated length for the 1.0 slm condition. It is clear that the length of the modified area increases until the gas flow rate of 2.0 slm (indicated by the dashed black line) after which it starts decreasing. Such a decrease in the modification length of the polymer surfaces for the increased gas flow rate was also observed for Olabanji and Bradley [31]. A similar trend is observed for the treated areas whose values increase approximately 15% when raising the gas flow rate from 1.0 to 1.5 slm or 2.0 slm. For pure helium flow, in such conditions, the obtained Reynolds number is quite low: Rn between 73 and 217 for 1.0 slm and 3.0 slm, respectively. The obtained values are also much lower than the critical Reynolds number of Rn = 2000 that indicates the laminar-turbulent transition [43]. Thus, the He flow is expected to be laminar for all chosen conditions. However, the target presence in the close vicinity to the He gas flow may affect it. Additionally, with plasma on, the fluid dynamics change completely, and a turbulent condition can be much more easily achieved. It was observed by Winter and coauthors [44] with shadowgraph images that a laminar He flow from a plasma jet device can become turbulent when the plasma is ignited. Thus, the shortening of the plasma plume length above 2.0 slm can be explained as the gas flow transition from laminar to turbulent regimes. Indeed, the transition seems to occur at 2.0 slm, where even though the modification length and area are as large as for 1.5 slm, an extra region of low WCAs can be observed further away along the jet axis. This pattern can also be observed in Figure 9c–e (2.0 slm, 2.5 slm, and 3.0 slm) and it arises from the unstable plasma plume typical for turbulent flows. Thus, the largest areas of lower contact angle values correspond to the laminar conditions (1.0–1.5 slm) and to the transition from laminar to turbulent flow (2.0 slm), where a reduction of WCA to almost 30° was obtained. The five PET modification areas presented in Figure 9 exhibited a similar plume-like longitudinal shape. This pattern can also be observed for the UV and VUV radiation profiles, where the transition to turbulent mode for 2.0 slm is defined by the diffuse plasma jet tip observed in both cases. Differently from the previous experiments (Section 3.2 and Section 3.3). Here the VUV profile matches the observed modification on the PET samples. Therefore, it is expected that the VUV radiation generated at the plasma plume plays a role only for very short distances (as in the case of side-on plasma jet application). For larger jet-to-target separations, as in the previous sections, the generated VUV is efficiently absorbed by the ambient air [34,41].

The same investigation as in Figure 8 was done for detecting the ROS spreading on starch-iodine-agar plates with the plasma jet in parallel with the agar surface at 1 mm distance and varying the He flow rate. The results are presented in the Figure 10 for different gas flow rates between 0.5 slm and 3.0 slm. The area of spreading ROS tends to steadily increase up to 1.5 slm. From 2.0 slm the ROS covered area is slightly reduced due to the gas flow transition from laminar to turbulent mode in accordance to the observations of the WCA measurements and UV/VUV profiles. Thus, for achieving the largest treated area with more homogeneous modification, the operation of the plasma jet at 0° should be within the laminar flow regime (up to 1.5 slm in this case) and operating in the grounded condition. Since the highest concentration of reactive species is close to the effluent region, the plasma jet operation in laminar mode, at 0° and 1 mm distance, appears to be the most attractive configuration for PET treatment in which most of the produced plasma species generated along the entire plasma plume extension can act on the surface leading to a more efficient WCA reduction within a short treatment time.

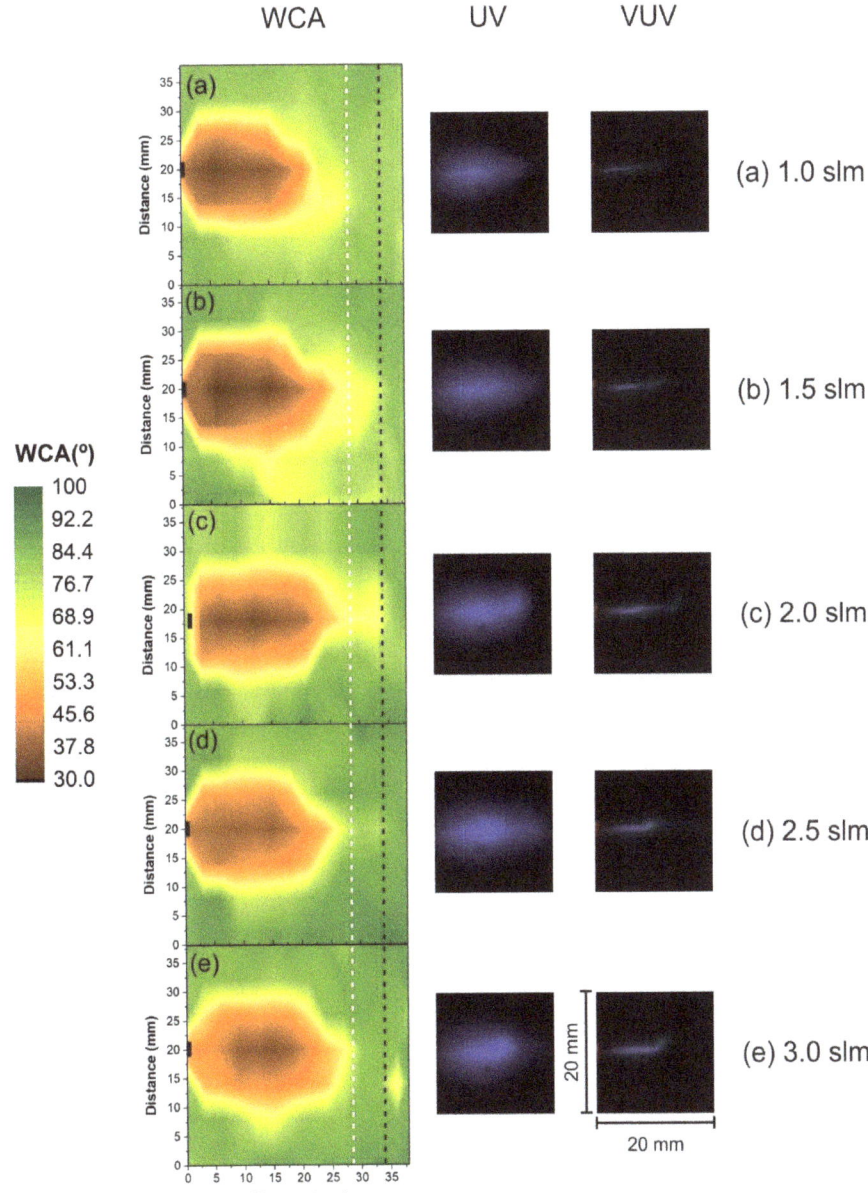

Figure 9. WCA reduction color map of PET samples treated by a He plasma jet placed parallel to the polymer surface for different applied gas flow rates: (**a**) 1.0 slm, (**b**) 1.5 slm, (**c**) 2.0 slm, (**d**) 2.5 slm and (**e**) 3.0 slm. The black rectangle identifies the position of the tube tip. UV (blue) and VUV (green) emission profiles are also displayed for each condition in scale.

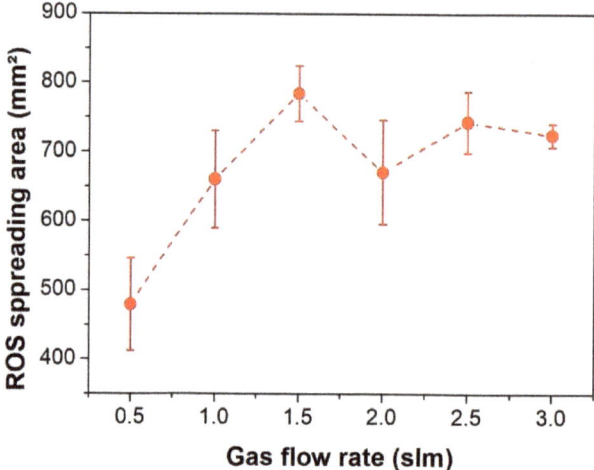

Figure 10. Effective spreading of ROS on starch-iodine-agar plates when exposed to plasma at 1 mm distance in parallel (0°) configuration for 60 s.

3.5. Combination of Different Processing Parameters: Treatment Optimization

Variation and combination of all the processing parameters presented previously make the surface modification by plasma a complex process. It is difficult to find a specific set of parameters that would lead to the best surface modification results once some of these parameters are not independent. For instance, increasing gas flow rate (within the limit of laminar flow) makes the plasma jet length increase, in this way, also affecting the jet-target distance. On the other side, the target presence can also affect the transition from laminar to turbulent flow. Moreover, in most cases the surface modification effect does not scale linearly with any of these parameters. For instance, the active species produced in the plasma plume have limited lifetimes and the increase of treatment time, especially at larger distances, does not result in a lower WCA. Additionally, in general, the treated area increases with decreasing the distance. However, at very short distances, gas turbulence and buoyancy effects affect the ROS distribution and hinders any further increase of the modification area. Therefore, choosing a very short distance and a high flow rate does not necessarily mean the best treatment condition (i.e., the lowest WCA for the shortest treatment time). Thus, to achieve the best surface modification results (including a large and more uniform modified area) the selection of an appropriate set of gas flow rate and jet-to-target distance are important. This depends upon some other experimental parameter like the kind of gas used (He, Ar, or some other), the jet exit dimension, the target surface condition (soft or solid), the ambient humidity and, of course, the particular treatment goal (a punctual or large surface area modification). These will establish the primary set of parameters for an optimal treatment. In this study, we focus on two additional and sometimes overlooked parameters—the jet incidence angle and the target grounding—which can also affect the surface modification effect, significantly increasing (in some cases by about 20%) the size of modification area. Their variation can establish a second order of process optimization. According to the presented study, in order to obtain a more efficient surface modification with a shorter treatment time, the use of incidence angles smaller than 90° combined with a grounded target condition are important. Finally, the jet incidence parallel to the target is an interesting case that needs further investigation because it seems to be very much affected by the target grounding condition.

4. Conclusions

The wettability of PET samples after plasma jet treatment, conducted under different conditions was investigated using the WCA mapping method, which allowed the detection of the entire modification zone. This study allowed identifying two important alterations in the experimental setup configuration that led to a larger treated area and enhanced treatment homogeneity: the use of a grounded substrate holder and the effect of different jet incidence angles. The latter was shown to promote a better ROS spreading on the surface, which contributes to increasing the modified area on PET samples and leads to the formation of larger zones of more efficient treatment. This feature can be especially interesting for applications like the treatment of polymeric packaging, thin films deposition, and biomedical applications where the treatment of regions with difficult access may be required.

For all studied cases, the evaluations of ROS and UV spreading were in reasonable agreement with the WCA measurements. Thus, the effective area of modification can be correlated to the activity of different plasma components that work synergistically. For the shortest tested distance of 1 mm, in the parallel jet arrangement, the VUV irradiation produced by the plasma jet also seems to play a role.

The spreading area of ROS over the starch-iodine-agar plate changes drastically when the angle formed between the plasma jet and the treated surface is diminished. Its shape varies from round (at 90°) to elliptical (at 30°), and finally to a more diffuse and elongated plume-like shape (0°). Thus, according to the desired application, the choice of a particular incidence angle can provide better results, being able to adjust the treatment from more punctual to rather diffuse.

Applying the plasma jet parallel to the surface has shown to be the most attractive condition because a relatively large and homogeneous modification zone can be obtained in the longitudinal direction, which means that a large sample can possibly be scanned only in one direction (transversal to the plasma plume). The variation of gas flow rate can be used to adjust the size of the modified area where the transition from laminar to turbulent plasma plume can be easily controlled. Additionally, simply adjusting the incidence angle of the plasma jet can efficiently vary the resultant modification areas allowing an advantageous control over the treatment for different desired applications.

In conclusion, surface modification of materials by plasma is a complex process where several processing parameters, such as, jet-to-target distance, gas flow rate, treatment time, target condition, and jet incidence angle, should be considered in order to control the resultant modification area and efficiency.

Author Contributions: Conceptualization: T.M.C.N.; data curation: T.M.C.N. and R.W.; formal analysis: T.M.C.N.; funding acquisition: K.G.K.; investigation: T.M.C.N.; methodology: T.M.C.N. and R.W.; project administration: K.G.K.; supervision: K.G.K.; writing—original draft: T.M.C.N. and R.W.; writing—review and editing: T.M.C.N. and K.G.K. All authors have read and agreed to the published version of the manuscript.

Funding: This research was supported by CAPES under grant 88881.132157/2016-01 and FAPESP under grant 2015/21989-6.

Acknowledgments: The authors gratefully acknowledge the fruitful discussions with Torsten Gerling and Ekaterina Makhneva.

Conflicts of Interest: The authors declare no conflict of interest.

References

1. Fricke, K.; Reuter, S.; Schroder, D.; Schulz-Von Der Gathen, V.; Weltmann, K.D.; Von Woedtke, T. Investigation of surface etching of poly(ether ether ketone) by atmospheric-pressure plasmas. *IEEE Trans. Plasma Sci.* **2012**, *40*, 2900–2911. [CrossRef]
2. Shaw, D.; West, A.; Bredin, J.; Wagenaars, E. Mechanisms behind surface modification of polypropylene film using an atmospheric-pressure plasma jet. *Plasma Sources Sci. Technol.* **2016**, *25*. [CrossRef]
3. Prysiazhnyi, V.; Saturnino, V.F.B.; Kostov, K.G. Transferred Plasma Jet as a Tool to Improve Wettability of Inner Surfaces of Polymer tubes. *Int. J. Polym. Anal. Charact.* **2017**, *22*, 215–221. [CrossRef]
4. Laroussi, M. Plasma Medicine: A Brief Introduction. *Plasma* **2018**, *1*, 47–60. [CrossRef]

5. von Woedtke, T.; Reuter, S.; Masur, K.; Weltmann, K.D. Plasmas for medicine. *Phys. Rep.* **2013**, *530*, 291–320. [CrossRef]
6. Penkov, O.V.; Khadem, M.; Lim, W.; Kim, D. A review of recent applications of atmospheric pressure plasma jets for materials processing. *J. Coat. Technol. Res.* **2015**, *12*, 225–235. [CrossRef]
7. Nagatsu, M.; Kinpara, M.; Abuzairi, T. Fluorescence Analysis of Micro-scale Surface Modification Using Ultrafine Capillary Atmospheric Pressure Plasma Jet for Biochip Fabrication Masaaki. *Recent Glob. Res. Educ. Technol. Chall.* **2017**, *519*, 246–254. [CrossRef]
8. Uchida, G.; Kawabata, K.; Ito, T.; Takenaka, K.; Setsuhara, Y. Development of a non-equilibrium 60 MHz plasma jet with a long discharge plume. *J. Appl. Phys.* **2017**, *122*. [CrossRef]
9. Lu, X.; Laroussi, M.; Puech, V. On atmospheric-pressure non-equilibrium plasma jets and plasma bullets. *Plasma Sources Sci. Technol.* **2012**, *21*. [CrossRef]
10. Vesel, A.; Mozetic, M. New developments in surface functionalization of polymers using controlled plasma treatments. *J. Phys. D Appl. Phys.* **2017**, *50*, 293001. [CrossRef]
11. Cools, P.; Astoreca, L.; Tabaei, P.S.E.; Thukkaram, M.; de Smet, H.; Morent, R.; de Geyter, N. Surface Treatment of Polymers by Plasma. In *Surface Modification of Polymers: Methods and Applications*, 1st ed.; Pinson, J., Thiry, D., Eds.; Wiley Online Library: Hoboken, NJ, USA, 2019; pp. 33–66. [CrossRef]
12. Onyshchenko, I.; de Geyter, N.; Morent, R. Improvement of the plasma treatment effect on PET with a newly designed atmospheric pressure plasma jet. *Plasma Process. Polym.* **2017**, *14*. [CrossRef]
13. Kostov, K.G.; Nishime, T.M.C.; Hein, L.R.O.; Toth, A. Study of polypropylene surface modification by air dielectric barrier discharge operated at two different frequencies. *Surf. Coat. Technol.* **2013**, *234*, 60–66. [CrossRef]
14. Musa, G.P.; Shaw, D.R.; West, A.T.; Wagenaar, E.; Momoh, M. An Atmospheric-Pressure Plasma Jet Treatment of Polyethylene Polymer Films for Wettability Enhancement. *Sci. World J.* **2016**, *11*, 27–29.
15. Van Deynse, A.; Cools, P.; Leys, C.; Morent, R.; De Geyter, N. Surface modification of polyethylene in an argon atmospheric pressure plasma jet. *Surf. Coat. Technol.* **2015**, *276*, 384–390. [CrossRef]
16. Kostov, K.G.; Nishime, T.M.C.; Castro, A.H.R.; Toth, A.; Hein, L.R.O. Surface modification of polymeric materials by cold atmospheric plasma jet. *Appl. Surf. Sci.* **2014**, *314*. [CrossRef]
17. Al-Sabagh, A.M.; Yehia, F.Z.; Eshaq, G.; Rabie, A.M.; ElMetwally, A.E. Greener routes for recycling of polyethylene terephthalate. *Egypt. J. Pet.* **2016**, *25*, 53–64. [CrossRef]
18. Vesel, A.; Zaplotnik, R.; Primc, G.; Mozetič, M. Evolution of the surface wettability of PET polymer upon treatment with an atmospheric-pressure plasma jet. *Polymers* **2020**, *12*, 87. [CrossRef]
19. Onyshchenko, I.; Nikiforov, A.Y.; de Geyter, N.; Morent, R. Local analysis of pet surface functionalization by an atmospheric pressure plasma jet. *Plasma Process. Polym.* **2015**, *12*, 466–476. [CrossRef]
20. Nie, Q.Y.; Cao, Z.; Ren, C.S.; Wang, D.Z.; Kong, M.G. A two-dimensional cold atmospheric plasma jet array for uniform treatment of large-area surfaces for plasma medicine. *New J. Phys.* **2009**, *11*. [CrossRef]
21. Ghasemi, M.; Olszewski, P.; Bradley, J.W.; Walsh, J.L. Interaction of multiple plasma plumes in an atmospheric pressure plasma jet array. *J. Phys. D Appl. Phys.* **2013**, *46*. [CrossRef]
22. Mui, T.S.M.; Mota, R.P.; Quade, A.; de Hein, L.R.O.; Kostov, K.G. Uniform surface modification of polyethylene terephthalate (PET) by atmospheric pressure plasma jet with a horn-like nozzle. *Surf. Coat. Technol.* **2018**, *352*, 338–347. [CrossRef]
23. Borges, A.C.; de Lima, G.M.G.; Nishime, T.M.C.; Gontijo, A.V.L.; Kostov, K.G.; Koga-Ito, C.Y. Amplitude-modulated cold atmospheric pressure plasma jet for treatment of oral candidiasis: In vivo study. *PLoS ONE* **2018**, *13*, 1–19. [CrossRef] [PubMed]
24. Borges, A.C.; Nishime, T.M.C.; de Rovetta, S.M.; de Lima, G.M.G.; Kostov, K.G.; Thim, G.P.; de Menezes, B.R.C.; Machado, J.P.B.; Koga-Ito, C.Y. Cold Atmospheric Pressure Plasma Jet Reduces *Trichophyton rubrum* Adherence and Infection Capacity. *Mycopathologia* **2019**. [CrossRef]
25. Kostov, K.G.; Machida, M.; Prysiazhnyi, V.; Honda, R.Y. Transfer of a cold atmospheric pressure plasma jet through a long flexible plastic tube. *Plasma Sources Sci. Technol.* **2015**, *24*. [CrossRef]
26. Kostov, K.G.; Borges, A.C.; Koga-Ito, C.Y.; Nishime, T.M.C.; Prysiazhnyi, V.; Honda, R.Y. Inactivation of Candida albicans by Cold Atmospheric Pressure Plasma Jet. *IEEE Trans. Plasma Sci.* **2015**, *43*. [CrossRef]
27. Kostov, K.G.; Nishime, T.M.C.; Machida, M.; Borges, A.C.; Prysiazhnyi, V.; Koga-Ito, C.Y. Study of Cold Atmospheric Plasma Jet at the End of Flexible Plastic Tube for Microbial Decontamination. *Plasma Process. Polym.* **2015**, *12*. [CrossRef]

28. Chen, W.K.; Huang, J.C.; Chen, Y.C.; Lee, M.T.; Juang, J.Y. Deposition of highly transparent and conductive Ga-doped zinc oxide films on tilted substrates by atmospheric pressure plasma jet. *J. Alloys Compd.* **2019**, *802*, 458–466. [CrossRef]
29. Damany, X.; Pasquiers, S.; Blin-Simiand, N.; Bauville, G.; Bournonville, B.; Fleury, M.; Jeanney, P.; Sousa, J.S. Impact of an atmospheric argon plasma jet on a dielectric surface and desorption of organic molecules. *EPJ Appl. Phys.* **2016**, *75*, 1–7. [CrossRef]
30. Hosseinpour, M.; Zendehnam, A.; Sangdehi, S.M.H.; Marzdashti, H.G. Effects of different gas flow rates and non-perpendicular incidence angles of argon cold atmospheric-pressure plasma jet on silver thin film treatment. *J. Theor. Appl. Phys.* **2019**. [CrossRef]
31. Olabanji, O.T.; Bradley, J.W. Side-on surface modification of polystyrene with an atmospheric pressure microplasma jet. *Plasma Process. Polym.* **2012**, *9*, 929–936. [CrossRef]
32. Kawasaki, T.; Eto, W.; Hamada, M.; Wakabayashi, Y.; Abe, Y.; Kihara, K. Detection of reactive oxygen species supplied into the water bottom by atmospheric nonthermal plasma jet using iodine-starch reaction. *Jpn. J. Appl. Phys.* **2015**, *54*, 1–7. [CrossRef]
33. Kawasaki, T.; Kawano, K.; Mizoguchi, H.; Yano, Y.; Yamashita, K.; Sakai, M. Visualization of the two-dimensional distribution of ROS supplied to a water-containing target by a non-thermal plasma jet. *Int. J. Plasma Environ. Sci. Technol.* **2016**, *10*, 41–46.
34. Golda, J.; Biskup, B.; Layes, V.; Winzer, T.; Benedikt, J. Vacuum ultraviolet spectroscopy of cold atmospheric pressure plasma jets. *Plasma Process. Polym.* **2020**, 1–12. [CrossRef]
35. Gerling, T.; Nastuta, A.V.; Bussiahn, R.; Kindel, E.; Weltmann, K.D. Back and forth directed plasma bullets in a helium atmospheric pressure needle-to-plane discharge with oxygen admixtures. *Plasma Sources Sci. Technol.* **2012**, *21*. [CrossRef]
36. Shao, X.J.; Zhang, G.J.; Zhan, J.Y.; Xu, G.M. Research on surface modification of polytetrafluoroethylene coupled with argon dielectric barrier discharge plasma jet characteristics. *IEEE Trans. Plasma Sci.* **2011**, *39*, 3095–3102. [CrossRef]
37. Darny, T.; Pouvesle, J.M.; Fontane, J.; Joly, L.; Dozias, S.; Robert, E. Plasma action on helium flow in cold atmospheric pressure plasma jet experiments. *Plasma Sources Sci. Technol.* **2017**, *26*. [CrossRef]
38. Goree, J.; Liu, B.; Drake, D. Gas flow dependence for plasma-needle disinfection of *S. mutans* bacteria. *J. Phys. D Appl. Phys.* **2006**, *39*, 3479–3486. [CrossRef]
39. Schmidt-Bleker, A.; Winter, J.; Iseni, S.; Dünnbier, M.; Weltmann, K.-D.; Reuter, S. Reactive species output of a plasma jet with a shielding gas device—Combination of FTIR absorption spectroscopy and gas phase modelling. *J. Phys. D Appl. Phys.* **2014**, *47*, 145201. [CrossRef]
40. Kaczmarek, H.; Kowalonek, J.; Szalla, A.; Sionkowska, A. Surface modification of thin polymeric films by air-plasma or UV-irradiation. *Surf. Sci.* **2002**, *507–510*, 883–888. [CrossRef]
41. Knoll, A.J.; Luan, P.; Bartis, E.A.J.; Kondeti, V.S.S.K.; Bruggeman, P.J.; Oehrlein, G.S. Cold Atmospheric Pressure Plasma VUV Interactions with Surfaces: Effect of Local Gas Environment and Source Design. *Plasma Process. Polym.* **2016**, *13*, 1067–1077. [CrossRef]
42. Borcia, C.; Borcia, G.; Dumitrascu, N. Surface treatment of polymers by plasma and UV radiation. *Rom. Rep. Phys.* **2011**, *56*, 224–232.
43. Jin, D.J.; Uhm, H.S.; Cho, G. Influence of the gas-flow Reynolds number on a plasma column in a glass tube. *Phys. Plasmas* **2013**, *20*, 083513. [CrossRef]
44. Winter, J.; Nishime, T.M.C.; Glitsch, S.; Lüder, H.; Weltmann, K.-D. On the development of a deployable cold plasma endoscope. *Contrib. Plasma Phys.* **2018**. [CrossRef]

© 2020 by the authors. Licensee MDPI, Basel, Switzerland. This article is an open access article distributed under the terms and conditions of the Creative Commons Attribution (CC BY) license (http://creativecommons.org/licenses/by/4.0/).

Article

Surface Thermo-Dynamic Characterization of Poly (Vinylidene Chloride-Co-Acrylonitrile) (P(VDC-co-AN)) Using Inverse-Gas Chromatography and Investigation of Visual Traits Using Computer Vision Image Processing Algorithms

Vijay Kakani [1], Hakil Kim [1], Praveen Kumar Basivi [2,*] and
Visweswara Rao Pasupuleti [3,4,*]

1. Information and Communication Engineering, Inha University, 100 inharo, Nam-gu Incheon 22212, Korea; vjkakani@inha.ac.kr (V.K.); hikim@inha.ac.kr (H.K.)
2. Department of Chemistry, Sri Venkateswara University, Tirupati, Andhra Pradesh 517502, India
3. Department of Biomedical Sciences & Therapeutics, Faculty of Medicine and Health Sciences, University Malaysia Sabah, Kota Kinabalu Sabah 88400, Malaysia
4. Department of Biochemistry, Faculty of Medicine and Health Sciences, Abdurrab University, Jl Riau Ujung No. 73, Pekanbaru 28292, Riau, Indonesia
* Correspondence: basivipraveen@gmail.com (P.K.B.); pvrao@ums.edu.my (V.R.P.); Tel.: +919494208338 (P.K.B.); +60189018799 (V.R.P.)

Received: 13 June 2020; Accepted: 25 June 2020; Published: 23 July 2020

Abstract: The Inverse Gas Chromatography (IGC) technique has been employed for the surface thermo-dynamic characterization of the polymer Poly(vinylidene chloride-co-acrylonitrile) (P(VDC-co-AN)) in its pure form. IGC attributes, such as London dispersive surface energy, Gibbs free energy, and Guttman Lewis acid-base parameters were analyzed for the polymer (P(VDC-co-AN)). The London dispersive surface free energy (γ_S^L) was calculated using the Schultz and Dorris–Gray method. The maximum surface energy value of (P(VDC-co-AN)) is found to be 29.93 mJ·m^{-2} and 24.15 mJ·m^{-2} in both methods respectively. In our analysis, it is observed that the γ_S^L values decline linearly with an increase in temperature. The Guttman–Lewis acid-base parameter K_a, K_b values were estimated to be 0.13 and 0.49. Additionally, the surface character S value and the correlation coefficient were estimated to be 3.77 and 0.98 respectively. After the thermo-dynamic surface characterization, the (P(VDC-co-AN)) polymer overall surface character is found to be basic. The substantial results revealed that the (P(VDC-co-AN)) polymer surface contains more basic sites than acidic sites and, hence, can closely associate in acidic media. Additionally, visual traits of the polymer (P(VDC-co-AN)) were investigated by employing Computer Vision and Image Processing (CVIP) techniques on Scanning Electron Microscopy (SEM) images captured at resolutions ×50, ×200 and ×500. Several visual traits, such as intricate patterns, surface morphology, texture/roughness, particle area distribution (D_A), directionality (D_P), mean average particle area (μ_{avg}) and mean average particle standard deviation (σ_{avg}), were investigated on the polymer's purest form. This collective study facilitates the researches to explore the pure form of the polymer Poly(vinylidene chloride-co-acrylonitrile) (P(VDC-co-AN)) in both chemical and visual perspective.

Keywords: Poly(vinylidene chloride-co-acrylonitrile) (P(VDC-co-AN)); thermo-dynamic surface characterization; surface free energy; inverse gas chromatography; visual traits; computer vision and image processing

1. Introduction

1.1. Purpose of Study

The primary goal of this study is to estimate the surface thermodynamic properties of the polymer (P(VDC-co-AN)) in its pure form using inverse gas chromatography techniques. In addition to that, the secondary intention is to explore the visual traits of polymer (P(VDC-co-AN)) using SEM images and CVIP techniques. Eventually, this study anticipates that the investigated aspects of this work might be obliging for future studies working on the polymer (P(VDC-co-AN)). Generally, polymers, such as Polyvinyl Chloride (PVC) and other variants of PVC, have been employed in industrial applications and other fields of research for decades[1–4]. Especially, when treated properly and shaped, the PVC and its variants can attain feasible properties that can be utilized in a variety of applications that are related to blood bags, healthcare, automobiles, electronics, and many more rigid and flexible products [5–8]. The variants of PVC can be used to blend with various other materials for better durability and effectiveness still being an affordable product for many applications [9,10]. One among such variants is Poly (vinylidene chloride-co-acrylonitrile (P(VDC-co-AN)), which has been considered for Inverse Gas Chromatography (IGC) based surface thermodynamic characterization in this study. (P(VDC-co-AN)) has been employed in many engineering applications related to optoelectronic devices, such as Organic Light-Emitting Diodes, Organic Thin-Film Transistors due to its outstanding barrier properties and transparency [11,12]. Applications, such as optoelectronic devices, require higher barrier properties, because the inner mechanics must avert the incursion of water and oxygen molecules into the light-emitting materials [13]. This vacuum system of using organic and inorganic materials is expensive and not robust against mechanical deformation and thermal shock. The use of (P(VDC-co-AN)) in the manufacturing of optoelectronic devices is more desirable due to its robust physical properties, which increases the lifespan and quality of the optoelectronic devices [14]. Furthermore, the polymer (P(VDC-co-AN)) incorporated with silica increases the adhesion to inorganic substrates and/or metal electrodes, making it substantial for optoelectronic devices and other engineering applications [15].

The major motivation of this study is to explore the polymer (P(VDC-co-AN)) primarily using the IGC techniques for surface thermodynamic properties. Besides, visual traits, such as intricate patterns, texture analysis, cross-sectional profiling, angular and radial spectra, particle area, directionality estimates were analyzed through the CVIP technique. Furthermore, in an economic stand-point, image analysis has its advantage, such as the solitary requirement of limited images with a decent resolution yet providing efficient analysis of visual traits with low cost and minimum human interference [16–19]. The usage of an expensive TEM imaging machinery would do the same by exhibiting the high definition imagery of the polymer. However, they lack in observing the intricate patterns and further visual traits, which, in most cases, are out shadowed by the imagery instrumental noise [20–23]. The real discussion is: how good is the exploration of this commercially useful polymer (P(VDC-co-AN)) through combined techniques from two relatively diverse fields of research CVIP & IGC? This question is partially elaborated through exhibiting the results and observations found by employing IGC and CVIP techniques on polymer (P(VDC-co-AN)) in its pure form. Eventually, the vicinity of the study is confined to characterize the surface thermodynamics of the polymer (P(VDC-co-AN)) using IGC techniques while also exhibiting a few of its visual traits when attempted to analyze using CVIP techniques. In the result-oriented perspective, this study aims to consolidate all of the aspects of IGC based polymer surface characterization as the main contribution. Nevertheless, this work also attempts to give some details of polymer's visual traits from the image processing point of view in the interest of contributing towards future polymer studies in terms of employing image processing techniques on not so clear images, yet exploring the pure polymer (P(VDC-co-AN)) from a visual standpoint.

1.2. Inverse Gas Chromatography

Inverse Gas Chromatography is an effective technique for characterizing materials to investigate their physicochemical properties [24]. IGC technique is considered as a feasible approach due to its ability to determine surface properties of solids in crystalline and amorphous structures within various forms, such as powder, films, and fibers [25]. Besides having similarity in the principle to that of analytical gas chromatography techniques, IGC bears a unique phenomenon of placing the material in a column with a known probe vapor used to facilitate information on the surface of the material [26,27]. The range of properties such as entropy and enthalpy of adsorption, dispersive surface energy, solubility parameters, and many more aspects can be characterized while using the IGC technique [28]. The information obtained from the IGC experimentation enables researchers to exploit the true potential of the materials and unveil its usefulness in industrial applications and other fields of research. In this paper, the dispersive component and specific component of surface free energy have been evaluated using Schultz and Dorris–Gray methods [29–31]. The surface character value and the Lewis acid-base parameters were also evaluated by the Schultz method to investigate the nature of the surface of the polymer.

1.3. Computer Vision and Image Processing

The advancements in image analysis and accessibility of image data contributed vast growth in the field of technologies, such as Computer Vision and Image Processing (CVIP) [32–34]. Analogous to this, improvements in instrumentations related imaging techniques, such as Scanning Electron Microscopy (SEM) and Transmission Electron Microscopy (TEM) added an unprecedented precision to the fields of research, such as Inverse Gas Chromatography (IGC) and Nanotechnology. Exploiting such technological advancements in favor of scientific research has always been a conventional practice for the researchers [35]. As of two decades, computer vision and image processing have made their way into every possible division of research, including medicine, artificial intelligence, chemistry, physics, automobile, and agriculture [36–40]. The crucial aspect that motivated this interdisciplinary collective study is the ability to handle images and observe significant traits of the physical substance from various perspectives. As far as the polymer materials are concerned, features, such as intricate patterns of the substance in its pure form, surface texture, area measurements of the lumps, and ridges, are the traits that can be explored while using the image processing techniques. In this study, the SEM image analysis of (P(VDC-co-AN)) polymer has been carried out in order to determine the visual patterns, particle area distribution, particle directionality, and cross-sectional profiling of particle in both overview (500 µm) and portrait modes (200 µm and 50 µm). All of these visual elements were observed through a series of CVIP techniques, such as segmentation, thresholding, detection of feature maxima, texture analysis, etc.

A survey of the literature revealed that the surface thermodynamic characterization using IGC, as well as SEM image analysis on a pure form of (P(VDC-co-AN)) using CVIP techniques has not been reported earlier. On a whole, the paper attempts to explore the polymer (P(VDC-co-AN)) primarily using the IGC techniques for surface thermodynamic properties and secondarily employing customized CVIP pipelines for investigation of visual traits. The Figure 1 below illustrates the scenario of exploring the polymer (P(VDC-co-AN)) using IGC for surface thermodynamic characterization and CVIP methodologies for possible visual traits. The major highlights of the paper are as follows:

- Surface thermo-dynamic characterization of the polymer (P(VDC-co-AN)) has been carried. IGC attributes, such as London dispersive surface energy, Gibbs free energy, and Guttman Lewis acid-base parameters, were estimated.
- Visual traits, such as intricate patterns, surface morphology, texture/roughness, particle area distribution (D_A), directionality (D_P), mean average particle area (μ_{avg}), and mean average particle standard deviation (σ_{avg}), were investigated using CVIP techniques on SEM images of the polymer in its purest form.

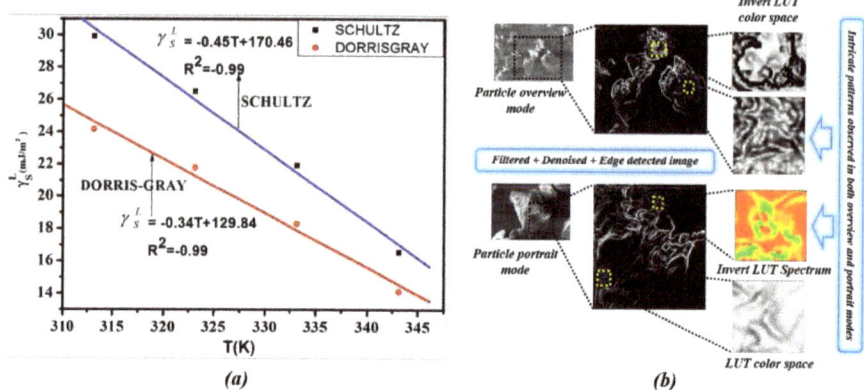

Figure 1. Inverse Gas Chromatography (IGC) surface thermodynamic characterization and Computer Vision and Image Processing (CVIP) Scanning Electron Microscopy (SEM) image analysis: (a) (P(VDC-co-AN)) surface free energy versus temperature using IGC methods. (b) Intricate visual patterns on the pure polymer surface observed using CVIP SEM image analysis.

The rest of this paper is organized, as follows. Section 2 elaborates on the IGC experimental setup and SEM image acquisition used in this study. Section 3 describes the detailed calculation of IGC attributes, such as London dispersive surface energy, Gibbs free energy, and Guttman Lewis acid-base parameters. Section 4 illustrates the experimental analysis of polymer visual traits while using CVIP techniques. Section 5 presents the results and discussions corresponding to the IGC Surface thermo-dynamic characterization and CVIP based SEM image analysis. Finally, this paper is concluded in Section 6 with a summary.

2. Experimental Setup

2.1. IGC Experimental Setup

The IGC experimental analysis has been studied by using the AIMIL (model 5700, AIMIL Ltd., New Delhi, India) gas chromatography with dual column setup and fixed with FID detector. The IGC reading measurements were carried out at constant oven temperature at 10 °C intervals and the range is set from 313.15–343.15 K. The P(VDC-co-AN) used in this study molecular weight, $\overline{M_n} = 80,000 \sim 150,000$ was purchased from Sigma–Aldrich Pvt. Ltd. (USA) and directly used for the preparation of the column packing. An exact amount of 0.8752 grams of polymer is packed directly into the column with inert component as helium gas. The probes n-alkanes (C5–C10), dichloromethane (DCM), Trichloro methane (TCM), acetone (AC), diethyl ether (DEE), ethyl acetate (EA), and tetrahydrofuran (THF) were analytical grade chemicals purchased from S.D Fine and Merck. The sieved P(VDC-co-AN) particles with 150–180 μm diameter were employed in column packing. The utilized stainless steel column (3 mm internal diameter and 30 cm length) was purchased from NUCON and the column was cleaned multiple times with acetone and methanol. The column is dried in the hot air oven. The sieved P(VDC-co-AN) was directly packed with the aid of mechanical vibrator and both ends of the column are filled with glass wool. Furthermore, the column was conditioned for 10 h under the continuous flow of nitrogen. The retention time (t_R) was measured using a Hamilton syringe 0.1 μL of each solute that was injected onto the P(VDC-co-AN) surface. The injection process for each solute was carried out three times on the P(VDC-co-AN) column. The average of all the three cycles was considered to calculate the net retention time.

2.2. SEM Image Acquisition

The SEM images were obtained using SNE-3000M mini-SEM instrument and they were analyzed using image processing software, such as OpenCV library [41], MATLAB [42], and Mathematica [43]. Languages, such as C++, C, and Wolfram, were employed to code the algorithms, such as thresholding, segmentation, and texture analysis on a Windows PC with NVIDIA GeForce GTX 980Ti Graphics Processor Unit (GPU). Properties, such as the distribution of area (D_A) and particle directionality (D_P) and cross-sectional profiling, were investigated using open-source image processing software, such as ImageJ [44] and Scanning Probe Image Processing (SPIP) [45]. Microsoft Excel and Origin7 [46] have been employed for pre-processing and post-calculations of polymer data. Various parameters involved in the texture analysis, directionality analysis, image enhancement, thresholding, and segmentation algorithms were fine-tuned to attain the best adaptive scaling on pixel-wise visual patterns that can reveal the finest patterns in the SEM image of the polymer (P(VDC-co-AN)). Certain arbitrary values have been chosen for better visualization to eliminate outliers and to enhance hidden patterns, under such circumstances a clear illustration has been documented in a pictorial representation regarding the effects imposed by those arbitrary values on the visual patterns. Figure 2 below illustrates the image acquisition of polymer (P(VDC-co-AN)) using mini-SEM SNE-3000M instrument at different resolutions (overview mode: ×50, portrait mode: ×200 and ×500).

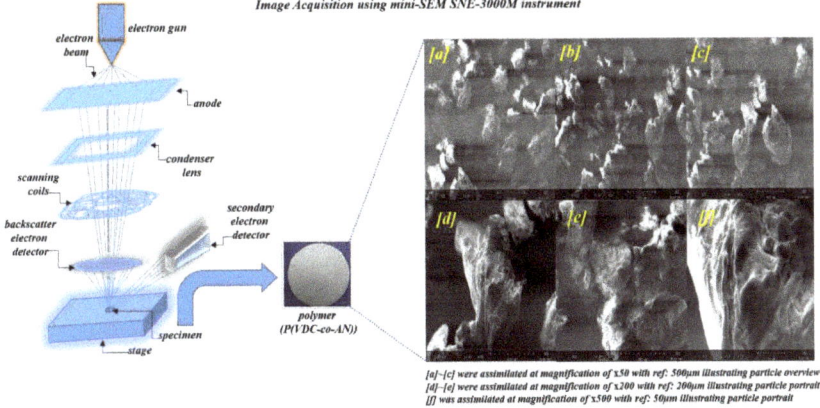

Figure 2. Image acquisition of polymer (P(VDC-co-AN)) using mini-SEM SNE-3000M instrument at different resolutions (overview mode: ×50, portrait mode: ×200 and ×500).

3. IGC Surface Thermo-Dynamic Characterization

IGC technique has been used to determine the net retention volume, V_N according to the following equation:

$$V_N = (t_R - t_0) \, FJ \left(\frac{P_o - P_W}{P_o} \right), \tag{1}$$

where t_R is the retention time of a probe, and to is the retention time of methane, F is the flow rate of carrier gas, P_W is the saturated vapor pressure of water at ambient temperature, P_o is the atmospheric pressure, and J is the James Martin correction factor.

The dispersive surface free energy, γ_S^d of (P(VDC-co-AN)) has been calculated following Dorris–Gray equation [30]:

$$\gamma_S^d = \frac{1}{4\gamma_{CH_2}} \left[\frac{RT \ln \left(\frac{V_{N,n+1}}{V_{N,n}} \right)}{N \cdot a_{CH_2}} \right]^2, \tag{2}$$

where γ_{CH_2} is the surface dispersive free energy of solid material containing only methylene groups (e.g., linear (P(VDC-co-AN))), and it is calculated at any temperature t (°C) with the Equation (3) stated in [47]:

$$\gamma_{CH_2} = 35.6 - 0.058t, \quad (3)$$

N is Avogadro's number, R is gas constant, T is column temperature, and $a_{CH_2} = 6 \text{ Å}^2$ is the cross-sectional area of adsorbed methylene group. $V_{N,n+1}$ and $V_{N,n}$ are the net retention volumes of n–alkanes with carbon number $n+1$ and n, respectively.

Alternatively, the dispersive surface free energy has been calculated following Equation (4) proposed by Schultz et al. [29].

$$RT \ln V_N = 2Na \left(\gamma_l^d\right)^{1/2} \left(\gamma_S^d\right)^{1/2} + K, \quad (4)$$

where a and γ_l^d are the cross-sectional area and dispersive free energy of the adsorbate. The γ_S^d of (P(VDC-co-AN)) can be evaluated using slope values obtained from a linear plot drawn between $RT \ln V_N$ versus $a \left(\gamma_l^d\right)^{1/2}$ of n-alkanes.

The specific component of surface free energy, ΔG_a^S values for the polar probes are evaluated while using Equation (5)

$$-\Delta G_a^S = RT \ln \left(\frac{V_N}{V_{N(ref)}}\right). \quad (5)$$

The specific components of enthalpy of adsorption ΔH_a^s, entropy of adsorption ΔS_a^S, the Lewis acid-base parameters K_a, and K_b are obtained using Equations (6)–(7)

$$-\frac{\Delta G_a^S}{T} = \frac{-\Delta H_a^S}{T} + \Delta S_a^S, \quad (6)$$

$$-\frac{\Delta H_a^s}{AN^*} = K_a \left(\frac{DN}{AN^*}\right) + K_b, \quad (7)$$

where AN^* and DN are Guttmann's modified acceptor and donor numbers, respectively. AN^* and DN values along with the physical property data for the probes are given in Appendix A, Table A1. The a and γ_l^d values for different probes were retrieved from previous IGC studies [47].

4. Image Analysis of (P(VDC-co-AN)) Visual Traits Using CVIP Techniques

In the customized pipeline of image analysis, the images obtained from any imaging instrument bear a certain amount of uncertainty in pixel accuracy due to instrument-induced noise [48]. To reduce such noise outliers, image enhancement techniques, such as filtering, denoising, feature enhancement, etc., were employed. These techniques act as a primary image purifier for later stages of image analysis techniques, such as contour detection, intricate pattern analysis, classification of structural lumps, and surface area distribution. Filtering techniques of three types, such as anisotropic diffusion [49] with total variation denoising [50], edge preserving guided filtering [51] with spatial neighboring noise removal, and Fast Fourier Transform (FFT) filtering [52] with Adaptive Histogram Equalization (ADE) were employed in a customized pipeline model for image enhancement and noise removal. The images were analyzed based on the various customized pipelines and each pipeline has its unique pros and cons in correspondence to various specimen properties. Example: use of customized anisotropic diffusion with variation denoising and outliers removal pipeline produces higher peak signal to noise ratio (PSNR) and good structural similarity index measure (SSIM) as compared to the other techniques, which imply the ability to view the specimen in less noise environment. In contrast, the same image might have a relatively minor consideration in the context of anomaly detection using directional transforms. Given these aspects, suitable approaches have been chosen, depending upon requirements and scenarios of certain analyses. The Figure 3 below illustrates the customized CVIP pipelines for SEM image analysis of (P(VDC-co-AN)) polymer for visual traits.

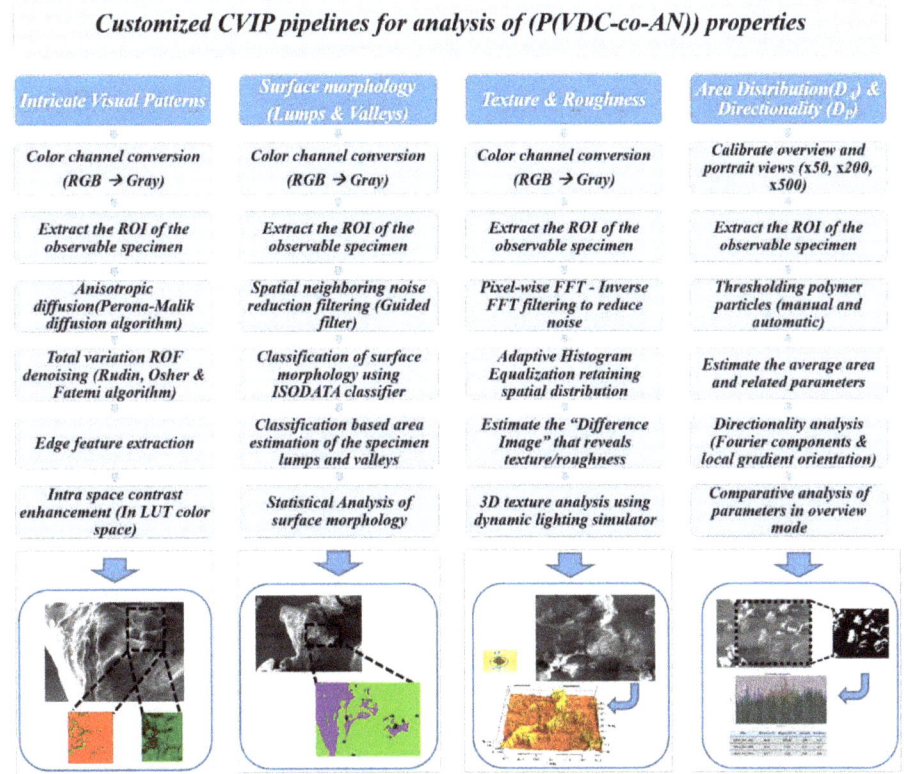

Figure 3. Customized CVIP pipelines for image analysis of (P(VDC-co-AN)) polymer for visual traits.

4.1. Intricate Visual Patterns

The visual traits can be characterized into various categories, such as surface roughness, structural visual patterns, texture forms, directionality, cross-sectional profiling and area distribution, etc. The intricate visual patterns of the polymer can be observed through the application of edge contouring on an anisotropic diffusion that nis filtered with a noise-reduced image. The pipeline comprises of denoising, filtering, and edge contour as pre-processing and intra color adjustments as post-processing. The anisotropic diffusion filtering proposed by Perona–Malik [53] was employed as a part of the pre-processing. Furthermore, post-processing techniques, such as intra color space contrast adjustments, were employed using the ImageJ toolbox for a better perception of the visual patterns on the surface of the polymer. The major pre-processing was carried out using the anisotropic diffusion filtering with a noise reduction step, as depicted in the Equation (8), below:

$$\frac{\partial I}{\partial t} = \mathrm{div}(c(x,y,t)\nabla I) = \nabla c \cdot \nabla I + c(x,y,t)\Delta I, \tag{8}$$

where Δ, ∇ denotes Laplacian and gradient, respectively. $\mathrm{div}()$ denotes the divergence operator and $c(x,y,t)$ represents the diffusion coefficient proposed by Perona–Malik, which controls the magnitude of image gradient, preserving the edges and details.

4.2. Surface Morphology-Lumps and Valleys

The surface morphology of the polymer includes several structural elements in the image domain, such as shape, surface voxel distribution, etc. These traits were observed through a series of pre-processing and post-processing techniques, such as spatial neighboring noise reduction using guided filtering: a pioneering technique of edge preservation filtering. The filtering technique that was proposed by K. He et al. [54] was employed to obtain the initial edge preserved imaging for later pipeline processing. The pre-processed image has been classified using an ISODATA classifier [55] based on classes observed on the surface of the specimen namely, lumps, valleys, and high-intensity flat surfaces. The underlying properties of any distributed model can be analyzed while using the ISODATA classifier due to its unsupervised clustering capabilities. The polymer specimen surface has been classified using the ISODATA classifier technique in ImageJ automated software and then the lump areas were analyzed and marked statistically. The major pre-processing was carried out using the edge-preserving guided filtering, as shown in the Equation (9), below:

$$q_i = \frac{1}{|\omega|} \sum_{k:i \in \omega_k} (a_k I_i + b_k) = \overline{a}_i I_i + \overline{b}_i, \tag{9}$$

where q_i is filtered output and I denotes guidance image. (a_k, b_k) are some linear coefficients and ω_k is window centered at k.

4.3. Texture and Roughness

The frequency-domain representation of a specimen exhibits the spatially distributed properties, such as texture and roughness. Image processing techniques such as Fast Fourier Transform (FFT) were employed to ensure the presence of frequency representation of the polymer specimen by purging the high-frequency intensities (noise outliers). Moreover, the image gets reverted from the frequency domain to the image domain using an inverse FFT technique and equalized using Adaptive Histogram Equalization (AHE). Additionally, a non-trivial yet novel concept of the difference image (I_{diff}) has been formulated while using Equation (10) to customize the analysis step even more substantial to observable texture and roughness features.

$$I_{diff} = P_{3D}\left[I_{xy} - FFT_{AHE}\left(I_{xy}\right)\right], \tag{10}$$

where I_{diff} represents the difference image, P_{3D} is three-dimensional (3D) projection operator, I_{xy} denotes the original image and $FFT_{AHE}(I_{xy})$ denotes frequency domain filtered and histogram equalized image.

4.4. Area Distribution (DA) and Particle Directionality (DP)

In the estimation of structural aspects, such as area distribution and particle directionality, SEM images were analyzed with open-source software, such as SPIP and ImageJ. The polymer SEM particle overview image has been considered for the estimation of area distribution and particle directionality [56]. The selected Region of Interest (ROI) was processed accordingly for the estimation of the thresholding as a stage-1 in the pipeline. The threshold classified image is then manually refined by color space edge enhancement for robust segmentation revealing the fine contours as a stage-2. The particle directionality analysis (DP) of the particle overview has been made while using the ImageJ toolkit in two different scenarios, such as directionality based on Fourier components and directionality based on local gradients orientation. Furthermore, the average distribution of particle area has been estimated and statistics were applied to retrieve the spatial distribution, such as average mean particle area, the standard deviation in terms of μm² were estimated using Equations (11)–(12):

$$\left(\mu_{avg}\right) = \frac{\sum_{k=1}^{N} \mu_k}{N}, \tag{11}$$

$$(\sigma_{avg}) = \sqrt{\frac{1}{N}\sum_{k=1}^{N}(x_k - \mu_k)^2}, \tag{12}$$

where μ_{avg} and σ_{avg} are average mean and average standard deviation of the particle area distribution with N particles ($N > k$), k being the particular index of the particle x of individual mean μ_k.

5. Results and Discussions

5.1. IGC Study on Polymer Surface Characterization

The $RT \ln V_N$ values of n- alkanes and polar probes were measured for the (P(VDC-co-AN)) at temperatures ranging from 313.15 K to 343.15 K. The $RT \ln V_N$ values are linearly decreased with the increase in temperature. The $RT \ln V_N$ values are shown in Table 1 and the variation of $RT \ln V_N$ versus $a\sqrt{\gamma_l^d}$ plot is illustrated in Figure 4, below.

Table 1. $RT \ln V_N$ values of n- alkanes and polar probes measured for the (P(VDC-co-AN)) at temperatures ranging from 313.15 K to 343.15 K.

Solutes	313.15K	323.15K	333.15K	343.15K
n-Pentane	7.40	7.61	7.41	–
n-Hexane	9.51	10.02	9.37	9.27
n-Heptane	10.92	11.50	12.22	10.42
n-Octane	12.48	13.10	13.84	11.82
n-Nonane	16.03	15.16	15.51	13.22
n-Decane	18.11	17.21	16.94	14.91
Acetone	17.88	18.48	18.61	16.28
Di-ethyl ether	17.70	17.87	17.84	16.02
Dichloromethane	17.26	17.51	17.56	15.61
Trichloromethane	19.82	19.62	19.27	16.66
Tetrahydrofuran	19.11	19.05	18.74	17.04
Ethyl acetate	19.32	18.45	18.37	16.98

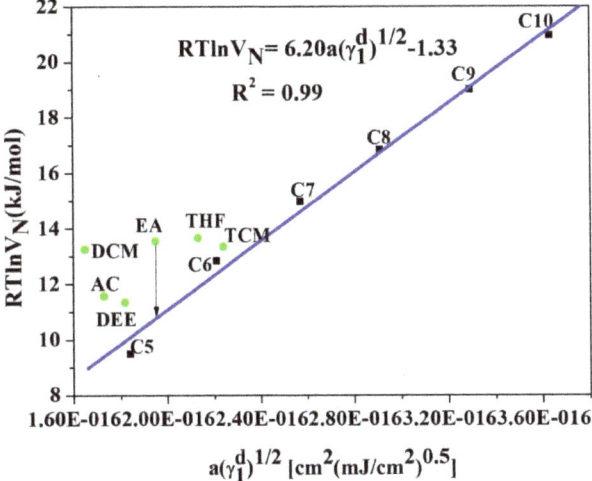

Figure 4. $RT \ln V_N$ versus $a\sqrt{\gamma_l^d}$ for n-alkanes and polar probes for (P(VDC-co-AN)) surface at 313.15 K. (Schultz).

The London dispersive surface free energy γ_S^L was calculated by Schultz and Dorris-Gray methods using Equations (2)–(4). The γ_S^L values are decreasing linearly with the increase of temperature in both methods, and the correlation coefficient is reliable. The maximum γ_S^L is 29.93 mJ/m² in Schultz and 24.15 mJ/m² in Dorris-Gray method. The γ_S^L values are shown Table 2 and the variation of temperature versus γ_S^L values are shown in Figure 5 below. The γ_S^L value in between the temperature range of (313.15–343.15 K) signifies the surface structural changes of the solid material and its allowance to the penetration of the probe molecules. As observed, the temperature gradient of γ_S^L is negative, which indicates the rise in the distance of accession between the (P(VDC-co-AN)) molecules with the escalation in the temperature. According to the London expression, the dispersive energy is inversely proportional to the sixth power of the distance of separation between the molecules. Therefore, with an escalation in the temperature, the dispersive energy increases, and the distance of dissolution between molecules decreases.

Table 2. Dispersive surface free energy, γ_S^L (mJm^{-2}) of (P(VDC-co-AN)).

$T(K)$	γ_S^L (Schultz)	γ_S^L (Dorris-Gray)
313.15	29.93	24.15
323.15	26.49	21.78
333.15	21.92	18.30
343.15	16.84	14.11

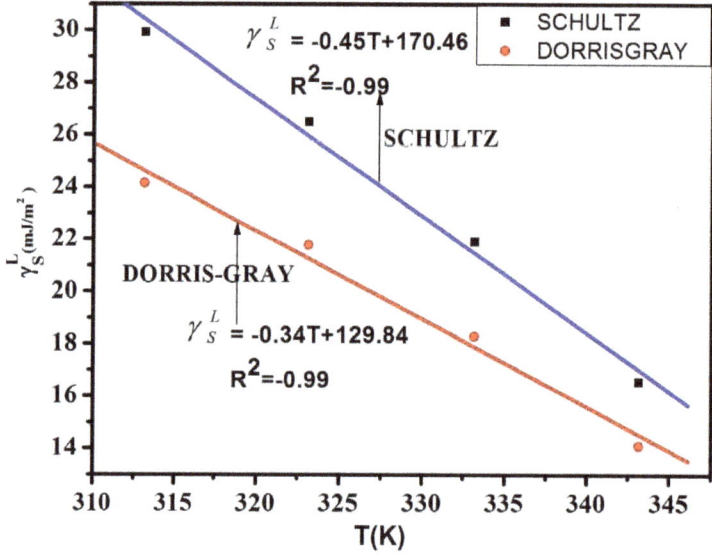

Figure 5. Comparison Plots of dispersive surface free energy versus temperature for (P(VDC-co-AN)) surface.

The results obtained from the Schultz method and Dorris–Gray method clearly illustrate the increase and gradual decrease of the London dispersive surface energy value with an increase in temperature. This discrepancy may be ascribed to the values of the dispersive surface tensions of the liquid probes γ_S^L. However, as compared to the Dorris–Gray method, the Schultz method is the most substantial for the estimation of London dispersive surface energy, γ_S^L at elevated temperatures [47]. Nevertheless, the Dorris-Gray method is applicable at ambient temperatures with a known temperature dependency on γ_S^L. Commonly, the strength of acid-base communication between the adsorbent and polar probe determines the numerical values of ΔG_a^S. Furthermore, the regular area of the probe's adsorption on the drug material has some influence on the ΔG_a^S values. The elevation in temperature effects Gibb's surface energy of adsorption to be further negative. The Gibb's surface free energy

ΔG_a^S values for the polar probes were obtained by the Schultz's method while using Equations (5)–(7) respectively. The ΔG_a^S versus temperature variation is shown in Figure 6.

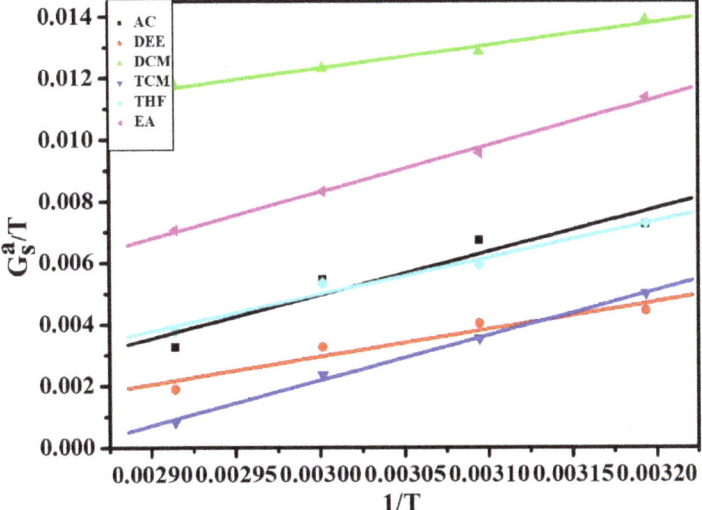

Figure 6. The plot of $-\Delta G_a^S/T$ versus $1/T$ on (P(VDC-co-AN)) (Schultz).

The ΔG_a^S values indicate that the polar probes interacted strongly with the polymer surface. In (P(VDC-co-AN)), the lowest ΔG_a^S value observed in the DCM solvent and the highest $-\Delta G_a^S$ value was observed in EA. The order of increase is shown below.

$$DCM < DEE < THF < AC < TCM < EA$$

Based on Equation (6), the polymer (P(VDC-co-AN)) variation of $-\Delta G_a^S/T$ versus $1/T$ is found to be linear and, hence, a statistical fit has been applied to evaluate ΔH_a^s and ΔS_a^S. The ΔH_a^s, ΔS_a^S values, and the correlation coefficient R^2 are given in Table 3. Furthermore, ΔH_a^s values along with Guttmann Lewis acid-base parameters have been used in Equation (7) to evaluate their Lewis acid-base parameters. The variation of $\Delta H_a^S/AN$ and DN/AN has been shown in Figure 7.

Table 3. The specific components of enthalpy of adsorption, ΔH_a^S and entropy of adsorption, ΔS_a^s for polar probes on (P(VDC-co-AN)).

Solutes	$-\Delta H_a^S$ (kJ mol^{-1})	$-\Delta S_a^s$ (kJ mol^{-1}K^{-1})	r
Acetone	14.07	−0.04	0.95
Di-ethyl ether	9.00	−0.02	0.96
Dichloromethane	7.49	−0.01	0.99
Trichloromethane	14.67	−0.04	0.99
Tetrahydrofuran	11.89	−0.03	0.99
Ethyl acetate	15.19	−0.03	0.99

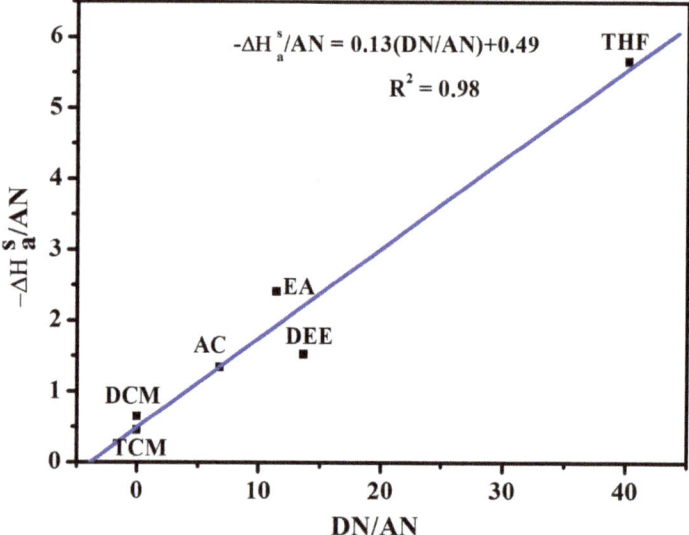

Figure 7. The plot of $\Delta H_a^S / AN$ versus DN/AN for the surface of (P(VDC-co-AN)) (Schultz).

The K_a and K_b values are obtained from the slope and intercept of the linear plot, respectively, and are shown in Table 4. The correlation coefficient (R^2) obtained from the calculation appears to be significant. The K_a and K_b values are related to the amount of acidic and basic sites on the surface of the polymer (P(VDC-co-AN)). The K_a value is 0.13, K_b value is 0.49 and the correlation coefficient R^2 is 0.98.

Table 4. Lewis acid-base parameters, surface character and correlation coefficient of the polymer (P(VDC-co-AN)).

Method	K_a	K_b	S	R^2
Schultz et al.	0.13	0.49	3.77	0.98

The surface character 'S' represents the overall character of the polymer surface. The surface S value is 3.77, which is higher than 1, that implies the overall surface character is considered to be basic, if less than 1, the surface is considered as acidic. Therefore the overall character of the polymer (P(VDC-co-AN)) is considered basic. Accordingly, the polymer (P(VDC-co-AN)) can interact with strongly acidic media. The K_a and K_b values were associated with the quality and nature (acidic or basic) of the sites of (P(VDC-co-AN)) surface. The results revealed that the (P(VDC-co-AN)) polymer surface contains more basic sites than acidic sites and, hence, can closely associate in acidic media.

5.2. CVIP Study on Polymer Visual Traits

The non-linear space-variant transformation of the polymer specimen was examined by using anisotropic diffusion filter. The retrieved image attributes were denoised and edge detected to observe the polymer's intricate patterns shown in Figure 8. The illustrated patterns were unable to observe in the presence of the instrumental noise and imaging anomalies. However, when customized pre-processing is applied, the patterns unveil the real surface perceptions of the polymer in both overview and portrait views. The patterns observed in Figure 8 illustrate the elaborate details of the polymer, such as quills and spikes, which otherwise cannot be seen using low-level imaging instruments. Furthermore, the visual intricate patterns on the pure form of polymer surface enable the chemical enthusiasts to obtain intuition regarding the binding possibilities in fusion with new materials.

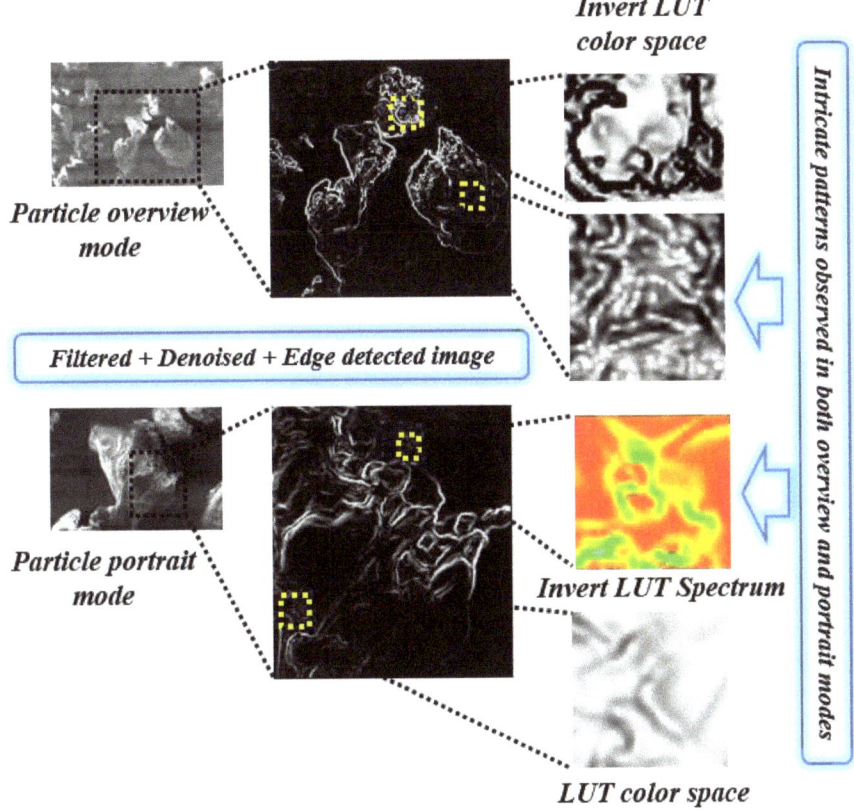

Figure 8. Customized image analysis pipeline for exploring visual intricate patterns in (P(VDC-co-AN)) polymer.

The ISODATA classification has been employed on both overview particle image and portrait image. The SEM was calibrated to set the dimensions as per their captured resolutions and reference scales, such as 500 μm, 200 μm, 50 μm. Moreover, chosen ROI dimensions of 375.00 × 340.91 μm and 265.09 × 257.55 μm were applied on the particle overview and particle portrait, respectively. Essential morphological features, such as edges, were preserved by utilizing filtering techniques, such as guided filters, and further reducing the neighborhood pixel-level noise. The surface of the specimen has been classified into lumps and valleys along with the estimation of their distribution attributes such as surface voxels, area, etc. The clear idea of distribution in the population of lumps and valleys can affect the surface chemical properties of the associating chemical in the context of composite research. As a statistical estimate, the mean lump area estimates were calculated for both overview and portrait modes as 123 μm^2 and 162 μm^2, respectively, which are depicted in Figure 9. Additionally, visual traits of the polymer were extracted using extended methodologies, such as gradient filters and ridge convolutions, which are used to reveal the elaborate lumps and contour details of the specimen. Additionally, the significant peaks in the polymer specimen can help the chemists to estimate the best possible use of the polymer for future composite reactions. The Hough transform has been employed to identify the significant peaks in the specimen and corresponding results were illustrated in Figure A1 in the Appendix B.

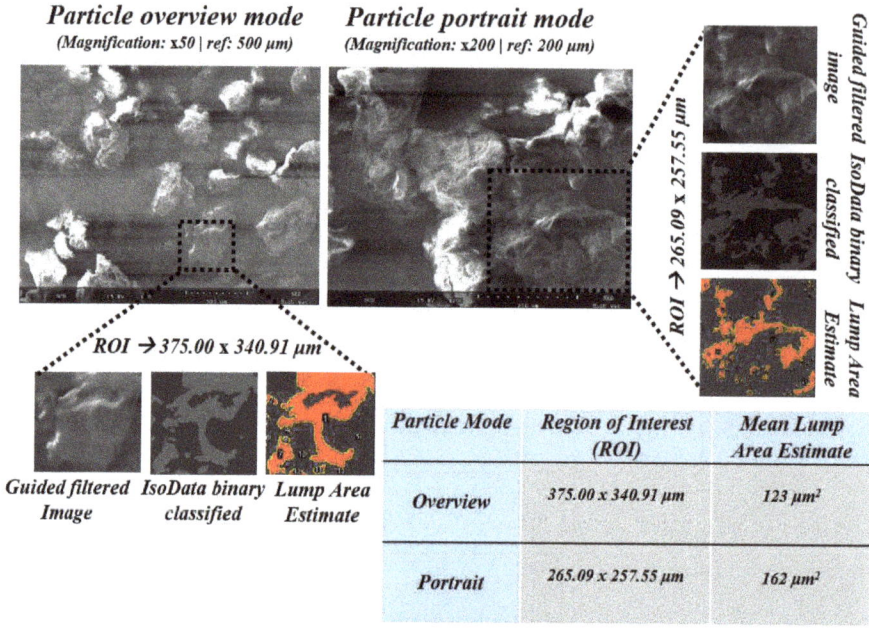

Figure 9. Customized image analysis pipeline for estimation of polymer surface morphology attributes (mean lump area).

Although the resultant visuals obtained from the inspection of the polymer in the image domain were prominent, it is essential to also examine the visual footprint of the polymer (P(VDC-co-AN)) in the frequency domain. In that context, the SEM images were processed while using a customized pipeline of consequent frequency-domain noise removal techniques. The images were filtered using the fast Fourier transform filter which yields the frequency domain component of the image, which comprises all of the frequency-wise distribution analogous to the pixel-wise in the image domain. This is followed by an inverse fast Fourier transform recovering back the image obtained from the frequency components which denoised the image in the frequency domain by retaining the most crucial part of the image. Additionally, to compensate for any spatial irregularities that are caused by the transition of image-to-frequency domain and back- the spatial arrangement of intensity values were balanced using adaptive histogram equalization. Also, the concept of difference image has been formulated from the pixel-wise subtraction of FFT filtered histogram equalized image from the original specimen. This pipeline modeling was illustrated in Figure 10 with additional techniques, such as post-color transformation to Inverse LUT domain and the application of customized lighting simulator. The results obtained using these frequency-based image processing with lighting simulator aids the clear observance of the polymer surface morphology in 3D format. The adaptive histogram equalization method enhanced the image pixel intensities and high-frequency noise components are removed for better visualization. This can be observed in Figure 10 within the visualizations of 3D projection with texture loading. These results might aid the researchers in visually exploring the pure form of the polymer's surface without employing expensive imaging machinery such as TEM.

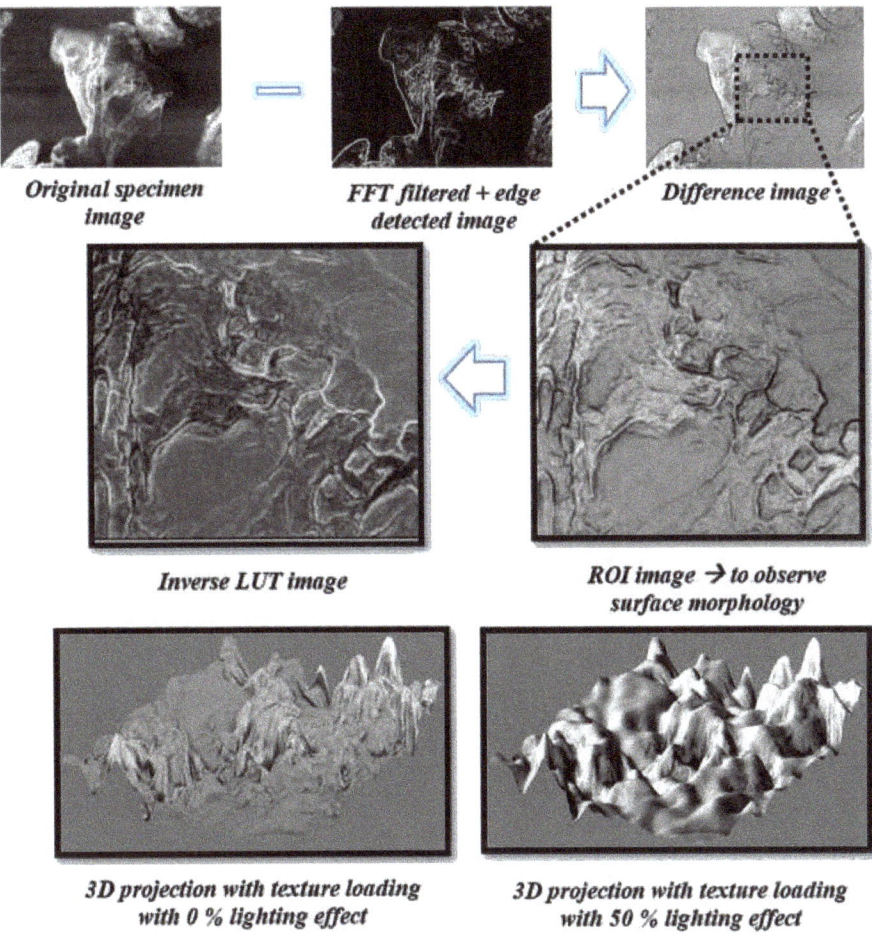

Figure 10. Customized difference image pipeline for investigating texture and three-dimensional (3D) roughness with lighting variations.

Furthermore, properties such as texture parameters were estimated using the SPIP texture analysis toolkit; the results are illustrated in Figures A2–A4 in the Appendix B. Visual observations such as dual axis polymer surface profiling, angular and radial spectrum, texture analysis were estimated from particle overview and portrait modes. These results are shown in Figures A2–A4 assist the researchers to speculate on the texture-based traits of the pure polymer (P(VDC-co-AN)). In terms of statistical analysis, investigation on particle overview mode has been carried out to estimate the particle distribution in terms of area distribution (D_A) and particle directionality (D_P). The refined thresholding pipeline yielded a detection of ($N = 7$) particles within a particle overview ROI of 381.68 × 442.75 µm. The average mean area of the ROI considered was estimated to be 157 µm^2 and the average mean standard deviation was estimated to be 37.2 µm^2.

Figure 11a depicts the clear analysis of the area distribution in the considered specimen along with the particle individual area estimates. The specimen's overview modes in (max = 3) view with resolution ×50 has been pre-processed in order to extract the ROI of 2303.85 × 1580.77 µm and directionality analysis has been made using Fourier components and local gradient orientation. Figure 11b illustrates the directionality of the particles depicted in a plot with histogram bins with

Direction (θ) in X-axis representing the center of the Gaussian fit and Amount in Y-axis represents the amount of directionality, which is calculated by the statistical parameter, such as standard deviation limit threshold. In the case of Fourier components, the peaks ranged between $-90°$ to $+90°$ among which major peaks can be observed between $+5°$ to $+10°$, which represents the polymer particle orientation at that bin in analogous to the overview specimen picture. In the case of local gradient orientation, the peaks can be observed between $0°$ to $+5°$ depicting the strong orientations in the particles. For adhesive applications and etching processes, the polymer surface directionality information could be of assistance towards employing polymer (P(VDC-co-AN)) in conjunction with other materials.

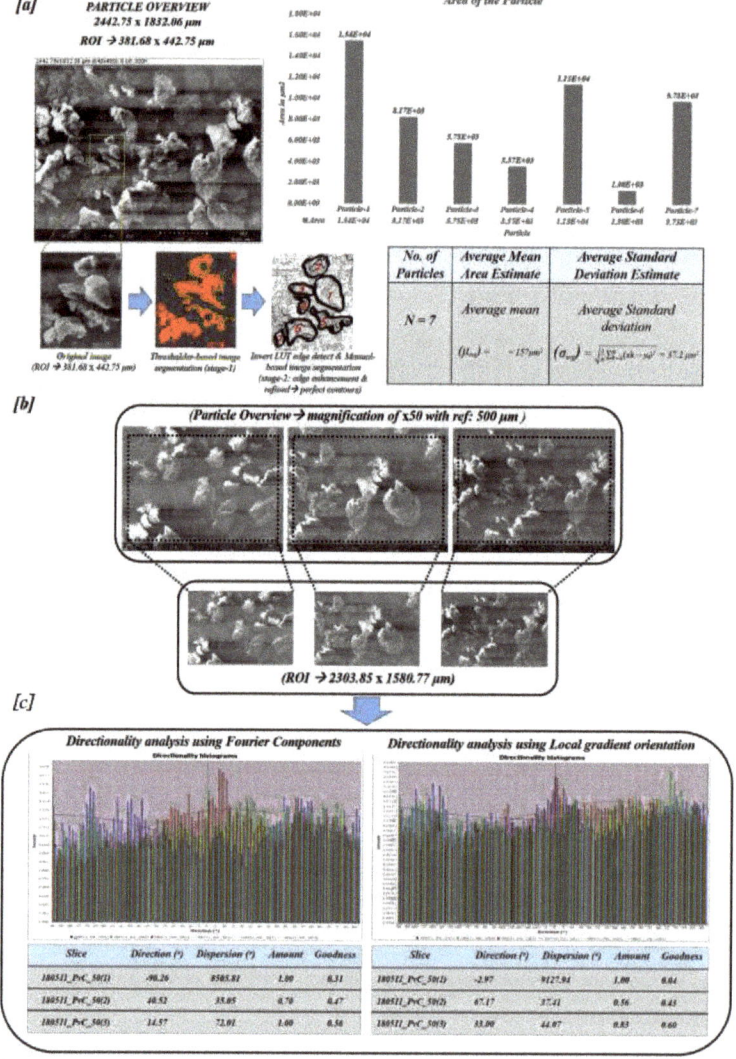

Figure 11. Particle area distribution and directionality analysis: (**a**) Statistical attributes (μ_{avg}, σ_{avg}) of particle area distribution. (**b**) Observation of polymer visual intricate patterns using ridge filter. (**c**) Directionality analysis using Fourier components and local gradient orientation.

The visual patterns of the polymer can also be observed using some standard filters such as gradient filters that enhance the features of the image specimen in certain desirable directions (along x-axis & y-axis). Unlike the gradient filters, the ridge filters can employ the convolutions to work on the range of luminosity in a given image. Both the gradient filter and ridge filters have their fine-tunable kernel size and ridge factor (σ) respectively. With the smaller kernel size in the gradient filter, the intricate patterns can be observed in the specimen and with the larger kernel size, contours can be observed clearly. Likewise, the ridge filter outputs the significant curvatures in the image specimen when applied with a larger ridge factor and all of the intricate patterns can be observed with the smaller ridge factor. This phenomenon can be observed in the Figure A1a,b. The image specimen contains various pixel-wise features and one way to identify the significant features in the image domain is through the application of transforms, such as Hough transform. The Hough parameters (H, ρ, θ) are tuned such that the marked region in Figure A1c of particle portrait region appears to be significant due to the presence of Hough peaks. The additional image processing techniques were employed by the usage of examples provided by MATLAB and Mathematica repositories.

The visual traits of the polymer related to texture and surface roughness were investigated using a SPIP toolbox. The surface morphology of the polymer along with its cross-sectional profiling can be better explored using this toolbox. Additionally, special features in an image, such as angular spectrum and radial spectrum, were plotted along with the brightness distribution of the image specimen. Finally, texture parameters, such as Texture Average (Z_a) and Root Mean Square (Z_q) were plotted on an arbitrary scale and 3D structural projections with optimum lighting conditions were plotted for better understanding of the surface morphology. The above-stated traits were calculated and plotted for overview ($\times 50$) and portrait ($\times 200$, $\times 500$) modes. The values of $Z_{td}[°]$, Z_{tdi}, Z_{tr20} and Z_{tr37} represents texture direction in degrees, texture direction index, aspect ratio at 20% and 37% of the highest peak respectively. The above readings are illustrated in Figures A2–A4 for specimen resolutions of $\times 50$, $\times 200$ and $\times 500$, respectively. These values represent the texture and roughness aspects of the polymer in various modes and resolutions and the angular spectrum represents the angular specification of high occurring peaks of texture. For better visualization, the 3D structural projection of the polymer can be observed in the respective depictions. The use of customized CVIP techniques aims to target and provide the insights of the polymer (P(VDC-co-AN)) surface observations, such as intricate patterns, surface morphology, area distribution, dimensionality, texture, and roughness analysis, which can be used to estimate the proper usage of the pure form of polymer (P(VDC-co-AN)) for future application and research purposes.

6. Conclusions

In this work, polymer (P(VDC-co-AN)) was systematically investigated using IGC and CVIP techniques. Using IGC attributes, such as London dispersive surface energy, Gibbs free energy, and Guttman Lewis acid-base parameters, were analyzed for the polymer. The London dispersive surface energy was calculated while using the Schultz and Dorris–Gray method. The values are decreased linearly with the increase of temperature. The maximum surface value of (P (VDC-co-AN)) is found to be 29.93 and 24.15 in both methods, respectively. The Guttman–Lewis acid-base parameter K_a, K_b values were estimated to be 0.13 and 0.49. The surface character S value is 3.77 and the correlation coefficient is 0.98, respectively. After the thermo-dynamic surface characterization, the (P(VDC-co-AN)) polymer overall surface character is found to be basic. The substantial results revealed that the (P(VDC-co-AN)) polymer surface contains more basic sites than acidic sites and, hence, can closely associate in acidic media. Additionally, the intricate visual patterns, texture traits, and particle distribution statistics were investigated through series of customized image processing pipelines on the pure polymer form an overview ($\times 50$) and portrait ($\times 200$, $\times 500$) modes. This combined study of exploring the (P(VDC-co-AN)) polymer facilitated the better representation of its properties and traits both in chemical and visual perspective.

Author Contributions: Conceptualization, V.K.; Methodology, V.K., P.K.B.; Validation, V.K., P.K.B., V.R.P. and H.K.; Formal Analysis, V.K., P.K.B., V.R.P. and H.K.; Writing—Original Draft Preparation, V.K.; Writing—Review & Editing, V.K., P.K.B. and V.R.P.; Visualization, V.K., P.K.B.; Supervision, H.K., V.R.P.; Project Administration, V.R.P. All authors have read and agreed to the published version of the manuscript.

Funding: This research received funding from Universitas Abdurrab Research Grant: 065/LPPM/SPKINT/VII/2019 and 066/LPPM/SPKINT/VII/2019.

Acknowledgments: The corresponding author Pasupuleti Visweswara Rao would like to thank Universitas Abdurrab for the research grant.

Conflicts of Interest: The authors declare no conflict of interest.

Appendix A. IGC Appendix

The values of AN^* and DN along with the physical property data for the probes used in the IGC experiment are given in Table A1.

Table A1. Physical constants for probes used in IGC experiments.

Solute	$a\left(\gamma_l^d\right)^{0.5} \times 10^{-16}$ cm^2 (mJ/cm^2)$^{0.5}$	a (nm^2)	AN (kJ/mol)	DN (kJ/mol)
n-Hexane	2.21	0.515	-	-
n-Heptane	2.57	0.570	-	-
n-Octane	2.91	0.630	-	-
n-Nonane	3.29	0.690	-	-
n-Decane	3.63	0.750	-	-
Acetone	1.73	0.425	10.5	71.4
Diethyl ether	1.82	0.470	5.88	80.6
Trichloromethane	2.24	0.440	22.7	0
Tetrahydrofuran	2.13	0.450	2.1	84.4
Ethyl acetate	1.95	0.480	6.3	71.8

Appendix B. CVIP Appendix

During the process of image analysis, an extra set of customized image processing pipelines were used from existing algorithms such as ridge filtering, Hough transforms to observe possible outcomes retrieved from a noisy SEM image. The impact of kernel size and gradients in the context of enhancement filtering is shown in Figure A1a. Similarly, the impact of the ridge factor in conjunction with luminosity on the surface curvatures in the polymer image can be observed in Figure A1b. Additionally, the usage of Hough transforms on the portrait mode identified the particle peaks in the surface morphology of the polymer in the observed context. The resulting peaks in the form of polar coordinates are depicted in Figure A1c. In general, techniques such as these can be used to unveil the structural analogies which are often overshadowed by the instrumental noise. The SPIP toolbox has been employed to investigate the polymer texture and surface roughness in conjunction with the polymer surface morphology concerning the cross-sectional profiling (horizontal and vertical). The brightness distributions of the polymer image are analyzed using spectrum in angular and radial modes. These resultant plots are depicted in Figure A2 which are useful for understanding the curvatures of the surface in overview mode. A similar set of attributes were estimated for the portrait modes which were shown in Figures A3 and A4 to have a side-by-side comparison of the cross-sectional profiling and spectra. The SPIP toolbox served as a great tool in calculating various numerical estimates for the texture and roughness of the polymer surface in overview and portrait modes. The Z_a and Z_q values estimated using the toolbox concerning portrait mode with resolution ×500 tend to have higher values indicating the roughness texture of the polymer surface. These supplementary visual observations retrieved from the SPIP image analysis toolbox appeared to offer additional information associated with that of the mainstream visual traits conversed in the main text and thus these supplementary results are organized into this CVIP related appendix section for further reference.

[a]. Finding Intricate patterns using local gradients over x & y

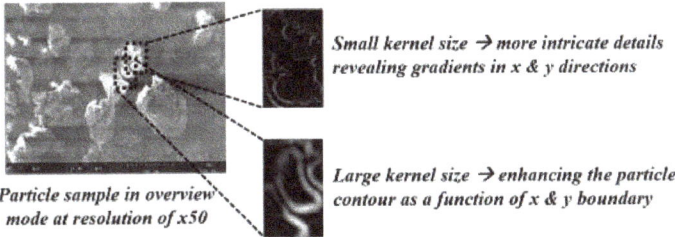

[b]. Finding ridges as a range of luminosity functions employing ridge convolutions (best for observing portrait mode visual properties)

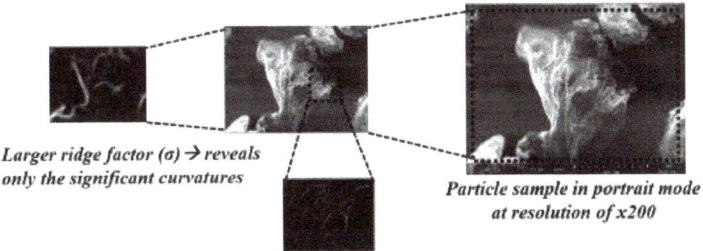

[c]. To identify significant peaks in a particle using Hough transform

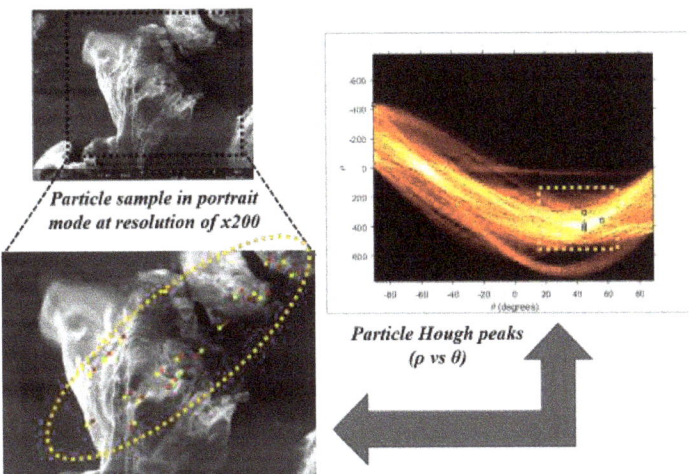

Figure A1. Detecting intricate patterns and substantial peaks: (**a**) Investigation of polymer visual intricate patterns using gradient filter. (**b**) Observation of polymer visual intricate patterns using ridge filter. (**c**) Identification of significant peaks in the polymer image specimen using Hough transform.

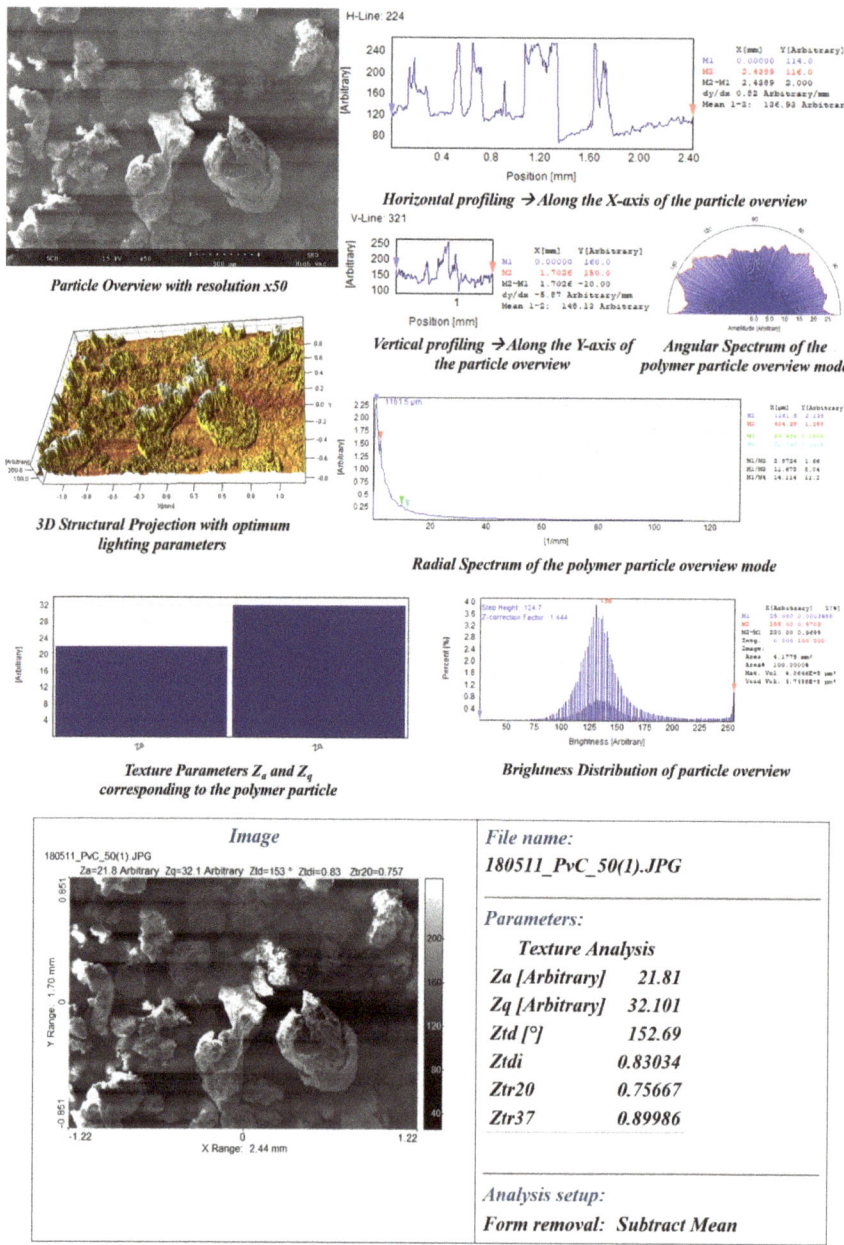

Figure A2. Investigation of texture properties, cross-sectional profiling, angular spectrum and 3D structural projection of the particle (in overview mode with resolution: ×50) using SPIP.

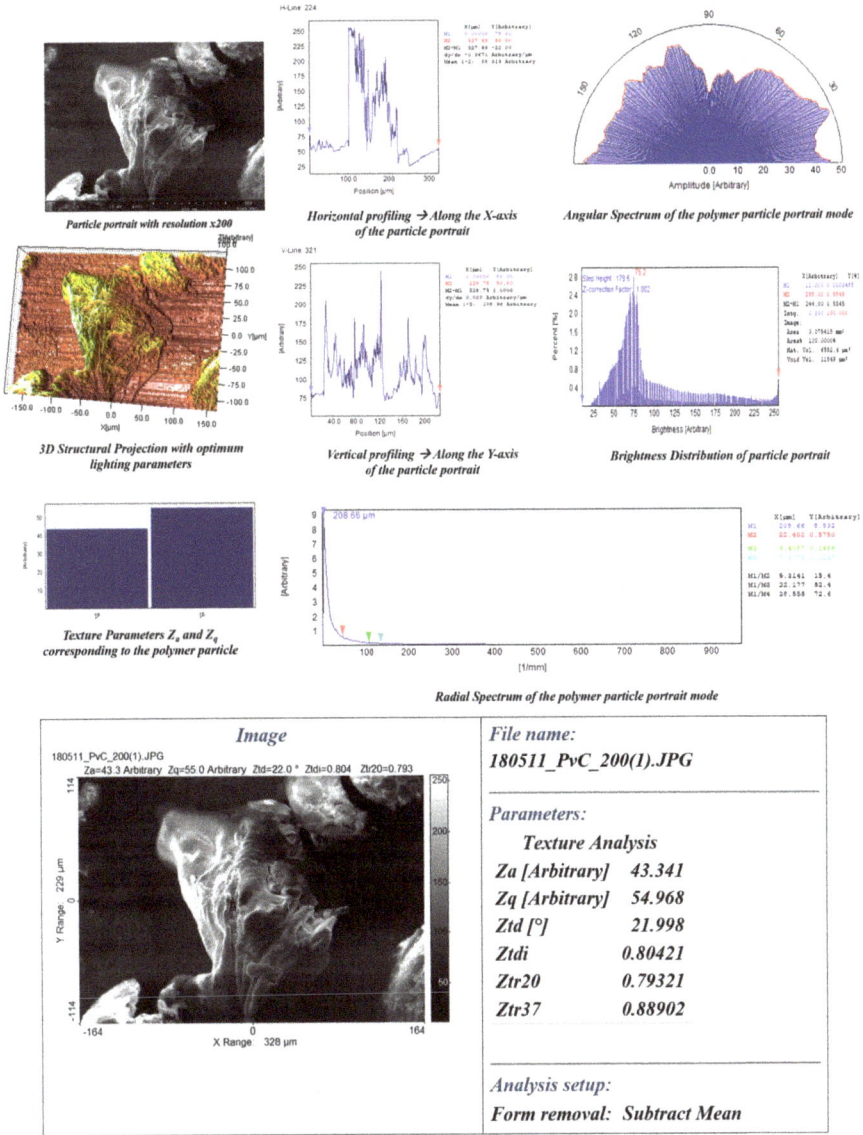

Figure A3. Investigation of texture properties, cross-sectional profiling, angular spectrum and 3D structural projection of the particle (in portrait mode with resolution: ×200) using SPIP.

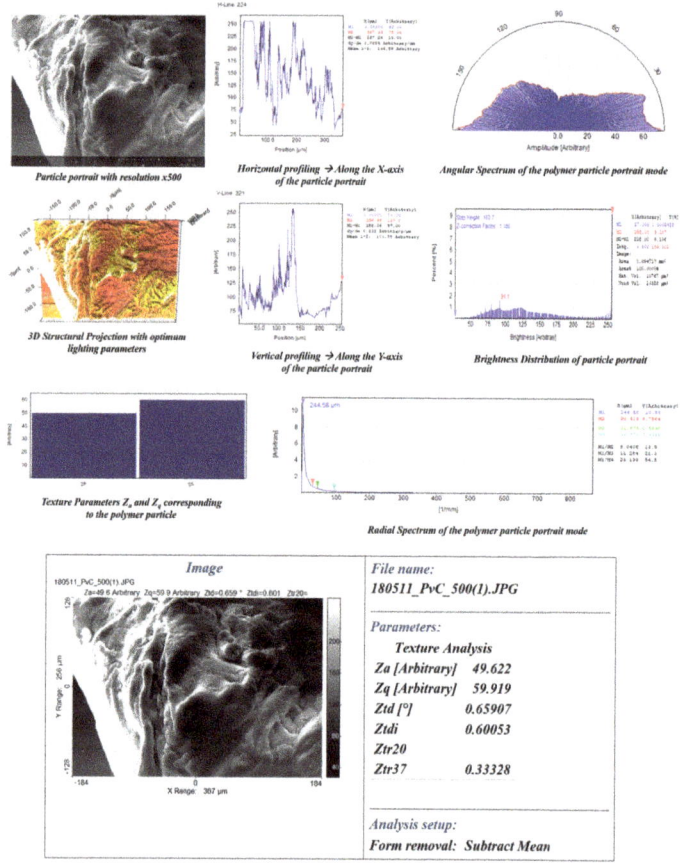

Figure A4. Investigation of texture properties, cross-sectional profiling, angular spectrum and 3D structural projection of the particle (in portrait mode with resolution: ×500) using SPIP.

References

1. Titow, W.V. *PVC Plastics: Properties, Processing, and Applications*; Springer Science & Business Media: Heidelberg, Germany, 2012.
2. Leblanc, J.L. *Filled Polymers: Science and Industrial Applications*; CRC Press: Boca Raton, FL, USA, 2009.
3. Chanda, M.; Roy, S.K. *Industrial Polymers, Specialty Polymers, and Their Applications*; CRC Press: Boca Raton, FL, USA, 2008; Volume 74.
4. Pleşa, I.; Noţingher, P.V.; Schlögl, S.; Sumereder, C.; Muhr, M. Properties of polymer composites used in high-voltage applications. *Polymers* **2016**, *8*, 173. [CrossRef] [PubMed]
5. Lee, J.H.; Kim, K.O.; Ju, Y.M. Polyethylene oxide additive-entrapped polyvinyl chloride as a new blood bag material. *J. Biomed. Mater. Res. Off. J. Soc. Biomater. Jpn. Soc. Biomater. Aust. Soc. Biomater.* **1999**, *48*, 328–334. [CrossRef]
6. Joshi, G.M.; Deshmukh, K. Optimized quality factor of graphene oxide-reinforced PVC nanocomposite. *J. Electron. Mater.* **2014**, *43*, 1161–1165. [CrossRef]
7. Luzio, A.; Canesi, E.V.; Bertarelli, C.; Caironi, M. Electrospun polymer fibers for electronic applications. *Materials* **2014**, *7*, 906–947. [CrossRef]
8. Shi, H.; Liu, X.; Zhang, Y. Fabrication of novel antimicrobial poly (vinyl chloride) plastic for automobile interior applications. *Iran. Polym. J.* **2014**, *23*, 297–305. [CrossRef]

9. Schmid, M.; Wollecke, F.; Levy, G. Long-term Durability of SLS Polymer Components under Automotive Application Environment. In Proceedings of the 23th Annual International Solid Freeform Fabrication (SFF) Symposium, Austin, TX, USA, 6–8 August 2012.
10. Köhl, M.; Jorgensen, G.; Brunold, S.; Carlsson, B.; Heck, M.; Möller, K. Durability of polymeric glazing materials for solar applications. *Sol. Energy* **2005**, *79*, 618–623. [CrossRef]
11. Subbu, C.; Rajendran, S.; Kesavan, K.; Mathew, C. Lithium ion conduction in PVdC-co-AN based polymer blend electrolytes doped with different lithium salts. *Int. Polym. Process.* **2015**, *30*, 476–486. [CrossRef]
12. Gonzalez, A.; Iriarte, M.; Iriondo, P.; Iruin, J. Miscibility and carbon dioxide transport properties of blends of bacterial poly (3-hydroxybutyrate) and a poly (vinylidene chloride-co-acrylonitrile) copolymer. *Polymer* **2002**, *43*, 6205–6211. [CrossRef]
13. Subbu, C.; Mathew, C.M.; Kesavan, K.; Rajendran, S. Electrochemical, structural and optical studies on poly (vinylidene chloride-co-acrylonitrile) based polymer blend membranes. *Int. J. Electrochem. Sci.* **2014**, *9*, 4944–4958.
14. Subbu, C.; Rajendran, S.; Kesavan, K.; Premila, R. The physical and electrochemical properties of poly (vinylidene chloride-co-acrylonitrile)-based polymer electrolytes prepared with different plasticizers. *Ionics* **2016**, *22*, 229–240. [CrossRef]
15. Hwang, T.; Pu, L.; Kim, S.W.; Oh, Y.S.; Nam, J.D. Synthesis and barrier properties of poly (vinylidene chloride-co-acrylonitrile)/SiO2 hybrid composites by sol–gel process. *J. Membr. Sci.* **2009**, *345*, 90–96. [CrossRef]
16. Tanaka, H.; Hayashi, T.; Nishi, T. Application of digital image analysis to pattern formation in polymer systems. *J. Appl. Phys.* **1986**, *59*, 3627–3643. [CrossRef]
17. Pouteau, C.; Baumberger, S.; Cathala, B.; Dole, P. Lignin–polymer blends: Evaluation of compatibility by image analysis. *Comptes Rendus Biol.* **2004**, *327*, 935–943. [CrossRef]
18. Gleeson, J.; Larson, R.; Mead, D.; Kiss, G.; Cladis, P. Image analysis of shear-induced textures in liquid-crystalline polymers. *Liq. Cryst.* **1992**, *11*, 341–364. [CrossRef]
19. Sasov, A.Y.; Ermakova, T.; Lotmentsev, Y.M. Quantitative analysis of images depicting structure of polymers and polymer composites. *Polym. Sci. USSR* **1991**, *33*, 597–602. [CrossRef]
20. Ruskin, R.S.; Yu, Z.; Grigorieff, N. Quantitative characterization of electron detectors for transmission electron microscopy. *J. Struct. Biol.* **2013**, *184*, 385–393. [CrossRef]
21. Uhlemann, S.; Müller, H.; Hartel, P.; Zach, J.; Haider, M. Thermal magnetic field noise limits resolution in transmission electron microscopy. *Phys. Rev. Lett.* **2013**, *111*, 046101. [CrossRef]
22. Misell, D. Conventional and scanning transmission electron microscopy: Image contrast and radiation damage. *J. Phys. D Appl. Phys.* **1977**, *10*, 1085. [CrossRef]
23. Kushwaha, H.S.; Tanwar, S.; Rathore, K.; Srivastava, S. De-noising filters for tem (transmission electron microscopy) image of nanomaterials. In Proceedings of the 2012 Second International Conference on Advanced Computing & Communication Technologies, Rohtak, India, 7–8 January 2012; pp. 276–281.
24. Lloyd, D.R.; Ward, T.C.; Schreiber, H.P. *Inverse Gas Chromatography*; Technical Report; American Chemical Society: Washington, DC, USA, 1989.
25. Voelkel, A. Inverse gas chromatography: Characterization of polymers, fibers, modified silicas, and surfactants. *Crit. Rev. Anal. Chem.* **1991**, *22*, 411–439. [CrossRef]
26. Mohammadi-Jam, S.; Waters, K. Inverse gas chromatography applications: A review. *Adv. Colloid Interface Sci.* **2014**, *212*, 21–44. [CrossRef]
27. Jennings, W.; Mittlefehldt, E.; Stremple, P. *Analytical Gas Chromatography*; Academic Press: San Diego, CA, USA, 1997.
28. Panzer, U.; Schreiber, H.P. On the evaluation of surface interactions by inverse gas chromatography. *Macromolecules* **1992**, *25*, 3633–3637. [CrossRef]
29. Schultz, J.; Lavielle, L.; Martin, C. The role of the interface in carbon fibre-epoxy composites. *J. Adhes.* **1987**, *23*, 45–60. [CrossRef]
30. Dorris, G.M.; Gray, D.G. Adsorption of n-alkanes at zero surface coverage on cellulose paper and wood fibers. *J. Colloid Interface Sci.* **1980**, *77*, 353–362. [CrossRef]
31. Shi, B.; Wang, Y.; Jia, L. Comparison of Dorris–Gray and Schultz methods for the calculation of surface dispersive free energy by inverse gas chromatography. *J. Chromatogr. A* **2011**, *1218*, 860–862. [CrossRef] [PubMed]

32. Umbaugh, S.E. *Computer Vision and Image Processing: A Practical Approach Using Cviptools with Cdrom*; Prentice Hall PTR: Upper Saddle River, NJ, USA, 1997.
33. Qahwaji, R.; Qahwaji, R.; Green, R.; Hines, E.L. *Applied Signal and Image Processing: Multidisciplinary Advancements*; IGI Publishing: Hershey, PA, USA, 2011.
34. Rosenfeld, A. From image analysis to computer vision: An annotated bibliography, 1955–1979. *Comput. Vis. Image Underst.* **2001**, *84*, 298–324. [CrossRef]
35. Kraut, R.; Egido, C.; Galegher, J. Patterns of contact and communication in scientific research collaboration. In Proceedings of the 1988 ACM Conference on Computer-Supported Cooperative Work, Portland, OR, USA, 26–29 September 1988; pp. 1–12.
36. Kakani, V.; Nguyen, V.H.; Kumar, B.P.; Kim, H.; Pasupuleti, V.R. A critical review on computer vision and artificial intelligence in food industry. *J. Agric. Food Res.* **2020**, *2*, 100033. [CrossRef]
37. Kakani, V.; Kim, H.; Kumbham, M.; Park, D.; Jin, C.B.; Nguyen, V.H. Feasible Self-Calibration of Larger Field-of-View (FOV) Camera Sensors for the Advanced Driver-Assistance System (ADAS). *Sensors* **2019**, *19*, 3369. [CrossRef]
38. Kakani, V.; Kim, H.; Lee, J.; Ryu, C.; Kumbham, M. Automatic Distortion Rectification of Wide-Angle Images Using Outlier Refinement for Streamlining Vision Tasks. *Sensors* **2020**, *20*, 894. [CrossRef]
39. Ayache, N. Medical computer vision, virtual reality and robotics. *Image Vis. Comput.* **1995**, *13*, 295–313. [CrossRef]
40. Capitán-Vallvey, L.F.; Lopez-Ruiz, N.; Martinez-Olmos, A.; Erenas, M.M.; Palma, A.J. Recent developments in computer vision-based analytical chemistry: A tutorial review. *Anal. Chim. Acta* **2015**, *899*, 23–56. [CrossRef]
41. Kaehler, A.; Bradski, G. *Learning OpenCV 3: Computer Vision in C++ with the OpenCV Library*; O'Reilly Media, Inc.: Sebastopol, CA, USA, 2016.
42. Higham, D.J.; Higham, N.J. *MATLAB Guide*; SIAM: Philadelphia, PA, USA, 2016.
43. Maeder, R.E. *Programming in Mathematica*; Addison-Wesley Longman Publishing Co., Inc.: Boston, USA, 1991.
44. Abràmoff, M.D.; Magalhães, P.J.; Ram, S.J. Image processing with ImageJ. *Biophotonics Int.* **2004**, *11*, 36–42.
45. Silly, F. A robust method for processing scanning probe microscopy images and determining nanoobject position and dimensions. *J. Microsc.* **2009**, *236*, 211–218. [CrossRef]
46. Hua-jun, L. Application of Origin7. 0 to Chemical Engineering Calculation. *J. Anhui Univ. Sci. Technol. Nat. Sci.* **2005**, *1*, 1–5.
47. Basivi, P.K.; Pasupuleti, V.R.; Seella, R.; Tukiakula, M.R.; Kalluru, S.R.; Park, S.J. Inverse gas chromatography study on London dispersive surface free energy and electron acceptor–donor of fluconazole drug. *J. Chem. Eng. Data* **2017**, *62*, 2090–2094. [CrossRef]
48. Perdigao, J.; Lambrechts, P.; Vanherle, G. Microscopy investigations: Techniques, results, limitations. *Am. J. Dent.* **2000**, *13*, 3D18D.
49. Weickert, J. *Anisotropic Diffusion In Image Processing*; Teubner Stuttgart: Copenhagen, Denmark, 1998; Volume 1.
50. Vogel, C.R.; Oman, M.E. Iterative methods for total variation denoising. *SIAM J. Sci. Comput.* **1996**, *17*, 227–238. [CrossRef]
51. Li, S.; Kang, X.; Hu, J. Image fusion with guided filtering. *IEEE Trans. Image Process.* **2013**, *22*, 2864–2875.
52. Buijs, H.; Pomerleau, A.; Fournier, M.; Tam, W. Implementation of a fast Fourier transform (FFT) for image processing applications. *IEEE Trans. Acoust. Speech Signal Process.* **1974**, *22*, 420–424. [CrossRef]
53. Perona, P.; Shiota, T.; Malik, J. Anisotropic diffusion. In *Geometry-Driven Diffusion in Computer Vision*; Springer: Heidelberg, Germany, 1994; pp. 73–92.
54. He, K.; Sun, J.; Tang, X. Guided image filtering. In Proceedings of the European Conference on Computer Vision, Crete, Greece, 5–11 September 2010; pp. 1–14.
55. Ball, G.H.; Hall, D.J. *ISODATA, A Novel Method of Data Analysis and Pattern Classification*; Technical Report; Stanford Research Institute: Menlo Park, CA, USA, 1965.
56. Liu, Z.Q. Scale space approach to directional analysis of images. *Appl. Opt.* **1991**, *30*, 1369–1373. [CrossRef]

© 2020 by the authors. Licensee MDPI, Basel, Switzerland. This article is an open access article distributed under the terms and conditions of the Creative Commons Attribution (CC BY) license (http://creativecommons.org/licenses/by/4.0/).

Article

Accelerated Shape Forming and Recovering, Induction, and Release of Adhesiveness of Conductive Carbon Nanotube/Epoxy Composites by Joule Heating

Petr Slobodian [1,*], Pavel Riha [2,*], Robert Olejnik [1] and Jiri Matyas [1]

1. Centre of Polymer Systems, University Institute, Tomas Bata University, Tr. T. Bati 5678, 760 01 Zlin, Czech Republic; olejnik@utb.cz (R.O.); matyas@utb.cz (J.M.)
2. The Czech Academy of Sciences, Institute of Hydrodynamics, Pod Patankou 5, 166 12 Prague 6, Czech Republic
* Correspondence: slobodian@ft.utb.cz (P.S.); riha@ih.cas.cz (P.R.)

Received: 30 March 2020; Accepted: 23 April 2020; Published: 1 May 2020

Abstract: The versatile properties of a nanopaper consisting of a porous network of multi-walled carbon nanotubes were applied to enhance the mechanical and electrical properties of a thermosetting epoxy polymer. The embedded nanopaper proved useful both in the monitoring of the curing process of the epoxy resin by the self-regulating Joule heating and in the supervising of tensile deformations of the composite by detecting changes in its electrical resistance. When heated by Joule heating above its glass transition temperature, the embedded carbon nanotube nanopaper accelerated not only the modelling of the composites into various shapes, but also the shape recovery process, wherein the stress in the nanopaper was released and the shape of the composite reverted to its original configuration. Lastly, in comparison with its respective epoxy adhesive, the internally heated electro-conductive carbon nanotube nanopaper/epoxy composite not only substantially shortened curing time while retaining comparable strength of the adhesive bonding of the steel surfaces, but also enabled a release of such bonds by repeated application of DC current.

Keywords: carbon nanotubes; epoxy; Joule heating; fast curing; accelerated forming; shape memory

---

1. Introduction

We recently used a nanopaper made of multi-walled carbon nanotubes (MWCNTs) to monitor infiltration of epoxy resin through a glass fiber textile and study the curing process and deformation of a glass fiber/epoxy composite [1]. The embedded MWCNT nanopaper offers further opportunities for technological applications. In this paper, we introduced curing of carbon nanotube/epoxy composites by Joule heating, and forming of the composites into various shapes. This was enabled by rapid tempering of the composite above its glass transition temperature by means of Joule heating, feasibility of repetition of this reheating and reshaping without losing the material properties, and efficient adhesive bonding of metals by these composites.

The idea of curing of carbon allotrope/epoxy composites by means of self-heating has proven successful in several composite arrangements. For instance, carbon fiber/epoxy composites can be cured by Joule heating. Such cured composites compare favorably with oven-cured ones, thanks to the lower energy consumption of the Joule heating process [2]. Similarly, epoxy resins with dispersed multiwall carbon nanotubes (MWCNTs) or graphene nanoplatelets can be cured well by Joule heating. Examples of practical applications of such curing processes are the use of an epoxy resin with MWCNTs as an adhesive to repair aerospace composite parts or soldering of an assembly of MWCNTs to a metallic surface [3]. Joule heating has been also used to cure an epoxy composite with an embedded single-wall carbon nanotube network and graphene nanoplates supported by glass fibers, which has been applied to deicing [4,5]. A self-regulating out-of-oven manufacturing of a fiber-reinforced epoxy by an embedded layer of graphene nanoplatelets is described in [6]. When compared to state-of-the-art out-of-autoclave oven curing, the curing by nanoplates consumes only 1% of the energy required for respective curing in the oven and has no negative effects on the mechanical performance and glass transition temperature of the final composite. Similarly, the curing by Joule heating of the graphene nanoplatelets/epoxy mixtures results in more compact composite structures with fewer micro-voids and enhanced electrical and mechanical properties than in the respective oven cured composites [7].

It has recently been shown that polypropylene with an embedded carbon fiber veil can be molded into a composite product by Joule heating, with properties comparable to that prepared by an oven heating process [8]. Similarly, since comparable bonding strengths of electrically cured and conventional adhesives has been achieved, Joule heating was successfully used to bond parts made of conductive or non-conductive materials with a MWCNT/epoxy composite [9,10] made of a conductive carbon black/epoxy-based shape memory polymer [11] or a carbon fiber mesh embedded in epoxy [12]. The controlled heating to achieve the desired bondline temperature for the out-of-autoclave adhesive bonding of carbon-fiber composite components has been tested in [13]. The control of the bondline temperature is arranged without an embedded sensor, which can cause a reduced bond strength. The conductive self-healing nanocomposites, which consists of multi-walled carbon nanotubes dispersed into crosslinked polyketons can repair emerged microcracks. The healing effect induced by electricity prolongs the service life of polymer nanocomposites and improves product performance [14]. The flexible piezoresistive sensors based on the RTV-silicone and milled carbon fibers change their resistance when an electric current is applied and Joule heating increases their temperature. The resistance change is caused by swelling of the polymer matrix, which reduces interconnections between the milled carbon fibers [15].

Different embedded reinforcing conductive structures in polymer matrices increase strength and add electrical properties to such polymer composites. These properties may be used for forming shapes of industrial products and parts or for an electrical-driven actuation, among others. The recent progress in the investigation of shape memory polymer composites with embedded carbon black, carbon nanotubes (CNTs), carbon nanofibers (CNFs), and graphene is reviewed in [16]. The carbon black distribution in a polymer matrix correlates with the shape recovery of the composite, which is triggered by applying voltage [17]. While uniform distribution of the spherical carbon black fillers diminishes the chance of their contacts, carbon black aggregates form bunches of microcircuits, which facilitate the shape recovery of such composites when applying a voltage. In addition, a shape memory polymer is employed to form a new composite with temperature and water sensing, as well as actuating capabilities. The shape memory function of this composite is characterized using a Joule heating-based activation method to understand shape recovery at different temperatures [18]. A low-voltage-driven transparent actuator, which is made from a polymer and single-layer carbon nanotube film composites, is used in variable-focus lens. The respective actuating mechanism depends on a large volume change of polymers when they are Joule-heated by electrical current [19].

The above examples show possible versatile uses of the various carbon allotrope structures embedded in epoxy polymers or in similar polymers. However, we think that the use of the carbon nanotube nanopaper embedded in the epoxy in various technological applications is still insufficient. In our previous paper [1], we described the monitoring of curing and deformation of a glass fiber reinforced epoxy by an integrated MWCNT nanopaper. Here, we showed the embedded MWCNT nanopaper in several further applications: induction of rapid Joule heating of the MWCNT nanopaper/epoxy composite, sped-up curing, facilitating of shape recovery of a modelled composite formation to its original configuration, and bonding of parts with shorter times and enabled release of such adhesive bonds by repeated application of DC current.

2. Materials and Methods

Purified MWCNTs produced by chemical vapor deposition of acetylene were supplied by Sun Nanotech Co. Ltd., Jiangxi, China. According to the supplier, the nanotube diameter was 10–30 nm, length 1–10 μm, purity ~ 90%, and volume resistivity 0.12 Ω cm. Further details about the nanotubes and results of the transmission electron microscope (TEM) analysis can be found in our previous papers [20,21].

The aqueous dispersion of MWCNTs (0.8 mg of nanotubes, 530 mL of water with a surfactant system based on a solution of 15.4 g sodium dodecyl sulphate as a surfactant agent and 8.5 mL of 1-pentanol as a co-surfactant agent) was prepared by sonication using the Dr Hielscher GmbH UP400St apparatus (ultrasonic horn S7, amplitude 88 μm, power density 300 W/cm^2, frequency 24 kHz, (Hielscher Ultrasonics GmbH, Teltow, Germany) for 15 min at 50% power of the apparatus, 50% pulse mode, and temperature about 50 °C. NaOH aqueous solution was added to adjust pH to 10.

To make an entangled MWCNT nanopaper (denoted as MWCNTnanopaper further on) from pristine nanotubes, the nanotubes were deposited on a porous polyurethane electrospun non-woven membrane by vacuum filtration. Two hundred and fifty mL of the homogenized MWCNT dispersion was filtered through a funnel 90 mm in diameter. Then, the resulting disk-shaped filtration cake was washed in situ several times with deionized water (at 65 °C) and afterwards with methanol. The filtering membrane was peeled off and dried between two filtration papers for 24 h at room temperature. The measured electrical conductivity of the MWCNT nanopaper was S = 11.97 S/cm.

The two-component epoxy resin Epox G 200 (Davex Chemical s.r.o., Prague, Czech Republic) is a transparent epoxy casting system with adjustable hardness and extended processing time. According to the supplier, the ratio of the epoxy components (A/B-hard/elastic) 100:50 yields a hard epoxy of hardness 79 Shore D, while the component ratio of 100:100 yields a flexible one (44 Shore A). The glass transition temperatures T_g, which was determined by differential scanning calorimetry (DSC) analysis (DSC 1, Perkin Elmer), were 18.4, 35.1 and 58.1 °C for the component ratios of 100:100, 100:75, and 100:50, respectively. The chosen mixing epoxy component ratio of 100:75 enabled manipulation and shape forming of experimental epoxy samples above T_g (35.1 °C) manually. The curing time was 48 h at room temperature. The exothermic heat of the curing was determined by the DSC analysis as −275 J/g and peak temperature 140 °C at a heating rate 10 °C/min. The cured epoxy was elastic and flexible (without any damage at the bending over radius of 70 mm); its ultimate tensile strain was 3.5% and hardness 66 Shore D.

The samples of MWCNT nanopaper/epoxy composite for the tests were fabricated in steps, which included at first the gluing of Cu strip electrodes to the opposite sides of a nanopaper plate. Then, the plate was put on a polytetrafluoroethylene (PTFE) foil, filled with epoxy resin, and its surface covered by the PTFE foil again. To avoid wrinkling in the course of curing by inner Joule heating at a temperature around 100 °C for 6 min, the plate was kept under a pressure load of 300 kPa. The measured electrical conductivity of the resulting composite was S = 3.16 S/cm. The glass transition temperature of the epoxy matrix with the embedded nanopaper, determined as peak temperature, was 53.9 °C.

To illustrate the shape recovery of the bent strips of the MWCNTnanopaper/epoxy composite induced by Joule heating, two electrodes were attached lengthwise to the opposite sides of the strip (length 40 mm, width 20 mm). The electrodes were made of Cuprexit (a thin copper layer supported by a 0.22 mm thick fiberglass plate). The composite strip was stapled to the conductive copper layer of the Cuprexit electrodes at 60 °C, to which in turn were soldered conductive wires. Subsequently, the composite strip was warmed up by Joule heating to about 60 °C and molded by hand alongside an edge to form a right-angled shape. The composite strip was held in that shape by hand until the material cooled to below its T_g and thus retained the imposed shape. Thereafter, the voltage was reapplied, and the deformed composite strip straightened against gravity, regaining its original flat shape.

An identical composite strip was warmed in an oven to 60 °C and a part of the strip was reeled onto a glass rod. It was then held by hand in that shape until the material cooled below its T_g and thus retained the imposed shape. The straight portion of the partially coiled composite strip was attached by double-sided tape to a vertical side of a box and the coiled portion was left protruding above. Next, the misshapen composite strip was reheated in an oven to 60 °C and its coiled portion straightened against gravity, regaining its original flat shape.

Thermogravimetric analysis (TGA) of the epoxy and MWCNT nanopaper/epoxy composite samples was carried out using the thermogravimeter SETARAM SETSYS Evolution 1200 (METTLER TOLEDO, Prague, Czech Republic). The samples were examined under inert atmosphere of helium (5.5 purity, SIAD TP s.r.o., Prague, Czech Republic), which minimized the oxidation-dependent weight fraction loss of carbon nanotubes. Gas flow of 30 cm^3/min at pressure of 101.325 kPa (i.e., 30 sccm) was set for all experiments. A platinum crucible was used to hold the sample, which weighted about 4 mg. The temperature was increased at a rate of 20 °C/min within the range from ambient room temperature to 1200 °C.

The structure of the MWCNT nanopaper and the cross-section of the nanopaper/epoxy composite were analyzed by a scanning electron microscope (SEM) (NOVA NanoSEM 450, FEI Co., Advex Instruments, Brno, Czech Republic). The surface temperature of the specimens was measured by means of a thermal camera (Flir E5-XT, FLIR Systems, Inc., Sally Gao, China), which was able to create a detailed temperature pattern. Sensor resistance was measured lengthwise by the two-point technique using the Multiplex datalogger 34980A (KEYSIGHT Technologies, Santa Rosa, CA, USA), which stored the readouts once per second. The DC power supply Metex AX 502 (AEMC Instruments, Dover, DE, USA) was used to power the Joule heating of composites. Tensile tests were carried out using the Testometric M350-5CT system (Testometric Co. Ltd., Rochdale, UK).

3. Results

The temperature progression induced by the imposed electric power on the MWCNT nanopaper and the dependence of the reached maximal temperature on the electric power are shown in Figure 1.

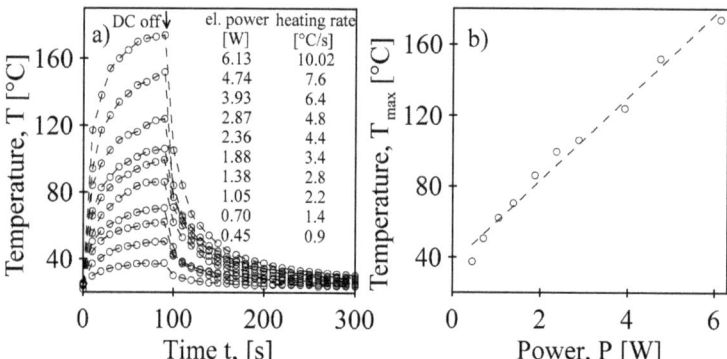

Figure 1. (a) Temperature progression of the MWCNT nanopaper through Joule heating at different electrical power rates and an initial heating rate. (b) The dependence of the reached maximal temperature on the applied DC electrical power through Joule heating.

The process of epoxy resin curing is shown in Figure 2a. The changing shades of grey in the photographs correspond to the advancing cross-linking of epoxy resin at the indicated times and temperatures. The temperature progression of the composite in Figure 2b indicated at first progressive resistive heating by DC voltage 4.6 V through inter-tube contacts, up to a temperature of 105 °C, which was reached in about 40 sec. The following temperature decrease was apparently a consequence of epoxy resin polymerization, which affected the number of inter-nanotube contacts and/or the contact resistance and resulted in a reduction of the applied power. On the other hand, when the same voltage was applied, the temperature of the MWCNT nanopaper continuously rose, thanks to the unaffected nanotube contacts until the temperature reached the arbitrarily chosen limit of 200 °C.

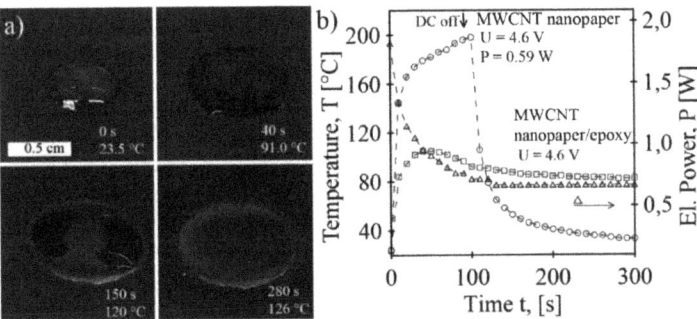

Figure 2. (a) Time-dependent curing of the epoxy resin droplet on the surface of the entangled MWCNT nanopaper by Joule heating at constant DC voltage. Respective times of curing and corresponding temperatures are indicated in the figure. (b) Temperature progression in the course of MWCNT nanopaper/epoxy curing by Joule heating at DC voltage 4.6 V is denoted by squares, the corresponding time dependence of the electrical power by triangles, and the time-dependence of the temperature during the MWCNT nanopaper heating at DC voltage of 4.6 V by circles.

The presence of MWCNTs affected the curing kinetics, which was obvious from the value of the glass transition temperatures of pure epoxy and of the MWCNTnanopaper/epoxy (100:75) composite. The measured T_g as peaks by the DSC was 35.1 °C for the former and 53.9 °C for the latter. The peak shift to the higher temperature was similar to the effect of the decreasing of the ratio between the resin components A:B in favor of component A. The peak shift was not due to creation of the second phase of epoxy as the immobilized rigid amorphous fracture near the MWCNT surface [22]. Only one enthalpy relaxation peak was measured and enthalpy relaxation, as expressed by the term of enthalpy losses [23], was not changed by the MWCNT presence. The single enthalpy relaxation peak represents transition energy from the glassy state to the rubbery state and is not affected by the embedded MWCNTs [23].

Component A was an epoxy resin prepolymer (hydrogenated bisphenol A polymer with epichlorohydrin [CAS: 30583-72-3]). The prepolymer was prepared with sufficient excess of epichlorohydrin so the termination was by free oxirane rings capable to react chemically with amine groups. Component B (Trimethylolpropane tris[poly(propylene glycol), amine terminated] ether [CAS: 39423-51-3]) was a hardener or a curing agent containing three polymeric chains of poly(propylene glycol) terminated by amine groups. The poly(propylene glycol) chains themselves are highly movable, capable to rotate around a single chemical bond. The chains contain oxygen, which has a relatively low rotational barrier, and therefore the conformation of the chains changes easily under, e.g., an applied mechanical stress. The reaction of components A and B leads to the formation of a three-dimensional polymer network, resulting in a final epoxy matrix. At an excess of component B at the 100:100 ratio, some poly(propylene glycol) branches remain unbounded, by a lack of oxirane rings with a high molecular mobility potential, and the epoxy is elastic. With a lack of component B at the 100:50 ratio, the proportion of functional groups for crosslinking is more equimolecular. The epoxy contains less free unbounded poly(propylene glycol) branches, resulting in resistance against conformation changes, stiffness, and a higher T_g. The epoxy curing in the presence of MWCNTs, which have oxygenated functional groups on their surface [24], increases epoxy matrix rigidity since the amine groups link to the carboxylic groups attached to the nanotubes, forming imide covalent bonds [25]. It was also confirmed independently that the curing of melamine-formaldehyde resin in the presence of the COOH-functionalized MWCNTs was affected by amine groups [26].

X-ray photoelectron spectroscopy (XPS) signals were measured to obtain information on functional groups attached to the nanotube surfaces [24]. The signals from MWCNTs were recorded by the Thermo Scientific K-Alpha XPS system (Thermo Fisher Scientific, UK) equipped with a micro-focused, monochromatic Al Kα X-ray source (1486.6 eV). The main binding energy peak (284.5 eV) in the XPS spectra of MWCNTs was assigned to the C1s–sp^2, while the other ones were assigned to the respective oxygenated functional groups C–O (286.2 eV), C=O (287.1 eV), O–C=O (288.6–289 eV), and C1s–π–π^* (291.1–291.5 eV).

A TGA analysis was performed on the MWCNT nanopaper, the MWCNT nanopaper/epoxy composite, and the cured epoxy to determine the weight fraction of MWCNTs, Figure 3. The TGA curves remained flat until the temperature reached about 300 °C when a loss of weight of the composite and the cured epoxy was observed owing to a decomposition of the epoxy. The weight percentages of the embedded MWCNTs in the composite after heating was (slightly) below 40 wt.%, which corresponded reasonably well to the weight fraction of the MWCNT nanopaper with the calculated porosity $\varphi = 0.67 = 1 - \rho_{nan}/\rho_{MWCNT}$, where $\rho_{nan} = 0.56 \pm 0.03$ g/cm^3 denoted the measured apparent density of the nanopaper and $\rho_{MWCNT} = 1.7$ g/cm^3 the measured average density of nanotubes, which was close to the theoretical value of 1.8 g/cm^3 [12].

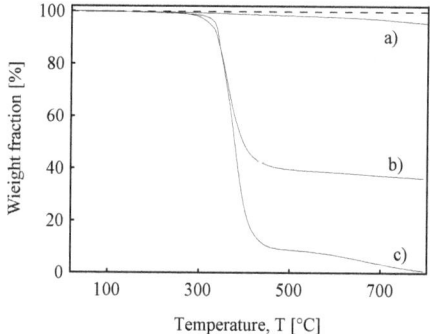

Figure 3. TGA curves of (**a**) the pure MWCNT nanopaper, (**b**) the MWCNT nanopaper/epoxy composite, and (**c**) the cured pure epoxy matrix.

A tensile test of the MWCNT nanopaper, the MWCNT nanopaper/epoxy composite, and the cured epoxy was carried out to determine their fracture tensile strength and strain. According to the test results in Figure 4, the MWCNT nanopaper had a sharp break at brittle fracture strength of about 1 MPa and strain 0.75%. The cured epoxy at first experienced strain hardening through plastic deformation and then a necking until fracture strain of 3.5% and tensile stress of 14 MPa was reached. On the other hand, the MWCNT nanopaper/epoxy composite strengthened until ultimate tensile stress of 24 MPa at fracture strain 2.9%.

A detailed view of the structure of an individual nanotube consisting of about 15 rolled sheets of graphene obtained by means of TEM is shown in Figure 5a. SEM micrographs depict the surface of a porous structure of the MWCNT nanopaper (Figure 5b), and the cross-section through the epoxy composite with the embedded conductive MWCNT nanopaper with some nanotubes protruding from the epoxy matrix surface together with cut nanotubes (bright spots on the surface), Figure 5c. Finally, Figure 5d depicts the temperature distribution of the MWCNT nanopaper/epoxy strip as measured by thermal camera Flir E5-XT in the course of the curing by means of the embedded Joule-heated carbon nanotubes. The temperature of the strip decreased towards the edges, apparently due to higher cooling in the marginal areas.

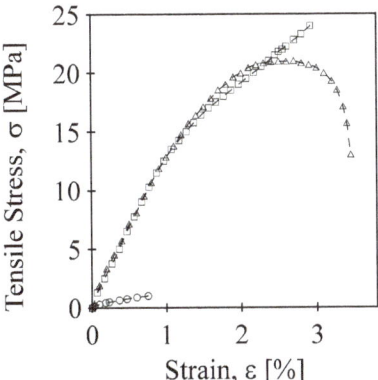

Figure 4. Tensile tests on the MWCNT nanopaper (circles), the MWCNT nanopaper/epoxy composite (squares), and the cured epoxy (triangles) at elongation rate of 5 mm/min.

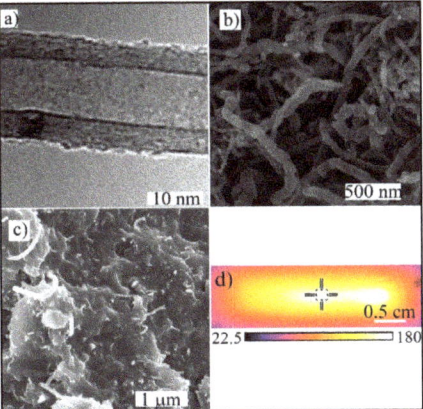

Figure 5. (**a**) A TEM image of an individual MWCNT with visible lines of rolled graphene layers at a distance of about 0.35 nm. (**b**) A SEM micrograph of the surface of the MWCNT nanopaper. (**c**) The cross-section of the MWCNT nanopaper/epoxy composite cured by Joule heating with cuts of individual MWCNTs, which protruded from the plane of the cross-section. (**d**) The thermographic image of the temperature distribution in the MWCNT nanopaper/epoxy strip in the course of curing at 4.6 V together with the temperature scale. The cross indicates the temperature of 173 °C.

The fabricated thermoset composites, which consisted of the epoxy matrix and the embedded conductive MWCNT nanopaper, offer a novel technique of formation by means of Joule heating. This self-heating, which was used for the curing of composite plates and strips, was repeated to quickly heat the composite plates (in 1–2 min) to the forming temperature of 70 °C, which was above their glass transition temperature T_g of 53.9 °C. At this temperature, the composites went from being rigid and glassy to rubbery and flexible, and thus highly formable. When a bending force was applied, the strips deformed easily to the chosen shapes, which they held after cooling below T_g, despite the deformed embedded MWCNT nanopaper (Figure 6).

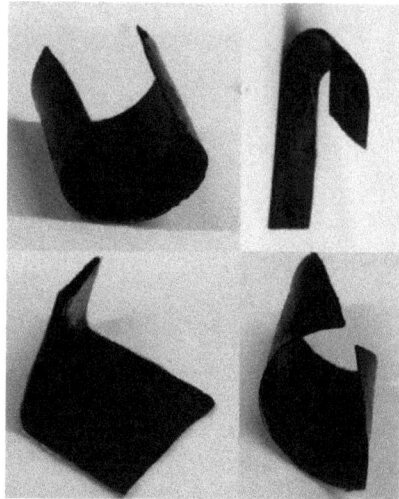

Figure 6. Different shapes of the MWCNT nanopaper/epoxy composites manipulated by hand after composite Joule heating to 70 °C and cooling below the T_g temperature of 53.9 °C to form new shapes.

Surprisingly, when the bent strips were reheated (as described in Section 2), their original shape recovered. The photos illustrating recovery by means of both nanopaper self-heating and heating in an oven at 60 °C are shown in Figure 7a,b, respectively (see Supplementary Video). After the composite is heated over T_g to the transformation temperature of the shape memory, shape recovery stress in the embedded MWCNT nanopaper was released due to the epoxy matrix softening. Apparently, the embedded nanotubes were compressed on the inside surface of the bent strip and stretched on the outside. When the nanotubes were released and were able to move within the epoxy matrix, the deformed arrangement of the stretched portion and compressed portion of the nanopaper recovered to its initial arrangement. Moreover, the strip's Joule heating through inter-nanotube contacts was more efficient than heating in an oven. Consequently, when Joule heating was applied to the bent composite strip attached to a box, the time of the strip shape's unbending reached about 20 s (Figure 7a), which was about one order of magnitude shorter than that of the strip coiled portion's recovery in the oven (Figure 7b).

Rapid heating of an embedded nanopaper may facilitate polymerization of the epoxy components of the composite and thus aid its adhesive properties. To test this surmise, we used two zinc-coated steel strips (length 100 mm, width 10 mm, thickness 0.5 mm, overlap 10 mm, overlapping area 100 mm^2) on which a few drops of the epoxy were deposited; the MWCNT nanopaper was attached, a few drops of the epoxy were added on top of the nanopaper surface, and then the second steel strip was overlaid. The overlapped strips were under pressure load of 300 kPa heated by electric power of 6 W to a temperature of about 160 °C for 5 min (Figure 8). In comparison, in the absence of the MWCNT nanopaper/epoxy composite, adhesive bonding of the identical zinc-coated steel strips with the ordinary epoxy resin under the same load of 300 kPa took 48 h: initially 24 h at room temperature, which was followed by curing at the bonding temperature of 60 °C in an oven for an additional 24 h. According to the results (Figure 8), the test of the adhesive strength of the bonding of the steel strips with both the MWCNT nanopaper/epoxy and epoxy adhesives gave ultimate shear stress values of 19.4 ± 0.4 N/mm^2 and 17.1 ± 0.3 N/mm^2, respectively. The corresponding ultimate strains were 7.4 and 10.3%. Thus, the MWCNT nanopaper/epoxy composite achieved comparable strength of bonding; yet, the required curing time was considerably shorter than when conventional epoxy adhesive was used. Moreover, the bonding retained its electrical conductivity and thus was able to be further modulated by the Joule heating. Once the temperature of the MWCNT nanopaper/epoxy composite, which glued together the steel strips, reached 72 °C within about 1 min, the bond of the steel strips was released at ultimate shear stress of 1.2 N/mm². Hence, the bond of the steel strips, which overlapped at 100 mm² and could withstand a total adhesive force of 1940 N (which is equivalent to about 200 kg of load under gravity) could be released by force of 120 N or about 12 kg of load under gravity, which can be readily exerted by human hand.

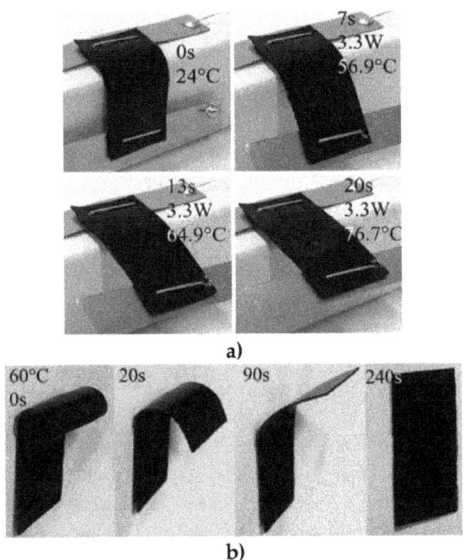

Figure 7. The time-dependent shape memory recovery of the MWCNT nanopaper/epoxy strip after (**a**) Joule heating to 76.7 °C (3.3 W), (**b**) heating in an oven to 60 °C. The respective temperatures, electric power, and times are included in the figure.

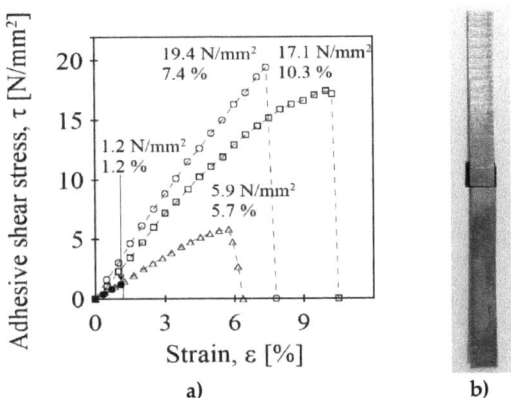

Figure 8. (**a**) The shear stress-strain relations for the bonded steel strips using the MWCNT nanopaper/epoxy composite (circles) and the epoxy adhesive (squares), and for the reheated adhesive bonding by the MWCNT nanopaper/epoxy composite at temperature of 55 °C (triangles) and 72 °C (filled circles). The respective ultimate stress and ultimate strain values are included in the figure. (**b**) The overlapped strips bonded by the MWCNT nanopaper/epoxy composite.

4. Discussion

We created an MWCNT nanopaper/epoxy composite consisting of electro-conductive MWCNT nanopaper, which was embedded in the epoxy matrix, and investigated its applications in several processing techniques. Previously, we showed that such an MWCNT nanopaper/epoxy composite can be used to monitor infiltration of epoxy resin through a glass fiber textile or a curing process of glass fiber/epoxy composite together with testing of its deformation [1]. In this paper, we focused on the process of epoxy resin curing by Joule heating. The results suggested that the composite was able to self-regulate the Joule heating (Figure 2). After initial rapid rise of the temperature of the cured prepolymer, a phase of decrease in temperature followed. It is probable that this decrease was due to volume changes resulting from polymerization. In the curing process, the volume of epoxy resin initially expands during polymerization [27]. As the curing continues, the volume shrinks. In this process, van der Walls interactions of the embedded nanotubes are apparently affected. This may change the nanopaper resistivity and/or number of inter-nanotube contacts, which in turn can be reflected in a reduction of applied power and a decrease in temperature.

One of the advantages of the MWCNT/nanopaper composite was the facility of its formation by means of Joule heating. While heated, the composite quickly (within 1–2 min) reached T_g and became easily modellable into various shapes, which were retained once the composite cooled down. Such a utilization of Joule heating obviates the necessity of using an oven to form the given composites and thus may be beneficial e.g., for such processes done ad hoc, in restricted spaces or in situ in complex systems.

The MWCNT nanopaper/epoxy composite could not only be easily formed, but also retained its shape memory. Namely, when Joule heating was reapplied to a molded MWCNT nanopaper/epoxy composite, it reverted to its original shape. Such a process was in an order of magnitude shorter when Joule heating was applied than when an oven was used. Thus, the shape recovery time as well as processability, light weight, and low manufacturing costs may be a highly desirable combination.

Facile forming and shape memory apart, another advantageous property of the MWCNT nanopaper/epoxy composite is its adhesiveness. Albeit no substantial difference in the strength of the bonding was observed when the composite was compared with a respective conventional epoxy adhesive, the time period necessary for curing and thus establishing a firm connection between the surfaces was substantially shorter in the nanomaterial, thanks to the possibility of application of Joule heating. In addition, repeated application of the DC current released these bonds, which suggests that materials based on the MWCNT nanopaper/epoxy composites might be used e.g., to form electrically controllable seams.

5. Conclusions

We combined a multi-walled carbon nanotube free-standing nanopaper and an epoxy resin into a reinforced thermosetting polymer matrix composite and demonstrated its versatile use in technological processes with advanced processing parameters, thus saving time and electrical energy. In our earlier papers [20,21,24], the nanopaper was introduced individually or embedded in thermoplastic polymers as a multifunctional sensor of ambient vapors or of a tensile and compressive stress. The nanopaper has been used as a built-in sensor in a glass fiber/epoxy composite to primarily measure composite deformation, as well to monitor epoxy resin infusion into layered glass fiber textiles and, subsequent curing process [1]. The nanopaper embedded in the epoxy matrix was investigated here as a thermosetting material in several practical applications listed in Table 1. However, at first, the properties and technological parameters of this conductive epoxy composite were measured to specify particulars of its manufacturing and use. In particular, we assessed the rise of nanopaper temperature in the course of its Joule heating at different voltages, temperature progression in the course of the MWCNT nanopaper/epoxy curing by Joule heating, TGA analysis of the MWCNT nanopaper/epoxy composite, tensile tests of the MWCNT nanopaper/epoxy composite, and shear stress-strain relations for the steel strips bonded by the MWCNT nanopaper/epoxy composite or conventional epoxy adhesive as well as for the reheated adhesive bonding by the MWCNT nanopaper/epoxy composite.

Table 1. Overview of introduced applications of the MWCNT nanopaper/epoxy composite.

Initial State of Material	Unit Operation	Particularity	Outcome
Nanopaper filled up with epoxy resin	Curing by Joule heating	Self-regulating Joule heating	MWCNTnanopaper/epoxy composite
MWCNTnanopaper/epoxy composite	Forming by Joule heating	Forming temperature reached in 1–2 min	Novel technique of forming
MWCNTnanopaper/epoxy composite	Shape recovery by Joule heating	Recovery stress release by matrix softening	Short shape recovery time
Nanopaper filled up with epoxy resin	Adhesive bonding by Joule heating	Bonding time within minutes	Higher adhesion then by epoxy adhesive
MWCNTnanopaper/epoxy composite	Debonding adhesive joint by Joule heating	Bond of steel strips released within 1 min	Easy release without destroying parts

An innovative finding was self-regulation of the epoxy curing temperature (Figure 2). After the initial rapid rise in temperature of the cured prepolymer, the phase of decrease in temperature followed, probably due to volume changes resulting from the polymerization, as discussed in Section 4. The shape recovery measurement accentuated the contribution of Joule heating in the shape recovery process. This transition from deformed to undeformed arrangement was an order of magnitude shorter when Joule heating was applied, compared to when an oven was used for heating of the MWCNT nanopaper/epoxy composite, thanks to efficient heating through the inter-nanotube contacts. The second innovative finding was the possibility of the efficient and fast debonding of objects glued by the MWCNTnanopaper/epoxy composite. Reheating of the composite over T_g decreased its adhesive stress and consequent release was done readily by human hand. The list of possible applications of the MWCNT nanopaper/epoxy composite and modulation of its characteristics by Joule heating, as presented in Table 1, need not be exhaustive, as further feasible practical applications are being evaluated.

Supplementary Materials: The following are available online at http://www.mdpi.com/2073-4360/12/5/1030/s1.

Author Contributions: P.S.: Coordinator of measurement, measurement, writing, figures designer, goals and ideas, literature search, and writing; P.R.: Writing, goals and ideas, literature search, proof reading; R.O.: sample preparation, help with measurement; J.M.: sample preparation, help with measurement. All authors have read and agreed to the published version of the manuscript.

Funding: This work was supported by the Ministry of Education, Youth and Sports of the Czech Republic-Program NPU I (grant number: LO1504); the Operational Program Research and Development for Innovations co-funded by the European Regional Development Fund; the national budget of the Czech Republic, within the framework of the project CPS-strengthening research capacity (reg. number: CZ.1.05/2.1.00/19.0409) and the Fund of the Institute of Hydrodynamics, grant number AV0Z20600510.

Conflicts of Interest: The authors declare no conflict of interest.

References

1. Slobodian, P.; Pertegás, S.L.; Riha, P.; Matyas, R.; Olejnik, R.; Schledjewski, R.; Kovar, M. Glass fiber/epoxy composites with integrated layer of carbon nanotubes for deformation detection. *Compos. Sci. Technol.* **2018**, *156*, 61–69. [CrossRef]
2. Joseph, C.; Viney, C. Electrical resistance curing of carbon-fibre/epoxy composites. *Compos. Sci. Technol.* **2000**, *60*, 315–319. [CrossRef]
3. Mas, B.; Fernandez-Blazquez, J.P.; Duval, J.; Bunyn, H.; Vilatela, J.J. Thermoset curing through Joule heating of nanocarbons for composite manufacture, repair and soldering. *Carbon* **2013**, *63*, 523–529. [CrossRef]
4. Chu, H.; Zhang, Z.; Liu, Y.; Leng, J. Self-heating fiber reinforced polymer composite using meso/macropore carbon nanotube paper and its application in deicing. *Carbon* **2014**, *66*, 154–163. [CrossRef]

5. Zhang, Q.; Yu, Y.; Yang, K.; Zhang, B.; Zha, K.; Xiong, G.; Zhang, X. Mechanically robust and electrically conductive graphene-paper/glass-fibers/epoxy composites for stimuli-responsive sensors and Joule heating deicer. *Carbon* **2017**, *124*, 296–307. [CrossRef]
6. Liu, Y.; Van Vliet, T.; Tao, Y.; Busfield, J.J.C.; Peijs, T.; Bilotti, E.; Zhang, H. Sustainable and self-regulating out-of-oven manufacturing of FRPs with integrated multifunctional capabilities. *Compos. Sci. Technol.* **2020**, *190*, 108032. [CrossRef]
7. Xia, T.; Zeng, D.; Li, Z.; Young, R.J.; Vallés, C.; Kinloch, I. Electrically conductive GNP/epoxy composites for out-of-autoclave thermoset curing through Joule heating. *Compos. Sci. Technol.* **2018**, *164*, 304–312. [CrossRef]
8. Liu, C.; Li, M.; Gu, Y.; Gong, Y.; Liang, J.; Wang, S.; Zhang, Z. Resistance heating forming process based on carbon fiber veil for continuous glass fiber reinforced polypropylene. *J. Reinf. Plast. Comp.* **2018**, *37*, 366–380. [CrossRef]
9. Sung, P.C.; Chang, S.C. The adhesive bonding with buckypaper-carbon nanotube/epoxy composite adhesives cured by Joule heating. *Carbon* **2015**, *91*, 215–223. [CrossRef]
10. Monreal-Bernal, A.; Mas, B.; Fernández-Blázquez, J.P.; Vilatela, J.J. Electric curing of nanocarbon/epoxy adhesives for composite repair. In Proceedings of the ECCM16—16th European Conference on Composite Materials, Seville, Spain, 22–26 June 2014.
11. Eisenhaure, J.; Kim, S. An internally heated shape memory polymer dry adhesive. *Polymers* **2014**, *6*, 2274–2286. [CrossRef]
12. Ashrafi, M.; Devasia, S.; Tuttle, M.E. Resistive embedded heating for homogeneous curing of adhesively bonded joints. *Int. J. Adhes. Adhes.* **2015**, *57*, 34–39. [CrossRef]
13. Smith, B.P.; Ashrafi, M.; Tuttle, M.E.; Devasia, S. Boundary control of embedded heaters for uniform bondline temperatures during composite joining. *ASME J. Manuf. Sci. Eng.* **2018**, *140*, 091013. [CrossRef]
14. Lima, M.G.R.; Orozco, F.; Picchioni, F.; Morreno-Villoslada, I.; Pucci, A.; Bose, R.K.; Araya-Hermosilla, R. Electrically Self-Healing Thermoset MWCNTs Composites Based on Diels-Alder and Hydrogen Bonds. *Polymers* **2019**, *11*, 1885. [CrossRef] [PubMed]
15. Sanchez-Gonzales, C.M.; Soriano-Pena, J.P.; Rubio-Avalos, J.C.; Pacheco-Ibarra, J.J. Fabrication of flexible piezoresistive sensors based on RTV-silicone and milled carbon fibers and the temperature´s effect on their electric resistance. *Sens. Actuators A Phys.* **2020**, *302*, 111311. [CrossRef]
16. Lei, M.; Chen, Z.; Lu, H.; Yu, K. Recent progress in shape memory polymer composites: Methods, properties, applications and prospects. *Nanotechnol. Rev.* **2019**, *8*, 327–351. [CrossRef]
17. Le, H.H.; Kolesov, I.; Ali, Z.A.; Uthardt, M.; Osazuwa, O.; Ilisch, S.; Radusch, H.J. Effect of filler dispersion degree on the Joule heating stimulated recovery behaviour of nanocomposites. *J. Mater. Sci.* **2010**, *45*, 5851–5859. [CrossRef]
18. Lu, H.; Liu, Y.; Gou, J.; Leng, J.; Du, S. Surface coating of multi-walled carbon nanotube nanopaper on shape-memory polymer for multifunctionalization. *Comp. Sci. Tech.* **2011**, *71*, 1427–1434. [CrossRef]
19. Weng, M.; Chen, L.; Huang, F.; Liu, C.; Zhang, W. Transparent actuator made by highly-oriented carbon nanotube film for bio-inspired optical systems. *Nanotechnology* **2020**, *31*, 065501. [CrossRef]
20. Slobodian, P.; Riha, P.; Saha, P. A highly-deformable composite composed of an entangled network of electrically-conductive carbon-nanotubes embedded in elastic polyurethane. *Carbon* **2012**, *50*, 3446–3453. [CrossRef]
21. Slobodian, P.; Riha, P.; Lengalova, A.; Svoboda, P.; Saha, P. Multi-wall carbon nanotube networks as potential resistive gas sensors for organic vapor detection. *Carbon* **2011**, *49*, 2499–2507. [CrossRef]
22. Slobodian, P. Rigid amorphous fraction in poly(ethylene terephthalate) determined by dilatometry. *J. Therm. Anal. Calorim.* **2008**, *94*, 545–551. [CrossRef]
23. Slobodian, P.; Riha, P.; Rychwalski, R.W.; Emri, I.; Saha, P.; Kubat, J. The relation between relaxed enthalpy and volume during physical aging of amorphous polymers and selenium. *Eur. Polym. J.* **2006**, *42*, 2824–2837. [CrossRef]
24. Slobodian, P.; Riha, P.; Olejnik, R.; Benlikaya, R. Analysis of sensing properties of thermoelectric vapor sensor made of carbon nanotubes/ethylene-octene copolymer composites. *Carbon* **2016**, *110*, 257–266. [CrossRef]
25. Slobodian, P.; Svoboda, P.; Riha, P.; Boruta, R.; Saha, P. Synthesis of PMMA-co-PMAA Copolymer Brush on Multi-Wall Carbon Nanotubes. *J. Surf. Eng. Mater. Adv. Technol.* **2012**, *2*, 221–226. [CrossRef]

26. Licea-Jimenez, L.; Henrio, P.Y.; Lund, A.; Laurie, T.M.; Perez-Garcia, S.A.; Nyborg, L.; Hassander, H.; Bertilsson, H.; Rychwalski, R.W. MWNT reinforced melamine-formaldehyde containing alpha-cellulose. *Compos. Sci. Technol.* **2007**, *67*, 844–854.
27. Wang, Z.P.; Zheng, S.R.; Zheng, Y.P. *Polymer Matrix Composites and Technology*; Woodhead Publishing Limited: Cambridge, UK, 2011; pp. 253–318.

© 2020 by the authors. Licensee MDPI, Basel, Switzerland. This article is an open access article distributed under the terms and conditions of the Creative Commons Attribution (CC BY) license (http://creativecommons.org/licenses/by/4.0/).

Article

Balanced Viscoelastic Properties of Pressure Sensitive Adhesives Made with Thermoplastic Polyurethanes Blends

Mónica Fuensanta [1], María Agostina Vallino-Moyano [2] and José Miguel Martín-Martínez [1,*]

[1] Adhesion and Adhesives Laboratory, University of Alicante, 03080 Alicante, Spain; monica.fuensanta@ua.es
[2] National University of Tucumán, San Miguel de Tucumán, Tucumán T4000, Argentina; agovallino@gmail.com
* Correspondence: jm.martin@ua.es; Tel.: +34-965-903-977; Fax: +34-965-909-416

Received: 24 August 2019; Accepted: 30 September 2019; Published: 3 October 2019

Abstract: Pressure sensitive adhesives made with blends of thermoplastic polyurethanes (TPUs PSAs) with satisfactory tack, cohesion, and adhesion have been developed. A simple procedure consisting of the physical blending of methyl ethyl ketone (MEK) solutions of two thermoplastic polyurethanes (TPUs) with very different properties—TPU1 and TPU2—was used, and two different blending procedures have been employed. The TPUs were characterized by infra-red spectroscopy in attenuated total reflectance mode (ATR-IR spectroscopy), differential scanning calorimetry, thermal gravimetric analysis, and plate-plate rheology (temperature and frequency sweeps). The TPUs PSAs were characterized by tack measurement, creep test, and the 180° peel test at 25 °C. The procedure for preparing the blends of the TPUs determined differently their viscoelastic properties, and the properties of the TPUs PSAs as well, the blending of separate MEK solutions of the two TPUs imparted higher tack and 180° peel strength than the blending of the two TPUs in MEK. TPU1 + TPU2 blends showed somewhat similar contributions of the free and hydrogen-bonded urethane groups and they had an almost similar degree of phase separation, irrespective of the composition of the blend. Two main thermal decompositions at 308–317 °C due to the urethane hard domains and another at 363–373 °C due to the soft domains could be distinguished in the TPU1 + TPU2 blends, the weight loss of the hard domains increased and the one of the soft domains decreased by increasing the amount of TPU2 in the blends. The storage moduli of the TPU1 + TPU2 blends were similar for temperatures lower than 20 °C and the moduli at the cross over of the moduli were lower than in the parent TPUs. The improved properties of the TPU1 + TPU2 blends derived from the creation of a higher number of hydrogen bonds upon removal of the MEK solvent, which lead to a lower degree of phase separation between the soft and the hard domains than in the parent TPUs. As a consequence, the properties of the TPU1 + TPU2 PSAs were improved because good tack, high 180° peel strength, and sufficient cohesion were obtained, particularly in 70 wt% TPU1 + 30 wt% TPU2 PSA.

Keywords: thermoplastic polyurethanes blends; pressure sensitive adhesives; viscoelastic properties; adhesion properties; tack; creep; cohesion properties

1. Introduction

Pressure sensitive adhesives (PSAs) are polymeric materials that can form an immediate bond without chemical reaction to a substrate upon brief contact by applying light pressure for short time [1]. Typical applications of PSAs include labels, sticky notes, packaging, diapers, auto/masking tapes, bandages, and decals. PSAs form physical bonds at a molecular level and sustain a minimum level of stress upon de-bonding. PSAs are soft and viscoelastic solids which have properties derived from the differences in the energy gained in forming van der Waals interactions with a substrate and the energy dissipated during the de-bonding.

Most PSAs are based on acrylics [2] and natural and synthetic rubbers [3], and, less commonly, silicones [4] and polyurethanes [5]. Rubber PSAs are composed of rubber polymer and low molecular-weight compatible tackifier, whereas acrylic PSAs are composed of mixtures of random acrylic copolymers of long side- and short-side chains, acrylic acid, and a tackifier can also be added. Silicone PSAs are used mainly when low temperature use or high-temperature stability is required and they do not contain tackifier [4]. Table 1 summarizes some properties of the typical polymers used as PSAs. Rubber PSAs show excellent tack, peel strength, and cohesive strength, but they have poor skin sensibility and low skin trauma, and low solvent resistance. On the contrary, acrylic and silicone PSAs show low to medium tack and peel strength but good solvent resistance. On the other hand, polyurethane PSAs show adequate resistance to the temperature and the solvents, and they have better low temperature performance than the acrylic or rubber PSAs. Nevertheless, polyurethane PSAs are limited by their low tack and low peel adhesive strength (Table 1).

Table 1. Main properties of typical polymers used in pressure sensitive adhesives.

	Natural and Synthetic Rubber	Acrylic	Silicone	Polyurethane
Tack	High	Low to high	Low to high	Low
Peel strength	High	Medium to high	Medium	Low to medium
Cohesive strength	High	Low to high	High	High
Oxidative resistance	Poor	Good	Excellent	Excellent
Solvent resistance	Fair	High	Excellent	Excellent
Low skin sensibility	Poor to good	Good	Excellent	Good
Low skin trauma	Poor	Poor	Excellent	Good
Repositionability	Poor	Poor	Excellent	Fair
Cost	Low	Medium	High	Medium

Medical tapes are one of the most exigent products based on PSAs because of the need to be compatible with the skin [6–8]. Acrylic PSAs used for medical tapes produce irritation of the skin due to the presence of unreacted monomers and residual solvent. Despite natural and synthetic rubber PSAs usually being employed for skin contact, their formulations contain tackifiers and additives that produce irritability and skin trauma with long-time application [9]. Silicone PSAs are preferred in medical tapes due to lower skin trauma, good biocompatibility, and low toxicity, but they have high cost [8,10]. Similar to silicone PSAs, polyurethane PSAs show high compatibility to the skin but they are less costly; however, they showed low tack and peel strength.

The performance of PSAs is tightly related to their viscoelastic properties, i.e., the viscous component (to wet the substrate for good contact during bonding) and the elastic component (to withstand shear stresses and peel forces during de-bonding). Copolymers with segmented structure are commonly used for controlling the viscoelastic properties of the PSAs because the hard phase imparts the elastic properties and the soft phase imparts the viscous properties [11]. In this sense, thermoplastic polyurethanes (TPUs) are potential versatile polymers for manufacturing PSAs because of their segmented structure, comprised of hard and soft segments which are thermodynamically incompatible, leading to microphase separation; on the other hand, the hydrogen bond interactions between the hard segments determine the final morphology of the TPUs [12]. Unfortunately, polyurethane PSAs are limited because of their inherent low tack and low peel adhesive strength [13], i.e., the pressure-sensitive adhesion property is not typical of polyurethanes. For improving the tack of the polyurethane PSAs, tackifier resins and/or plasticisers can be added to increase the glass transition temperature (T_g) and decrease the modulus at room temperature [14–16]. Different tackifier resins (rosin esters, coumarone-indene resins, unsaturated aliphatic hydrocarbon resins, polyterpene resins) have been added in polyurethane PSAs, and even they displayed high peel strength, the substrate can be damaged during de-bonding due to the migration of the tackifier resin to the surface with time. Additional strategies have been proposed for balancing the properties of the polyurethane PSAs. Thus, the use of low NCO/OH ratios, i.e., insufficient equivalents of isocyanate with respect to the equivalents of high functionality polyol, has been proposed but the low degree of cross-linking in the polyurethane PSAs produced poor cohesion [17–20]. Nakamura et al. [21] have shown that the addition of one crosslinking agent increased the peel strength of polyurethane PSAs but the tack did not increase sufficiently. On the other hand, more recently, thermoplastic polyurethanes (TPUs) with pressure sensitive adhesion properties (TPU PSAs) with good tack but insufficient peel strength and poor cohesion have been synthesized by reacting 4,4-diphenylmethane diisocyanate (MDI) with 1,4-butanediol chain extender and mixtures of polypropylene glycols (PPGs) of different molecular weights (1000 and 2000 Da) [22]. TPU PSAs synthesized with mixtures of PPGs containing 50 wt% or more PPG of higher molecular weight showed good tack at 10–37 °C and their pressure sensitive adhesion was related to their minor content of bonded urethane groups and important degree of phase separation [22]. Furthermore, these TPU PSAs followed the Dahlquist criterion and they showed low glass transition temperatures, but they had low cohesion and low 180° peel strength.

For balancing the adhesion and cohesion properties, in a recent approach, TPU PSAs with different hard segments content were synthesized by using different mixtures of PPGs with molecular weights of 450 and 2000 Da [23]. The hard segments contents and the degrees of phase separation of the TPUs affected their pressure sensitive adhesion properties. The increase in the hard segments content increased the percentage of the hydrogen bonded urethane groups and produced a lower degree of phase separation in the TPUs, the storage moduli of the TPUs increased, and high shear PSAs were obtained. On the other hand, TPU PSAs with lower hard segments content showed high tack and adequate de-bonding properties, whereas the increase in the hard segments content increased the cohesive strength, the storage moduli of the TPUs, and the 180° peel strength values. Despite the change in the hard segments content of the TPUs producing adjustable properties in the PSAs, an adequate balance of tack, cohesion, and adhesion was not achieved, i.e., the TPU PSAs showing high tack and sufficient peel strength had low cohesion and the ones with low tack and peel strength showed high cohesion.

In this study, TPU PSAs with satisfactory tack, cohesion, and adhesion were prepared. A simple procedure for balancing the adhesion and cohesion of the TPU PSAs was used, consisting of the physical blending of two TPUs with very different properties, i.e., one TPU with excellent tack and poor cohesion (TPU1) and another TPU with good cohesion and poor tack (TPU2). For preparing the TPUs PSAs, the TPUs were dissolved in methyl ethyl ketone (MEK) and the solutions were mixed; blends of TPU1 and TPU2 containing 60–80 wt% TPU1 were prepared and characterized.

2. Materials and Methods

2.1. Materials

4,4'-Diphenylmethane diisocyanate (MDI) flakes—Desmodur® 44MC (Covestro, Leverkusen, Germany)—was used. Polypropylene glycols with molecular weights of 450 Da (PPG450), Alcupol® D0511, and 2000 Da (PPG2000), Alcupol® D2021, both supplied by Repsol (Madrid, Spain), were used as polyols. Before use, the polyols were melted and dried at 80 °C under reduced pressure (300 mbar) for 2 h. 1,4-butanediol (BD) was used as a chain extender and dibutyl tin dilaurate (DBTDL) was used as a catalyst, both were supplied by Sigma Aldrich Co. LLC (St. Louis, MO, USA). Methyl ethyl ketone (MEK) (Jaber Industrias Químicas, Madrid, Spain) was used as the solvent for the TPUs.

2.2. Synthesis of the Thermoplastic Polyurethanes (TPUs)

The thermoplastic polyurethanes (TPU1 and TPU2) were synthesized using the prepolymer method and an NCO/OH ratio of 1.10 was selected. The polyurethane prepolymer was synthesized in a 500 cm^3 four-neck round-bottom glass reactor under nitrogen atmosphere (flow: 50 mL/min) by reacting the melted MDI with the mixtures of PPGs under stirring with an anchor shaped stirrer in a Heidolph overhead stirrer RZR-2000 (Kelheim, Germany) at 80 °C and 250 rpm for 30 min. Then, 0.04 mmol catalyst (DBTDL) was added and the stirring was carried out at 80 °C and 80 rpm for 2 h. The amount of free NCO in the prepolymer was monitored by dibutylamine titration. Once the desired free NCO content was obtained, the chain extender (BD) was added and stirred at 80 °C and 80 rpm for 5 min. Figure 1 shows the scheme of the synthesis of the TPUs.

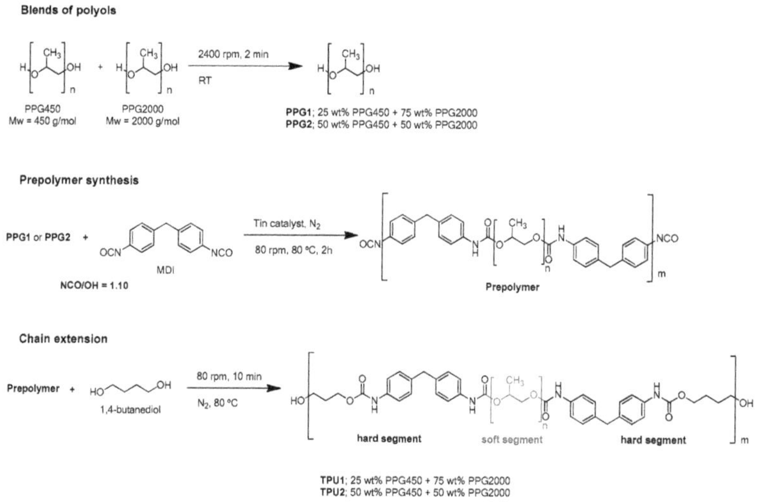

Figure 1. Scheme of the synthesis of the thermoplastic polyurethanes (TPUs).

Two TPUs were synthesized. TPU1 was synthesized with a mixture of 75 wt% PPG2000 and 25 wt% PPG450, and TPU2 was synthesized with a mixture of 50 wt% PPG2000 and 50 wt% PPG450. The same amount of 1,4 butanediol chain extender was added. The hard segments contents of TPU1 and TPU2 were calculated as the ratio of the amount of MDI by the amounts of MDI, polyols, and chain extender, and they were 21% and 28%, respectively. TPU1 was selected for its excellent tack and poor cohesion, whereas TPU2 was chosen for its good cohesion and poor tack [23].

2.3. Preparation of the Blends of TPUs in MEK Solutions

Methyl ethyl ketone (MEK) solutions of solid TPU1 and solid TPU2 were prepared for making the blends. Two different methods for obtaining the TPU1 + TPU2 blends were used.

In method A, 18 wt% solid TPU (15 g) was added to MEK (68.33 g) in a hermetically closed polypropylene container of 58 mm length and 68 mm diameter. The top of the container was sealed with parafilm (Parafilm, Bemis, Oshkosh, USA) and the mixture was magnetically stirred with a magnetic Teflon® cylindrical stirrer of 20 mm length and 8 mm diameter in an IKA C-MAG HS 7 stirrer (IKA, Staufen, Germany) at 25 °C and 60 rpm for at least 30 min. Because of the solvent evaporation during the preparation of the solutions, additional MEK was added to adjust the solids content to 18 wt%. Afterwards, 60–80 wt% TPU1 solution in MEK and 20–40 wt% TPU2 solution in MEK were added in a hermetically closed polypropylene container of 58 mm length and 68 mm diameter and stirred with a magnetic Teflon® cylindrical stirrer of 20 mm length and 8 mm diameter in an IKA C-MAG HS 7 stirrer (IKA, Staufen, Germany) at 25 °C and 60 rpm for at least 30 min. Figure 2 shows the scheme of the procedure employed to prepare the blends of TPU1 and TPU2 (TPU1 + TPU2) by using method A. The nomenclature of the blends consists of the amount of each TPU in wt% followed by the capital letters "TPU" and "/", ending with the capital letter "A" between brackets. For example, 80TPU1/20TPU2 (A) corresponds to the blend made with 80 wt% TPU1 solution in MEK and 20 wt% TPU2 solution in MEK by using method A.

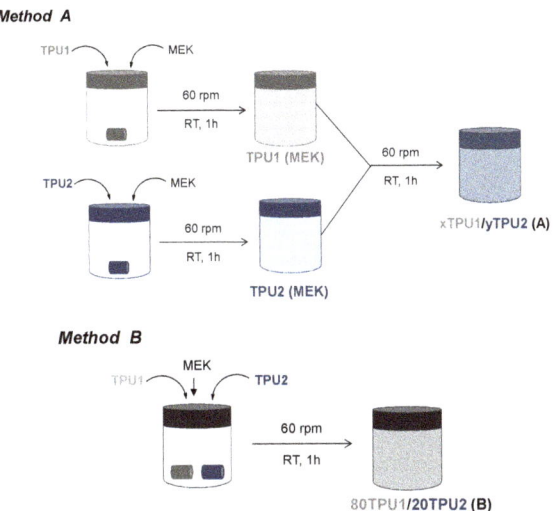

Figure 2. Schemes of the procedures used to prepare the blends of TPU1 and TPU2 in methyl ethyl ketone (MEK) solutions.

In method B, 80 wt% solid TPU1 and 20 wt% solid TPU2 were added together in a closed cylindrical polypropylene container (58 mm length and 68 mm diameter) and MEK was added to obtain a final solids content of 18 wt%. The top of the container was sealed with parafilm (Parafilm, Bemis, Oshkosh, USA). The mixture was magnetically stirred with a magnetic Teflon® cylindrical stirrer of 20 mm length and 8 mm diameter in an IKA C-MAG HS 7 stirrer (IKA, Staufen, Germany) at 25 °C and 60 rpm for at least 60 min. The resulting blend was named 80TPU1/20TPU2 (B). Figure 2 shows the scheme of the procedure employed to obtain the 80TPU1/20TPU2 (B) blend.

MEK from the solutions of TPU1, TPU2, and TPU1 + TPU2 blends was removed for preparing solid films. The solid films were obtained by placing 4 g of TPU solution in MEK in an open cylindrical polypropylene mold of 30 mm height and 25 mm diameter, allowing the solvent removal at room temperature for 72 h.

2.4. Preparation of the TPU PSAs

PSAs consist of an adhesive supported on a thin substrate. TPU PSAs were prepared by placing the TPU solution in MEK on a 50 μm thick polyethylene terephthalate (PET) film; before applying the TPU solution, the PET film was wiped with MEK. The TPU solution was applied to the PET film with a pipette and spread by means of a metering rod of 400 μm. Then, the solvent was removed at room temperature for 24 h to obtain a dry TPU film on 30–40 μm thick PET film.

2.5. Experimental Techniques

2.5.1. Solids Content

The solids contents of the TPUs were obtained by the difference in the weights before and after the evaporation of the solvent. Approximately 0.5 g TPU solution was placed and spread by means of a

Pasteur pipette on aluminum foil, and the solvent was evaporated at 50 °C in an oven until a constant weight was obtained. Two replicates were tested and averaged for each TPU solution.

2.5.2. Attenuated Total Reflection Infrared Spectroscopy (ATR-IR)

The ATR-IR spectra of TPU1, TPU2, and TPU1 + TPU2 blends were obtained in a Tensor 27 FT-IR spectrometer (Bruker Optik GmbH, Erlinger, Germany) by using a Golden Gate single reflection diamond ATR accessory. The angle of the incident beam was 45° and 64 scans were recorded in absorbance mode with a resolution of 4 cm^{-1} in the wavenumber range of 4000 to 400 cm^{-1}.

2.5.3. Differential Scanning Calorimetry (DSC)

The thermal and structural properties of TPU1, TPU2, and TPU1 + TPU2 blends were determined in a DSC Q100 calorimeter (TA Instruments, New Castle, DE, USA) under nitrogen atmosphere (flow: 50 mL/min). A total of 8–9 mg of TPU film was placed in a hermetically sealed aluminum pan and placed in the oven of the DSC equipment. For removing the thermal history, the TPU film was heated from −80 to 100 °C using a heating rate of 10 °C/min. Then, the TPU was cooled down to −80 °C using a cooling rate of 10 °C/min and, finally, the TPU was heated again from −80 to 150 °C using a heating rate of 10 °C/min. The glass transition temperature (T_g) of TPU1, TPU2, and TPU1 + TPU2 blends were obtained from the second DSC heating run.

2.5.4. Thermal Gravimetric Analysis (TGA)

The thermal degradation and the structure of TPU1, TPU2, and TPU1 + TPU2 blends were determined in a TGA Q500 equipment (TA Instruments, New Castle, USA) under nitrogen atmosphere (flow: 50 mL/min). A total of 8–9 mg of TPU was placed in platinum crucible and then heated from 35 to 800 °C using a heating rate of 10 °C/min.

2.5.5. Plate-Plate Rheology

The rheological and viscoelastic properties of TPU1, TPU2, and TPU1 + TPU2 blends were assessed in a DHR-2 rheometer (TA Instruments, New Castle, DE, USA) using parallel plate-plate geometry. The gap selected was 0.40 mm, and 20 mm diameter stainless steel parallel plates were used. Temperature sweep experiments were carried out from −10 to 120 °C, a frequency of 1 Hz and a heating rate of 5 °C/min were used. Oscillatory frequency sweep experiments were also performed at 25 °C using 2.5% strain amplitude in the angular frequency range from 0.01 to 100 rad s^{-1}. All rheological experiments were carried out in the region of linear viscoelasticity.

2.5.6. Probe Tack

The probe tack of the TPU PSAs (TPU on PET films) was measured in the range of 15 to 50 °C with an interval of 5 °C in a TA.XT2i Texture Analyzer (Stable Micro Systems, Surrey, UK). The TPU PSAs were attached to square stainless steel 304 plates of 6 cm × 6 cm × 0.1 cm by means of double-sided tape (Miarco, Paterna, Spain). A flat end cylindrical stainless-steel probe of 3 mm diameter was used. The probe was brought into contact with the TPU PSA surface for 1 s under a load of 5 N. Then, the probe was pulled out at a constant rate of 10 mm/s and a stress–strain curve was obtained. The maximum of the stress–strain curve was taken at the tack of the TPU PSA. At least five replicates were carried out and averaged.

2.5.7. Creep Test under Shear

Pieces of the TPU PSAs (TPU on PET films) were cut to obtain strips of 2.4 cm × 20 cm. On the other hand, rectangular pieces of polished stainless steel 304 of 77 cm × 51 cm × 1.5 cm were wiped with MEK to remove surface contaminants, allowing the solvent to evaporate for 15 min. Then, the TPU PSA strip was placed in the central area of the clean polished stainless steel 304 piece, an area of 2.4 cm

× 2.4 cm was joined, and a rubber coated roller of 2 kg was passed 30 times over the joint. Afterwards, a piece of stainless steel was placed at the bottom of the TPU PSA strip, which was plied and fixed with a staple at a distance of 4 cm from the polished stainless-steel plate. The coupon was placed on the holder of a Shear-10 equipment (ChemInstruments, Fairfell, OH, USA) and 1 Kg weight was held at the bottom (Figure 3). The creep resistance at 25 °C, which is related to the cohesion, was obtained as the "holding time", i.e., the time needed for the TPU PSA strip to fall down. Three replicates were tested for each TPU PSA and the results obtained were averaged.

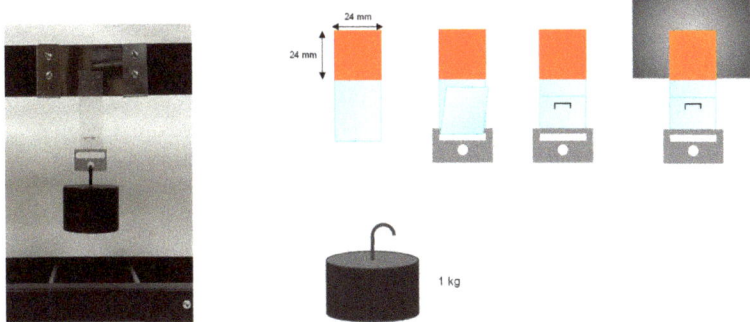

Figure 3. Scheme of the manufacturing of the coupons and the creep test of TPU PSA.

2.5.8. 180° Peel Test

The adhesion properties of the TPU PSAs (TPU on PET films) were determined by 180° peel tests of stainless steel 304/TPU PSA joints. TPU PSA strips of 30 mm × 180 mm × 0.50 mm were placed on a stainless-steel 304 plate of 30 mm × 150 mm × 1 mm and a 2 Kg rubber coated roller was passed 30 times over the joint. After 30 and 72 min of the joints formation, the 180° peel tests (Figure 4) were carried out in an Instron 4411 universal testing machine (Instron Ltd., Buckinghamshire, UK) using a pulling rate of 152 mm/min. A length of 7 cm of each joint was peeled, and the initial values of 180° peel strength were discarded. Five replicates were tested and averaged for each joint.

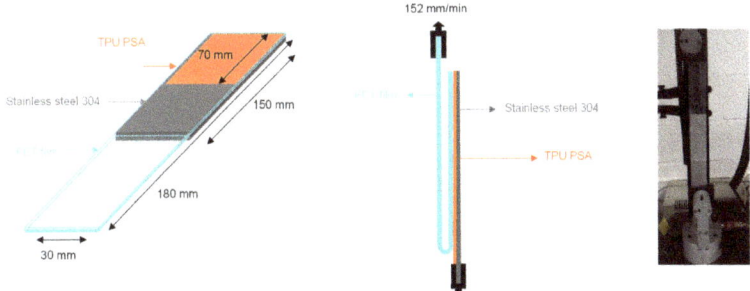

Figure 4. Scheme of the 180° peel strength test of a stainless-steel 304/TPU PSA joint.

3. Results and Discussion

3.1. Influence of the Procedure for Preparing the Blends of TPUs on Their Properties

In a recent study [23], TPU1 PSA synthetized with 75 wt% PPG polyol with molecular weight 2000 Da (PPG2000) and 25 wt% PPG with molecular weight 450 Da (PPG450) showed high tack (752 kPa) but low cohesion at 25 °C, whereas TPU2 PSA synthetized with 50 wt% PPG2000 and 50 wt% PPG450 showed low tack (295 kPa) but high cohesion at 25 °C. For balancing the tack and the cohesion of the TPU PSAs, different blends of TPU1 and TPU2 were prepared. Two different procedures for preparing the blends were used (Figure 2): (i) Method A—different amounts of TPU1 solution in MEK and TPU2 solution in MEK were blended; (ii) Method B—80 wt% solid TPU1 and 20 wt% solid TPU2 were dissolved together in MEK. For determining the influence of the procedure for preparing the blends of the TPUs on their properties, the blend of 80 wt% TPU1 and 20 wt% TPU2 was selected.

The chemical structure of the blends of the TPUs was assessed by ATR-IR spectroscopy. Figure 5a shows the ATR-IR spectra of the blends prepared with the two procedures. Both blends showed the same chemical structure. The ATR-IR spectra show the bands of the PPG soft segments due to the asymmetric and symmetric C–H stretching of the hydrocarbon chains at 2971 and 2869 cm^{-1}, CH$_3$ and CH$_2$ bands at 1373—scissor and rocking CH$_3$ (sym), 1453—scissor and rocking CH$_3$ + scissor and rocking CH$_2$, and 927 cm^{-1}—CH$_2$ bending, and the strong band at 1084 cm^{-1} due to the asymmetric stretching of C–O–C. The bands corresponding to the hard segments (urethane groups) appeared at 3300 cm^{-1}—symmetric and asymmetric N–H stretching, 1598 cm^{-1}—in plane N–H bending, and 1727 cm^{-1}—C=O stretching. Furthermore, the typical C=C stretching and bending in the benzene ring of MDI at 1412, 818, and 512 cm^{-1} were also observed.

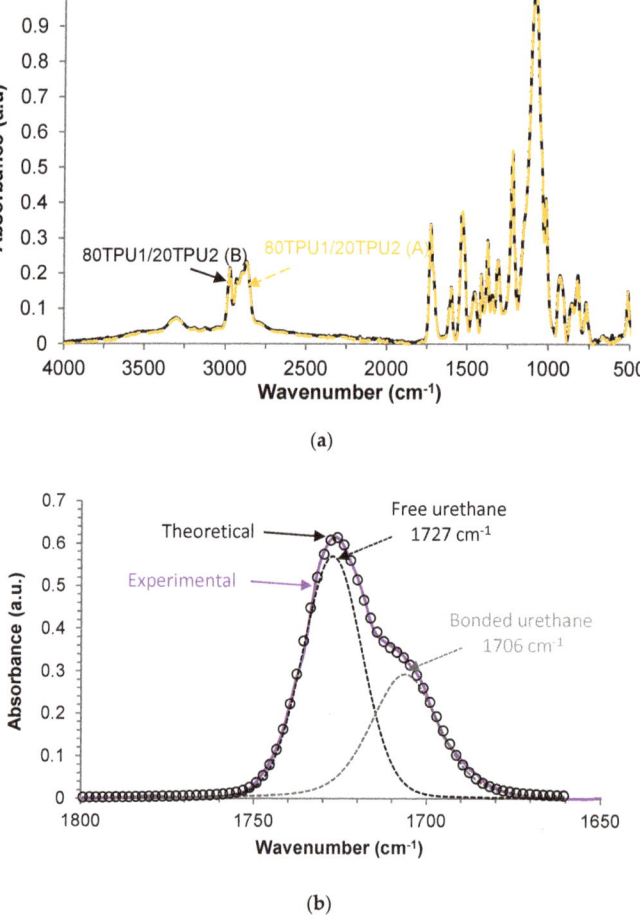

Figure 5. (a) ATR-IR spectra of 80TPU1/20TPU2 (A) and 80TPU1/20TPU2 (B) blends. (b) Curve fitting of the carbonyl region (1650–1800 cm^{-1}) of the ATR-IR spectrum of the 80TPU1/20TPU2 (A) blend.

The existence of hydrogen bond formation in polyurethanes by IR spectroscopy has been extensively studied in the existing literature [24,25]. The hydrogen bond interactions between the N–H and C=O groups of the urethane groups in the hard segments (associated urethanes) were assessed from the ATR-IR spectra of Figure 5a. The relative percentages of the free and associated urethane groups were assessed by curve fitting of the C=O band of the ATR-IR spectra of the blends. Figure 5b shows a typical example of the curve fitting of the carbonyl band of 80TPU1/20TPU2 (A) blend; the curve fitting was carried out by assuming a Gaussian distribution. The free urethane groups were fitted at 1727 cm^{-1} and the associated urethane groups were fitted at 1706–1705 cm^{-1}. According to Table 2, the free urethane groups were dominant in the structure of both TPU1 + TPU2 blends, and similar contributions of the free and associated urethane groups were evidenced in both blends.

Table 2. Relative contribution of the free and hydrogen bonded urethane groups of 80TPU1/20TPU2 (A) and 80TPU1/20TPU2 (B) blends. Curve fitting of the C=O band of the ATR-IR spectra.

Wavenumber (cm^{-1})	Relative Contribution of Species (%)	
	80TPU1/20TPU2 (A)	80TPU1/20TPU2 (B)
1727 cm^{-1} (Free urethane)	61	62
1706–1705 cm^{-1} (H-bonded urethane)	39	38

The structure of the blends was also determined by DSC. The DSC thermograms of the 80TPU1/20TPU2 (A) and 80TPU1/20TPU2 (B) blends are shown in Figure 6, and they exhibit the glass transition temperatures due to the soft segments at −30 °C and −32 °C, respectively. The small differences between the glass transition temperatures of the blends suggest slight differences in the degree of phase separation between the soft and the hard segments.

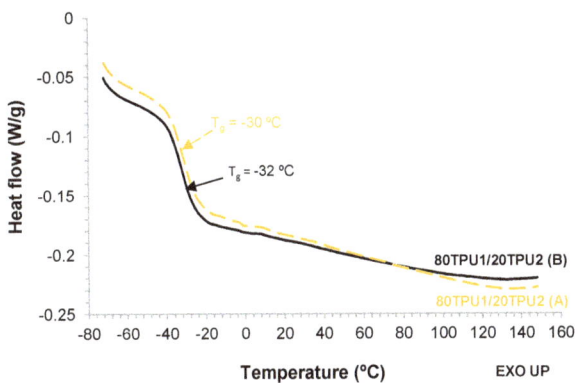

Figure 6. DSC thermograms of 80TPU1/20TPU2 (A) and 80TPU1/20TPU2 (B) blends. Second heating run.

The viscoelastic properties of 80TPU1/20TPU2 (A) and 80TPU1/20TPU2 (B) blends were determined by plate-plate rheology (temperature sweep experiments). Figure 7a shows the variation in the storage modulus (G') as a function of the temperature for the blends; at any temperature, the storage modulus was higher in the 80TPU1/20TPU2 (A) blend. On the other hand, a cross-over between the storage (G') and the loss (G'') moduli was found in both blends (Figure 7b,c). Figure 7b,c show that below 53 °C or 46 °C, respectively, the elastic rheological regime was dominant in the blends, whereas above 53 °C or 46 °C the viscous rheological regime was prevailing. The viscous behavior of the blends is mainly determined by their soft segments and, consequently, their content in the blend will determine the values of the temperature and the modulus at the cross-over of G' and G''. According to Figure 7b,c, and Table 3, the 80TPU1/20TPU2 (B) blend had a slightly lower cross-over temperature than the

80TPU1/20TPU2 (A) blend, likely due to a slightly different degree of phase separation; however, both blends showed similar moduli at the cross-over of G' and G''.

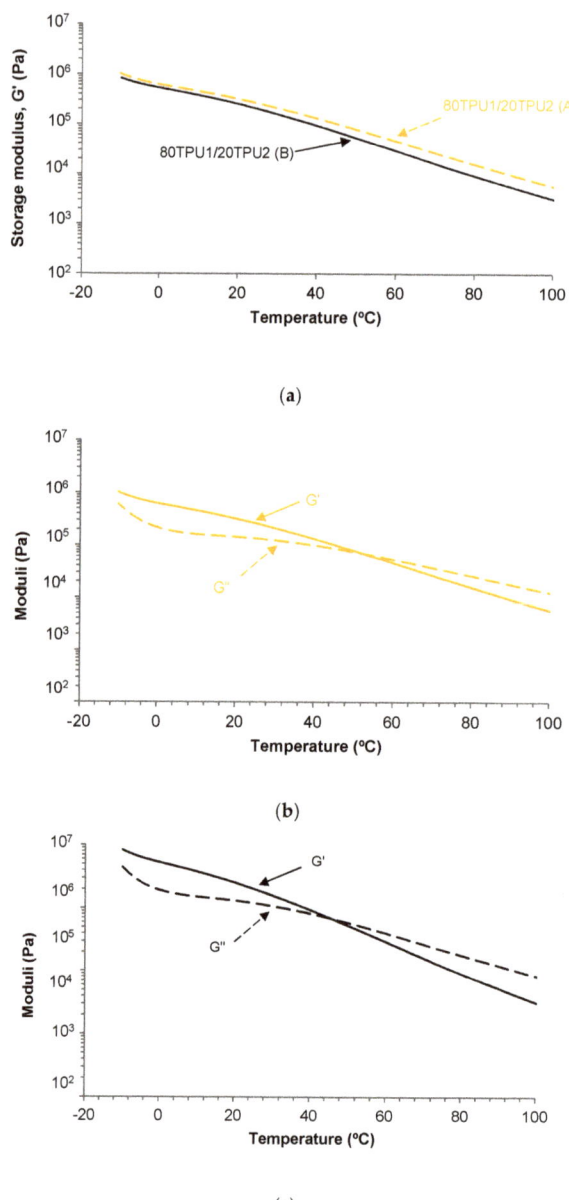

Figure 7. (**a**) Variation in the storage modulus (G') as a function of the temperature for 80TPU1/20TPU2 (A) and 80TPU1/20TPU2 (B) blends. (**b**) Variation in the storage (G') and loss (G'') moduli as a function of the temperature for the 80TPU1/20TPU2 (A) blend. (**c**) Variation in the storage (G') and loss (G'') moduli as a function of the temperature for 80TPU1/20TPU2 (B) blend. Plate-plate rheology experiments.

Table 3. Values of the temperature ($T_{\text{cross-over}}$) and the modulus ($G_{\text{cross-over}}$) at the cross-over of the storage and loss moduli of 80TPU1/20TPU2 (A) and 80TPU1/20TPU2 (B) blends. Plate-plate rheology experiments.

Blend	$T_{\text{cross-over}}$ (°C)	$G_{\text{cross-over}}$ (Pa)
80TPU1/20TPU2 (A)	53	6.8·10⁴
80TPU1/20TPU2 (B)	46	6.8·10⁴

The properties of the TPU PSAs made with 80 wt% TPU1 + 20 wt% TPU2 blends on PET film were characterized by tack measurements at different temperatures, cohesion (holding time) at 25 °C, and 180° peel strength of stainless steel/TPU PSA joints at 25 °C.

Figure 8 shows the variation in the tack as a function of the temperature for 80TPU1/20TPU2 (A) and 80TPU1/20TPU2 (B) PSAs. The tack of both TPU PSAs was higher than 300 kPa and decreased by decreasing the temperature from 50 to 15 °C; furthermore, the tack values at any temperature were higher in the 80TPU1/20TPU2 (A) PSA, indicating that the procedure to prepare the blend determines its tack. The higher tack of 80TPU1/20TPU2 (A) PSA can be ascribed to its slightly higher degree of phase separation, and its higher storage modulus and $T_{\text{cross-over}}$. On the other hand, both TPU PSAs showed good cohesion (i.e., high holding time) and acceptable 180° peel strength (Table 4), and the 80TPU1/20TPU2 (A) PSA had higher 180° peel strength and lower cohesion than 80TPU1/20TPU2 (B) PSA. However, the failed surfaces after the 180° peel test show a cohesive failure in the blend, which is not desirable. Therefore, TPU1 + TPU2 blends with different compositions were prepared and characterized.

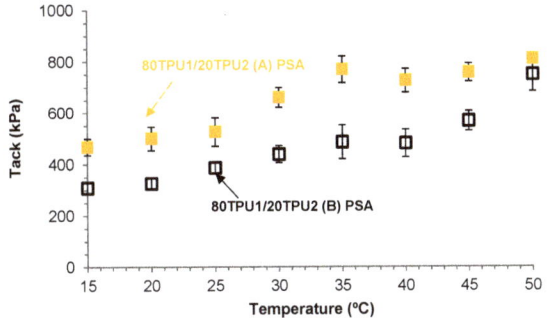

Figure 8. Variation in the tack of the 80TPU1/20TPU2 (**A**) PSA and 80TPU1/20TPU2 (**B**) PSA as a function of the temperature.

Table 4. Holding time at 25 °C and 180° peel strength at 25 °C of stainless steel/TPU PSA joints.

PSA	Holding Time (min)	180° Peel Strength After 30 min (kN/m)	180° Peel Strength After 72 h (kN/m)
80TPU1/20TPU2 (A) PSA	442 ± 134	1.29 ± 0.06 (CA) [a]	1.61 ± 0.07 (CA) [a]
80TPU1/20TPU2 (B) PSA	845 ± 75	0.95 ± 0.03 (CA) [a]	1.07 ± 0.02 (CA) [a]

[a] Locus of failure: CA, cohesive failure of the blend.

3.2. Characterization of the TPU1 + TPU2 Blends

The main target of this study is the development of TPU PSAs with balanced adhesion and cohesion properties, and because the best balance between tack, 180° peel strength, and cohesion was obtained in 80TPU1/20TPU2 (A) PSA, the procedure selected for preparing other TPU1 + TPU2 blends was method A. Table 5 shows the composition of the TPU1 + TPU2 blends (TPU1 content between 60 and 80 wt%). The solids contents of TPU1 and TPU2 were 18.8 wt% and 19.9 wt%, respectively, and the contents of the TPU1 + TPU2 blends were 17.8–19.3 wt%.

Table 5. Nomenclature and composition of the TPU1 + TPU2 blends (method A).

Sample Code	TPU1 (wt%)	TPU2 (wt%)	Solids Content (wt%)
TPU1	100	-	18.8 ± 1.7
80TPU1/20TPU2	80	20	17.9 ± 0.6
70TPU1/30TPU2	70	30	17.8 ± 0.7
60TPU1/40TPU2	60	40	19.3 ± 1.5
TPU2	-	100	19.9 ± 1.9

The chemical structures of the TPU1 + TPU2 blends were characterized by ATR-IR spectroscopy (Figure 9a). The absorption bands of the hard segments can be distinguished at 3310–3255 (N–H stretching), 1727–1726 (C=O stretching of urethane), 1533–1532 (C–N stretching), and 1598 cm^{-1} (N–H bending in plane), whereas the absorption bands of the soft segments were located at 2972–2971 and 2869–2868 cm^{-1} (asymmetric and symmetric C–H stretching, respectively), and at 927, 1084–1017, 1373, and 1454–1453 cm^{-1} (–C–O–C– group of PPG polyol).

In order to assess the contributions of the free and hydrogen-bonded urethane groups in TPU1, TPU2, and TPU1 + TPU2 blends, the carbonyl region of the ATR-IR spectra was curve fitted (Figure 9b). According to Table 6, the free urethane (fitted at 1727–1726 cm^{-1}) was dominant in TPU1, whereas the hydrogen-bonded urethane (fitted at 1706–1704 cm^{-1}) was the major contribution in TPU2. Interestingly, the TPU1 + TPU2 blends showed somewhat similar contributions of the free and hydrogen-bonded urethane groups, indicating that they have almost similar degrees of phase separation.

Table 6. Relative contributions of the free and hydrogen-bonded urethane groups of TPU1, TPU2, and TPU1 + TPU2 blends. Curve fitting of the C=O region of the ATR-IR spectra.

Wavenumber (cm^{-1})	Relative Contribution of Species (%)				
	TPU1	80TPU1/20TPU2	70TPU1/30TPU2	60TPU1/40TPU2	TPU2
1726–1727 cm^{-1} (Free urethane)	66	61	59	58	46
1706–1704 cm^{-1} (H-bonded urethane)	34	39	41	42	54

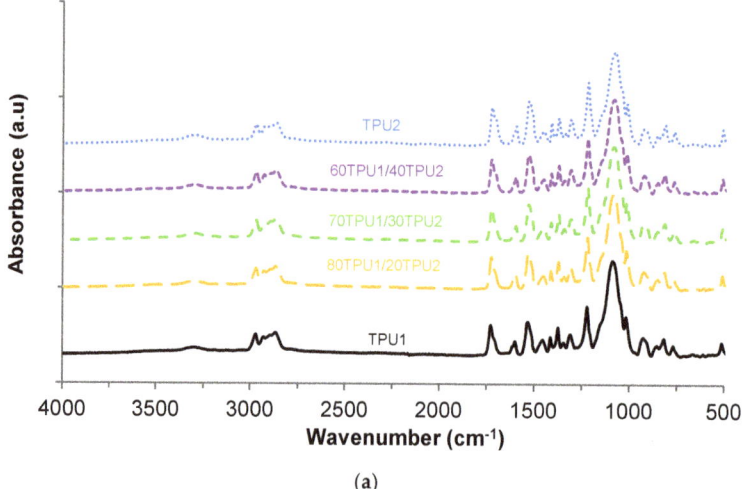

(a)

Figure 9. *Cont.*

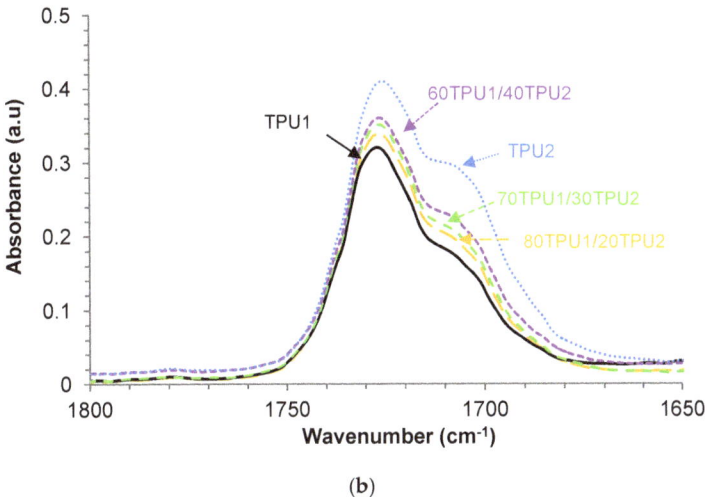

(b)

Figure 9. (a) ATR-IR spectra of TPU1, TPU2, and TPU1 + TPU2 blends. (b) Carbonyl region (1650–1800 cm^{-1}) of the ATR-IR spectra of TPU1, TPU2, and TPU1 + TPU2 blends.

The structural changes and the degree of phase separation in TPU1, TPU2, and TPU1 + TPU2 blends were determined by DSC. Figure 10 shows the DSC thermograms (second heating run) of TPU1, TPU2, and TPU1 + TPU2 blends. At low temperature, all TPUs showed the glass transition temperature of the soft segments (T_g), the lowest T_g corresponds to TPU1 (−36 °C) and the highest to TPU2 (−16 °C). Interestingly, all TPU1 + TPU2 blends had similar T_g values (near −30 °C), irrespective of the composition of the blend, indicating that the structure of the soft segments was similar in all blends and it was somewhat similar to the one of TPU1. In agreement with previous findings [23], the increase in the hard segment content in the TPUs increased their T_g values and the extent of mixing of the hard and soft segments, i.e., the degree of microphase separation, decreased. Therefore, similar structure of the soft segments and analogous degree of microphase separation of the soft and hard segments was obtained in the TPU1 + TPU2 blends, both were different than the ones in the parent TPUs, in agreement with the evidence provided by ATR-IR spectroscopy.

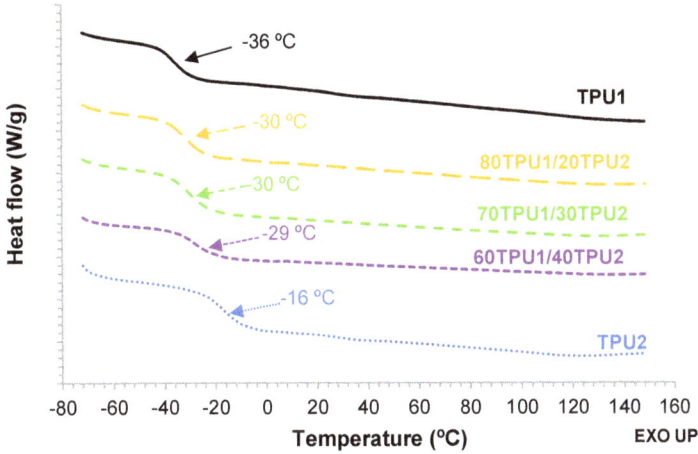

Figure 10. DSC thermograms of TPU1, TPU2, and TPU1 + TPU2 blends. Second DSC heating run.

The thermal stabilities and the structure of TPU1, TPU2, and TPU1 + TPU2 blends were analyzed by TGA. Figure 11a shows that TPU1 had the highest thermal stability and TPU2 the lowest. The thermal stabilities of TPU1, TPU2, and TPU1 + TPU2 blends were quantified by the values of the temperatures at which 5 ($T_{5\%}$) and 50 wt% ($T_{50\%}$) were lost. According to Table 7, the values of $T_{5\%}$ and $T_{50\%}$ of the blends decrease by increasing their TPU2 content. On the other hand, two main thermal decompositions can be distinguished in the TPUs and their blends (Figure 11b), one at 308–317 °C due to the urethane hard domains and another at 363–373 °C due to the soft domains [26]. The weight loss of the hard domains increased and the one of the soft domains decreased by increasing the amount of TPU2 in the blends, and the temperatures of the thermal decompositions were similar in 80TPU1/20TPU2 and 60TPU1/40TPU2; however, the temperatures of decomposition of the hard and soft domains were higher in 70TPU1/30TPU2 (Table 7).

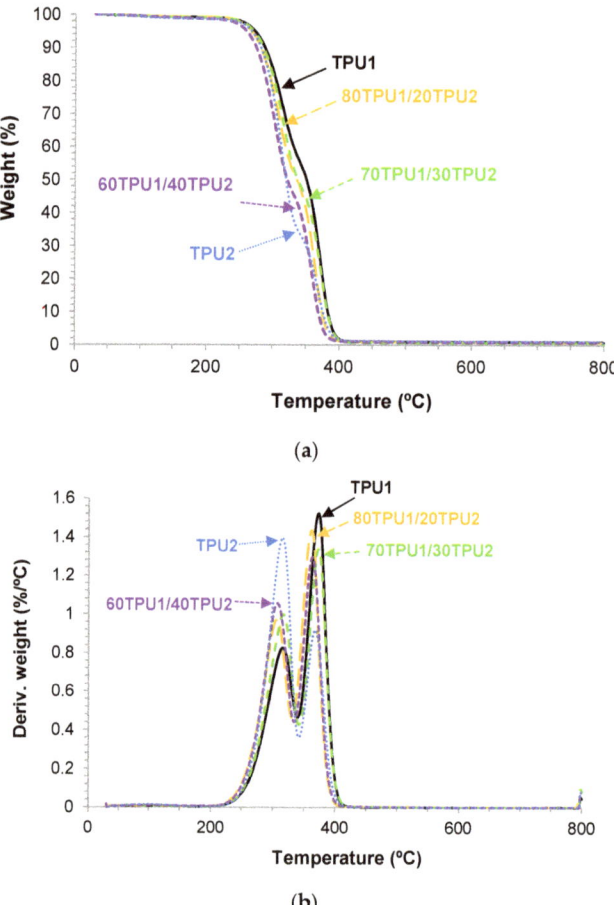

Figure 11. Variation in the (**a**) weight and (**b**) derivative of the weight as a function of the temperature for TPU1, TPU2, and TPU1 + TPU2 blends. TGA experiments.

Table 7. Temperatures at which 5 wt% ($T_{5\%}$) and 50 wt% ($T_{50\%}$) were lost, and temperatures and weight losses of the two thermal decompositions of TPU1, TPU2, and TPU1 + TPU2 blends. TGA experiments.

Sample Code.	$T_{5\%}$ (°C)	$T_{50\%}$ (°C)	1st Decomposition		2nd Decomposition		Residue (wt%)
			T_1 (°C)	Weight Loss$_1$ (%)	T_2 (°C)	Weight Loss$_2$ (%)	
TPU1	272	351	309	44	373	54	2
80TPU1/20TPU2	267	336	309	48	365	51	1
70TPU1/30TPU2	271	338	317	52	373	48	0
60TPU1/40TPU2	263	326	308	53	363	45	2
TPU2	267	320	315	66	368	32	2

The structure of the TPUs affects their viscoelastic properties. The viscoelastic properties of TPU1, TPU2, and TPU1 + TPU2 blends were determined by plate-plate rheology experiments. Figure 12a,b show the variation in the storage modulus (G') and the loss modulus (G'') as a function of the temperature for TPU1, TPU2, and TPU1 + TPU2 blends. The storage and the loss moduli decreased by increasing the temperature, more noticeably in TPU1 than in TPU2, and the storage and loss moduli of the TPU1 + TPU2 blends were intermediate between the ones of TPU1 and TPU2. For temperatures below 20 °C, the storage and loss moduli of the TPU1 + TPU2 blends were similar, but above 20 °C the moduli were higher in the blends with higher content of TPU2. All TPUs and TPU1 + TPU2 blends showed a cross-over of the storage (G') and loss (G'') moduli (Figure 7b) and the values of the temperatures and the moduli at the cross-over are given in Table 8. The temperature at the cross-over of G' and G'' was lower in TPU1 than in TPU2 because of the lower content of PPG450 polyol, and the temperatures at the cross-over in the blends increased by increasing their TPU2 content. Interestingly, the moduli at the cross-over were higher in TPU1 and TPU2 than in the blends, this can be related to their lower degree of phase separation.

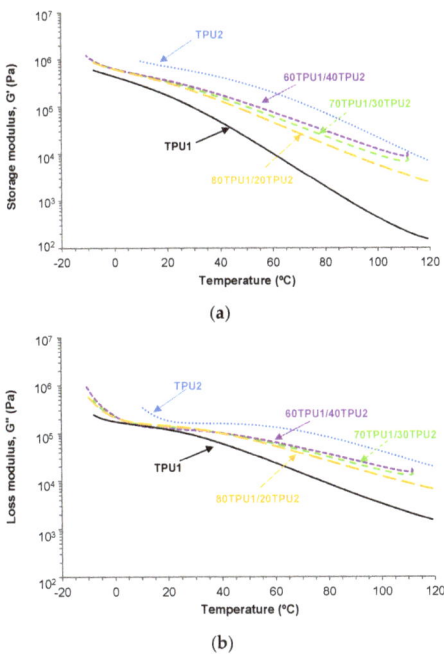

Figure 12. Variation in the (**a**) storage modulus (G') and (**b**) loss modulus (G'') as a function of the temperature for TPU1, TPU2, and TPU1 + TPU2 blends. Plate-plate rheology experiments. Temperature sweep.

Table 8. Values of temperature ($T_{\text{cross-over}}$) and modulus ($G_{\text{cross-over}}$) at the cross-over of the storage and loss moduli of TPU1, TPU2, and TPU1 + TPU2 blends. Plate-plate rheology experiments.

Sample Code	$T_{\text{cross-over}}$ (°C)	$G_{\text{cross-over}}$ (Pa)
TPU1	32	$8.3 \cdot 10^4$
80TPU1/20TPU2	53	$6.8 \cdot 10^4$
70TPU1/30TPU2	61	$5.7 \cdot 10^4$
60TPU1/40TPU2	69	$5.0 \cdot 10^4$
TPU2	75	$9.6 \cdot 10^4$

The unexpected particular structures of the TPU1 + TPU2 blends with respect to the ones of the parent TPUs leading to lower degree of phase separation should derive from the structural changes produced when the solvent (MEK) in the solutions is removed. It has been shown [27–29] that the interactions by hydrogen bonds between the polymer chains in the TPUs can be reversibly destroyed by increasing the temperature or by adding organic solvents (particularly ketones). In this study, the solid TPU1 and solid TPU2 were dissolved in MEK, which caused the rupture of the hydrogen bonds between the hard segments (Figure 13a,b). The structures of TPU1 and TPU2 were re-formed upon MEK removal, i.e., the hydrogen bonds between the hard segments were created. However, when the MEK solutions of TPU1 and TPU2 were mixed and the solvent was removed, the structure was different because the interactions between the hard domains were more complex and a higher number of hydrogen bonds were formed (Figure 13c), this led to a lower degree of phase separation between the soft and the hard domains. As a consequence, the structures of the TPU1 + TPU2 blends were different than the ones in TPU1 and TPU2.

The experimental results shown above indicate that the most efficient TPU1 + TPU2 blends were obtained by adding 20–30 wt% TPU2, likely due to easy mobility of the polymeric chains of TPU2 during MEK removal. Because the number of hydrogen bond interactions in the TPUs are tightly related to their cohesion, higher cohesion in the TPU1 + TPU2 blends than in TPU1 can be anticipated; however, at the same time the mobility of the polymeric chains of TPU2 should be reduced, so a decrease in tack can be expected. The properties of the TPU PSAs made with TPU1, TPU2, and TPU1 + TPU2 blends are studied in the next section.

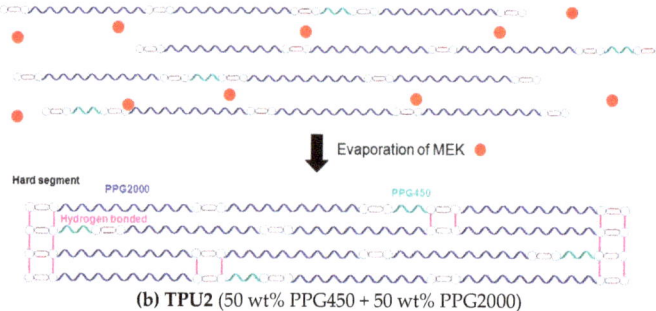

Figure 13. Cont.

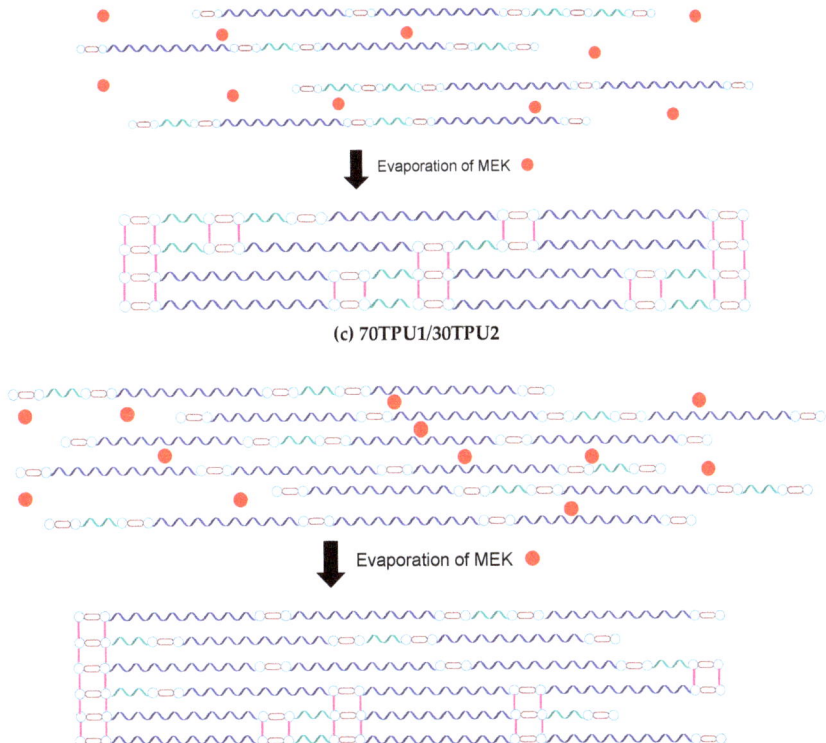

Figure 13. Scheme of the structure of TPU1, TPU2, and 70TPU1/30TPU2 solutions in MEK and after solvent removal.

3.3. Characterization of the TPU1 + TPU2 PSAs

The performance of the PSAs is tightly related to their viscoelastic properties. The TPU PSAs were made by placing the MEK solutions of TPU1, TPU2, and TPU1 + TPU2 blends on PET film. The most PSAs are used at ambient temperature, so the viscoelastic properties of the TPU PSAs were studied at 25 °C by oscillatory frequency sweep plate-plate rheology experiments. The storage modulus (G') at low frequency of the TPU PSA is related to its tack and shear resistance, whereas at higher frequencies the G' is associated with its peel strength. An excellent PSA must have low G' value at high frequency (high shear and easy peel) and high G' value at low frequency (good resistance to creep). Figure 14a shows that the TPU1 PSA had low G' values at low frequencies, anticipating poor cohesion, but the TPU2 PSA had high G' values in all range of frequencies, anticipating high cohesion; the G' values of the TPU1 + TPU2 PSAs showed reasonable values and they must have good cohesion. On the other hand, all TPU PSAs had high G' values at high frequencies, anticipating that they should have easy peel. Furthermore, the variation in the loss modulus (G'') as a function of the frequency (Figure 14b) shows the same trend as the ones of the storage modulus (Figure 14a), but the differences in the loss moduli are less marked.

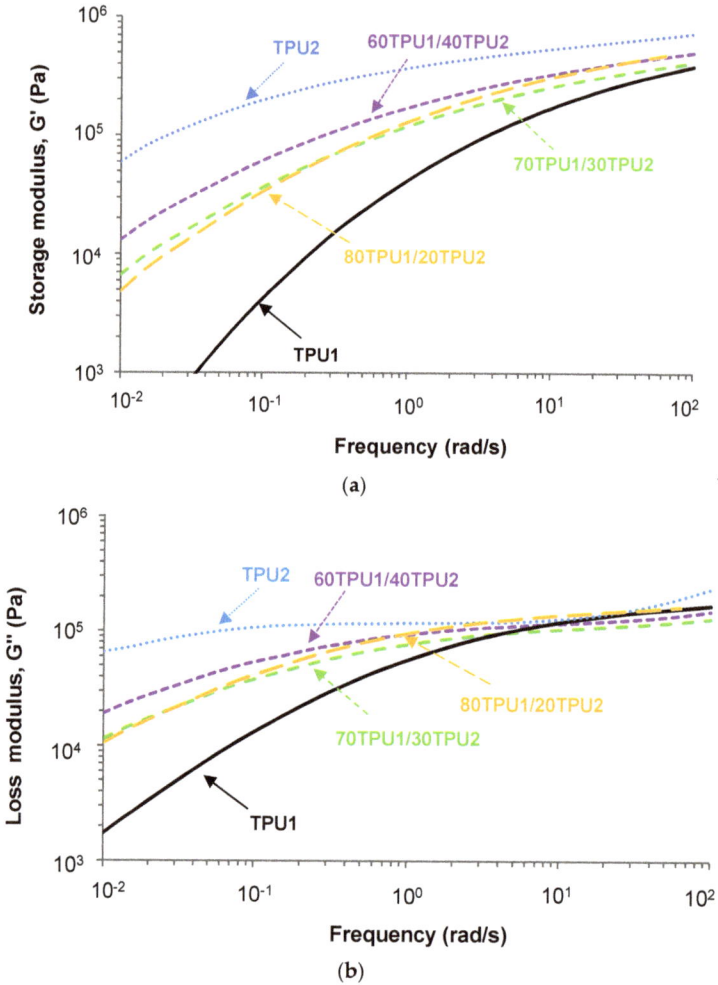

Figure 14. Variation in the (**a**) storage modulus (G') and (**b**) loss modulus (G'') at 25 °C as a function of the frequency of TPU1, TPU2, and TPU1 + TPU2 blends. Plate-plate rheology experiments. Frequency sweep.

Table 9 compiles the values of G' at 0.1 and 100 rad/s. TPU1 PSA had low G' value at 0.1 rad/s and high G' (0.1 rad/s)/G' (100 rad/s) value, which is typical of PSAs with high tack [30]. On the contrary, the G' value at 0.1 rad/s was high and the G' (0.1 rad/s)/G' (100 rad/s) ratio was low in TPU2 PSA, and its tack should be low. The G' value at 0.1 rad/s of the TPU1 + TPU2 PSAs increased and their G' (0.1 rad/s)/G' (100 rad/s) ratios decreased by increasing their TPU2 content; according to the values in Table 9, 60TPU1/40TPU2 PSA should have low tack, easy peel, and good creep resistance, but 70TPU1/30TPU2 PSA and 80TPU1/20TPU2 PSA should have a good balance of tack, cohesion, and peel.

Table 9. Values of the storage moduli (G') at 25 °C and different frequencies for TPU PSAs. Frequency sweep plate-plate rheology experiments.

Sample Code	G' (kPa)-0.1 rad/s	G' (kPa)-100 rad/s	G' (0.1 rad/s)/G' (100 rad/s)
TPU 1	0.42	38.68	92.7
80TPU1/20TPU2	3.33	52.50	15.8
70TPU1/30TPU2	3.64	41.66	11.4
60TPU1/40TPU2	6.13	50.63	8.3
TPU2	10.68	23.55	2.2

Chang's viscoelastic window is a simple tool to determine the balance of the viscoelastic properties of the PSAs [31]. Chang's viscoelastic windows at 25 °C of TPU1, TPU2, and TPU1 + TPU2 PSAs are given in Figure 15. The differences between the storage (G') and loss (G") moduli were lower in the TPU1 + TPU2 PSAs with respect to TPU1 PSA and TPU2 PSA, thus anticipating a better compromise between tack and cohesion. All TPU1 + TPU2 PSAs had G' values at 0.01 rad/s lower than $3 \cdot 10^5$ Pa, anticipating an efficient contact with the substrate, and they followed the Dahlquist criterion—Dahlquist suggested that the tack of the PSAs requires a storage modulus value lower than $3 \cdot 10^5$ Pa at 25 °C and 1Hz [32]. Furthermore, the TPU1 + TPU2 PSAs are good general-purpose PSAs (i.e., they are located in the center of the Chang's viscoelastic window).

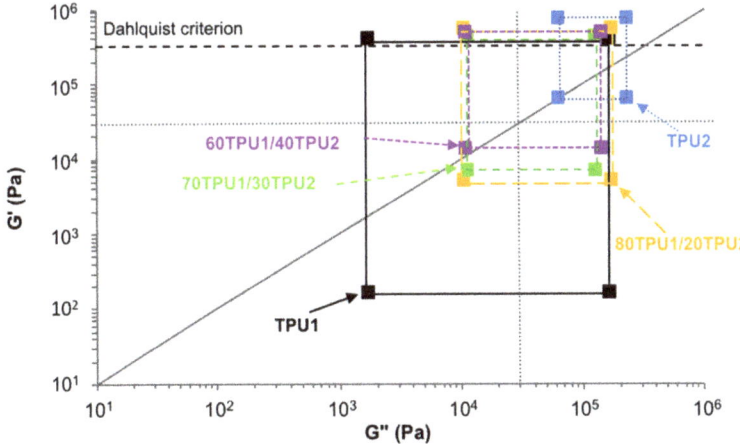

Figure 15. Chang's viscoelastic windows at 25 °C of TPU1, TPU2, and TPU1 + TPU2 blends. Dotted lines indicate the four regions of Chang's viscoelastic window. Solid line corresponds to G' = G" (tan delta = 1). Dashed line indicates the Dahlquist criterion.

Figure 16 shows the variation of the tack of the TPU PSAs as a function of the temperature. TPU1 PSA showed the highest tack at any temperature and the tack was maintained between 15 and 30 °C, and an increase in production above 30 °C. A similar trend and slightly lower tack values were obtained in 80TPU1/20TPU2 PSA, whereas the tack was lower and similar between 15 and 40 °C in 70TPU1/30TPU2 PSA. The lowest tack values corresponded to TPU2 PSA and the maximum tack was obtained at 25–30 °C. At 25 °C, the tack ranged between 391 and 634 kPa and was somewhat similar in all TPU1 + TPU2 PSAs with TPU1 content of 70 wt% or lower; however, at 15 and 30 °C the tack of the TPU PSAs decreased by increasing their content in TPU2, which can be related to the decrease in the soft segments content. In summary, the tack of the TPU PSAs can be designed by changing their soft segments contents and its variation with the temperature is related to their viscoelastic properties.

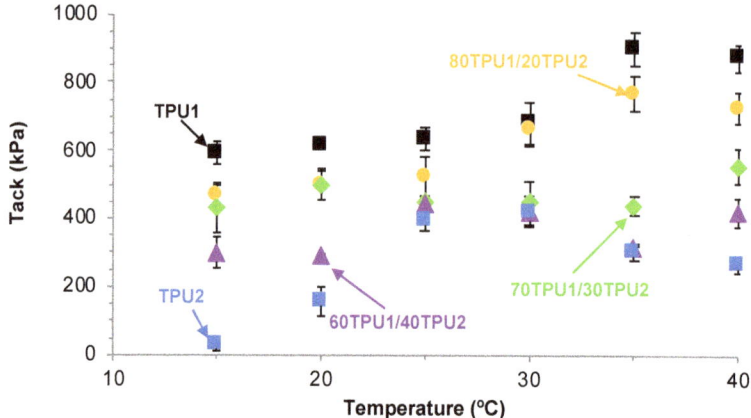

Figure 16. Variation of the tack of TPU1, TPU2, and TPU1 + TPU2 PSAs as a function of the temperature.

Apart from an adequate tack, the PSAs must possess an adequate peel strength and sufficient cohesion. Table 10 shows the 180° peel strength values of the stainless steel 304/TPU PSA film joints measured after 30 min of joint formation. The 180° peel strengths of the joints made with TPU1 PSA and TPU2 PSA were low and their loci of failure were different, i.e., the joint made with TPU1 PSA failed cohesively in the adhesive, which is not acceptable in PSAs, whereas an adhesion failure to the stainless-steel substrate was obtained in the joint made with TPU2 PSA. The loci of failure of the joints made with TPU1 and TPU2 PSAs are related to their cohesion or holding time, which was quite low in TPU1 PSA and quite high in TPU2 PSA. Interestingly, the 180° peel strength values of the joints made with all TPU1 + TPU2 PSAs were higher than the ones of TPU1 and TPU2 PSAs and the highest 180° peel strengths corresponded to the joints made with 80TPU1/20TPU2 and 70TPU1/30TPU2 PSAs. Because the holding time was higher in 70TPU1/30TPU2 PSA, the joints made with this TPU PSA showed an adhesion failure to the stainless-steel. In summary, the 70TPU1/30TPU2 PSA showed an excellent balance of tack, peel, and cohesion for being used as general purpose pressure sensitive adhesive. This balance of properties is due to the adequate soft segments content and the formation of new hydrogen bonded urethane interactions, which causes a lower degree of phase separation.

Table 10. Tack values and holding times at 25 °C of TPU1, TPU2, and TPU1 + TPU2 PSAs, and 180° peel strength values at 25 °C of stainless steel/TPU PSA joints.

Sample Code	Tack at 25 °C (kPa)	180° Peel Strength (kN/m) [a]	Holding Time at 25 °C (min)
TPU 1	634 ± 33	0.35 ± 0.04 (CA)	152 ± 46
80TPU1/20TPU2	525 ± 55	1.29 ± 0.06 (CA)	442 ± 134
70TPU1/30TPU2	450 ± 5	1.43 ± 0.25 (A)	847 ± 55
60TPU1/40TPU2	440 ± 21	0.85 ± 0.08 (A)	2115 ± 128
TPU2	391 ± 30	0.22 ± 0.04 (A)	4211 ± 10

[a] Locus of failure: CA, cohesive failure of the blend; A, adhesion failure.

In order to demonstrate the potential of the novel TPU PSAs, their pressure sensitive adhesive properties were compared to the ones of a commercial polyurethane pressure sensitive adhesive (SPUR PSA 3.0, Momentive, UK) of unknown formulation supported on PET film. For the commercial polyurethane PSA, the tack at 25 °C was 504 ± 22 kPa, the 180° peel strength of stainless steel/commercial polyurethane PSA joints was 0.41 ± 0.03 N/m, and the holding time was 359 ± 3 min. Therefore, the TPU PSAs developed in this study show comparable tack to commercial polyurethane PSA but higher 180° peel strength and higher creep resistance.

4. Conclusions

Pressure sensitive adhesives with balanced adhesion and cohesion properties were prepared by blending thermoplastic polyurethanes with different properties. The procedure for preparing the blends of the TPUs determined their different viscoelastic properties, and the properties of the TPU PSAs as well, the blending of separate MEK solutions of the two TPUs imparted higher tack and 180° peel strength, and adequate cohesion.

The TPU1 + TPU2 blends showed somewhat similar contributions of the free and hydrogen-bonded urethane groups, indicating that they had almost similar degrees of phase separation, which was lower than in the parent TPUs. All TPU1 + TPU2 blends showed the glass transition temperature of the soft segments at about −30 °C, which were similar irrespective of the composition of the blend, confirming that they showed a similar degree of microphase separation. Furthermore, two main thermal decompositions at 308–317 °C due to the urethane hard domains and another at 363–373 °C due to the soft domains could be distinguished in the TPU1 + TPU2 blends, the weight loss of the hard domains increased and the one of the soft domains decreased by increasing the amount of TPU2 in the blends. The storage moduli of the TPU1 + TPU2 blends were intermediate between the ones of TPU1 and TPU2 and they were similar for temperatures lower than 20 °C, and the moduli at the cross-over were lower than in the parent TPUs, which can be related to their lower degree of phase separation.

The improved properties of the TPU1 + TPU2 blends derived from the removal of the hydrogen bonds between the hard segments in the MEK solutions of TPU1 and TPU2 that were re-formed upon MEK removal, producing a different structure because the interactions between the hard domains were more complex and a higher number of hydrogen bonds were formed, which led to a lower degree of phase separation between the soft and the hard domains. The most efficient TPU1 + TPU2 blends were obtained by adding 20–30 wt% TPU2, likely due to the easy mobility of the polymeric chains of TPU2 during MEK removal. As a consequence, the properties of the TPU1 + TPU2 PSAs were improved because good tack, high 180° peel strength, and sufficient cohesion were obtained, particularly in 70TPU1/30TPU2 PSA. Therefore, the novel TPU PSAs can be used for manufacturing labels and tapes for medical and automotive applications in which tack can be maintained in a wide range of temperatures without sacrificing the cohesion and the peel strength.

Author Contributions: M.F. carried out some experiments, synthesize the parent TPUs and write the original draft, M.A.V.-M. carried out the experiments related to TPU PSA characterization, and J.M.M.-M. conceptualize and supervise the study, and write the article.

Funding: This research received no external funding.

Conflicts of Interest: The authors declare no conflict of interest.

References

1. Feldstein, M.M.; Dormidontova, E.E.; Khokhlov, A.R. Pressure sensitive adhesives based on interpolymer complexes. *Prog. Polym. Sci.* **2015**, *42*, 79–153. [CrossRef]
2. Mehravar, E.; Gross, M.A.; Agirre, A.; Reck, B.; Leiza, J.R.; Asua, J.M. Importance of film morphology on the performance of thermo-responsive waterborne pressure sensitive adhesives. *Eur. Polym. J.* **2018**, *98*, 63–71. [CrossRef]
3. Deng, X. Progress on rubber-based pressure-sensitive adhesives. *J. Adhes.* **2018**, *94*, 77–96. [CrossRef]
4. Mecham, S.; Sentman, A.; Sambasivam, M. Amphiphilic silicone copolymers for pressure sensitive adhesive applications. *J. Appl. Polym. Sci.* **2010**, *116*, 3265–3270. [CrossRef]
5. Czech, Z.; Milker, R.; Malec, A. Crosslinking of PUR-PSA water-borne systems. *Rev. Adv. Sci.* **2006**, *12*, 189–199.
6. Venkatraman, S.; Gale, R. Skin adhesives and skin adhesion. 1. Transdermal drug delivery systems. *Biomaterials* **1998**, *19*, 1119–1136. [CrossRef]
7. Lobo, S.; Sachadeva, S.; Goswami, T. Role of pressure sensitive adhesives in transdermal drug delivery systems. *Ther. Deliv.* **2016**, *7*, 33–48. [CrossRef]

8. Ho, K.Y.; Dodou, K. Rheological studies on pressure-sensitive silicone adhesives and drug-in-adhesive layers as a means to characterize adhesive performance. *Int. J. Pharm.* **2007**, *333*, 24–33. [CrossRef]
9. Wolf, H.-M.; Dodou, K. Investigations on the viscoelastic performance of pressure sensitive adhesives in drug-in-adhesive type transdermal films. *Pharm. Res.* **2014**, *31*, 2186–2202. [CrossRef]
10. Tombs, E.L.; Nikolaou, V.; Nurumbetov, G.; Haddleton, D.M. Trandermal delivery of ibuprofen utilizing a novel solvent-free pressure sensitive adhesive (PSA): TEPI® Technology. *J. Pharm. Innov.* **2018**, *13*, 48–57. [CrossRef]
11. Feldstein, M.M.; Siegel, R.A. Molecular and nanoscale factors governing pressure-sensitive adhesion strength of viscoelastic polymers. *J. Polym. Sci. Part B Polym. Phys.* **2012**, *50*, 739–772. [CrossRef]
12. Yilgör, I.; Yilgör, E.; Wilker, G.L. Critical parameters in designing segmented polyurethanes and their effect on morphology and properties: A comprehensive review. *Polymer* **2015**, *58*, A1–A36. [CrossRef]
13. Czech, Z.; Hinterwaldner, R. Pressure-sensitive adhesives based on polyurethanes. In *Handbook of Pressure Sensitive Adhesives and Products. Fundamentals of Pressure Sensitive Adhesives and Products Applications of Pressure–Sensitive Products*, 1st ed.; Benedek, I., Feldstein, M.M., Eds.; CRC Press, Taylor and Francis: Boca Raton, FL, USA; London, UK; New York, NY, USA, 2009; Chapter 11; pp. 1–19.
14. Dahl, R. Polyurethane Pressure-sensitive Adhesive. US Patent No. 3,437,622, 8 April 1969.
15. Tushaus, L. Pressure-sensitive Adhesives Based on Cyclic Terpene Urethane Resin. US Patent No. 3,718,712 A, 27 February 1973.
16. Hartmann, H.; Druschke, W.; Eisentraeger, K.; Muellern, H. Manufacture of Pressure-Sensitive Adhesives. US Patent No. 4,087,392 A, 2 May 1978.
17. Müller, H.; Szonn, B. Adhesive Sheets and Webs and Process for Their Manufacture. U.S. Patent No. 3,930,102 A, 11 April 1974.
18. Orr, R.B.; Del, W. Pressure-sensitive Adhesive Formulation Comprising Underindexed Isocyanate to Active Hidrogen Composition. US Patent No. 5,157101 A, 4 June 1990.
19. Kydonieus, A.; Bastar, L.; Shah, K.; Jamshidi, K.; Chang, T.-L.; Kuo, S.-H. Polyurethane Pressure-sensitive Adhesives. US Patent No. 5,591,820 A, 7 January 1997.
20. Shah, K.R.; Chang, T.-L.; Kydonieus, A. Water Soluble Polymer Additives for Polyurethane-based Pressure Sensitive Adhesives. US Patent No. 5,714,543 A, 3 February 1998.
21. Nakamura, Y.; Nakanoa, S.; Itoa, K.; Imamuraa, K.; Fujiia, S.; Sasakic, M.; Urahamad, Y. Adhesion properties of polyurethane pressure-sensitive adhesive. *J. Adhes. Sci. Technol.* **2013**, *27*, 263–277. [CrossRef]
22. Fuensanta, M.; Martín-Martínez, J.M. Thermoplastic polyurethane coatings made with mixtures of polyethers of different molecular weights with pressure sensitive adhesion property. *Prog. Org. Coat.* **2018**, *118*, 148–156. [CrossRef]
23. Fuensanta, M.; Martín-Martínez, J.M. Thermoplastic polyurethane pressure sensitive adhesives made with T mixtures of polypropylene glycols of different molecular weights. *Int. J. Adhes. Adhes.* **2019**, *88*, 81–90. [CrossRef]
24. Yılgör, E.; Yılgör, I.; Yurtsever, E. Hydrogen bonding and polyurethane morphology. I. Quantum mechanical calculations of hydrogen bond energies and vibrational spectroscopy of model compounds. *Polymer* **2002**, *43*, 6551–6559. [CrossRef]
25. Mattia, J.; Painter, P. A comparison of hydrogen bonding and order in a polyurethane and poly(urethane-urea) and their blends with poly(ethylene glycol). *Macromolecules* **2007**, *40*, 1546–1554. [CrossRef]
26. Fuensanta, M.; Jofre-Reche, J.A.; Rodríguez-Llansola, F.; Costa, V.; Iglesias, J.I.; Martín-Martínez, J.M. Structural characterization of polyurethane ureas and waterborne polyurethane urea dispersions made with mixtures of polyester polyol and polycarbonate diol. *Prog. Org. Coat.* **2017**, *112*, 141–152. [CrossRef]
27. Bagde, K.; Molnár, K.; Sajó, I.; Pukánszky, B. Specific interactions, structure and properties in segmented polyurethane elastomers. *Express Polym. Lett.* **2011**, *5*, 417–427. [CrossRef]
28. Wilkes, G.L.; Bagrodia, S.; Humphries, W.; Wildnauer, R. The time dependence of the Thermal and mechanical properties of segmented urethanes following Thermal treatment. *J. Polym. Sci. Polym. Lett.* **1975**, *13*, 321–327. [CrossRef]
29. Wilkes, G.L.; Wildnauer, R. The kinetic behavior of the thermal and mechanical properties of segmented urethanes. *J. Appl. Phys.* **1975**, *46*, 4148–4152. [CrossRef]
30. Chu, S.G. Dynamic mechanical properties of pressure-sensitive adhesives. In *Adhesive Bonding*; Lee, L.H., Ed.; Plenum Press: New York, NY, USA, 1991; pp. 97–138.

31. Chang, E.-P. Viscoelastic properties of pressure sensitive adhesives. *J. Adhes.* **1997**, *60*, 233–248. [CrossRef]
32. Dahlquist, C.A. Pressure sensitive adhesives. In *Treatise on Adhesion and Adhesives*; Patrick, R.L., Ed.; Marcel Dekker, Inc.: New York, NY, USA, 1969; pp. 219–260.

© 2019 by the authors. Licensee MDPI, Basel, Switzerland. This article is an open access article distributed under the terms and conditions of the Creative Commons Attribution (CC BY) license (http://creativecommons.org/licenses/by/4.0/).

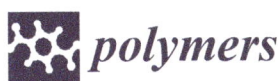

Article

PA6 and Halloysite Nanotubes Composites with Improved Hydrothermal Ageing Resistance: Role of Filler Physicochemical Properties, Functionalization and Dispersion Technique

Valentina Sabatini [1,2,*], Tommaso Taroni [1,2], Riccardo Rampazzo [1,2,3], Marco Bompieri [1], Daniela Maggioni [1], Daniela Meroni [1,2], Marco Aldo Ortenzi [1,2,3] and Silvia Ardizzone [1,2,3]

[1] Dipartimento di Chimica, Università degli Studi di Milano, Via Golgi 19, 20133 Milano, Italy; tommaso.taroni@unimi.it (T.T.); riccardo.rampa@gmail.com (R.R.); marco.bompieri@studenti.unimi.it (M.B.); daniela.maggioni@unimi.it (D.M.); daniela.meroni@unimi.it (D.M.); marco.ortenzi@unimi.it (M.A.O.); silvia.ardizzone@unimi.it (S.A.)
[2] Consorzio Interuniversitario per la Scienza e Tecnologia dei Materiali (INSTM), Via Giusti 9, 50121 Firenze, Italy
[3] CRC Materiali Polimerici "LaMPo", Dipartimento di Chimica, Università degli Studi di Milano, Via Golgi 19, 20133 Milano, Italy
* Correspondence: valentina.sabatini@unimi.it; Tel.: +39-02-503-14115

Received: 22 November 2019; Accepted: 13 January 2020; Published: 15 January 2020

Abstract: Polyamide 6 (PA6) suffers from fast degradation in humid conditions due to hydrolysis of amide bonds, which limits its durability. The addition of nanotubular fillers represents a viable strategy for overcoming this issue, although the additive/polymer interface at high filler content can become privileged site for moisture accumulation. As a cost-effective and versatile material, halloysite nanotubes (HNT) were investigated to prepare PA6 nanocomposites with very low loadings (1–45% w/w). The roles of the physicochemical properties of two differently sourced HNT, of filler functionalization with (3-aminopropyl)triethoxysilane and of dispersion techniques (*in situ* polymerization vs. melt blending) were investigated. The aspect ratio (5 vs. 15) and surface charge (−31 vs. −59 mV) of the two HNT proved crucial in determining their distribution within the polymer matrix. *In situ* polymerization of functionalized HNT leads to enclosed and well-penetrated filler within the polymer matrix. PA6 nanocomposites crystal growth and nucleation type were studied according to Avrami theory, as well as the formation of different crystalline structures (α and γ forms). After 1680 h of ageing, functionalized HNT reduced the diffusion of water into polymer, lowering water uptake after 600 h up to 90%, increasing the materials durability also regarding molecular weights and rheological behavior.

Keywords: polyamide 6; halloysite nanotube; functionalizing agent; nanocomposite; *in situ* polymerization; melt blending; polymorphism; hydrothermal ageing

1. Introduction

In recent decades, polyamide 6 (PA6) has received considerable attention [1] for a broad range of applications including in the automotive sector [2] and the textile industry [3], owing to its good mechanical performances and high thermal resistance, which are the result of the strong hydrogen bonds in its chemical structure.

However, in demanding environmental conditions (i.e., extremely high/low temperatures, high humidity, solar exposure and salts absorption), different degradation mechanisms occur and decrease

PA6 material performance and durability [4–6]. In particular, in humid conditions, water can interpenetrate between –NH and C=O groups of PA6 and break the pre-existent bonds, leading to a loss in mechanical cohesion [7,8]. Because the prediction and extension of PA6 durability is a very important issue for chemists, engineers and final users, numerous studies have been devoted to investigate its hydrothermal ageing.

In this context, PA6 properties can be preserved or modified to suit engineering requirements for several applications, using fillers and fiber additives such as silicate clays [9], glass fiber [10], black carbon [11], graphite [12], silica [13] and graphene [14]. However, it should be underlined that in some composites, such as PA6 and glass fibers-based materials, the additive/matrix interface is considered a privileged site for water accumulation during diffusion processes. Since this behavior can become catastrophic as the additive content increases [15,16], the use of low filler amounts represents a preferable strategy.

By using fillers with nanotubular shapes, significant improvements in the composite properties can be obtained using very low loadings, even down to a range of 1–5% w/w, maintaining in this way processing conditions similar to the ones of the neat material [9,17,18]. This is because their high surface/volume ratio favors their dispersion into the polymer matrix. Thus far, numerous investigations have focused on PA6 functionalization with carbon nanotubes (CNTs) [19–21] and boron nitride nanotubes (BNNTs) [22,23], often with very high loadings (up to 40% w/w [24,25]), due to the interesting mechanical properties of the resulting nanocomposites. However, both of these additives are technologically demanding to produce in bulk and functionalize, making them quite expensive and suitable only for limited high added value applications, e.g., robotics and organic electronics [26,27]. Thus, there is still a need to develop new types of high-performance PA6-nanocomposites that can be prepared via large scale, low cost fabrication routes.

In this framework, a phyllosilicate clay, halloysite nanotubes (HNT), represents a novel 1D natural nanomaterial morphologically similar to CNTs and BNNTs, with a unique combination of tubular nanostructure, natural availability, low cost, ease of functionalization, good biocompatibility, light color and high mechanical strength [28–30]. These features can impart useful mechanical, thermal, and biological properties to HNT-polymer nanocomposites, while keeping costs low [31,32]. Nanocomposites of polyamides and HNT have been reported in the literature to exhibit higher thermal stability [33] and flame inhibition properties [34], as well as enhanced strength without loss of ductility [35], even at very low filler content (<5%) [36–38]. However, to the authors' best knowledge, no previous reports about the effect of HNT filler on the durability of polyamide materials in hydrothermal conditions can be found in the literature. The present study aims at shedding light on this aspect by exploiting the effects of low content (1% and 4% w/w) of HNT fillers on the hydrothermal ageing of PA6 nanocomposites.

As halloysite is an extractive material, it shows a large morphological variance (from dimension to specific surface area and aspect ratio) depending on the extraction site. This aspect can have a great impact on the final properties of nanocomposites, but it has often been overlooked in previous literature reports. In this regard, here we thoroughly compare two different sources of HNT with several structural, morphological, optical and surface differences.

Moreover, we here compare the two different incorporation strategies more often used for fillers incorporation in PA6: in situ polymerization and melt blending processes. In situ polymerization involves modifiers dispersion into monomer followed by polymerization, while melt extrusion proceeds via a mechanical blending of filler in the molten polymer. The melt blending approach is more cost effective and straightforward than the in situ polymerization technique, but full filler intercalation is not always achievable [39]. On the other hand, relatively little work has been reported on the in situ polymerization technique, particularly concerning the relationship between the filler dispersion efficiency and PA6-based composites resultant macromolecular properties [39], and no previous work can be found in relation to HNT-polyamide composites using this incorporation strategy.

One of the main advantages of HNT is their ease of functionalization, which enables the tailoring of their surface properties. Guo et al. [33] reported the covalent incorporation of HNT functionalized with 3-(trimethoxysilyl)propyl methacrylate within a PA6 matrix, showing good incorporation and enhanced thermal and mechanical properties of the resulting composite. Here we compare HNT fillers both in their pristine state and after functionalization with (3-aminopropyl)triethoxysilane (APTES); this surface modification was adopted to increase compatibility between the filler surface and the organic matrix, as well as to reduce the surface hydrophilicity.

For all of the prepared composites, the effect of HNT incorporation on the composite molecular weights, molecular weight distribution, thermal and rheological properties, together with the filler dispersion state inside the polymer matrix, was investigated with respect to standard PA6. In particular, the comparison between in situ and melt blending methods highlighted the role of HNT surface charge with respect to the filler dispersion efficiency. Finally, the composite resistance to humidity was determined via accelerated ageing tests. It is well known that the macroscopic properties of most polyamides are affected by hydrothermal ageing after few months, or even weeks [40–42]. In the present case, notable differences between the neat PA6 and the composite materials are observed even after just 70 days.

2. Materials and Methods

2.1. Materials

ε-caprolactam (99%), 6-aminocaproic acid (≥99%), distilled water Chromasolv® (≥99.9%), toluene (anhydrous, ≥99.8%), potassium nitrate (KNO_3), 2-propanol, (3-aminopropyl)triethoxysilane (APTES, ≥98%) and dichloromethane (DCM anhydrous, ≥99.8%) were supplied by Sigma Aldrich (Milan, Italy and used without further purification. The commercial polyamide 6 "Technyl® C 246 natural" (PA6TECH) was purchased from Solvay Chimica Italia S.p.A (Livorno, Italy). Two different kinds of HNT nano-clays were used to prepare nanocomposites: HNT_H supplied by iMineral Inc. (commercially known as Ultra HalloPure®, Vancouver, Canada) and HNT_S received from Sigma Aldrich (Milan, Italy).

2.2. HNT Functionalization

2 g of HNT was suspended in 100 mL of toluene. Then, 1.4 g (6.3 mmol) of APTES was added and the suspension was refluxed under stirring for 7 h. Heating was stopped during the night (15 h) and the retrieved powder was washed with toluene by centrifugation/resuspension cycles (5 × 40 mL). The solid was then dried in the oven at 90 °C for 24 h.

2.3. HNT Characterization

The specific surface area (SSA) of the samples was measured through N_2 adsorption-desorption isotherms in subcritical conditions, utilizing a Coulter SA 3100 apparatus (Beckman Coulter Life Sciences, Indianapolis, Indiana), analyzed according to the Brunauer Emmett Teller (BET) method.

X-ray diffraction (XRD) (Shimadzu, Milan, Italy) patterns of the samples were acquired using a Siemens D5000 diffractometer equipped with a Cu Kα source (λ = 0.15406 nm), working at 40 kV × 40 mA nominal X-ray power, at room temperature. θ:2θ scans were performed between 10° and 80° with a step of 0.02°.

The oxide ζ-potential was measured using a Zetasizer Nano ZS (Malvern Instruments, Malvern, United Kingdom), operating at λ = 633 nm with a solid-state He-Ne laser at a scattering angle of 173°, using the dip-cell kit. Powders were suspended in a 0.01 M KNO_3 aqueous solution. Each ζ-potential value was averaged from at least three measurements.

Thermogravimetric analyses (TGA) (Mettler Toledo, Milan, Italy) were performed on bare and modified HNT using a Mettler-Toledo TGA/DSC 3+ STAR System, between 30 and 900 °C with a rate of 3 °C/min under air flux.

Transmission electron microscopy (TEM) images were obtained with a Zeiss LEO 912ab Energy Filtering TEM (Zeiss, Oberkochen, Germany), equipped with a CCD-BM/1K system, operating at an acceleration voltage of 120 kV. Samples were suspended in 2-propanol or water (1 mg/mL) and deposited on Cu holey carbon grids (200 mesh).

Fourier transform infrared (FTIR) spectra were acquired using a Perkin Elmer Spectrum 100 spectrophotometer in Attenuated Total Reflectance (ATR) mode (Perkin Elmer, Milan, Italy), registering 12 scans between 4000 and 400 cm^{-1} with a resolution of 4.0 cm^{-1}. A single-bounce diamond crystal was used with an incidence angle of 45°.

Diffuse reflectance spectra (DRS) were collected using a Shimadzu UV2600 spectrophotometer (Shimadzu, Milan, Italy).

2.4. PA6 and HNT Nanocomposites (PA6_HNT) Preparation and Characterization

2.4.1. In Situ Polymerization

A 250 cm^3 one-necked round bottom flask was loaded with ε-caprolactam (50.0 g), 6-aminocaproic acid (2.5 g) and the selected amount of HNT (either 1 or 4% w/w), and equipped with a polymerization candelabrum having nitrogen inlet and outlet adapters and an overhead mechanical stirrer. The flask was flushed with nitrogen, heated at 270 °C in a polymerization oven and then stirred (150–180 rpm) for 6 h. Afterwards, the reaction mixture was gradually brought back to room temperature. Flowing nitrogen atmosphere was maintained throughout the polymerization reaction. The resulting product was washed from unreacted ε-caprolactam with boiling water for 24 h, and dried in vacuum (about 4 mbar) at 100 °C for at least 12 h. For the sake comparison, neat PA6 was synthesized using the same conditions.

2.4.2. Melt Processing

Melt blended PA6-HNT nanocomposites (with filler content 1 or 4% w/w) were prepared using a ThermoScientific HAAKE Process 11 co-rotating twin screw extruder (Thermo Fisher Scientific, Rodano, Italy), with extrusion temperatures set at 235, 235, 235, 235, 235, 230, 217 and 210 °C from hopper to die respectively, a screw speed of 160 rpm, and a feed rate of 3 rpm. The extruded strands were quenched in water, air dried and then pelletized. Neat PA6, used as reference, was prepared by extruding PA6TECH under the same processing conditions. Prior to the use, nano-clays and PA6TECH pellets were dried under nitrogen atmosphere for 12 h at 60 °C in an oven to remove physisorbed water.

2.4.3. PA6 Nanocomposites Characterization

The HNT actual content in PA6-nanocomposites was determined via dynamic TGA analyses performed from 30 to 900 °C at 10 °C/min, under nitrogen flow, using a Perkin Elmer TGA 4000 Instrument (Perkin Elmer, Milan, Italy).

Scanning Electron Microscopy (SEM) studies of PA6 nanocomposites fracture surfaces were carried out on a Leica Electron Optics 435 VP microscope (Leica, Milan, Italy), working at an acceleration voltage of 15 kV and 50 pA of current probe and equipped with an energy dispersive X-ray (EDX) spectrometer (Philips, XL-30W/TMP), to investigate the HNT dispersion in PA6 polymer matrix. The samples were manually fractured after cooling in liquid nitrogen and sputter-coated with a 20 nm thick gold layer in rarefied argon, using an Emitech K550 Sputter Coater, with a current of 20 mA for 180 s.

The isothermal crystallization behavior of PA6 and its nanocomposites was determined using a Mettler Toledo DSC1 calorimeter (Mettler Toledo, Milan, Italy); analyses were conducted under nitrogen atmosphere, weighting 5–10 mg of each sample in a standard 40 μL aluminum crucible and using an empty 40 μL crucible as reference.

In addition, non-isothermal behavior was studied using the following program: (i) heating at a rate of 10 °C/min from 25 to 250 °C; (ii) 5 min of isotherm at 250 °C; (iii) cooling from 250 to 25 °C at

10 °C/min (crystallization temperature, T_c and crystallization enthalpy, ΔH_c, were determined during cooling); (iv) 5 min of isotherm at 25 °C; and (v) heating from 25 to 250 °C at 10 °C/min (melting temperature, T_m, and heat of fusion, ΔH_f, were measured during heating). ΔH_f was used to calculate the level of crystallinity, defined by the ratio of ΔH_f to the heat of fusion of the purely crystalline forms of PA6 (ΔH_{f°). Since the ΔH_{f° values of α and γ PA6 crystalline forms are nearly identical, the level of crystallinity was calculated using the average of the two, i.e., 240 J/g [35].

Rheological analyses, conducted using frequency sweep experiments, were performed with a Physica MCR 300 rotational rheometer with a parallel plate geometry (φ = 25 mm; 1 mm distance between plates) (Anton Paar, Rivoli, Italy). Neat PA6 and PA6-HNT nanocomposites were dried in a vacuum oven (around 4 mbar) at 100 °C for 12 h and linear viscoelastic regimes were studied; strain was set equal to 5% and curves of complex viscosity as function of frequency were recorded, taking 30 points ranging from 100 Hz to 0.1 Hz with a logarithmic progression, at 250 °C.

The molecular weight properties were evaluated using a Size Exclusion Chromatography (SEC) system having a Waters 1515 Isocratic HPLC pump, four Waters Styragel (103Å-104Å-105Å-500Å) columns and a UV detector (Waters 2487 Dual λ Absorbance Detector) at 244 nm, using a flow rate of 1 $cm^3 \cdot min^{-1}$ and 15 μL as injection volume (Waters, Sesto San Giovanni, Italy). Samples were prepared dissolving 10–15 mg of polymer in 1 cm^3 of anhydrous DCM after N-trifluoroacetilation [43]; before the analysis, the solution was filtered with 0.45 μm filters. Molecular weight data were expressed in polystyrene (PS) equivalents. The calibration was built using monodispersed PS standards having the following nominal peak molecular weight (Mp) and molecular weight distribution (D): Mp = 1600 kDa (D ≤ 1.13), Mp = 1150 kDa (D ≤ 1.09), Mp = 900 kDa (D ≤ 1.06), Mp = 400 kDa (D ≤ 1.06), Mp = 200 kDa (D ≤ 1.05), Mp = 90 kDa (D ≤ 1.04), Mp = 50,400 Da (D = 1.03), Mp = 37,000 Da (D = 1.02), Mp = 17,800 Da (D = 1.03), Mp = 6520 Da (D = 1.03), Mp = 5460 Da (D = 1.03), Mp = 2950 Da (D = 1.06), Mp = 2032 Da (D = 1.06), Mp = 1241 Da (D= 1.07), Mp = 906 Da (D = 1.12); ethyl benzene (molecular weight = 106 g/mol). For all analyses, 1,2-dichlorobenzene was used as internal reference.

2.5. Hydrothermal Ageing Tests

Samples for the accelerated ageing test were prepared via injection molding with a BABYPLAST 610P instrument (BABYPLAST, Vicenza, Italy) and were immersed in distilled water at a set temperature of 90 °C for 70 days in glass jar containers. The main parameters of the injection molding are reported in Table S1. The ageing temperature was chosen to accelerate the diffusion process as well as to accentuate the materials degradation mechanism [7,44]. Samples were fully immersed and periodically, i.e., after 25 (600 h), 50 (1200 h) and 70 days (1680 h), weighted using a digital balance of 0.0001 g accuracy. The moisture uptake (M_t) was measured through a gravimetric method, as reported in Equation (1).

$$M_t (\%) = (m_t - m_0)/m_0 \times 100 \tag{1}$$

where "m_t" and "m_0" are the wet and dry weights of the sample after any specific time t. Aged samples were analyzed via rheological, SEC and DSC analyses, according to the experimental procedures described in Section 2.4.3.

3. Results and Discussion

3.1. HNT Characterization and Functionalization

As previously stated, HNT_H and HNT_S fillers were acquired from different suppliers, and in order to investigate the differences between the two types of nanotubes, they were analyzed via SSA, XRD, TGA and TEM analyses. The two powders present different colors, HNT_S being white and HNT_H brownish (Supplementary Information, Figure S1). As reported in Table 1, HNT_S powder possesses a specific surface area of 53 m^2/g, in good agreement with literature data [45,46]; on the other hand, HNT_H sample shows a lower surface area of 34 m^2/g. According to XRD patterns reported in Figure 1A, HNT_S filler presents several impurity phases alongside halloysite, including kaolinite, quartz and gibbsite. These findings are in good agreement with previous literature reports on HNT samples from the same supplier [45,47]. Instead, HNT_H powder appears purer, as apparent from the absence or much lower intensity of impurity peaks, and more crystalline, as shown by the sharper halloysite (10Å) peaks. Our diffractogram is in good agreement with data from the supplier, showing only minor crystalline impurities, such as quartz (<1%) and kaolinite (<10%).

Table 1. Physicochemical properties of bare and functionalized HNT: specific surface area (SSA), average length and aspect ratio, ς-potential at spontaneous pH, and APTES loading from TGA.

Sample	SSA (m^2/g)	Avg. Length (nm)	Avg. Aspect Ratio	ς-Potential (Bare HNT) (mV)	ς-Potential (HNT_APTES) (mV)	APTES Loading (%)
HNT_S	53	400 ± 300	5 ± 3	−31 ± 2	+5.9 ± 0.6	4.6
HNT_H	34	1100 ± 800	15 ± 9	−59 ± 3	−44 ± 2	5.1

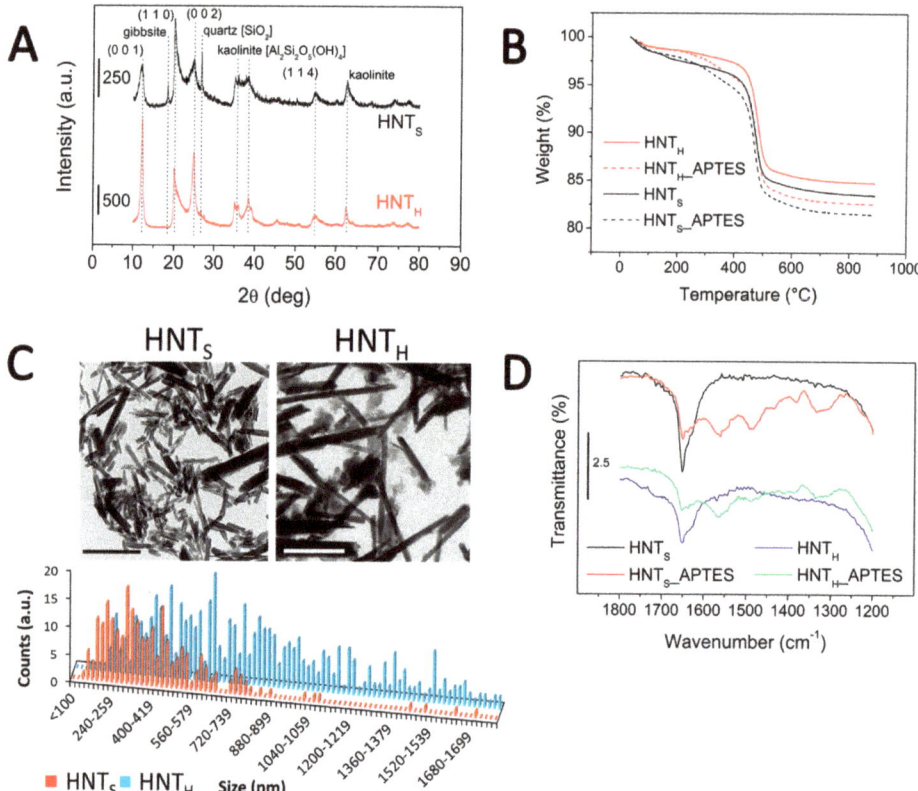

Figure 1. Characterization of adopted HNT powders: (**A**) XRD patterns with indexing of the halloysite peaks and main impurities; (**B**) TGA curves; (**C**) TEM images (scale bar: 1 μm); and (**D**) FTIR spectra of bare and APTES-modified fillers.

TGA measurements are reported in Figure 1B. The initial mass loss, between 50 and 300 °C, can be associated with the removal of physisorbed and interlayer water, in this order. The most severe weight loss, at around 500 °C, can be attributed to the dehydroxylation of the inner surface [46]. The two sources of HNT show a slight difference in weight loss at lower temperatures. This can be related to the amount of physisorbed water on their surface: as the HNT_S sample has a higher surface area (Table 1), it also carries more water, resulting in a greater weight loss. Moreover, upon APTES addition a weight loss at around 250 °C can be appreciated, related to the degradation of the aminopropyl chain. Overall, in both cases loading is around 5% (Table 1), notwithstanding HNT$_H$ lower surface area. This suggests that in this case functionalization was more efficient.

TEM images of untreated HNT were acquired to determine the difference in morphology between the two sources (Figure 1C). As can be seen in Table 1, the two families of HNT show greatly different aspect ratios: HNT$_H$ appear much more elongated than HNT$_S$, as their aspect ratio is three times larger, and also much less fragmented.

The FTIR spectra of HNT samples are reported in Figure 1D, in the 1800–1200 cm^{-1} range. As can be seen, HNT$_S$ filler shows a more intense water bending peak at 1650 cm^{-1}, which is in line with what reported by the TGA. After the functionalization with APTES, few signals between 1600 and 1300 cm^{-1} become appreciable: the peak at 1561 cm^{-1} can be attributed to –NH$_2$ scissoring modes of APTES amino groups, while that at 1490 cm^{-1} can be related to the symmetric deformation of –NH$_3^+$ moieties. Finally, the band at 1335 cm^{-1} can be assigned to C–N stretching modes [48].

Interestingly, one of the most blatant differences between the two sources of HNT resides in their surface charge: as reported in Table 1, the recorded ς-potential at spontaneous pH for HNT$_H$ is much more negative than that of HNT$_S$, and this difference persists even after surface functionalization with APTES, which, due to its positively charged –NH$_3^+$ groups, raises the value of surface charge.

3.2. Synthesis of PA6 and PA6-HNT Nanocomposites via In Situ Polymerization and Melt Blending Processes

The dispersion of HNT$_H$ and HNT$_S$ inside PA6 polymer matrix was performed comparing two different incorporation routes, i.e., in situ polymerization and melt blending processes. The fillers were tested at 1 and 4% w/w, both unmodified and functionalized with APTES. Table 2 lists all of the prepared samples, along with the actual content of HNT inside the PA6 nanocomposites, as determined via TGA analyses. The latter are, in all cases, very close to the theoretical amount.

Table 2. PA6 nanocomposites prepared via in situ polymerization and melt blending processes: theoretical and actual loading from TGA analyses, and kinetic analysis of isothermal crystallization data obtained via Avrami equation.

Preparation Route	Sample	HNT$_{theoretical}$ (% w/w)	HNT$_{real}$ (% w/w)	Avrami Data		
				onset T_c (°C)	n	$t_{1/2}$ (s)
In situ polymerization	PA6$_{in\ situ}$	-	-	190	1.6	63
				199	1.8	408
	PA6$_{in\ situ}$_HNT$_H$_1	1	0.8	190	1.0	135
	PA6$_{in\ situ}$_HNT$_S$_1	1	0.9	199	2.3	408
	PA6$_{in\ situ}$_HNT$_H$_4	4	3.3	190	1.1	163
	PA6$_{in\ situ}$_HNT$_S$_4	4	4.0	199	1.6	408
	PA6$_{in\ situ}$_HNT$_H$_1_APTES	1	0.7	190	1.0	137
	PA6$_{in\ situ}$_HNT$_S$_1_APTES	1	1.0	199	2.4	420
	PA6$_{in\ situ}$_HNT$_H$_4_APTES	4	4.0	190	1.2	174
	PA6$_{in\ situ}$_HNT$_S$_4_APTES	4	3.9	199	1.6	408
Melt extrusion	PA6$_{melt}$	-	-	200	1.8	135
	PA6$_{melt}$_HNT$_H$_1	1	1.0	200	2.3	135
	PA6$_{melt}$_HNT$_S$_1	1	0.6	200	2.6	135
	PA6$_{melt}$_HNT$_H$_4	4	4.0	200	2.3	135
	PA6$_{melt}$_HNT$_S$_4	4	4.0	200	2.1	135
	PA6$_{melt}$_HNT$_H$_1_APTES	1	0.8	200	2.3	126
	PA6$_{melt}$_HNT$_S$_1_APTES	1	0.7	200	2.1	126
	PA6$_{melt}$_HNT$_H$_4_APTES	4	3.8	200	2.0	135
	PA6$_{melt}$_HNT$_S$_4_APTES	4	3.6	200	2.0	135

It should be noted that at the highest filler content (4% w/w) HNT$_H$ noticeably alters the original white color of neat PA6 (Figure S1), as also shown in DRS spectra (Figure S2).

To investigate the dispersion efficiency of HNT$_H$ and HNT$_S$ by the two preparation methods, the morphology of cryo-fractured PA6-HNT nanocomposite samples prepared was studied via SEM analyses; the presence of HNT-based aggregates was also investigated via EDX measurements. Representative images of 4% w/w composites are reported in Figures 2 and 3.

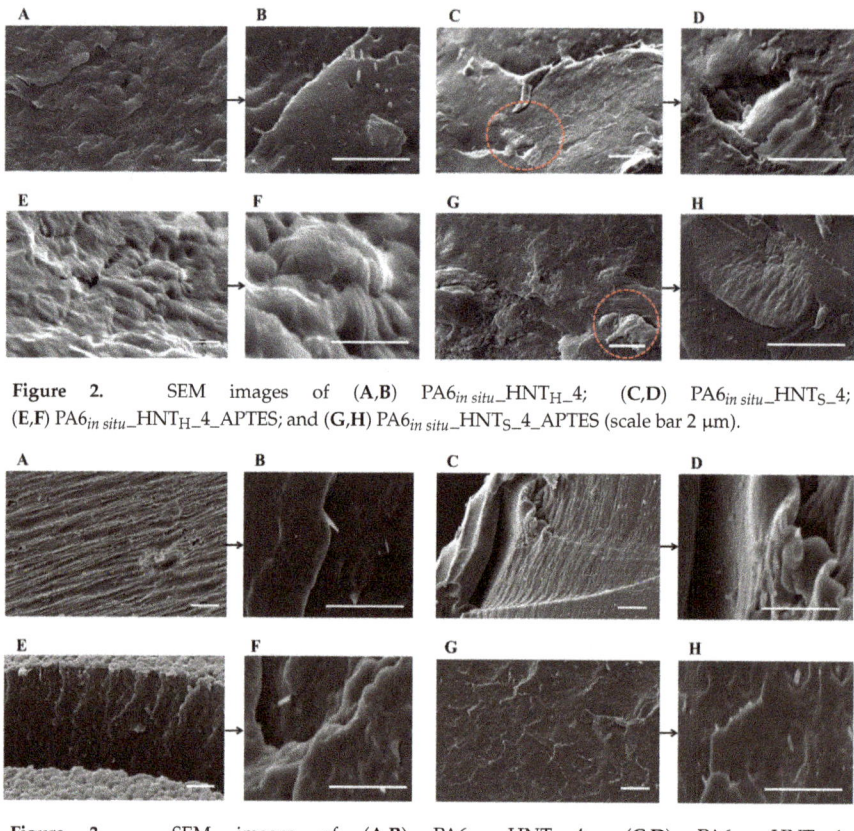

Figure 2. SEM images of (**A,B**) PA6$_{in\,situ}$_HNT$_H$_4; (**C,D**) PA6$_{in\,situ}$_HNT$_S$_4; (**E,F**) PA6$_{in\,situ}$_HNT$_H$_4_APTES; and (**G,H**) PA6$_{in\,situ}$_HNT$_S$_4_APTES (scale bar 2 µm).

Figure 3. SEM images of (**A,B**) PA6$_{melt}$_HNT$_H$_4; (**C,D**) PA6$_{melt}$_HNT$_S$_4; (**E,F**) PA6$_{melt}$_HNT$_H$_4_APTES; and (**G,H**) PA6$_{melt}$_HNT$_S$_4_APTES (scale bar: 2 µm).

In in situ polymerization samples, both unmodified and silane-modified HNT$_S$ form micrometric aggregates (Figure 2C,G red circles), indicating that the filler has not been efficiently dispersed in the polymer matrix due to its poor interaction with PA6 [37]. On the other side, both pristine and functionalized HNT$_H$ fillers are homogeneously dispersed in the polymer and no aggregates are present (Figure 2A,E). The reason behind this different behavior is probably the greater surface charge of HNT$_H$, which more effectively separates nanotubes due to same-charge repulsing interactions. This effect, together with the higher aspect ratio leading to a more pronounced self-ordering, might favor a more homogeneous distribution of the filler in PA6_HNT$_H$ composites, whereas HNT$_S$ aggregates, possibly already present in the untreated powder, remain in the polymer matrix.

The dispersion of HNT$_H$ further improves with APTES functionalization: APTES-modified HNT$_H$ are enclosed and well-penetrated within the fractured PA6 surface. This observation can be explained considering that, while the surface charge remains high upon functionalization (Table 1), the amino functional groups of APTES promote the compatibility between the functionalized filler and PA6 matrix [49].

In the case of nanocomposites prepared via melt extrusion process (Figure 3), no significant difference related to the use of HNT$_S$/HNT$_H$ and APTES is detectable: in fact, the nanotubes distribution appears to be uniform across all the specimens and the good interaction between HNT and PA6 matrix seems to be independent of surface functionalization. However, it should be noted that in the case of melt blending, the HNT dispersion, although homogeneous, is essentially close to the surface, as also supported by the final performance (see Sections 3.4 and 3.5). On the other hand, the in situ polymerization guarantees a bulk penetration of the filler. This difference may be explained by the fact that, as opposed to extrusion, during in situ preparation HNT fillers have more time to disperse in the medium, allowing the filler to spread evenly.

3.3. Isothermal Crystallization Behavior of PA6 Nanocomposites

The crystallization behavior of a crystalline polymer can be distinguished in two main processes, nucleation and crystal growth. The initial stage of crystallization, i.e., primary crystallization, can be analyzed by the Avrami kinetic equation (Equation (2)).

$$1 - X(t) = \exp(-K \cdot t^n) \qquad (2)$$

where $X(t)$ is the fraction of crystallized polymer at time t, n is the parameter that details nucleation and growth processes, and K is the crystallization rate constant. Linear regression was adopted to obtain n and K from plots of the data in the form of Equation (3).

$$\ln[1 - \ln(1 - X(t))] = \ln K + n \ln t \qquad (3)$$

Based on the Avrami theory, n is related to the geometry of the crystal growth and on the type of nucleation.

Here, isothermal crystallization behavior was studied by heating the sample from 25 to 250 °C at 10 °C/min, holding for 1 min to ensure melting, and then rapidly cooling to (i) 199 °C in the case of PA6_HNT$_S$ nanocomposites prepared via in situ polymerization, (ii) 190 °C for PA6_HNT$_H$ synthesized in situ and (iii) 200 °C for all samples prepared via melt processing. Neat PA6 was consequently analyzed at 190, 199 and 200 °C. The chosen crystallization temperatures (190/199/200 °C) were determined via a series of experiments conducted at various crystallization temperatures. Below the indicated onset crystallization temperature, the crystallization peak overlapped with the initial transient portion of heat flow curve. Setting the onset temperature as shown above deletes this problem, while it enables the completion of the crystallization process in a reasonable time. According to Fornes et al. [50], the different crystallization behavior is probably due to the different molecular weight of PA6 in the different composites. Moreover, in the case of PA6$_{in\ situ}$ nanocomposites, there can be an effect of the different interaction of the two used HNT with the polymer matrix, as previously described in SEM results. As reported in Table 2 (6th column), n values for pure PA6$_{in\ situ}$ and PA6$_{melt}$ samples range from $1.6_{T=190\ °C}$ to $1.8_{T=199\ and\ 200\ °C}$ and are consistent with values reported in the literature [51]. In PA6$_{in\ situ}$ nanocomposites, n shows a decrease with the addition of HNT$_H$ filler and is not significantly affected by the presence of the functionalizing agent. According to Gurato et al. [52], this behavior suggests that small amounts of filler (lower than 5–10% w/w) can act as homogeneous nucleation sites followed by mono and bi-dimensional growth. On the other side, HNT$_S$ samples overall show higher n values, possibly due to the not satisfactory HNT$_S$ dispersion in PA6 matrix that enhances the formation of heterogeneous nucleation sites [52].

The n values for extruded materials are generally higher than the ones obtained by in situ polymerization. Also, in this case, functionalization with APTES does not significantly change the PA6 nanocomposites crystallization behavior. As shown by Khanna et al. [53], the increase in n can be related to memory effects associated with the processing of PA6. As a matter of fact, numerous studies have shown that the processing history of the polymer, e.g., melting, cooling, pelletizing and so on, are often not fully erased when the polymer is annealed at high melting temperatures [50,52,54]. In this way, memory effects associated with thermal and stress histories that remain present in the material after annealing in the melt may also lead to an increased formation of heterogeneous crystallization sites.

The half-time of crystallization ($t_{1/2}$) was calculated according to Equation (4):

$$t_{1/2} = (\ln(2/K))^{1/n} \quad (4)$$

The resulting values for all samples are reported in Table 2 (7th column). PA6$_{in\ situ}$ nanocomposites with bare and APTES-modified HNT$_H$ fillers noticeably decrease their crystallization rate than neat PA6; this effect is due to the good adhesion between PA6 and the filler that impedes the motion of PA6 molecular chains, hampering the crystallization of PA6 matrix [13]. On the other hand, pristine and modified HNTs nanotubes lead to a less pronounced decrease of their crystallization rate with respect to neat PA6. Lastly, the crystallization rate of the samples obtained via melt extrusion process is not significantly influenced by the presence of bare and modified HNT.

Overall, the results obtained in the case of samples prepared via melt blending are consistent with those reported in the literature [33,50]. On the basis of SEM results, the inefficient distribution of HNTs filler inside PA6 matrix by in situ polymerization results in a non-homogenous crystallization behavior of the corresponding composites. Moreover, the different dispersion of HNT$_H$ during in situ polymerization and melt blending process is reflected in the crystallization behavior of the two types of composite: n and $t_{1/2}$ are lower for the PA6$_{in\ situ}$ samples, consistently with the better filler incorporation within the bulk polymer matrix.

3.4. Non-Isothermal Crystallization Behavior of PA6 Nanocomposites

Non-isothermal crystallization studies were carried out to highlight the effect of pristine and modified fillers onto the crystallization behavior and crystal structure of the prepared PA6-HNT nanocomposites. Table 3 and Table S2 report the DSC data of PA6 nanocomposites in non-isothermal heating and cooling conditions.

Table 3. DSC data of PA6-HNT nanocomposites collected during cooling and second heating thermal steps.

Sample	Cooling	2nd Heating	
	T_c (°C)	T_m (°C)	χ_c (%)
PA6$_{in\ situ}$	188.1	220.8	30.2
PA6$_{in\ situ}$_HNT$_H$_1	187.2	220.8	23.8
PA6$_{in\ situ}$_HNT$_S$_1	185.5	220.4	22.3
PA6$_{in\ situ}$_HNT$_H$_4	182.1	219.0	23.5
PA6$_{in\ situ}$_HNT$_S$_4	182.2	219.1	22.4
PA6$_{in\ situ}$_HNT$_H$_1_APTES	187.2	220.3	21.5
PA6$_{in\ situ}$_HNT$_S$_1_APTES	187.0	220.3	18.3
PA6$_{in\ situ}$_HNT$_H$_4_APTES	184.6	214.6	14.0
PA6$_{in\ situ}$_HNT$_S$_4_APTES	185.0	214.4	14.9
PA6$_{melt}$	189.4	221.1	24.9
PA6$_{melt}$_HNT$_H$_1	188.8	221.2	20.1
PA6$_{melt}$_HNT$_S$_1	188.9	221.6	24.3
PA6$_{melt}$_HNT$_H$_4	186.2	220.1	17.1
PA6$_{melt}$_HNT$_S$_4	185.9	220.4	21.8
PA6$_{melt}$_HNT$_H$_1_APTES	188.6	221.4	24.4
PA6$_{melt}$_HNT$_S$_1_APTES	188.4	221.4	26.3
PA6$_{melt}$_HNT$_H$_4_APTES	187.1	219.8	21.7
PA6$_{melt}$_HNT$_S$_4_APTES	188.3	219.6	19.3

By comparing the thermal behavior of each composite with respect to the relative neat PA6, a decrease in T_c (Table 3, 2nd column) and ΔH_c (Table S2, 2nd column) values is observed upon addition of both pristine and APTES-modified nanotubes by in situ polymerization. In particular, T_c values consistently show a decreasing trend at increasing HNT content. As described by Wurm et al. [55], the filler introduction in a polymer matrix hinders its macromolecular chains mobility, thus delaying crystal growth. Further evidence of slower kinetics for HNT$_H$ and HNT$_S$-based nanocomposites, can be seen in the χ_c values obtained upon subsequent heating (Table 3, 4th column). ΔH_f values used for χ_c calculations are reported in the Supporting Information (Table S2, 3rd column). These effects are instead much less appreciable in the case of melt blending composites, in good agreement with isothermal crystallization studies (see Section 3.2) and with the literature [50,56].

Furthermore, it is interesting to note that in the case of the samples prepared via in situ and melt-blending processes, with 4% w/w of both pristine and modified HNT$_H$/HNT$_S$ the peak temperature related to the melting transition (T_m) slightly decreases; the above observation may arise from several factors, such as the use of high filler loadings that could disturb the crystalline structures formation [35,57,58], but in particular, the formation of a small shoulder prior and close to the endothermic peak, associated with the PA6 γ crystalline form (Figure 4), may be a reason [59]. Several works [50,56,60,61] explain that the lower melting point associated with the presence of the γ form could be due to the lower crystalline density and increased melting entropy associated with the reduction in trans-conformation bonding as compared to the α counterpart. On the other hand, the presence of the γ crystalline form and the melting temperature decrease may simply reflect changes and imperfections in crystallites thickness and their distribution. In other words, the lower crystallinity could be due to the inability of polymer chains to be fully incorporated in to growing crystalline lamella; this leads to a smaller crystallite structures and more defects ridden crystalline lamella. Such imperfections in crystalline structure could contribute to the reduction of nanocomposites T_m [62]. Furthermore, several works [33,35,56] report that HNT addition in polyamide-based composites favors polymer polymorphism, i.e., the formation of γ crystalline form. In particular, Guo et al. [56] suggested that the Si–O and –OH groups present along HNT surface are capable of forming hydrogen bonds with PA6 amide groups. The interfacial interaction could restrain the mobility of PA6 chains and weaken the hydrogen bonding interactions between the polymer chains, hindering the formation α-phase crystals.

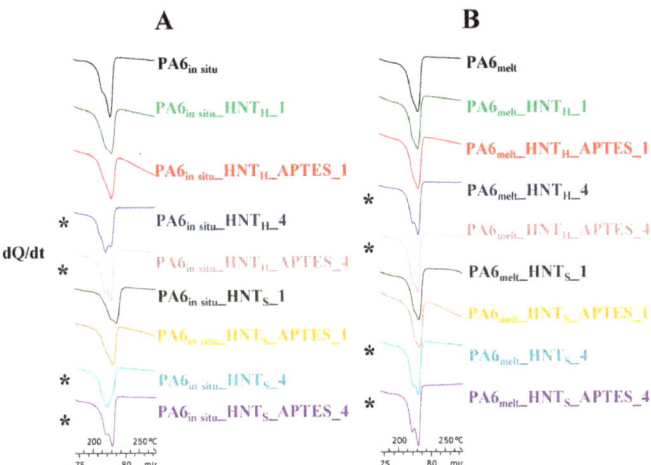

Figure 4. Heating scans of (**A**) PA6_HNT$_{in\ situ}$ and (**B**) PA6_HNT$_{melt}$ nanocomposites. Samples characterized by the presence of γ form are highlighted with a star (*).

3.5. Water Uptake and Rheological, SEC and DSC Characterization of Aged Nanocomposites

Figure 5 reports the absolute mass variation, calculated according to Equation (1), of PA6 and PA6-based nanocomposites conditioned in distilled water at 90 °C for 70 days. As widely reported in the literature [7,8,44,63], the PA6 hydrothermal stability can be studied at different ageing temperatures, e.g., at room temperature or close to water boiling point (~70–90 °C). It is generally accepted that temperature acts like an activator of eater diffusion within the polymer matrix [64]. As a result, a much shorter duration is needed to reach the equilibrium moisture content values at high temperature, such as 70–90 °C, than at room temperature [64]. Very recently, Sang et al. [8] compared the hydrothermal ageing at room temperature to that at temperatures higher than T_g, showing similar trends in terms of water uptake at the different temperatures, although higher temperatures gave rise to a much faster initial water uptake and more appreciable surface and internal damage. Hence, in the present study, an ageing temperature of 90 °C was chosen to accelerate the diffusion processes that occur at lower ageing temperatures, as well as to accentuate degradation phenomena that otherwise would take weeks or months to happen [65,66].

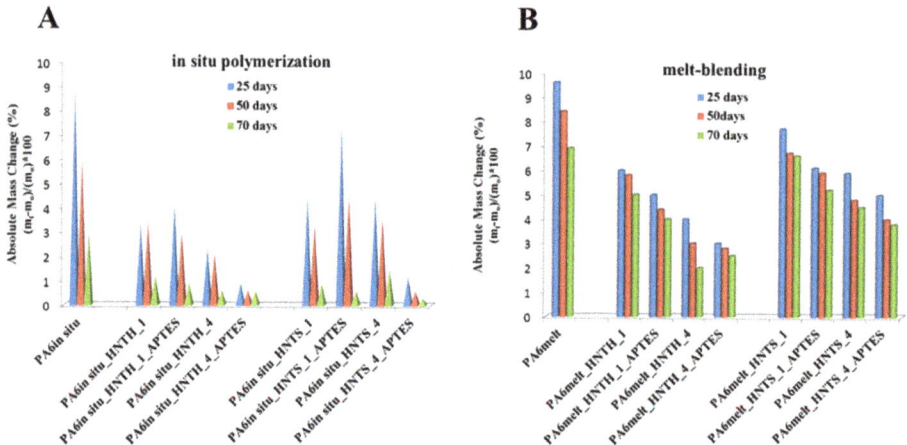

Figure 5. Water absorption in (**A**) PA6$_{in\ situ}$_HNT and (**B**) PA6$_{melt}$_HNT nanocomposites.

The weight of aged samples after 25, 50 and 70 days of accelerated ageing test is reported in the Supporting Information, Table S3. The plotted data show the same trend involving two main steps, corresponding to the ageing periods of 25 and 50–70 days, respectively. In the first 25 days, an initial mass increase is observed as a consequence of a fast initial water uptake, associated with water penetration within PA6 amorphous regions; the mass increase continues until the equilibrium saturation level of the moisture uptake is reached [7]. Then, the equilibrium previously described disappears and a tendency of mass loss is noticed (25–70 days). The latter behavior can be ascribed to the polymer degradation consequent to hydrolysis phenomena [67].

The moisture uptake differs depending on the tested material. Overall, under the same ageing conditions, neat PA6 absorbs more water than modified PA6. Moreover, the decrease of moisture absorption gets higher with the increase of the modified-filler loading. This latter behavior is consistent with those reported in previous studies related to the combined use of functionalizing agents and polymers [68,69]. The improved decrease of water uptake in APTES-modified samples can be explained by the fact that amino functional groups present in the polymer matrix decrease PA6-HNT nanocomposites surface free energy [49]. Therefore, the use of modified-HNT fillers hinders and decelerates the mechanism of moisture sorption.

Furthermore, HNT$_S$ seems to be more prone to water uptake than HNT$_H$; as reported in Section 3.1, this behavior can be related to the HNT$_S$ higher surface area that results in a greater capability of moisture sorption than HNT$_H$ filler.

The effect of hydrothermal ageing on the rheological properties of PA6_HNT nanocomposites was studied. According to Figure S3A,B, PA6$_{in\ situ}$ nanocomposites with the 4% w/w of modified filler have higher complex viscosity values than all other samples thanks to the combined effect of the higher HNT loading and APTES presence that favors the dispersion of the inorganic filler inside the polymer matrix [70]. In the case of PA6$_{melt}$ nanocomposites (Figure S4A,B), their complex viscosity is not significantly affected by the addition of bare and modified HNT.

After ageing, the complex viscosity decreases in all the samples prepared (Figures S3C,D and S4C,D), with the exception of the PA6$_{in\ situ}$ samples prepared with the 4% of modified HNT (Figure 6). The complex viscosity decrease for PA6 and other polyamides after long ageing has been widely reported in the literature [71] and can be ascribed to the plasticization effect of water: during hydrothermal ageing, water molecules are mainly absorbed by the amorphous phase where they interfere with the hydrogen bonds and break the secondary bonds between the PA6 groups. This induces an increase in the chain mobility and decreases consequently the complex viscosity. However, Figure 6 clearly shows that the decrease of PA6$_{in\ situ}$ nanocomposites complex viscosities is only slightly present in the samples with modified HNT$_H$/HNT$_S$ at 4% w/w, in good agreement with water uptake results. The observed differenced in terms of complex viscosity between unaged and aged samples are not significant.

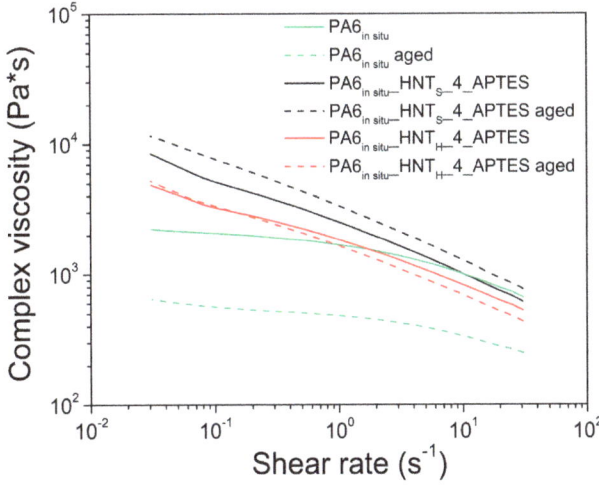

Figure 6. Rheological behavior of neat and APTES-functionalized HNT composites from in situ polymerization before and after hydrothermal ageing test.

Furthermore, the effect of the hydrothermal ageing on PA6-based nanocomposites can be highlighted by the evaluation of the number average molecular weight ($\overline{Mn}$) and molecular weight distribution (D) determined by SEC analyses (Table 4). As described in Section 2.4.3, and in agreement with our previous works [9,18], samples were filtered with a 0.45 μm filter; therefore, molecular weight data refer only to the part of the polymer present in solution. In particular, no significant decrease in molecular weight can be observed in samples from melt blending pointing out to the absence of hydrolysis of the polymer chains due to the filler surface hydroxylation. The lack of any significant effect might be explained, on the one hand, with the drying step performed on the filler before incorporation, which removes physisorbed water that could lead to undesired hydrolysis reactions. On the other hand, the surface hydroxylation of HNT is mainly related to the inner lumen, exposing Al(OH)$_x$ groups, while the outer surface of the nanotubes mainly exposes Si–O–Si groups and silanol groups are present only as defects [72]. The lack of accessible surface hydroxyl groups limits the occurrence of hydrolysis phenomena of the polymer matrix. However, in good agreement with water uptake behavior, a slight decrease in $\overline{Mn}$ data and broader D values are generally observed upon hydrothermal ageing as a result of degradation of PA6 chains [67]. This behavior is shown by both neat PA6 and HNT-PA6 composite, with the notable exception of the nanocomposites prepared with high loadings of modified fillers exhibiting almost unaltered parameters. For the latter samples, the sorption mechanism of water molecules is impeded by the presence of just 4% w/w of modified filler, which limits the diffusion of water molecules into the nanocomposites. As representative examples, Figure 7 compares the SEC curves obtained before and after the hydrothermal test of neat PA6 with those of PA6 nanocomposites prepared with the 4% w/w of modified filler.

Table 4. Number average molecular weight ($\overline{Mn}$) and molecular weight distribution (D) data of PA6_HNT nanocomposites collected before and after the hydrothermal ageing test.

Sample	SEC$_{unaged}$		SEC$_{aged}$	
	$\overline{Mn}$ (Da)	D	$\overline{Mn}$ (Da)	D
PA6$_{in\ situ}$	57,200	2.1	45,800	2.4
PA6$_{in\ situ}$_HNT$_H$_1	48,600	2.1	32,400	2.5
PA6$_{in\ situ}$_HNT$_S$_1	56,700	2.1	49,300	2.5
PA6$_{in\ situ}$_HNT$_H$_4	60,400	2.0	42,000	2.2
PA6$_{in\ situ}$_HNT$_S$_4	62,600	2.1	50,300	2.6
PA6$_{in\ situ}$_HNT$_H$_1_APTES	55,600	2.1	45,400	2.4
PA6$_{in\ situ}$_HNT$_S$_1_APTES	54,600	2.2	49,900	2.6
PA6$_{in\ situ}$_HNT$_H$_4_APTES	47,800	2.3	41,800	2.3
PA6$_{in\ situ}$_HNT$_S$_4_APTES	60,600	2.3	59,000	2.3
PA6$_{melt}$	43,600	1.9	33,100	2.2
PA6$_{melt}$_HNT$_H$_1	40,300	2.0	19,300	2.5
PA6$_{melt}$_HNT$_S$_1	40,400	2.0	19,700	2.5
PA6$_{melt}$_HNT$_H$_4	42,400	2.0	21,000	2.3
PA6$_{melt}$_HNT$_S$_4	42,300	2.0	22,100	2.3
PA6$_{melt}$_HNT$_H$_1_APTES	40,400	2.0	33,100	2.2
PA6$_{melt}$_HNT$_S$_1_APTES	39,700	1.9	22,300	2.2
PA6$_{melt}$_HNT$_H$_4_APTES	42,500	2.0	42,000	2.0
PA6$_{melt}$_HNT$_S$_4_APTES	39,900	2.0	39,800	2.0

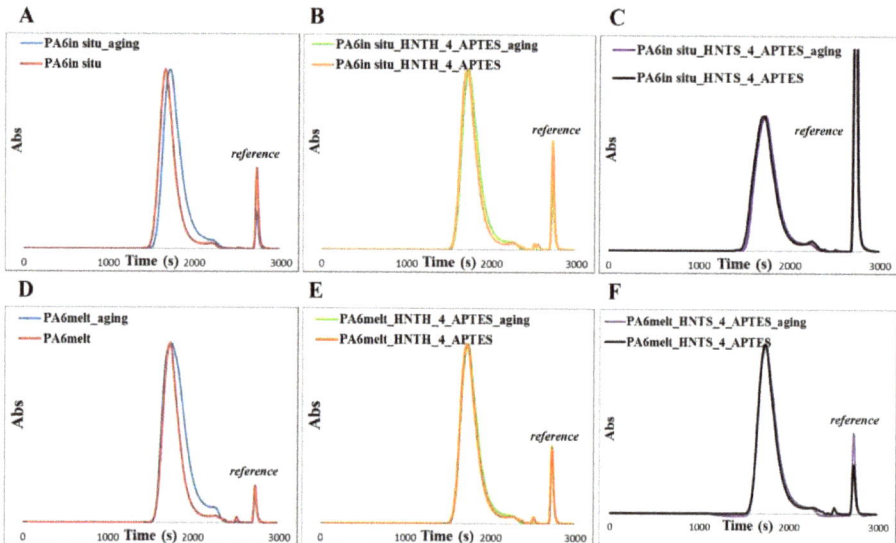

Figure 7. SEC curves before and after the ageing test of (**A**) PA6$_{in\ situ}$, (**B**) PA6$_{in\ situ}$_HNT$_H$_4_APTES, (**C**) PA6$_{in\ situ}$_HNT$_S$_4_APTES, (**D**) PA6$_{melt}$, (**E**) PA6$_{melt}$_HNT$_H$_4_APTES, and (**F**) PA6$_{melt}$_HNT$_S$_4_APTES nanocomposites.

Lastly, to evaluate the consequences of harsh ageing conditions on PA6 nanocomposites crystallization and melting temperatures, DSC measurements were obtained (Table S4). Comparing the thermal data obtained before and after the ageing test, it is worth noting that only the following samples showed minor changes in crystallization kinetics:

- PA6$_{in\ situ}$_HNT$_H$_4_APTES: χ_{c_unaged}: 14.0% vs. χ_{c_aged}: 13.3%
- PA6$_{in\ situ}$_HNT$_S$_4_APTES: χ_{c_unaged}: 14.9% vs. χ_{c_aged}: 14.9%
- PA6$_{melt}$_HNT$_H$_4_APTES: χ_{c_unaged}: 21.7% vs. χ_{c_aged}: 21.1%
- PA6$_{melt}$_HNT$_S$_4_APTES: χ_{c_unaged}: 19.3% vs. χ_{c_aged}: 20.3%.

Together with the reported presence of PA6 γ crystalline even after the ageing test (Figure 8), this is further evidence of the high hydrothermal stability of the nanocomposites prepared with 4% w/w of modified HNT.

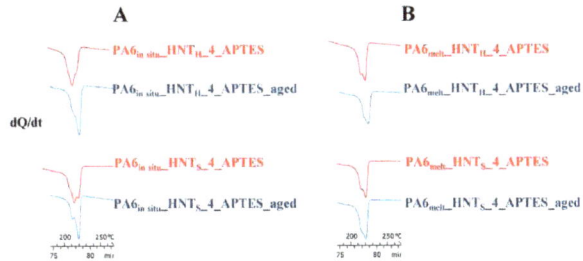

Figure 8. Heating scans of (**A**) PA6$_{in\ situ}$_HNT, and (**B**) PA6$_{melt}$_HNT nanocomposites prepared with the 4% w/w of modified HNT before and after the ageing test.

4. Conclusions

In the field of shelf-life improvement for polymer materials, the development of PA6 nanocomposites characterized by enhanced moisture absorption resistance is always in high demand due to the irreversible decay mechanisms subsequent to the accumulation of water molecules in the polymer matrix. In this context, the amino-silanization of HNT used as PA6 additives could be a way to create tailor-made materials with enhanced durability.

The purpose of this study was to investigate the influence of very low amounts, i.e., 1% and 4% w/w, of HNT on the hydrothermal ageing resistance of PA6 nanocomposites. The use of very low filler loadings provides multiple advantages, including lower alteration of the polymer appearance and a less extended polymer/filler interface at which moisture accumulation may occur.

Crystal growth and type of nucleation in PA6 nanocomposites were studied on the basis of Avrami theory via isothermal crystallization studies, and of dynamic DSC evaluations, showing that the use of low HNT loading does not influence the homogenous PA6 crystal growth and, at the same time, adding HNT promotes the PA6 polymorphism, i.e., the formation of γ counterpart.

Here we employed a multifaceted approach, by studying the physicochemical properties of differently sourced HNT, the filler functionalization, and, last but not least, the dispersion technique, i.e., the in situ polymerization and melt extrusion processes.

The HNT dispersion efficiency into the PA6 matrix was monitored by means of SEM/EDX analyses, showing clear differences between the two preparation methods: while HNT is homogeneously distributed via melt blending, in situ polymerization guarantees a better incorporation within the polymer matrix.

Both in situ and melt blending dispersion techniques showed that a lower surface area, a higher aspect ratio and a larger surface charge of HNT enhance filler incorporation and improve the final composite performances. It should be noted that the tested HNT samples also showed different structural properties in terms of crystallinity and phase impurity, which might affect the filler incorporation. These observations highlight the importance of a preliminary assessment of the physicochemical features of naturally sourced fillers, as relatively minor changes can notably affect the final composite properties.

The filler surface modification with aminosilanes plays a major role in the durability of the PA6_HNT nanocomposites, as determined by very long hydrothermal ageing exposure (1680 h). Notably, the nanocomposites with 4% w/w of modified HNT_H and HNT_S appear to be the most durable ones in terms of water uptake and rheological, molecular weights, molecular weight distribution and thermal properties change before and after the ageing test. This might be related to the ability of amino groups present on HNT surface to avoid the diffusion of water molecules into the nanocomposites, enhancing the overall durability of the resulting nanocomposites.

The present research has led to a deep comprehension of the macromolecular, thermal and rheological properties of HNT-based nanocomposites that can be successfully adopted as composite materials with improved resistance to hydrothermal ageing. Future work will study the natural ageing of the most promising samples, i.e., $PA6_{in\ situ}$_HNT_H_4_APTES and $PA6_{in\ situ}$_HNT_S_4_APTES samples, in a real environment.

Supplementary Materials: The following are available online at http://www.mdpi.com/2073-4360/12/1/211/s1, Figure S1: Visual comparison of HNT_S (A) and HNT_H (B) and their respective composites prepared via injection moulding: $PA6_{melt}$_HNT_S_4 (C) and $PA6_{melt}$_HNT_H_4 (D); Figure S2: DRS spectra of neat, $PA6$_HNT_H_4_APTES and $PA6$_HNT_S_4_APTES nanocomposites prepared via melt blending; Figure S3: Rheological curves of (A) $PA6_{in\ situ}$_HNTH, (B) $PA6_{in\ situ}$_HNT_S, (C) $PA6_{in\ situ}$_HNT_H_aged and (D) $PA6_{in\ situ}$_HNT_S_aged samples; Figure S4: Rheological curves of (A) $PA6_{melt}$_HNT_H, (B) $PA6_{melt}$_HNT_S, (C) $PA6_{melt}$_HNT_H_aged and (D) $PA6_{melt}$_HNT_S_aged samples; Table S1: Injection moulding parameters for PA6 nanocomposites specimen preparation; Table S2: Crystallization enthalpy (ΔH_c) and heat of fusion (ΔH_f) of unaged and aged PA6_HNT nanocomposites collected during cooling and second heating thermal steps; Table S3: Weight of the PA6_HNT nanocomposites collected during the hydrothermal ageing test at different times; Table S4: Differential scanning calorimetry data of PA6_HNT nanocomposites collected during cooling and second heating thermal steps after the hydrothermal ageing test.

Author Contributions: Investigation, M.B., R.R., V.S., T.T. and D.M. (Daniela Maggioni); Writing—Original Draft Preparation, V.S.; Writing—Review & Editing, S.A., D.M. (Daniela Meroni), M.A.O., V.S. and T.T.; Supervision, S.A., D.M. (Daniela Meroni) and M.A.O. All authors have read and agreed to the published version of the manuscript.

Funding: This research received no external funding.

Acknowledgments: Nadia Santo and UNITECH COSPECT (Università degli Studi di Milano, Piattaforme Tecnologiche di Ateneo, Via Golgi 19, 20133 Milano) are kindly acknowledged for SEM-EDX analyses. The authors wish to thank iMinerals Inc. for providing complimentary samples.

Conflicts of Interest: The authors declare no conflict of interest.

References

1. Jacob, A. Carbon fibre and cars—2013 in review. *Reinf. Plast.* **2014**, *58*, 18–19. [CrossRef]
2. Stewart, R. Rebounding automotive industry welcome news for FRP. *Reinf. Plast.* **2011**, *55*, 38–44. [CrossRef]
3. Horrocks, A.R.; Kandola, B.K.; Davies, P.J.; Zhang, S.; Padbury, S.A. Developments in flame retardant textiles—A review. *Polym. Degrad. Stab.* **2005**, *88*, 3–12. [CrossRef]
4. Dong, W.; Gijsman, P. Influence of temperature on the thermo-oxidative degradation of polyamide 6 films. *Polym. Degrad. Stab.* **2010**, *95*, 1054–1062. [CrossRef]
5. Gijsman, P.; Meijers, G.; Vitarelli, G. Comparison of the UV-degradation chemistry of polypropylene, polyethylene, polyamide 6 and polybutylene terephthalate. *Polym. Degrad. Stab.* **1999**, *65*, 433–441. [CrossRef]
6. Forsström, D.; Terselius, B. Thermo oxidative stability of polyamide 6 films I. Mechanical and chemical characterisation. *Polym. Degrad. Stab.* **2000**, *67*, 69–78. [CrossRef]
7. Ksouri, I.; De Almeida, O.; Haddar, N. Long term ageing of polyamide 6 and polyamide 6 reinforced with 30% of glass fibers: Physicochemical, mechanical and morphological characterization. *J. Polym. Res.* **2017**, *24*, 133. [CrossRef]
8. Lin, S.; Wang, C.; Wang, Y.; Hou, W. Effects of hydrothermal aging on moisture absorption and property prediction of short carbon fiber reinforced polyamide 6 composites. *Compos. Part B Eng.* **2018**, *153*, 306–314. [CrossRef]
9. Sabatini, V.; Farina, H.; Basilissi, L.; Di Silvestro, G.; Ortenzi, M.A. The Use of Epoxy Silanes on Montmorillonite: An Effective Way to Improve Thermal and Rheological Properties of PLA/MMT Nanocomposites Obtained via "in Situ" Polymerization. *J. Nanomater.* **2015**, *16*, 213. [CrossRef]

10. Akkapeddi, M.K. Glass fiber reinforced polyamide-6 nanocomposites. *Polym. Compos.* **2000**, *21*, 576–585. [CrossRef]
11. Wu, G.; Li, B.; Jiang, J. Carbon black self-networking induced co-continuity of immiscible polymer blends. *Polymer* **2010**, *51*, 2077–2083. [CrossRef]
12. Uhl, F.M.; Yao, Q.; Nakajima, H.; Manias, E.; Wilkie, C.A. Expandable graphite/polyamide-6 nanocomposites. *Polym. Degrad. Stab.* **2005**, *89*, 70–84. [CrossRef]
13. Yang, F.; Ou, Y.; Yu, Z. Polyamide 6/silica nanocomposites prepared by in situ polymerization. *J. Appl. Polym. Sci.* **1998**, *69*, 355–361. [CrossRef]
14. Xu, Z.; Gao, C. In situ polymerization approach to graphene-reinforced nylon-6 composites. *Macromolecules* **2010**, *43*, 6716–6723. [CrossRef]
15. Pramoda, K.P.; Liu, T. Effect of moisture on the dynamic mechanical relaxation of polyamide-6/clay nanocomposites. *J. Polym. Sci. Part B Polym. Phys.* **2004**, *42*, 1823–1830. [CrossRef]
16. Hassan, A.; Rahman, N.A.; Yahya, R. Moisture absorption effect on thermal, dynamic mechanical and mechanical properties of injection-molded short glass-fiber/polyamide 6,6 composites. *Fibers Polym.* **2012**, *13*, 899–906. [CrossRef]
17. Nazir, M.S.; Mohamad Kassim, M.H.; Mohapatra, L.; Gilani, M.A.; Raza, M.R.; Majeed, K. Characteristic Properties of Nanoclays and Characterization of Nanoparticulates and Nanocomposites. In *Nanoclay Reinforced Polymer Composites*; Springer: Singapore, 2016; pp. 35–55. ISBN 978-981-10-1952-4.
18. Basilissi, L.; Silvestro, G.D.; Farina, H.; Ortenzi, M.A. Synthesis and characterization of PLA nanocomposites containing nanosilica modified with different organosilanes I. Effect of the organosilanes on the properties of nanocomposites: Macromolecular, morphological, and rheologic characterization. *J. Appl. Polym. Sci.* **2013**, *128*, 1575–1582. [CrossRef]
19. Kodgire, P.V.; Bhattacharyya, A.R.; Bose, S.; Gupta, N.; Kulkarni, A.R.; Misra, A. Control of multiwall carbon nanotubes dispersion in polyamide6 matrix: An assessment through electrical conductivity. *Chem. Phys. Lett.* **2006**, *432*, 480–485. [CrossRef]
20. Zhang, W.D.; Shen, L.; Phang, I.Y.; Liu, T. Carbon nanotubes reinforced nylon-6 composite prepared by simple melt-compounding. *Macromolecules* **2004**, *37*, 256–259. [CrossRef]
21. Ayatollahi, M.R.; Shadlou, S.; Shokrieh, M.M.; Chitsazzadeh, M. Effect of multi-walled carbon nanotube aspect ratio on mechanical and electrical properties of epoxy-based nanocomposites. *Polym. Test.* **2011**, *30*, 548–556. [CrossRef]
22. Pan, B.; Li, N.; Chu, G.; Wei, F.; Liu, J.; Zhang, J.; Zhang, Y. Tribological investigation of MC PA6 reinforced by boron nitride of single layer. *Tribol. Lett.* **2014**, *54*, 161–170. [CrossRef]
23. Song, W.L.; Wang, P.; Cao, L.; Anderson, A.; Meziani, M.J.; Farr, A.J.; Sun, Y.P. Polymer/boron nitride nanocomposite materials for superior thermal transport performance. *Angew. Chem.-Int. Ed.* **2012**, *51*, 6498–6501. [CrossRef] [PubMed]
24. De Monte, M.; Moosbrugger, E.; Quaresimin, M. Influence of temperature and thickness on the off-axis behaviour of short glass fibre reinforced polyamide 6.6—Cyclic loading. *Compos. Part A Appl. Sci. Manuf.* **2010**, *41*, 1368–1379. [CrossRef]
25. Ozkoc, G.; Bayram, G.; Bayramli, E. Effects of polyamide 6 incorporation to the short glass fiber reinforced ABS composites: An interfacial approach. *Polymer* **2004**, *45*, 8957–8966. [CrossRef]
26. Reynaud, E.; Jouen, T.; Gauthier, C.; Vigier, G.; Varlet, J. Nanofillers in polymeric matrix: A study on silica reinforced PA6. *Polymer* **2001**, *42*, 8759–8768. [CrossRef]
27. Gensler, R.; Gröppel, P.; Muhrer, V.; Müller, N. Application of nanoparticles in polymers for electronics and electrical engineering. *Part. Part. Syst. Charact.* **2002**, *19*, 293–299. [CrossRef]
28. Liu, M.; Jia, Z.; Jia, D.; Zhou, C. Recent Advance in Research on Halloysite Nanotubes-Polymer Nanocomposite. *Prog. Polym. Sci.* **2014**, *39*, 1498–1525. [CrossRef]
29. Lvov, Y.; Wang, W.; Zhang, L.; Fakhrullin, R. Halloysite Clay Nanotubes for Loading and Sustained Release of Functional Compounds. *Adv. Mater.* **2016**, *28*, 1227–1250. [CrossRef]
30. Tarì, G.; Bobos, I.; Gomes, C.S.F.; Ferreira, J.M.F. Modification of surface charge properties during kaolinite to halloysite-7Å transformation. *J. Colloid Interface Sci.* **1999**, *210*, 360–366. [CrossRef]
31. Rawtani, D.; Agrawal, Y.K. Multifarious applications of halloysite nanotubes: A review. *Rev. Adv. Mater. Sci.* **2012**, *30*, 282–295.

32. Shchukin, D.G.; Sukhorukov, G.B.; Price, R.R.; Lvov, Y.M. Halloysite nanotubes as biomimetic nanoreactors. *Small* **2005**, *1*, 510–513. [CrossRef] [PubMed]
33. Guo, B.; Zou, Q.; Lei, Y.; Jia, D. Structure and performance of polyamide 6/halloysite nanotubes nanocomposites. *Polym. J.* **2009**, *41*, 835–842. [CrossRef]
34. Marney, D.C.O.; Russell, L.J.; Wu, D.Y.; Nguyen, T.; Cramm, D.; Rigopoulos, N.; Wright, N.; Greaves, M. The suitability of halloysite nanotubes as a fire retardant for nylon 6. *Polym. Degrad. Stab.* **2008**, *93*, 1971–1978. [CrossRef]
35. Handge, U.A.; Hedicke-Höchstötter, K.; Altstädt, V. Composites of polyamide 6 and silicate nanotubes of the mineral halloysite: Influence of molecular weight on thermal, mechanical and rheological properties. *Polymer* **2010**, *51*, 2690–2699. [CrossRef]
36. Prashantha, K.; Schmitt, H.; Lacrampe, M.F.; Krawczak, P. Mechanical behaviour and essential work of fracture of halloysite nanotubes filled polyamide 6 nanocomposites. *Compos. Sci. Technol.* **2011**, *71*, 1859–1866. [CrossRef]
37. Lecouvet, B.; Gutierrez, J.G.; Sclavons, M.; Bailly, C. Structure-property relationships in polyamide 12/halloysite nanotube nanocomposites. *Polym. Degrad. Stab.* **2011**, *96*, 226–235. [CrossRef]
38. Prashantha, K.; Lacrampe, M.F.; Krawczak, P. Highly dispersed polyamide-11/halloysite nanocomposites: Thermal, rheological, optical, dielectric, and mechanical properties. *J. Appl. Polym. Sci.* **2013**, *130*, 313–321. [CrossRef]
39. Tung, J.; Gupta, R.K.; Simon, G.P.; Edward, G.H.; Bhattacharya, S.N. Rheological and mechanical comparative study of in situ polymerized and melt-blended nylon 6 nanocomposites. *Polymer* **2005**, *46*, 10405–10418. [CrossRef]
40. Bergeret, A.; Ferry, L.; Ienny, P. Influence of the fibre/matrix interface on ageing mechanisms of glass fibre reinforced thermoplastic composites (PA-6,6, PET, PBT) in a hygrothermal environment. *Polym. Degrad. Stab.* **2009**, *94*, 1315–1324. [CrossRef]
41. Bergeret, A.; Pires, I.; Foulc, M.P.; Abadie, B.; Ferry, L.; Crespy, A. The hygrothermal behaviour of glass-fibre-reinforced thermoplastic composites: A prediction of the composite lifetime. *Polym. Test.* **2001**, *20*, 753–763. [CrossRef]
42. Seltzer, R.; Frontini, P.M.; Mai, Y.W. Effect of hygrothermal ageing on morphology and indentation modulus of injection moulded nylon 6/organoclay nanocomposites. *Compos. Sci. Technol.* **2009**, *69*, 1093–1100. [CrossRef]
43. Jacobi, E.; Schuttenberg, H.; Schulz, R.C. A new method for gel permeation chromatography of polyamides. *Macromol. Rapid Commun.* **1980**, *1*, 397–402. [CrossRef]
44. Piao, H. Influence of water absorption on the mechanical properties of discontinuous carbon fiber reinforced polyamide 6. *J. Polym. Res.* **2019**, *44*, 353–358. [CrossRef]
45. Falcón, J.M.; Sawczen, T.; Aoki, I.V. Dodecylamine-Loaded Halloysite Nanocontainers for Active Anticorrosion Coatings. *Front. Mater.* **2015**, *2*, 69.
46. Joussein, E.; Petit, S.; Churchman, J.; Theng, B.; Righi, D.; Delvaux, B. Halloysite clay minerals—A review. *Clay Miner.* **2005**, *40*. [CrossRef]
47. Taroni, T.; Meroni, D.; Fidecka, K.; Maggioni, D.; Longhi, M.; Ardizzone, S. Halloysite nanotubes functionalization with phosphonic acids: Role of surface charge on molecule localization and reversibility. *Appl. Surf. Sci.* **2019**, *486*, 466–473. [CrossRef]
48. Meroni, D.; Lo Presti, L.; Di Liberto, G.; Ceotto, M.; Acres, R.G.; Prince, K.C.; Bellani, R.; Soliveri, G.; Ardizzone, S. A close look at the structure of the TiO_2-APTES interface in hybrid nanomaterials and its degradation pathway: An experimental and theoretical study. *J. Phys. Chem. C* **2017**, *121*, 430–440. [CrossRef]
49. Prashantha, K.; Lacrampe, M.F.; Krawczak, P. Processing and characterization of halloysite nanotubes filled polypropylene nanocomposites based on a masterbatch route: Effect of halloysites treatment on structural and mechanical properties. *Express Polym. Lett.* **2011**, *5*, 295–307. [CrossRef]
50. Fornes, T.D.; Paul, D.R. Crystallization behavior of nylon 6 nanocomposites. *Polymer* **2003**, *44*, 3945–3961. [CrossRef]
51. Turska, E.; Gogolewski, S. Study on crystallization of nylon-6 (polycapramide): Part 2. Effect of molecular weight on isothermal crystallization kinetics. *Polymer* **1971**, *12*, 629–641. [CrossRef]
52. Gurato, G.; Gaidano, D.; Zannetti, R. Influence of nucleating agents on the crystallization of 6-polyamide. *Die Makromol. Chem.* **1978**, *179*, 231–245. [CrossRef]

53. Khanna, Y.P.; Reimschuessel, A.C.; Banerjie, A.; Altman, C.; Reimschuessel, A.; Banerjie, C. Altman Memory effects in polymers. II. Processing history vs. crystallization rate of nylon 6—Observation of phenomenon and product behavior. *Polym. Eng. Sci.* **1988**, *28*, 1600–1606. [CrossRef]
54. Reid, B.O.; Vadlamudi, M.; Mamun, A.; Janani, H.; Gao, H.; Hu, W.; Alamo, R.G. Strong memory effect of crystallization above the equilibrium melting point of random copolymers. *Macromolecules* **2013**, *46*, 6485–6497. [CrossRef]
55. Wurm, A.; Ismail, M.; Kretzschmar, B.; Pospiech, D.; Schick, C. Retarded crystallization in polyamide/layered silicates nanocomposites caused by an immobilized interphase. *Macromolecules* **2010**, *43*, 1480–1487. [CrossRef]
56. Guo, B.; Zou, Q.; Lei, Y.; Du, M.; Liu, M.; Jia, D. Crystallization behavior of polyamide 6/halloysite nanotubes nanocomposites. *Thermochim. Acta* **2009**, *484*, 48–56. [CrossRef]
57. Caamaño, C.; Grady, B.; Resasco, D.E. Influence of nanotube characteristics on electrical and thermal properties of MWCNT/polyamide 6,6 composites prepared by melt mixing. *Carbon N. Y.* **2012**, *50*, 3694–3707. [CrossRef]
58. Kiziltas, A.; Gardner, D.J.; Han, Y.; Yang, H.S. Dynamic mechanical behavior and thermal properties of microcrystalline cellulose (MCC)-filled nylon 6 composites. *Thermochim. Acta* **2011**, *519*, 38–43. [CrossRef]
59. Bradbury, E.M.; Brown, L.; Elliott, A.; Parry, D.A.D. The structure of the gamma form of polycaproamide (Nylon 6). *Polymer* **1965**, *6*, 465–482. [CrossRef]
60. Lincoln, D.M.; Vaia, R.A.; Wang, Z.G.; Hsiao, B.S.; Krishnamoorti, R. Temperature dependence of polymer crystalline morphology in nylon 6/montmorillonite nanocomposites. *Polymer* **2001**, *42*, 09975–09985. [CrossRef]
61. Porubská, M.; Szöllös, O.; Kóňová, A.; Janigová, I.; Jašková, M.; Jomová, K.; Chodák, I. FTIR spectroscopy study of polyamide-6 irradiated by electron and proton beams. *Polym. Degrad. Stab.* **2012**, *97*, 523–531. [CrossRef]
62. Zammarano, M.; Bellayer, S.; Gilman, J.W.; Franceschi, M.; Beyer, F.L.; Harris, R.H.; Meriani, S. Delamination of organo-modified layered double hydroxides in polyamide 6 by melt processing. *Polymer* **2006**, *47*, 652–662. [CrossRef]
63. Lei, Y.; Zhang, J.; Zhang, T.; Li, H. Water diffusion in carbon fiber reinforced polyamide 6 composites: Experimental, theoretical, and numerical approaches. *J. Reinf. Plast. Compos.* **2019**, *38*, 578–587. [CrossRef]
64. Taktak, R.; Guermazi, N.; Derbeli, J.; Haddar, N. Effect of hygrothermal aging on the mechanical properties and ductile fracture of polyamide 6: Experimental and numerical approaches. *Eng. Fract. Mech.* **2015**, *148*, 122–133. [CrossRef]
65. Gao, L.; Ye, L.; Li, G. Long-Term Hydrothermal Aging Behavior and Life-Time Prediction of Polyamide 6. *J. Macromol. Sci. Part B* **2015**, *54*, 239–252. [CrossRef]
66. Li, R.; Ye, L.; Li, G. Long-Term Hydrothermal Aging Behavior and Aging Mechanism of Glass Fibre Reinforced Polyamide 6 Composites. *J. Macromol. Sci. Part B* **2017**, *57*, 67–82. [CrossRef]
67. Chaupart, N.; Serpe, G.; Verdu, J. Molecular weight distribution and mass changes during polyamide hydrolysis. *Polymer* **1998**, *39*, 1375–1380. [CrossRef]
68. Kim, N.; Shin, D.H.; Lee, Y.T. Effect of silane coupling agents on the performance of RO membranes. *J. Memb. Sci.* **2007**, *300*, 224–231. [CrossRef]
69. Shallenberger, J.R.; Metwalli, E.E.; Pantano, C.G.; Tuller, F.N.; Fry, D.F. Adsorption of polyamides and polyamide-silane mixtures at glass surfaces. *Surf. Interface Anal.* **2003**, *35*, 667–672. [CrossRef]
70. Pardo, S.G.; Bernal, C.; Ares, A.; Abad, M.J.; Cano, J. Rheological, Thermal, and Mechanical Characterization of Fly Ash-Thermoplastic Composites With Different Coupling Agents. *Polym. Compos.* **2012**, *31*, 1722–1730. [CrossRef]
71. Berland, S.; Launay, B. Rheological properties of wheat flour doughs in steady and dynamic shear: Effect of water content and some additives. *Cereal Chem.* **1995**, *72*, 48–52.
72. Yuan, P.; Tan, D.; Annabi-Bergaya, F. Properties and applications of halloysite nanotubes: Recent research advances and future prospects. *Appl. Clay Sci.* **2015**, *112–113*, 75–93. [CrossRef]

 © 2020 by the authors. Licensee MDPI, Basel, Switzerland. This article is an open access article distributed under the terms and conditions of the Creative Commons Attribution (CC BY) license (http://creativecommons.org/licenses/by/4.0/).

Article

Abrasion Wear Resistance of Polymer Constructional Materials for Rapid Prototyping and Tool-Making Industry

Janusz Musiał [1], Serhiy Horiashchenko [2], Robert Polasik [1,*], Jakub Musiał [1], Tomasz Kałaczyński [1], Maciej Matuszewski [1] and Mścisław Srutek [3]

[1] Faculty of Mechanical Engineering, University of Science and Technology, Kaliskiego 7 Street, 85-789 Bydgoszcz, Poland; janusz.musial@utp.edu.pl (J.M.); jakmus@onet.pl (J.M.); tomasz.kalaczynski@utp.edu.pl (T.K.); maciej.matuszewski@utp.edu.pl (M.M.)
[2] Khmelnitskiy National University, st Institutskaya st. 11, 29016 Khmelnitskiy, Ukraine; gsl7@ukr.net
[3] Faculty of Telecommunications, Computer Science and Electrical Engineering, University of Science and Technology, Kaliskiego 7 Street, 85-789 Bydgoszcz, Poland; mscislaw.srutek@utp.edu.pl
* Correspondence: robert.polasik@utp.edu.pl; Tel.: +48-52-340-82-96

Received: 31 March 2020; Accepted: 8 April 2020; Published: 10 April 2020

Abstract: The original test results of abrasive wear resistance of different type of construction polymer materials were presented and discussed in this article. Tests were made on an adapted test stand (surface grinder for form and finish grinding). Test samples were made of different types of polymer board materials including RenShape®, Cibatool®and phenolic cotton laminated plastic laminate (TCF). An original methodology based on a grinding experimental set-up of abrasion wear resistance of polymer construction materials was presented. Equations describing relations between material type and wear resistance were presented and discussed. Micro and macro structures were investigated and used in wear resistance prediction.

Keywords: construction composite; friction resistance; surface state

1. Introduction

Modern constructional polymers, especially composite materials, are often used in structural applications where the ability to create properties such as stiffness and strength make them attractive compared to traditional engineering materials [1–5]. In addition to structural applications, such materials are also used in applications where both thermal and structural properties are important. Silicon wafers, used in the electronics industry, are such an example. Consequently, coupled thermal–structural analysis of thin structures is becoming increasingly important from a simulation standpoint [6].

Currently, the results of research in the field of materials wear under different conditions of mechanical and thermal loads are dynamically changing [7–9]. This is a serious obstacle to the selection or development of constructive technological conditions to increase the durability of products.

The results of many friction and wear mechanisms and phenomena studies of various polymeric materials have been published in recent years. Some of them concerned the determination of various techniques and applications [10,11] for the assessment of wear resistance or the influence of conditions on friction and wear [12–15]. A significant part of the investigations concerns research on the impact of the material structure, additives or fillers used and coatings on the tribological behavior or wear of cooperating elements [16–22]. These works relate to the classic approach in determining the wear mechanism, determining wear resistance or friction. In these works, wear is usually determined on the basis of changes in the micro-geometric structure of the surface or based on the object weight loss. This study presents an original method of determining wear resistance in conditions of intensive friction using the geometric structure of the surface, constituted during the test, to determine the tribological properties of materials.

The characteristics of the external heat, lubrication and wear are directly related to the surface geometric structure of cooperating parts. The surface of the part is the outer layer, which differs from the inner part by its structure and other physical properties. The general concept of "surface quality" can be described as set of properties acquired by the surface layer during forming processes. The quality of the surface of the machine parts influences such service properties as the contact fatigue, wear, erosion and crossover resistance. The complex properties of surfaces are formed after processing and determine the concept of "the quality of the surface". Surface roughness, material structure, physical–chemical–mechanical upper-layer properties and general stress can determine the state (quality) of the surfaces of the machine parts and can be considered in the initial and working (operational) state (Figure 1) [6,23].

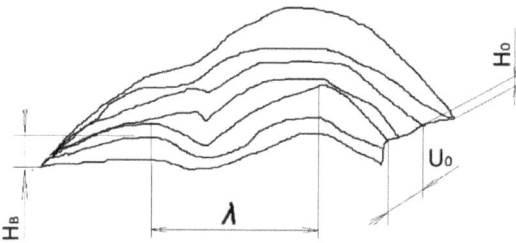

Figure 1. Scheme of surface waviness.

The final state of a workpiece surface can be determined by technological processes. Its state (value) is generally satisfactory at the beginning of its cooperation with other surfaces. The nature of the contact of two solids, except for the geometry of the surfaces, it determined by the mechanical, physical and chemical properties of the thin surface layers and their stress states. Such surface layers generally have a different structure and different properties than the materials inside the material. The thickness of upper surface layer ranges from tenths to hundreds of angstroms and, rarely, tenths of a millimeter. The difference between the properties of thin surface layers and the properties of the core is due to the following three main factors [6,24]:

- The state of the metal atoms in the surface layers, which is different from the state of the atoms in the bulk of the material. The presence of free surface energy and high adsorption activity is a consequence;
- The sum of mechanical, thermal and physicochemical effects on the metal surface during the final and preliminary processing operations;
- The sum of repeated cyclic, mechanical, thermal and physicochemical effects on the metal surface in case of friction load in operation.

The outgoing technological terrain quickly disappears in the process of operation. The chemical composition and geometry of the surfaces is completely changed. New surface qualities are formed.

The estimation of geometrical parameters of the surface of machine parts includes the estimation of macro-, micro- and submicron deviations, taking into account the nature and mechanisms of formation of geometric deviations. Deviations distributed into components are caused by machining, internal structure and loading during operation.

Microgeometric deviations can be technological and operational. Technological macro deviations are caused by insufficient precision of the machine, tool, processing modes, temperature stresses and deformations [25–27]. Operational macro-deviations usually are caused by uneven wear resulting from misalignment of the moving parts, vibrations and overloads during operation.

The durability of materials can be determined by the following:

- The combination of material properties and the type (state) of contact surfaces (surface cleanliness, lubrication);
- The nature of the movement (sliding, rolling, bumps, fluidity);
- The speed of mutual movement;
- The load level;
- The removal of particles that separate or the presence of particles of some other material that complicates friction, etc.

Universal indicator *Ra* used in most cases for measurements, which gives the most complete characteristic with all points of the profile. The value of the average height *Rz* used in case of difficulties is associated with the use of instruments for determining *Ra*. Such characteristics affect the resistance and vibration resistance, as well as the electrical conductivity of materials.

A direct correlation determines the characteristics of the processed surface; the higher the class index, the less important the height of the measured surface is and the better the quality of processing.

The arithmetic mean deviation Ra of the profile, called the arithmetic mean of the absolute values of the profile deviations within the base length *l*—Figure 2 [1,24,28], can be calculated from the following equation:

$$Ra = \frac{1}{n} \cdot \sum_{i=1}^{n} |y_i| \qquad (1)$$

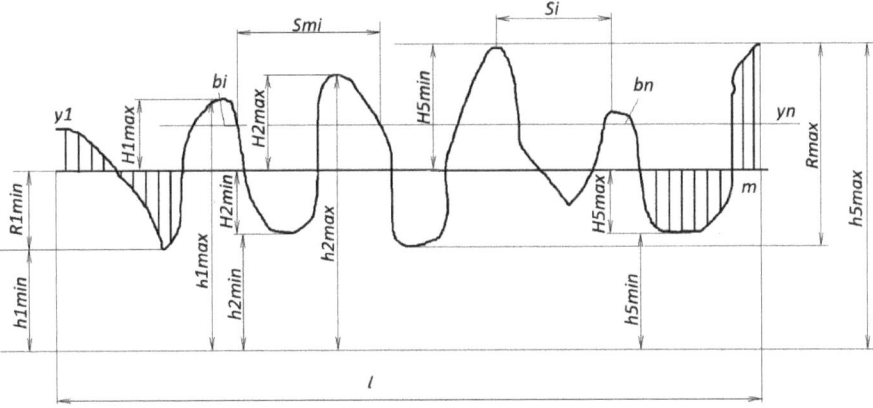

Figure 2. The initial roughness profile.

The middle line of the profile is defined so that areas of the projections and troughs of the contour of the profile *F* on both sides of it are equal.

The height of the irregularities of the profile at 10 points *Rz* is the sum of the mean absolute heights of the five largest projections of the profile and the depths of the five largest depressions of the profile within the base length.

$$Rz = \frac{1}{5} \cdot \left(\sum_{i=1}^{5} |y_{pi}| + \sum_{i=1}^{5} |y_{vi}| \right) \qquad (2)$$

where y_{pi} is the height of the highest *i*-th protrusion of the profile; and y_{vi} is the depth of the *i*-th largest recess of the profile.

Relative reference length of the profile t_p is the reference length of the profile ratio that is equal to the sum of the lengths of the sections. Segments cut off at a given level in the material of the profile by a line equivalent to the average line within the base length to the base length.

$$t_p = \sum_{i=1}^{n} \frac{b_i}{l} \tag{3}$$

The parameter t_p characterizes the shape of the profile irregularities and gives the notion of the distribution of the height of the irregularities across the levels of the profile cross section.

The wear and tear processes are uneven and multi-stage. Three periods of wear and tear are observed when parts and assemblies of machines are operating, namely working (1), permanent (normal) wear (2) and catastrophic (emergency) wear (3) (Figure 3). The working out period is characterized by increased wear rate, which is gradually decreasing. The condition of cooperating friction pairs gradually changes, because the initial stages of the process of wear and tear begin to eliminate irregularities on the part's surface. New relief forms, which are characteristic to specific loading conditions and structural changes of materials, occur. The area of actual contact thus changes, and the coefficient of friction and temperature in the contact zone decrease markedly (Figure 3). As soon as the structure and the relief on the surface of the materials become optimal for these friction conditions, their wear rate decreases to the minimum values.

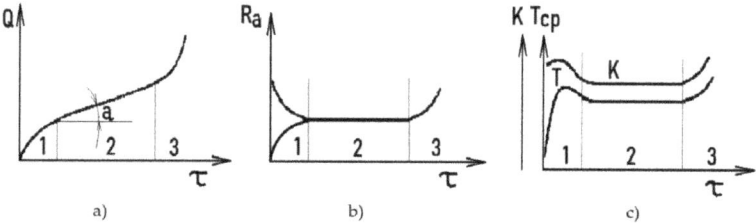

Figure 3. Dependence of the wear rate Q (**a**), the height of the micro-irregularities on the friction surface R_a (**b**), the coefficient of friction K and the average bulk temperature T of the durability elements of the work (**c**); 1,2,3 refer to stages of the wear process.

Period 2 of permanent wear is characterized by the relative constancy of the friction conditions and the wear rate. The coefficient of friction does not change. During this period, a dynamic equilibrium is established in the surface layers of the contacting bodies between the processes of strengthening and weakening, the formation of new structures and their destruction. In the surface layers of materials, the optimum structure formed during the working period and the corresponding relief is preserved during the period of permanent wear, as well as during the beginning of catastrophic wear 3. The wear resistance of machine parts depends essentially on the nature of the relief and the structure formed on the surface of the materials during the working-in period. Therefore, it is important to be able to control the processes of relief formation and structure on the surface of the machine parts in the initial period of wear, that is, during the period of wear.

2. Modeling Polymer Constructional Material Wear

An original function based on the analysis of physical processes of resistance of forms to wear and the formation of their resistance was made. This function describes the dynamics of wear for the period of its operation. The dependence of wear W on the amount of abrasion N is described by the following function [28–30]:

$$W(N) = \frac{1}{A \cdot b_0} \left[\frac{(A - N_0) \cdot N}{N_0 \cdot (A - N_0)} \right] \tag{4}$$

where A, b_0, N_0 are determined using the least squares method.

The wear rate ϑ_W is a derivative of the function $W(N)$ and has the form

$$\vartheta_W(N) = \frac{N_0 \cdot S}{A \cdot b_0 \cdot (A - N_0)}[(A - N_0) - (A - N_0) \cdot N] \tag{5}$$

where A is the number of cycles, and S is the wear area.

Since there is mass loss, $W_m = \frac{\Delta m}{S}$.

The wear rate, obtained during the experiment, is determined by the following formula:

$$\vartheta_W(N) = \frac{\Delta m_{i+1} - \Delta m_{i-1}}{N_{i+1} - N_{i-1}} \tag{6}$$

Since the wear rate depends on the stage of wear, it can be noticed that it decreases with time, but then it can rise again. The wear rate I_v is defined as the ratio of wear to the friction path L. For composition materials, specific wear by volume can be described by W_v.

$$W_v = \frac{\Delta m k}{\rho \Delta h} \tag{7}$$

where k is a correction coefficient taking into account the wear of the rod, $k = 0.98$; ρ is material density; Δh is the height of the micro-irregularities on the friction surface.

The friction path is determined through the number of friction cycles.

$$L = k \cdot l_p \cdot N \tag{8}$$

where l_p is the path when executing one cycle.

Permissible wear I_v can be determined by setting the number of friction cycles.

$$I_v = \frac{W_v}{k \cdot l_p \cdot N} = \frac{\Delta m \cdot k}{\rho \cdot \Delta h \cdot k \cdot l_p \cdot N} \tag{9}$$

The value of the change in the maximum height of surface is ΔRz and the specific volume can be defined as follows:

$$W_v = \Delta Rz \cdot S \cdot \alpha^2 \tag{10}$$

where α is the fill factor of the physical area.

The use of polymer constructional materials and composites in the manufacturing industry is rapidly increasing. Compared to traditional metallic engineering materials, polymer constructional materials and composites are lighter and more corrosion resistant, and properties like strength, stiffness and toughness can often be tailored to a specific application. The composite material is typically a laminate of individual layers, where the fibers in each layer are unidirectional.

3. Experimental Research Parameters of the Sub-Microrelief

Sub-microgeometry characterizes the type of irregularities, which is a geometric reflection of the structure of surface layers of metal and its imperfections. Submicroscopic relief is considered in the areas of the surface from one to several micrometers. The parameters of the sub-microrelief are quantitatively determined by means of electronic fractography, and in particular the following parameters: the relative surface area covered by films of secondary structures, the area of sections and juvenile areas, the thickness of secondary structures and the height of sub-micro-irregularities. The initial surface is characterized by the presence of a developed sub-relief, which is formed in the process of final technological processing of grinding. The coatings of films of secondary structures have a weakly expressed sub microrelief. When worn in the mode of structural and energy adaptation of the friction surface, the sub-relief of the friction surfaces is covered by secondary structures of its type and is caused mainly by the presence of smoothed areas. The sub-micro-roughness of the patches is negligible [31–34].

Seven samples of different polymer materials were used for the experiment. Their appearance is presented in Figure 4, and the characteristics are summarized in Table 1. TCF, phenolic cotton laminated plastic (laminate, e.g., tekstolit, tufnol, rezoteks, turbax, nowotex), is commonly used for high durability parts and structures, especially wear resistant ones. RenShape®materials are mostly used for prototyping processes and light structure construction. Cibatool®polymer materials are used for mold making, tool making and modelling.

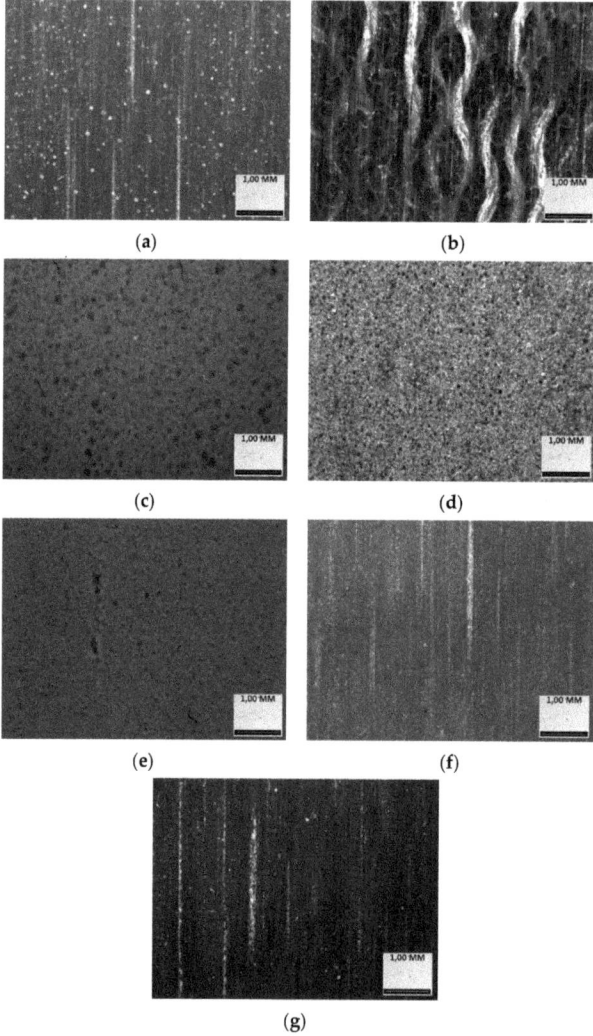

Figure 4. Experimental sample macro structures: (**a**) RenShape®BM5273 (No. 1), (**b**) TCF (Tekstolit, No. 2), (**c**) RenShape®BM5035 (No. 3), (**d**) Cibatool®BM5005 (No. 4), (**e**) RenShape®BM5185 (No. 5), (**f**) Cibatool®BM5272 (No. 6), (**g**) Cibatool®BM5168 (No. 7).

Table 1. Material properties [35,36].

No.	Material	ρ, g/cm^3	σ, MPa	Shore D
1	RenShape®BM5273	1.4	90	120
2	TCF (Tekstolit)	1.5	80	85
3	RenShape®BM5035	0.45	0,02	48
4	Cibatool®BM5005	0.56	25	68
5	RenShape®BM5185	0.5	15	-
6	Cibatool®BM5272	1.4	80	85
7	Cibatool®BM5168	1.4	90	85

The number of friction cycles selected for the experiment, calculated from grinding wheel speed and real contact time, was $N = 10{,}000$. The experiment was carried out for grinding depths of 0.005, 0.075 and 0.01 mm. Grinding wheel characteristic were 1-250x25x76 99A 24 M5B – 50.

To determine the quality of wear resistance, Q_R, the parameter of relative wear of the material, was determined. It can be calculated from the following equation:

$$Q_R = \frac{(Ra_{i+1} - Ra_i)}{Ra_i} \tag{11}$$

where Ra_i is the arithmetic mean deviation in area i and Ra_{i+1} is the arithmetic mean deviation in the next area.

Thus, the wear resistance is determined by the following formula:

$$I_v = \frac{\Delta m}{\rho \cdot \Delta Ra \cdot l_p \cdot N} \tag{12}$$

where Δm is the change of mass in area and ΔRa is surface deviation, $\Delta Ra = (Ra_{i+1} - Ra_i) = Q_R Ra_i$.

The proposed model allows the wear resistance of composite materials to be predicted when they formed.

For roughness models of the sub-microrelief we can use nonlinear regression, namely

$$Y = a_0 + b_1 \cdot Ra + b_2 \cdot Ra^2 \tag{13}$$

where the parameter a_0 is the empirical coefficient that describes the initial surface deviation, and b_1, b_2 are empirical coefficients that describe wear sub microrelief.

The results of the experiment are summarized in Table 2.

Table 2. Results of experiment.

No.	Formula	ΔRa Surface Deviation, μm	Roughness Class
1	$Y = -0.066 \cdot Ra^2 + 0.72 \cdot Ra + 0.12$	0.005264	7
2	$Y = -0.17 \cdot Ra^2 + 0.72 \cdot Ra + 0.12$	0.24025	6
3	$Y = 0.566 \cdot Ra^2 - 3.22 \cdot Ra + 9.92$	0.32089	3
4	$Y = 0.44 \cdot Ra^2 - 0.997 \cdot Ra + 4.92$	0.227079	4
5	$Y = 0.73 \cdot Ra^2 - 2.71 \cdot Ra + 6.53$	0.231453	4
6	$Y = -0.49 \cdot Ra^2 + 2.14 \cdot Ra - 0.84$	0.09569	5
7	$Y = -0.14 \cdot Ra^2 + 0.53 \cdot Ra + 0.46$	0.51174	6

Results of experimental formation of roughness are shown in Figure 5.

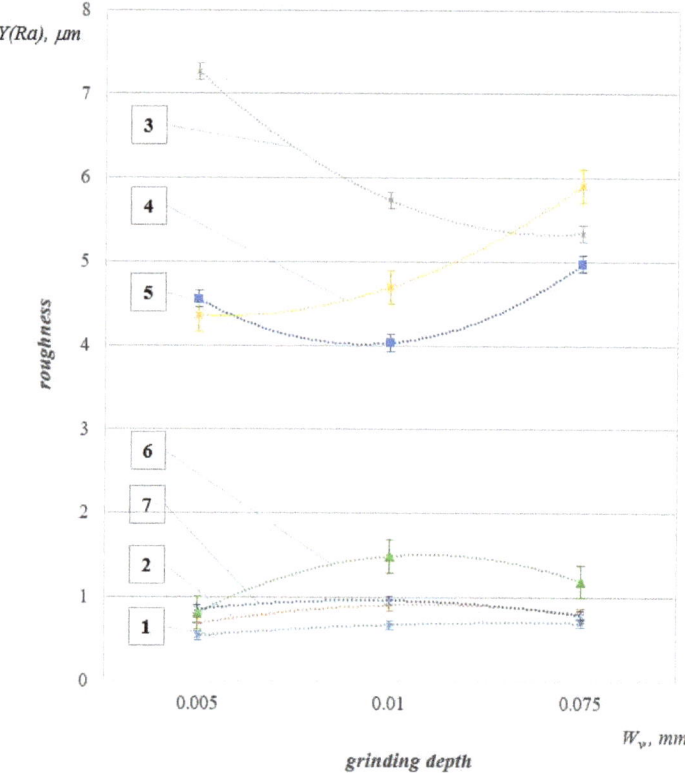

Figure 5. Graphs of experimental formation of roughness.

The dependence of Y on the parameter was studied. The parabolic trend was selected at the specification stage. Its parameters were least squares estimated. The statistical significance of the equation was verified using the coefficient of determination and the Fisher test. It was also established that the model parameters were statistically significant.

The following equation is used to describe and assess our proposed model of the roughness based on the correlation of multiple regression:

$$Y = a_0 + a_1 \cdot Ra + a_2 \cdot Rz \tag{14}$$

where the parameter a_0 describes the initial roughness, a_1 describes wear rate I_v and a_2 describes the change in the maximum height of surface.

Thus, the formation of roughness can describe by the following equation:

$$Y = a_0 + \Delta Rz \frac{N}{S \cdot \alpha^2} Ra + \frac{\Delta m \cdot Rz}{\rho \cdot \Delta Ra \cdot l_p \cdot N} \tag{15}$$

The results of the experiment are summarized in Table 3.

Table 3. Results of experiment.

No.	Formula	Roughness Class
1	$Y = 0.3575 + 0.4852 \cdot Ra - 0.00254 \cdot Rz$	7
2	$Y = 0.5273 - 2.1965 \cdot Ra + 0.1526 \cdot Rz$	6
3	$Y = -0.2199 + 0.04706 \cdot Ra - 0.000981 \cdot Rz$	2
4	$Y = 0.5758 + 0.02053 \cdot Ra - 0.0215 \cdot Rz$	2
5	$Y = 0.2429 - 0.194 \cdot Ra + 0.02197 \cdot Rz$	3
6	$Y = 0.3059 + 0.2019 \cdot Ra - 0.05005 \cdot Rz$	4
7	$Y = -26.8319 + 61.1495 \cdot Ra - 3.5595 \cdot Rz$	6

The partial correlation coefficient measures the correlation of the corresponding features (and Ra, Rz), provided that the influence of other factors on them is eliminated. Conclusions can be made based on partial coefficients about the validity of including variables in the regression model. If the coefficient is small or insignificant, this means that the relationship between this factor and the effective variable is very weak. Therefore, the factor can be excluded from the model.

4. Conclusions

Basic formulas that describe surface roughness were identified. The minimal error of the formula was 2% for sample 1 and the maximum was 8% for sample 7.

Seven samples of different polymer board materials and phenolic cotton laminated plastic laminate were used in experiment. The research conducted allowed us to determine that the best roughness was obtained after sample tests of the sample No. 1, the RenShape BM5273.

The dynamics of surface roughness change during intensive friction was the lowest in the case of test sample 1.

Samples 2 and 7 (TCF and Cibatool®BM5168) also obtained sufficient surface quality after abrasive wear resistance tests. Light structure (porous) materials, namely RenShape®BM5035 (No. 3), Cibatool®BM5005 (No. 4) and RenShape®BM5185 (No. 5), were characterized by reduced resistance to abrasive wear and had traces of internal structure, reflected in the surface roughness. These materials are unsuitable for applications in systems where high friction occurs. High density materials, namely RenShape®BM5273 (No. 1), TCF (No. 2), Cibatool®BM5272 (No. 6) and Cibatool®BM5168 (No. 7), had high abrasive wear resistance, which predisposed these materials to applications in elements subjected to a significant number of friction cycles.

As a result of the calculations, multiple regression equations were obtained for all samples.

The statistical significance of the equations was verified using the coefficient of determination and the Fisher test. They established that in the studied situation, 100% of the total variability of Y was explained by a change in factors Xj. It was also established that the model parameters were statistically significant.

Author Contributions: Conceptualization, J.M. (Jakub Musiał); Formal analysis, J.M. (Janusz Musiał); Investigation, R.P.; Methodology, S.H. and T.K.; Project administration, J.M. (Janusz Musiał); Validation, M.M.; Visualization, M.Ś.; Writing—original draft, S.H. and R.P. All authors have read and agreed to the published version of the manuscript.

Funding: This research received no external funding.

Conflicts of Interest: The authors declare no conflicts of interest.

References

1. Tuttle, M.E. *Structural Analysis of Polymeric Composite Materials*; Chapman and Hall/CRC: London, UK, 2019.
2. Chung, D.D.L. *Composite Materials: Science and Applications*; Springer: Berlin/Heidelberg, Germany, 2004.
3. Mangino, E.; Carruthers, J.; Pitarresi, G. The Future Use of Structural Composite Materials in the Automotive Industry. *Int. J. Veh. Des.* **2007**, *44*, 211–232. [CrossRef]
4. Vollrath, F.; Porter, D. Silks as Ancient Models for Modern Polymers. *Polymer* **2009**, *50*, 5623–5632. [CrossRef]

5. Horiashchenko, S.; Horiashchenko, K.; Musiał, J. Methodology of measuring spraying the droplet flow of polymers from nozzle. *Mechanika* **2020**, *26*, 82–86. [CrossRef]
6. Zakalov, O.V. *Fundamentals of Friction and Wear in Machines*; TNTU Publishing House: Ternopil, Ukraine, 2011; p. 322.
7. Muhandes, H.; Kalacska, G.; Kadi, N.; Skrifvars, M. Pin-on-Plate Abrasive Wear Test for Sevel Composite Materials. *Proceedings* **2018**, *2*, 469. [CrossRef]
8. Zsidai, L.; Katai, L. Abrasive Wear and Abrasion Testing of PA 6 and PEEK Composite in Small-Scale Model System. *Acta Polytech. Hung.* **2016**, *13*, 197–214.
9. Burris, D.L.; Sawyer, G. A low friction and ultra low wear rate PEEK/PTFE composite. *Wear* **2006**, *261*, 410–418. [CrossRef]
10. Sudeepan, J.; Kumar, K.; Barman, T.K.; Sahoo, P. Study of friction and wear of ABS/Zno polymer composite using Taguchi technique. *Procedia Mater. Sci.* **2014**, *6*, 391–400. [CrossRef]
11. Sudeepan, J.; Kumar, K.; Barman, T.K.; Sahoo, P. Study of friction and wear properties of ABS/Kaolin polymer composite using grey relational technique. *Procedia Technol.* **2014**, *14*, 196–203. [CrossRef]
12. Myshkin, N.K.; Petrokovets, M.I.; Kovalev, A.V. Tribology of polymers: Adhesion, friction, wear, and mass-transfer. *Tribol. Int.* **2005**, *38*, 910–921. [CrossRef]
13. Greco, A.C.; Erck, R.; Ajayi, O.; Fenske, G. Effect of reinforcement morphology on high-speed sliding friction and wear of PEEK polymers. *Wear* **2011**, *271*, 2222–2229. [CrossRef]
14. Samyn, P.; Schoukens, G. Experimental extrapolation model for friction and wear of polymers different testing scales. *Int. J. Mech. Sci.* **2008**, *50*, 1390–1403. [CrossRef]
15. Agrawal, S.; Singh, K.K.; Sarkar, P.K. Comparative investigation on the wear and friction behaviors of carbon fiber reinforced polymer composites under dry sliding, oil lubrication and inert gas environment. *Mater. Today Proc.* **2018**, *5*, 1250–1256. [CrossRef]
16. Yousif, B.F.; Nirmal, U. Wear and frictional performance of polymeric composites aged in various solutions. *Wear* **2011**, *272*, 97–104. [CrossRef]
17. Katiyar, J.K.; Sinha, S.K.; Kumar, A. Friction an wear durability study of epoxy-base polymer (SU-8) composite coatings with talc and graphite as fillers. *Wear* **2016**, *362–363*, 199–208. [CrossRef]
18. Voyer, J.; Jiang, Y.; Pakkanen, T.A.; Diem, A. Adhesive friction and wear of micro-pillared polymers in dry contact. *Polym. Test.* **2019**, *73*, 258–267. [CrossRef]
19. Karuppiah, K.S.K.; Bruck, A.L.; Sundararajan, S.; Wang, J.; Lin, Z.; Xu, Z.; Li, X. Friction and wear behavior of ultra-high molecular weight polyethylene as a function of polymer crystallinity. *Acta Biomater.* **2008**, *4*, 1401–1410. [CrossRef]
20. Iyer, S.B.; Dube, A.; Dube, N.M.; Roy, P.; Sailaja, R.R.N. Sliding wear and friction characteristics of polymer nanocomposite PAEK-PDMS with nano-hydroxyapatite and nano carbon fibres as fillers. *J. Mech. Behav. Biomed. Mater.* **2018**, *86*, 23032. [CrossRef]
21. Yadav, P.S.; Purohit, R.; Kothari, A. Study of friction and wear behavior of epoxy/nano SiO_2 based polymer matrix composites—A review. *Mater. Today Proc.* **2019**, *18*, 5530–5539. [CrossRef]
22. Wang, Q.; Xianqiang, P. Chapter 4—The influence of nanoparticle fillers on friction and wear behavior of polymer martices. In *Tribology of Polymeric Nanocomposites*; Elsevier: Amsterdam, The Netherlands, 2013; pp. 91–118.
23. Lakshminarayana, R.T. Tribological Performance of Different Crankshaft Bearings in Conjunction with Textured Shaft Surfaces. Master's Thesis, Luleå University of Technology, Department of Engineering Sciences and Mathematics, Luleå, Sweden, 2017.
24. Sorokatiy, R.V. Scientific Bases and Implementation of the Method of Calculation of Wear of Friction Units by the Method of Trioelements. Ph.D. Thesis, Khmelnytskyi National University, Khmelnitsky, Ukraine, 2009.
25. Musiał, J.; Polasik, R.; Kałaczyński, T.; Szczutkowski, M.; Łukasiewicz, M. Milling efficiency aspects during machining of 7075 aluminium alloy with reference to the surface geometrical structure. In Proceedings of the 24th International Conference Engineering Mechanics, Svratka, Czech Republic, 14–17 May 2018; pp. 569–572.
26. Musiał, J. *The Importance of Surface Topography in the Transformation of the Cylindrical Surface Layer of the Rolling Pairs*; Publishing House of University of Technology and Life Sciences: Bydgoszcz, Poland, 2014.
27. Musiał, J.; Szczutkowski, M.; Polasik, R.; Kałaczyński, T. The influence of hardness of cooperating elements on performance parameters of rolling kinematic pairs. In Proceedings of the 58th International Conference of Machine Design Departments—ICMD 2017, Prague, Czech Republic, 6–8 September 2017; pp. 260–265.
28. Naik, N.K.; Shrirao, P. Composite structures under ballistic impact. *Compos. Struct.* **2004**, *66*, 579–590. [CrossRef]

29. Nunes, L.M.; Paciornik, S.; d'Almeida, J.R.M. Evaluation of the damaged area of glass-fiber-reinforced epoxy-matrix composite materials submitted to ballistic impacts. *Compos. Sci. Technol.* **2004**, *64*, 945–954. [CrossRef]
30. Horiashchenko, S.; Golinka, I.; Bubulis, A.; Jurenas, V. Simulation and Research of the Nozzle with an Ultrasonic Resonator for Spraying Polymeric. *Mechanika* **2018**, *24*, 61–64. [CrossRef]
31. Nam, Y.-W.; Sathish Kumar, S.K.; Ankem, V.A.; Kim, C.G. Multi-functional aramid/epoxy composite for stealth space hypervelocity impact shielding system. *Compos. Struct.* **2018**, *193*, 113–120. [CrossRef]
32. Marx, J.; Portanova, M.; Rabiei, A. A study on blast and fragment resistance of composite metal foams through experimental and modeling approaches. *Compos. Struct.* **2018**, *194*, 652–661. [CrossRef]
33. Rajput, M.S.; Burman, M.; Segalini, A.; Hallstrom, S. Design and evaluation of a novel instrumented drop-weight rig for controlled impact testing of polymer composites. *Polym. Test.* **2018**, *68*, 446–455. [CrossRef]
34. González, E.V.; Maimi, P.; Martin-Santos, E.; Soto, A.; Cruz, P.; de la Escalera, M.F.; Sainz de Aja, J.R. Simulating drop-weight impact and compression after impact tests on composite laminates using conventional shell finite elements. *Int. J. Solids Struct.* **2018**, *144–145*, 230–247.
35. Available online: https://www.freemansupply.com/products/machinable-media/renshape-modeling-and-styling-boards (accessed on 2 February 2020).
36. Available online: https://www.obo-werke.de/en/products/renshaper-pu-boards.html (accessed on 2 February 2020).

© 2020 by the authors. Licensee MDPI, Basel, Switzerland. This article is an open access article distributed under the terms and conditions of the Creative Commons Attribution (CC BY) license (http://creativecommons.org/licenses/by/4.0/).

Article

Study of the Sound Absorption Properties of 3D-Printed Open-Porous ABS Material Structures

Martin Vasina [1,3,*], Katarina Monkova [2,3,*], Peter Pavol Monka [2,3], Drazan Kozak [4] and Jozef Tkac [2]

[1] Faculty of Mechanical Engineering, VŠB-Technical University of Ostrava, 17. listopadu 15/2172, 708 33 Ostrava-Poruba, Czech Republic
[2] Faculty of Manufacturing Technologies, Technical University in Kosice, 080 01 Presov, Slovakia; peter.pavol.monka@tuke.sk (P.P.M.); jozef.tkac@tuke.sk (J.T.)
[3] Faculty of Technology, Tomas Bata University in Zlin, Nam. T.G. Masaryka 275, 760 01 Zlin, Czech Republic
[4] Mechanical Engineering Faculty in Slavonski Brod, Josip Juraj Strossmayer University of Osijek, Trg Ivane Brlic-Mazuranic 2, HR-35000 Slavonski Brod, Croatia; dkozak@sfsb.hr
* Correspondence: vasina@utb.cz (M.V.); katarina.monkova@tuke.sk (K.M.); Tel.: +420-57-603-5112 (M.V.); +421-55-602-6370 (K.M.)

Received: 2 April 2020; Accepted: 2 May 2020; Published: 6 May 2020

Abstract: Noise pollution is a negative factor that affects our environment. It is, therefore, necessary to take appropriate measures to minimize it. This article deals with the sound absorption properties of open-porous Acrylonitrile Butadiene Styrene (ABS) material structures that were produced using 3D printing technology. The material's ability to damp sound was evaluated based on the normal incidence sound absorption coefficient and the noise reduction coefficient, which were experimentally measured by the transfer function method using an acoustic impedance tube. The different factors that affect the sound absorption behavior of the studied ABS specimens are presented in this work. In this study, it was discovered that the sound absorption properties of the tested ABS samples are significantly influenced by many factors, namely by the type of 3D-printed, open-porous material structure, the excitation frequency, the sample thickness, and the air gap size behind the sound-absorbing materials inside the acoustic impedance tube.

Keywords: Acrylonitrile Butadiene Styrene; sound absorption; 3D printing technology; frequency; thickness; air gap

1. Introduction

Noise pollution currently has a major impact on the environment, human health, and the economy. In general, noise control includes active and passive management [1]. An active noise control (ANC) system cancels undesirable sound based on its superposition attitude [2]. An ANC system features electroacoustic equipment, which is built based on destructive intervention by creating an antinoise of equal amplitude and an opposite phase to the unwanted sound [3]. ANC systems that use no noise-absorbing materials [4] are effective for low-frequency noise. In contrast, passive noise control (PNC) systems that use sound-absorbing materials are relatively large, expensive, and ineffective with low frequencies [2,5]. Regarding low frequencies, this is because the thickness of the absorbers is insufficient compared to the acoustic wavelength [4]. For this reason, PNC systems are effective only for high-frequency noise [6]. Porous materials (e.g., foam or glass wool) are usually applied to reduce undue noise [7]. New types of suitable sound-absorbing materials are continuously being developed. Compared to conventional manufacturing technologies, it is possible to produce sound-absorbing materials with advanced open-pore structures using a 3D printing technique.

Fakrul et al. [8] studied the use of a 3D-printed lattice structure as a sound absorber. Their paper investigates the sound absorption characteristics of a lattice structure produced by additive manufacturing to improve the acoustic characteristics that cannot be fulfilled by a typical sound absorber. The study confirmed that lattice structures can absorb sound mostly above 1 kHz. Sim et al. [9] were involved in the preparation of nanocomposite films of polycarbonate/polyacrylonitrile-butadiene-styrene (PC/ABS), and investigated not only the rheological and mechanical properties of this material, but also its ability to absorb sound. The PC/ABS thin films were made by a simple solvent casting method using phenylene modified-mesoporous silica materials as additives and dichloromethane as a solvent. The tensile properties of these materials were experimentally tested at ambient temperature according to ASTM D638. The authors discovered that depending on the production process, the PC/ABS nanocomposites containing mesoporous materials could also be used as sound insulation materials. Wang et al. [10] investigated the production of a lattice structure by an additive method, the geometry of which was optimized through graded mesostructures. The design of the geometry of the global cell structure, as well as the microstructure distribution in the material and its scale, reflect load bearing. Cao et al. [11] studied the application of porous materials for sound absorption. This review describes not only the process of sound absorption and the predictive behavioral models of porous materials that are characterized by their sound-absorption properties, but also analyses the development of the principles for designing foams and fibrous sound-absorbing materials, including their production. Some scenarios and perspectives on sound-absorbing structures and improvements to their properties are introduced in the conclusions of the paper. Gulia and Gupta [12] studied the damping of noise in a triple board for which they used a porous material and a sonic crystal with a specific resonant area. To suppress the sound transmission in the triple panel, they combined the characteristics of the material with porosity and a locally-resonant sonic crystal. An analytical method was used for this research. The results highlight two types of sudden decreases in the loss of sound transmission: Bragg's dip and the dip caused by panel oscillation. To avoid a sudden drop in sound loss, the cavities of a porous material were filled with a medium. Yoon et al. [13] studied porous materials and optimized their topology to improve their noise absorption properties. The goal of their research was to develop a numerical method to appropriately distribute the solid elements in a layer of a porous structure with a constant thickness so that the noise would be completely absorbed. No special process was used for the initial distribution of the solid inclusions. Only then were the systematic multiple distributions of the solid inclusions implemented in a given porous layer, which was realized using a topology optimization formula. The results of the finite element-based numerical method showed that different types of resonances appear in the optimized layer, which absorbs noise for all considered frequencies simultaneously. The influence of processing conditions on the quality of the geometry of the lattice structures was assessed by Rosli et al. [14], whose research investigated the effects of the processing parameters of the fused deposition modelling (FDM) technique on the geometrical quality of an ABS polymer lattice structure. The variations in the layer thickness of the FDM machine used in this study were 70 µm, 200 µm, and 300 µm. Examinations of the diameter circularity of the printed lattice blocks were conducted via direct measurements and formulations from a previous study. It was found that a layer thickness of 200 µm produced a more accurate strut diameter with a reliable mechanical response. Kim et al. [15] developed experiments to investigate the semi-active control of a sophisticated porous material related to its ability to absorb noise. The authors developed a smart structure that can reduce sound in a wide range of frequencies when semi-active regulation is realized through the application of a magnetic reactive material. Jimbo and Tateno [16] focused on creating a lattice structure with isotropic tensile strength that can be produced in an additive way. ABS material and geometry with a body-centered cubic (BCC) structure were selected to be studied. This type of cell was proposed to eliminate the anisotropic qualities caused by additive manufacturing. The specimens were produced via FFM technology, and the hypothesis of the isotropic tensile strength properties of such a structure was confirmed by experimental tests. Salomons et al. [17] studied the operation of soundwaves for fluid flow in a structure using the Lattice Boltzmann method (LBM). In the published manuscript, this method was applied to the analysis of noise propagation in several different environments, such as a free field, porous and nonporous structures, the sound barrier, and a windy atmosphere. The results obtained using the LBM were compared with the corresponding results computed based on the acoustic equations. It was concluded that the LBM is an appropriate method for soundwave investigations, but that it is necessary to take into account that the dispersion of soundwaves within the LBM is usually much greater than their dissipation in the air. Zvolensky et al. [18] investigated the behavior of noise propagation through a porous material and the specific parameters of noise pollution. The authors focused on a computer simulation of the sound transmitted through wagon walls and on the usage of noise analyses during the operation of a train. In the article, the acoustic characteristics of the primary materials were compared with those of the new ones. Chen et al. [19] were involved in the development of an acoustic superlens based on single-phase lattice metamaterials in the shape of a star made of steel. The authors invented low-density superlens which can reach a sound focus over the diffraction limit. These lattice structures offer rich resonances to induce abnormal dispersion effects, as specified by the negative parameter indices. Their study showed that this type of structure has double negative index properties that can mediate these effects for sound in water.

The present work aims to investigate the different factors that affect the sound absorption properties of 3D-printed ABS materials with various open-pore structures. Although many studies have been published related to the sound properties of porous materials, the proposed types of structures (produced from an ABS material) have not yet been investigated.

2. Materials and Methods

2.1. Materials

The specimens for the experimental tests were produced from an ABS (acrylonitrile-butadiene-styrene) material (Smart Materials 3D, Alcala la Real, Spain), which is one of the most commonly-used thermoplastic polymers for 3D printing technology. Due to its physical properties, this material is very often chosen to produce a wide range of components for use in relatively safe machines that are easy to operate. It is resistant to abrasion and high temperatures (its melting point is 200 °C/392 °F), and it is lightweight [20].

Its mechanical properties, such as impact and tensile strength, as well as stiffness, are very good. ABS is soluble in acetone. However, ABS also resists many chemical formulas. Moreover, this plastic has good surface quality and flame retardancy. It is recyclable and available to both professionals and the general public. Consequently, ABS is an ideal material for the production of inexpensive prototypes and architectural models for engineers or research departments, as well as for the creation of inexpensive material handling equipment [21,22].

2.2. Samples Production

ABS is a material that is primarily used by 3D printers based on the FDM (Fused Deposition modelling) technique (see Figure 1).

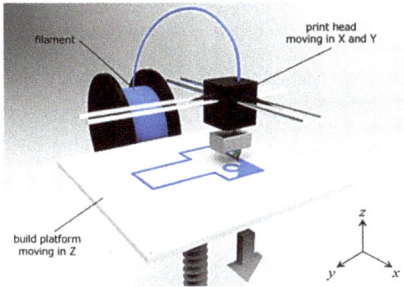

Figure 1. Principle of the Fused Deposition modelling (FDM) technique.

The material for 3D printing is prepared in the form of a long filament that is wound on a spool. This spool enters through the print head, which can move in the x and y axes (i.e., the plane of the printing bed). The movement of the head is digitally controlled based on the shape of the produced body in the given layer. The molten material is applied in a thin layer on the building platform to which it adheres. After the deposition of the first layer, the platform moves down, and a new layer is applied. This process repeats until the new component is finished. One of the advantages of 3D printing technology is the possibility to make parts with complex shapes that are non-manufacturable (or very difficult) with other methods. Such products include lightweight components with internal porous structures. The mechanical properties of porous structures are significantly affected by the material used and by the geometry of the structure. A very important parameter of porous materials is the "relative density", or "Volume ratio V_r". This parameter expresses how much cell space is filled with material, and is described by the Equation (1) [23]:

$$V_r(\%) = \frac{V_S}{V_T} \times 100 \tag{1}$$

where V_S is the volume of the solid phase and V_T is the total body volume. In this study, three types of lattice structures (Cartesian, Starlit, and Octagonal) were modelled and produced from the plastic material ABSplus-P430 Ivory using the FDM technique. All the samples had a cylindrical shape with an outer diameter of φ29 mm (based on the testing device requirements) and three different lengths (thicknesses), i.e., 10, 20, and 30 mm. The core of every sample was filled with a lattice structure with a volume ratio V_r = 57%, while the thickness of the outside cylindrical shell (fully filled by the material) was 2 mm. All basic cell types were modelled with the outer horizontal, outer vertical, and inner angular beams/struts. The z axis is the building axis, normal to the building platform (plane xy). The volume ratio was controlled by using a strut diameter of 1.4 mm. The structure types and their characteristics are presented in Table 1, while the produced samples are shown in Figure 2.

Table 1. Types of 3D-printed structures.

Structure Type	Volume Ratio (%)	Label	Front View	Strut Diameter	Cell Sizes (mm)
Cartesian	57	C_57		1.4	x = 5 y = 5 z = 5
Starlit	57	S_57		1.4	x = 9 y = 9 z = 5
Octagonal	57	O_57		1.4	x = 7 y = 7 z = 5

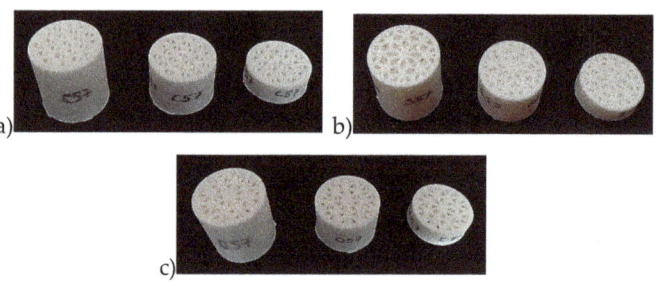

Figure 2. Tested samples: (**a**) Cartesian structure, (**b**) Starlit structure, and (**c**) Octagonal structure.

2.3. Measurement Methodology

2.3.1. Sound Absorption Coefficient

If the incident acoustic energy E_I is propagated from a noise source to a material surface, some of this incident energy is reflected from the surface. The remainder of the incident acoustic energy is absorbed by the tested acoustical material. The material's ability to damp noise is expressed by the sound absorption coefficient α, which is defined by the following equation [24]:

$$\alpha = \frac{E_A}{E_I} = 1 - \frac{E_R}{E_I} \qquad (2)$$

where E_A represents the absorbed acoustic energy and E_R is the reflected acoustic energy (see Figure 3b). The basic function of sound-absorbing materials is to transform the incident acoustic energy into heat. There are two mechanisms by which the acoustic energy is dissipated: viscous-flow losses and internal friction [25].

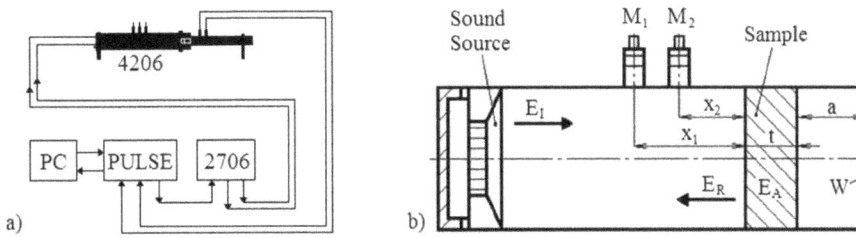

Figure 3. Schematic diagram of the apparatus for measuring the sound absorption coefficient (**a**) and a schematic of the impedance tube equipment (**b**). Legend of the abbreviations: a—air gap size; E_A—absorbed acoustic energy; E_I—incident acoustic energy; E_R—reflected acoustic energy; M_1, M_2—measuring microphones; t—sample thickness; W—solid wall; x_1, x_2—microphone distances from the tested ABS sample.

2.3.2. Noise Reduction Coefficient

The material's ability to absorb sound depends on several factors, such as the material type, thickness, density, structure, excitation frequency, and temperature. The noise reduction coefficient (NRC) includes the effect of the excitation frequency on the sound absorption coefficient and is a single number, which is defined as the arithmetical average of the sound absorption coefficients at the frequencies 250, 500, 1000, and 2000 Hz [26,27]:

$$NRC = \frac{\alpha_{250} + \alpha_{500} + \alpha_{1000} + \alpha_{2000}}{4}. \qquad (3)$$

2.3.3. Sound Absorption Properties

The sound absorption behavior of the investigated 3D-printed ABS materials was measured using a two-microphone acoustic impedance tube (BK 4206) in combination with a signal PULSE multi-analyzer (BK 3560-B-030) and a power amplifier (BK 2706) in the frequency range of 200–6400 Hz (Brüel & Kjær, Nærum, Denmark). A schematic diagram of the experimental apparatus is shown in Figure 3a. The normal incidence soundwave absorption of the tested samples of a given thickness t (from 10 to 30 mm) was experimentally determined for different air gap sizes a (ranging from 0 to 100 mm) behind the investigated ABS specimens, as shown in Figure 3b. All experiments were carried out at the ambient temperature of 23 °C.

The frequency dependencies of the sound absorption coefficient of the investigated materials were obtained by the transfer function method ISO 10534-2 [28], which is based on the partial standing wave principle. In this case, the normal incidence sound absorption coefficient α is expressed by the following formula [29,30]:

$$\alpha = 1 - |r|^2 = 1 - r_r^2 - r_i^2 \tag{4}$$

where r is the normal incidence reflection factor, and r_r and r_i are the real and imaginary components of the factor r, which is given by

$$r = r_r + i r_i = \frac{H_{12} - H_I}{H_R - H_{12}} \cdot e^{2 k_0 \cdot x_1 i} \tag{5}$$

where H_{12} is the complex acoustic transfer function, H_I is the transfer function for the incident wave, H_R is the transfer function for the reflection wave, k_0 is the wave number, and x_1 is the distance between the investigated material sample and the microphone M_1 (see Figure 3b). The transfer functions are expressed as follows:

$$H_{12} = \frac{p_2}{p_1} = \frac{e^{k_0 \cdot x_2 i} + r \cdot e^{-k_0 \cdot x_2 i}}{e^{k_0 \cdot x_1 i} + r \cdot e^{k_0 \cdot x_1 i}} \tag{6}$$

$$H_I = e^{-k_0 \cdot (x_1 \cdot x_2) i} \tag{7}$$

$$H_R = e^{k_0 \cdot (x_1 \cdot x_2) i} \tag{8}$$

where p_1 and p_2 are the complex sound pressures at the two microphone positions, and x_2 is the distance between the investigated material sample and the microphone M_2 (see Figure 3b).

3. Results and Discussion

This section explores the different factors that influence the sound absorption properties of the investigated 3D-printed, open-pore ABS materials, whose designations are as follows: The sample designation is first described by its structure type (see Table 1). Subsequently, the sample designation consists of two numbers. The first number represents the sample thickness (in mm). The second number defines the air gap size a (in mm) behind the sample inside the impedance tube (see Figure 3b).

3.1. Frequency Dependencies of the Sound Absorption Coefficient

3.1.1. Effect of Structure Type

The effect of the structure type on the sound absorption properties of the investigated 3D-printed ABS materials is demonstrated in Figure 4. Examples of the frequency dependencies of the sound absorption coefficient for the ABS samples of the same thickness (i.e., $t = 20$ mm) are shown in Figure 4a. In this case, the samples were mounted directly on the solid wall W (i.e., with $a = 0$ mm) inside the impedance tube (see Figure 3b). Similarly, the structural effect of the tested 3D-printed ABS materials (with $t = 10$ mm and $a = 30$ mm) is presented in Figure 4b. It is evident from these comparisons that the structural effect on sound absorption is negligible at low excitation frequencies, namely at $f < 2$ kHz (see Figure 4a) and at $f < 700$ Hz (see Figure 4b). Subsequently, better sound absorption properties were obtained for the ABS samples, which were produced with the Starlit structure. The ability of open-porous material structures to damp sound is related to the airflow resistivity of these structures. Generally, increasing the airflow resistivity improves sound absorption properties [31,32] in the whole frequency range but only up to an intermediate value. If the porous materials are too acoustically resistive and their airflow resistivity is very high, the sound absorption properties of these materials are very low during the propagation of an acoustic wave through their porous structures. The Starlit structure (see Table 1) is characterized by more complicated pore shapes compared to the Cartesian and Octagonal structures, resulting in higher airflow resistivity of the open-porous Starlit structure. For these reasons, the ABS samples produced with the Starlit structure exhibit better sound absorption performance than the other open-porous structures. This phenomenon is accompanied by higher internal friction during the propagation of an acoustic wave through this structure, and by greater transformation of incident acoustic energy into heat. This effect was observed in the frequency ranges of 2–3 kHz (see Figure 4a) and 0.7–1.9 kHz (see Figure 4b). Conversely, the 3D-printed ABS samples produced with the Cartesian and Octagonal structures, whose frequency dependencies of sound absorption are very similar, are more suitable for damping sound at higher excitation frequencies, namely in the frequency ranges $f = \langle 3.0; 6.4 \rangle$ kHz (see Figure 4a) and $f = \langle 1.9; 6.4 \rangle$ kHz (see Figure 4b). Similar structural effect results were found for the 3D-printed ABS samples further tested independently of the sample thickness and the air gap size.

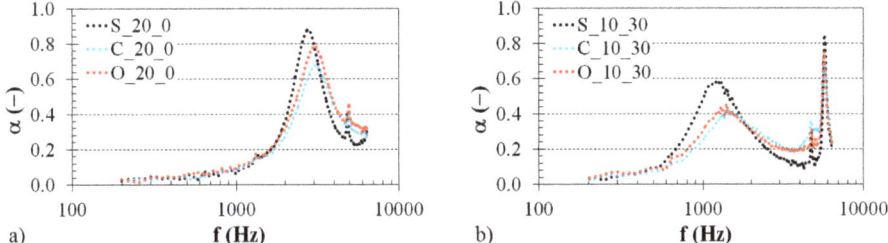

Figure 4. Effect of the 3D-printed ABS material structure on the frequency dependencies of the sound absorption coefficient; (**a**) sample thickness $t = 20$ mm, air gap size $a = 0$ mm, (**b**) sample thickness $t = 10$ mm, air gap size $a = 30$ mm.

3.1.2. Effect of Material Thickness

A material's ability to reduce noise is significantly influenced by its thickness. Figure 5 demonstrates the effect of varying the sample thickness when the tested sample is mounted directly (see Figure 5a) on a solid wall or placed at a distance of 10 mm from the wall (see Figure 5b). The sample thickness significantly improves sound absorption towards lower excitation frequencies. However, increasing the sample thickness leads to higher production costs for the 3D-printed ABS samples. The production costs of the 3D printing of components are generally influenced by various factors, such as the type of 3D-printed material, the printing layer thickness, the printing speed, the dimensions of the 3D-printed components, and the printing time. Under the same 3D printing process parameters of a given structure type, the production costs increase by increasing printing time. Because the ground plan dimensions of the investigated open-porous ABS samples are identical, an increase in sample thickness (or sample volume) increases the printing time and thus the production costs of the 3D printing process [33,34]. Therefore, this method is not effective in terms of noise reduction.

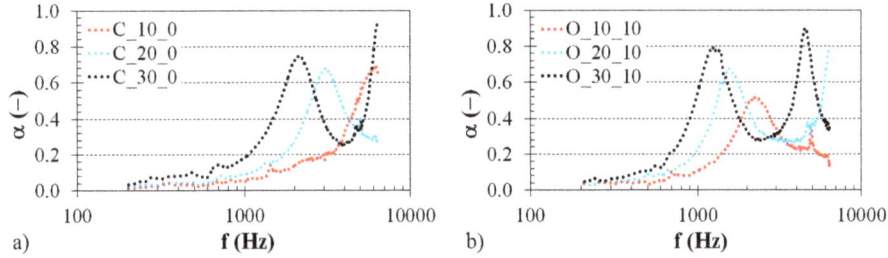

Figure 5. Effect of the sample thickness on the frequency dependencies of the sound absorption coefficient for the investigated ABS samples with a (**a**) Cartesian and (**b**) Octagonal structures.

3.1.3. Effect of the Air Gap Size

The air gap size a between the tested ABS sample and the solid wall also has a significant effect on sound damping. As shown in Figure 6, it is possible to observe a certain number of maxima and minima of the sound absorption coefficient over the whole frequency range depending on the sample type. These effects are explained by the sound reflections from the solid wall inside the impedance tube and by the wavelength of sound λ, which is defined as the speed of sound divided by the frequency [35].

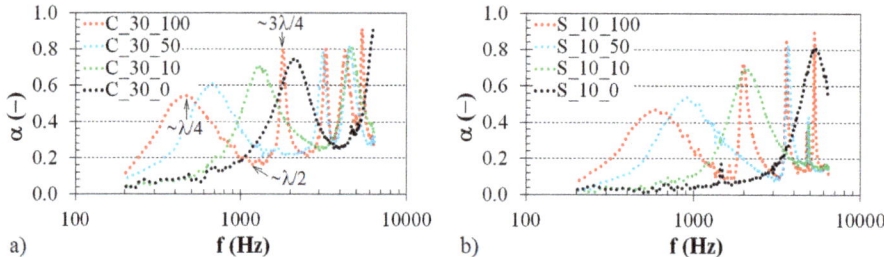

Figure 6. Effect of the air gap size on the frequency dependencies of the sound absorption coefficient for the investigated ABS samples with a (**a**) Cartesian and (**b**) Starlit structure.

At the wall surface, the acoustic pressure reaches its maximum value, but the air particle velocity is zero. Conversely, in the case of a quarter wavelength (see Figure 6a) from the wall W (see Figure 3b), the acoustic pressure is zero, and the air particle velocity is maximum. When the investigated porous material is placed at a quarter-wavelength distance from the solid wall, it is possible to obtain the maximum sound absorption because the air particle velocity is maximum. Similarly, at a half-wavelength, the air particle velocity is minimum, and the sound absorption coefficient is also minimum [35,36]. For these reasons, sound absorption maxima occur at odd multiples of quarter wavelengths in the standing-wave antinodes at the excitation frequencies:

$$f = \frac{c \cdot (2n+1)}{4l} \tag{9}$$

where c is the speed of sound, n is an integer ($n = 0, 1, 2 \ldots$), and l is the sample distance from the solid wall ($l = a + t/2$). Similarly, the sound absorption minima are obtained at even multiples of quarter wavelengths in standing-wave nodes at the excitation frequencies:

$$f = \frac{c \cdot n}{2l} \tag{10}$$

Table 2 presents the values of the primary sound absorption maxima α_{max1} (corresponding to a quarter-wavelength, i.e., $\lambda/4$), the primary sound absorption minima α_{min1} (corresponding to a half-wavelength, i.e., $\lambda/2$), and the corresponding excitation frequencies f_{max1} and f_{min1} of the ABS samples whose frequency dependencies of the sound absorption coefficient are depicted in Figure 6.

Here, the frequencies f_{max1} and f_{min1} generally decrease with an increase in the air gap size a. Therefore, it can be concluded that open-porous structures with air gaps are effective for improving sound absorption properties at low excitation frequencies instead of increasing the thickness of the sound absorber, which requires more materials [37].

Table 2. Primary sound absorption maxima and minima and their corresponding frequencies depending on the air gap size for the investigated ABS samples with Cartesian and Starlit structures.

Structure Type	t (mm)	a (mm)	f_{max1} (Hz)	α_{max1} (−)	f_{min1} (Hz)	α_{min1} (−)
Cartesian	30	0	2096	0.76	4184	0.25
		10	1304	0.73	3000	0.23
		50	656	0.63	1864	0.22
		100	432	0.55	1136	0.17
Starlit	10	0	5328	0.81	-	-
		10	2064	0.71	5912	0.15
		50	904	0.55	2992	0.08
		100	552	0.48	1312	0.09

3.1.4. Effect of Excitation Frequency

The excitation frequency f is another important factor that influences the sound absorption properties of the investigated 3D-printed ABS samples. It is evident (see Figures 4–6) that the highest value of the sound absorption coefficient is obtained at a given frequency depending on the sample type. For example, in the case of the sample with the Starlit structure that was mounted directly on the solid wall (i.e., without the air gap size) inside the impedance tube, the maximum value of the sound absorption coefficient $\alpha_{max} = 0.81$ was observed at the frequency $f = 5328$ Hz (see Figure 6b) for a sample thickness $t = 10$ mm. The same sample type (with a thickness $t = 20$ mm) that was directly mounted on the solid wall had the best sound absorption properties ($\alpha_{max} = 0.88$) at the frequency $f = 2760$ Hz (see Figure 4a). Similarly, the sample with a thickness $t = 10$ mm and the Starlit structure placed at a distance of 50 mm from the solid wall had the greatest ability to absorb sound ($\alpha_{max} = 0.84$) at the frequency $f = 3648$ Hz (see Figure 6b).

Generally, a very low material ability to absorb sound was observed at low excitation frequencies. In these cases, it is possible to increase the sample thickness (see Figure 5) to improve the sound damping of the open-porous materials. This method of noise elimination is not effective because the increase in material thickness is connected with the higher production costs of the investigated 3D-printed ABS materials. It is also possible to improve the sound absorption properties of the open-porous materials at low excitation frequencies via air-gap magnification (see Figure 6) behind the tested material sample.

The influence of the excitation frequency on sound absorption is considered based on the noise reduction coefficient defined according to Equation (3). The different factors that influence the noise reduction coefficient of the tested open-porous ABS samples are evaluated in the following section.

3.2. Noise Reduction Coefficient

As mentioned above, the noise reduction coefficient NRC is used to describe the average sound absorption performance of a given soundproofing material.

The dependence of the noise reduction coefficient on the air space size behind the 3D-printed ABS samples inside the impedance tube (see Figure 3b) is demonstrated in Figures 7 and 8.

The effect of the ABS sample structure on the noise reduction coefficient vs. the air gap size for the two various sample thicknesses is shown in Figure 7. The graphs in Figure 7 clearly show that the samples produced with the Starlit structure had higher noise reduction coefficient values, and thus, better sound absorption properties than the Cartesian and Octagonal structures. For example, for the ABS sample with a thickness $t = 10$ mm placed at a distance of 75 mm from the solid wall, a noise reduction coefficient NRC = 0.235 was observed for the ABS sample produced with the Starlit structure. However, when the ABS sample was produced with a Cartesian or Octagonal structure, the value of the noise coefficient (NRC $\cong$ 0.190) was lower (see Figure 7a). Similarly, for the ABS sample with a thickness of 20 mm that was placed at a distance of 100 mm from the solid wall, a higher noise reduction coefficient value was observed for the ABS sample produced with the Starlit structure

(NRC = 0.352) compared to the ABS samples produced with Cartesian (NRC = 0.312) and Octagonal (NRC = 0.299) structures (see Figure 7b). It was found in this study that the lowest value of the noise reduction coefficient (NRC = 0.047) was observed for the ABS sample with a thickness t = 10 mm that was produced with a Cartesian lattice structure and was mounted directly on the solid wall (i.e., without the air gap size) inside the impedance tube. Conversely, the maximum value of the noise reduction coefficient (NRC = 0.406) was observed for the ABS sample of a thickness t = 30 mm, which was produced with the Starlit lattice structure and placed at a distance of 100 mm from the solid wall of the impedance tube. The above phenomenon is caused by the higher internal friction during the propagation of an acoustic wave through the Starlit structure, which leads to a higher transformation of the incident acoustic energy into heat during sound propagation through this structure compared to the Cartesian and Octagonal structures, whose noise reduction is very similar (see Figure 7). It was also found that the noise reduction coefficient generally increases with an increase in the material thickness t and the air gap size a. For example (see Figure 7a,b), if the ABS sample produced with the Octagonal structure is placed at a distance of 30 mm from the solid wall, the noise reduction coefficient increases from 0.189 (for t = 10 mm) to 0.250 (for t = 20 mm). For the ABS sample of thickness t = 10 mm with a Cartesian structure, the noise reduction coefficient increased (see Figure 7a) from 0.047 (for a = 0 mm) to 0.268 (for a = 100 mm). Likewise (see Figure 8a), the minimum value of the noise reduction coefficient (NRC = 0.108) for the ABS sample of thickness t = 20 mm with a Cartesian lattice structure was observed in such a case when the tested sample was mounted directly (i.e., with a = 0 mm) on the solid wall. When the same ABS sample was placed at a distance of 100 mm from the wall, the maximum value of the noise reduction coefficient (NRC = 0.312) was found. If this ABS sample type was mounted directly on the wall, the noise reduction coefficient increased from 0.047 (for t = 10 mm) to 0.260 (for t = 30 mm). Similarly, if this sample was placed at a distance of 100 mm from the wall, the noise reduction coefficient increased from 0.268 (for t = 10 mm) to 0.312 (for t = 30 mm). Similar results were observed for the ABS sample produced with the Octagonal lattice structure (see Figure 8b). For the ABS specimen with a thickness t = 20 mm, the noise reduction coefficient increases from 0.121 (for a = 0 mm) to 0.299 (for a = 100 mm). When this specimen is mounted directly on the solid wall, the noise reduction coefficient rises from 0.051 (for t = 10 mm) to 0.271 (for t = 30 mm). In the case of the maximum air gap size (i.e., with a = 100 mm) inside the impedance tube (see Figure 8b), the noise reduction coefficient of this ABS specimen increased from 0.287 (for t = 10 mm) to 0.379 (for t = 30 mm).

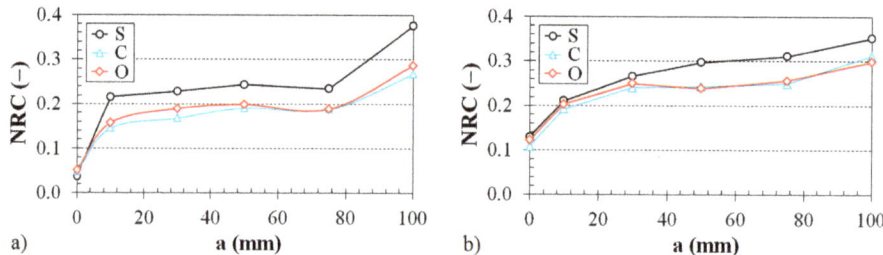

Figure 7. Effect of material structure on the noise reduction coefficient vs. air gap size dependencies for the investigated ABS samples with the following thicknesses: (**a**) 10 mm and (**b**) 20 mm.

The effect of the ABS sample thickness on the noise reduction coefficient vs. the air gap size for two different types of 3D-printed structures (Cartesian and Octagonal) is presented in Figure 8. It is evident from Figure 8 that the noise reduction coefficient increases by increasing the sample thickness independently of the structure type and the air gap size. For this reason, sample thickness has a positive influence on sound absorption. It is also obvious that the noise coefficient generally increases with an increase in the air gap size.

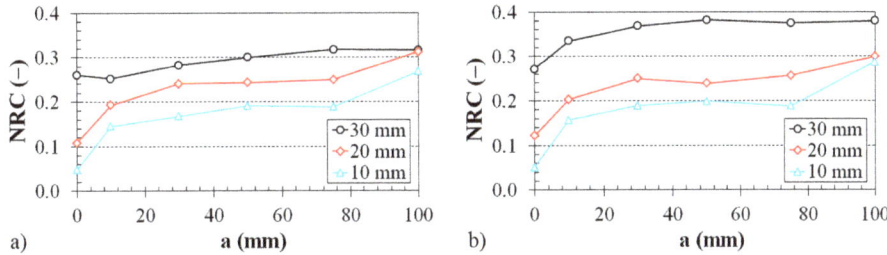

Figure 8. Noise reduction coefficient vs. air gap size dependencies for the investigated ABS samples with (**a**) Cartesian and (**b**) Octagonal structures. Inset legend: sample thickness.

The dependence of the noise reduction coefficient on the thickness of the 3D-printed ABS samples is shown in Figures 9 and 10.

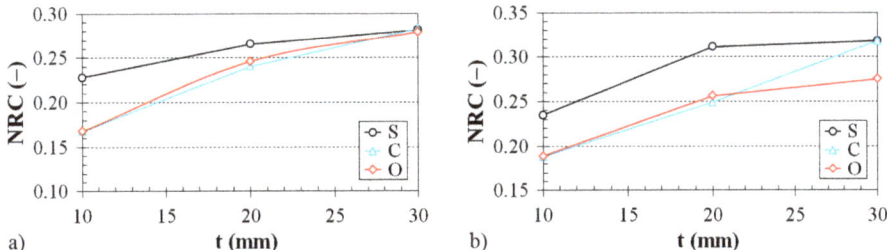

Figure 9. Effect of the material structure on the noise reduction coefficient vs. the material thickness dependencies for the investigated ABS samples placed at a distance from the wall inside the impedance tube: (**a**) 30 mm and (**b**) 75 mm.

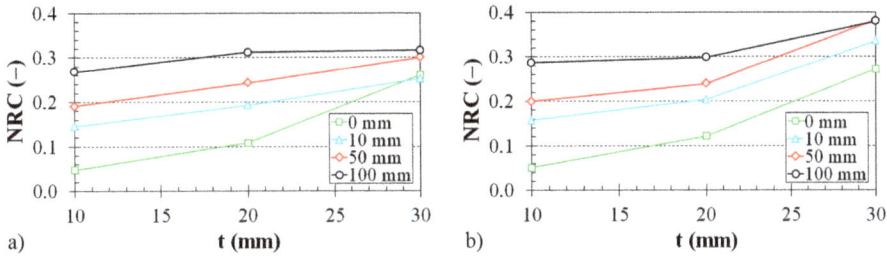

Figure 10. Noise reduction coefficient vs. material thickness dependencies for the investigated ABS samples with (**a**) Cartesian and (**b**) Octagonal structures. Inset legend: air gap size.

The effect of the ABS sample structure on the noise reduction coefficient vs. the material thickness for two various air gap sizes behind the tested ABS samples inside the impedance tube is shown in Figure 9. It was found again (as with the effect of the air gap size on the noise reduction coefficient; see Figure 7) that the samples produced with the Starlit structure have better sound absorption properties than those with Cartesian and Octagonal structures. This effect is especially significant for smaller sample thicknesses (i.e., for $t < 20$ mm). It was also found that the noise reduction coefficient generally increases with an increase in the sample thickness and air gap size (see Figure 9).

The effect of the air gap size on the noise reduction coefficient vs. the material thickness for two different types of 3D-printed structures is demonstrated in Figure 10. The noise reduction coefficient increases with an increase in the air gap size and the sample thickness independently of the structure type. Therefore, the sound absorption properties of the investigated open-porous, 3D-printed ABS samples generally increase with an increase in the air gap size and material thickness.

4. Conclusions

In this work, we investigated the sound absorption properties of open-porous, 3D-printed ABS samples manufactured using Cartesian, Starlit, and Octagonal structures with the same volume ratios. It was found that the ABS samples produced with the Starlit structure exhibited a greater ability to damp noise compared to the other ABS structures examined.

Depending on the ABS sample thickness and the air gap size behind the tested samples inside the impedance tube, the noise reduction coefficient of the samples produced with Starlit structures was higher (the increment was between 0 and 0.1) than those with Cartesian and Octagonal structures. This result is related to the pore spaces and sizes, which have a notable effect on airflow resistivity and thus on the sound absorption performance of the investigated open-porous structures.

It was also found that low sound damping properties are generally observable at the low excitation frequencies of an acoustic wave. Hence, it was necessary to increase the material thickness, which was reflected by the improved sound absorption in the low-frequency region. However, increasing the sample thickness results in higher manufacturing costs for the 3D-printed ABS samples. Therefore, this method is not effective for noise reduction. For this reason, we recommend applying open-porous sound absorbers with air gaps to improve sound damping at low excitation frequencies. This effect was particularly significant when increasing the air gap size, which resulted in a shift of the primary sound absorption maxima to lower excitation frequencies. Depending on the ABS sample thickness and structure type, the frequency of the primary sound absorption maxima decreased from $f_{max1} \cong \langle 2.0; 6.1 \rangle$ kHz for the ABS samples mounted directly (without an air gap) on the solid wall of the impedance tube to $f_{max1} = \langle 416; 656 \rangle$ Hz for the ABS samples placed at a distance of 100 mm from the wall.

Author Contributions: Conceptualization, M.V. and K.M.; methodology, K.M. and M.V.; software, M.V.; validation, P.P.M., K.M., and D.K.; formal analysis, P.P.M.; investigation, M.V.; resources, J.T.; data curation, J.T.; writing—original draft preparation, M.V. and K.M.; writing—review and editing, P.P.M.; visualization, D.K.; supervision, M.V. and K.M.; project administration, K.M. and M.V.; funding acquisition, M.V. and K.M. All authors have read and agreed to the published version of the manuscript.

Funding: This work was supported by the European Regional Development Fund in the Research Centre of Advanced Mechatronic Systems project, project number CZ.02.1.01/0.0/0.0/16_019/0000867 within the Operational Programme Research, Development and Education; as well as with direct support from the Ministry of Education, Science, Research, and Sport of the Slovak Republic through the projects KEGA 007TUKE-4/2018 and APVV-19-0550.

Acknowledgments: The authors would like to thank the Ministry of Education, Science, Research, and Sport of the Slovak Republic for its direct support of this research through the projects KEGA 007TUKE-4/2018 and APVV-19-0550, as well as the European Regional Development Fund in the Research Centre of Advanced Mechatronic Systems for its financial support of this research through the project number CZ.02.1.01/0.0/0.0/16_019/0000867 within the Operational Programme Research, Development, and Education.

Conflicts of Interest: The authors declare no conflict of interest.

References

1. Yang, W.D.; Li, Y. Sound absorption performance of natural fibers and their composites. *Sci. China Technol. Sci.* **2012**, *55*, 2278–2283. [CrossRef]
2. Gan, W.S.; Kuo, S.M. An integrated audio and active noise control headsets. *IEEE Trans. Consum. Electron.* **2002**, *48*, 242–247. [CrossRef]
3. Luo, L.; Sun, J.W.; Huang, B.Y. A novel feedback active noise control for broadband chaotic noise and random noise. *Appl. Acoust.* **2017**, *116*, 229–237. [CrossRef]
4. Chang, C.Y. Simple Approaches to Improve the Performance of Noise Cancellation. *JVC/J. Vib. Control* **2009**, *15*, 1875–1883. [CrossRef]
5. Kuo, S.M.; Mitra, S.; Gan, W.S. Active Noise control System for Headphone Applications. *IEEE Trans. Control Syst. Technol.* **2006**, *14*, 331–335. [CrossRef]

6. Chang, C.Y. Efficient active noise controller using a fixed-point DSP. *Signal Process.* **2009**, *89*, 843–850. [CrossRef]
7. Utsuno, H.; Tanaka, T.; Fujikawa, T.; Seibert, A.F. Transfer-function method for measuring characteristic impedance and propagation constant of porous materials. *J. Acoust. Soc. Am.* **1989**, *86*, 637–643. [CrossRef]
8. Fakrul, M.; Che, S.; Putra, A.; Kassim, D.H.; Alkahari, M.R. 3D-printed lattice structure as sound absorber. *Proc. Mech. Eng. Res. Day* **2019**, *2019*, 287–288.
9. Sim, S.; Kwon, O.M.; Ahn, K.H.; Lee, H.R.; Kang, Y.J.; Cho, E.B. Preparation of polycarbonate/poly(acrylonitrile -butadiene-styrene)/mesoporous silica nanocomposite films and its rheological, mechanical, and sound absorption properties. *J. Appl. Polym. Sci.* **2018**, *135*, 1–14. [CrossRef]
10. Wang, Y.; Zhang, L.; Daynes, S.; Zhang, H.; Feih, S.; Wang, M.Y. Design of graded lattice structure with optimized mesostructures for additive manufacturing. *Mater. Des.* **2018**, *142*, 114–123. [CrossRef]
11. Cao, L.; Fu, Q.; Si, Y.; Ding, B.; Yu, J. Porous materials for sound absorption. *Compos. Commun.* **2018**, *10*, 25–35. [CrossRef]
12. Gulia, P.; Gupta, A. Sound attenuation in triple panel using locally resonant sonic crystal and porous material. *Appl. Acoust.* **2019**, *156*, 113–119. [CrossRef]
13. Yoon, W.U.; Park, J.H.; Lee, J.S.; Kim, Y.Y. Topology optimization design for total sound absorption in porous media. *Comput. Methods Appl. Mech. Eng.* **2020**, *360*, 112723. [CrossRef]
14. Rosli, N.A.; Hasan, R.; Alkahari, M.R.; Tokoroyama, T. Effect of process parameters on the geometrical quality of ABS polymer lattice structure. *Proc. SAKURA Symp. Mech. Sci. Eng.* **2017**, *2017*, 3–5.
15. Kim, B.S.; Kwon, S.; Jeong, S.; Park, J. Semi-active control of smart porous structure for sound absorption enhancement. *J. Intell. Mater. Syst. Struct.* **2019**, *30*, 2575–2580. [CrossRef]
16. Jimbo, K.; Tateno, T. Design of isotropic-tensile-strength lattice structure fabricated by AMAMによる製作を想定した引張強度に関して等方性を有する格子構造の設計. *Trans. JSME* **2019**, *85*, 18-00098. (In Japanese)
17. Salomons, E.M.; Lohman, W.J.A.; Zhou, H. Simulation of sound waves using the lattice Boltzmann method for fluid flow: Benchmark cases for outdoor sound propagation. *PLoS ONE* **2016**, *11*, e0147206. [CrossRef]
18. Zvolensky, P.; Grencik, J.; Kasiar, L.; Stuchly, V. Simulation of sound transmission through the porous material, determining the parameters of acoustic absorption and sound reduction. *MATEC Web Conf.* **2018**, *157*, 2058. [CrossRef]
19. Chen, M.; Jiang, H.; Zhang, H.; Li, D.; Wang, Y. Design of an acoustic superlens using single-phase metamaterials with a star-shaped lattice structure. *Sci. Rep.* **2018**, *8*, 2–9. [CrossRef]
20. Stansbury, J.W.; Idacavage, M.J. 3D printing with polymers: Challenges among expanding options and opportunities. *Dent. Mater.* **2016**, *32*, 54. [CrossRef]
21. Bourell, D.; Pierre, J.; Leu, M.; Levy, G.; Rosen, D.; Beese, A.M.; Clare, A. Materials for additive manufacturing. *CIRP Ann.* **2017**, *66*, 659–681. [CrossRef]
22. Pantazopoulos, G.A. A Short Review on Fracture Mechanisms of Mechanical Components Operated under Industrial Process Conditions: Fractographic Analysis and Selected Prevention Strategies. *Metals* **2019**, *9*, 148. [CrossRef]
23. Brenne, F.; Niendorf, T.; Maier, H.J. Additively manufactured cellular structures: Impact of microstructure and local strains on the monotonic and cyclic behavior under uniaxial and bending load. *J. Mater. Process. Technol.* **2013**, *213*, 1558–1564. [CrossRef]
24. Koizumi, T.; Tsujiuchi, N.; Adachi, A. The development of sound absorbing materials using natural bamboo fibers. *High Perform. Struct. Mater.* **2002**, *4*, 157–166.
25. Borlea, A.; Rusu, T.; Ionescu, S.; Cretu, M.; Innescu, A. Acoustical materials—Sound absoring materials made of pine sawdust. *Rom. J. Acoust. Vib.* **2011**, *8*, 95–98.
26. Tiwari, W.; Shukla, A.; Bose, A. Acoustic properties of cenosphere reinforced cement and asphalt concrete. *Appl. Acoust.* **2004**, *65*, 263–275. [CrossRef]
27. Buratti, C. Indoor noise reduction index with an open window (Part II). *Appl. Acoust.* **2006**, *67*, 383–401. [CrossRef]
28. International Organization for Standardization. *ISO 10534-2, Acoustics-Determination of Sound Absorption Coefficient and Impedance in Impedance Tubes-Part 2: Transfer-Function Method*; ISO/TC 43/SC2 Building Acoustics; CEN, European Committee for Standardization: Brussels, Belgium, 1998; pp. 10534–10542.
29. Han, F.S.; Seiffert, G.; Zhao, Y.Y.; Gibbs, B. Acoustic absorption behaviour of an open-celled alluminium foam. *J. Phys. D Appl. Phys.* **2003**, *36*, 294–302. [CrossRef]

30. Cheerier, O.; Pommier-Budinger, V.; Simon, F. Panel of resonators with variable resonance frequency for noise control. *Appl. Acoust.* **2012**, *73*, 781–790. [CrossRef]
31. Doutres, O.; Atalla, N.; Dong, K. Effect of the microstructure closed pore content on the acoustic behavior of polyurethane foams. *J. Appl. Phys.* **2011**, *110*, 064901. [CrossRef]
32. Zhai, W.; Yu, X.; Song, X.; Ang, L.Y.L.; Cui, F.; Lee, H.P.; Li, T. Microstructure-based experimental and numerical investigations on the sound absorption property of open-cell metallic foams manufactured by a template replication technique. *Mater. Des.* **2018**, *137*, 108–116. [CrossRef]
33. Baich, L.; Monogharan, G.; Marie, H. Study of infill print design on production cost-time of 3D printed ABS parts. *Int. J. Rapid Manuf.* **2015**, *5*, 308–319. [CrossRef]
34. Derossi, A.; Caporizzi, R.; Azzollini, D.; Severini, C. Application of 3D printing for customized food. A case on the development of a fruit-based snack for children. *J. Food Eng.* **2018**, *220*, 65–75. [CrossRef]
35. Everest, F.A. Absorption of sound. In *Master Handbook of Acoustics*, 4th ed.; McGraw-Hill: New York, NY, USA, 2001; pp. 179–233.
36. Lim, Z.Y.; Putra, A.; Nor, M.J.M.; Yaakob, M.Y. Sound absorption performance of natural kenaf fibres. *Appl. Acoust.* **2018**, *130*, 107–114. [CrossRef]
37. Putra, A.; Or, K.H.; Selamat, M.Z.; Nor, M.J.M.; Hassan, M.H.; Prasetiyo, I. Sound absorption of extracted pineapple-leaf fibres. *Appl. Acoust.* **2018**, *136*, 9–15. [CrossRef]

© 2020 by the authors. Licensee MDPI, Basel, Switzerland. This article is an open access article distributed under the terms and conditions of the Creative Commons Attribution (CC BY) license (http://creativecommons.org/licenses/by/4.0/).

Article

The Piezoresistive Highly Elastic Sensor Based on Carbon Nanotubes for the Detection of Breath

Romana Daňová *[⬤], Robert Olejnik, Petr Slobodian and Jiri Matyas

Centre of Polymer Systems, University Institute, Tomas Bata University in Zlin, 760 01 Zlin, Czech Republic; olejnik@utb.cz (R.O.); slobodian@utb.cz (P.S.); matyas@utb.cz (J.M.)
* Correspondence: danova@utb.cz; Tel.: +420-576-038-120

Received: 4 February 2020; Accepted: 18 March 2020; Published: 23 March 2020

Abstract: Wearable electronic sensor was prepared on a light and flexible substrate. The breathing sensor has a broad assumption and great potential for portable devices in wearable technology. In the present work, the application of a flexible thermoplastic polyurethane/multiwalled carbon nanotubes (TPU/MWCNTs) strain sensor was demonstrated. This composite was prepared by a novel technique using a thermoplastic filtering membrane based on electrospinning technology. Aqueous dispersion of MWCNTs was filtered through membrane, dried and then welded directly on a T-shirt and encapsulated by a thin silicone layer. The sensing layer was also equipped by electrodes. A polymer composite sensor is capable of detecting a deformation by changing its electrical resistance. A T-shirt was capable of analyzing a type, frequency and intensity of human breathing. The sensitivity to the applied strain of the sensor was improved by the oxidation of MWCNTs by potassium permanganate ($KMnO_4$) and also by subsequent application of the prestrain.

Keywords: elastic sensor; carbon nanotubes; wearable electronics; monitoring of breathing; strain sensor; polymer composite; CNTs

1. Introduction

Significant attention is devoted to wearable and flexible electronic technologies in the past decade. It is very important not to confuse the term "wearable technology" with "wearable device" or "smart device". Wearable technology comes from describing the integration of electronics and computers into clothing or accessories that are comfortable on the body [1]. The research in an electronic textile has exposed to numeracy smart clothing applications for wearable electronic and healthcare monitoring. The base of success is an electrically conductive layer, which is flexible, stretchable, and lightweight [2].

Human breath is one of the basic natural bodily processes. Breathing can generally be divided into two basic types: diaphragm and thoracic. In the first case, the chest does not move, only the abdominal wall rises and falls. In the second type, in particular, the rib cage opens and contracts. [3].

The general trend of a healthy lifestyle is extended to areas that have the need to detect, record and analyze data from various human activities. There are already many advanced technologies that are also wearable but do not necessarily have integrated electronic or computing components. On the contrary, these belong to the range of smart textiles and clothing. Smart textiles are soft material with flexibility and drapability that are able to sense external conditions or stimulus, to respond and adapt to the behavior to them in an intelligent way. The stimulus that is detected by the smart textiles may have thermal, mechanical, chemical, electrical, magnetic and also optical properties. Similar formulations include intelligent, interactive, sensitive and adaptive. Smart textiles describe a novel category of textiles, which have the capability to detect and react to an external stimulus. In the current situation they can provide required information and help to master everyday life more effectively. They present a challenge in several fields of application such as health and sport. Among the first areas in which they appeared is military and health care [4,5]. In contrast, ultra-strong fire-resistant or breathing fabrics are not considered smart functionalized materials. In terms of the degree of intelligence, it is possible to classify intelligent textiles according to whether they can perform one or more of the following functions:

I. Passive smart textiles—ability to sense environmental conditions or an external element. The first generation of smart textiles that integrates sensors as conductive materials or optical fiber.
II. Active smart textiles—ability to respond after capture. The textiles consist of sensors and actuators as membranes, hydrogels and chromatic materials that provide the ability to sense and actuate or move a part of their environment.
III. Very smart textiles—ability to sense, react and adapt based on the learned experience from what it sensed and reacted to previously as thermo regulating clothing, space suits and health monitoring apparel for example [4,5].

Moreover, it is important to distinguish 'intelligent' clothing in terms of the degree of integration that relates to the extent to which the part that performs the intelligent function is embedded into the textile. At the lowest degree of integration, the smart material is attached to the surface of textile. In the second generation of intelligence, the material is integrated directly into the textile structure, for example, weaving or knitting conductive yarns to form a textile pressure sensor. The third generation of intelligence, which is the highest degree of integration is manifested innately as part of a yarn or fiber. It is thus able to be integrated much more discreetly without compromising user comfort. [1].

The term of electrically conductive fabrics are fabrics that can conduct electricity. They are designed for a wide range of textile fibber products woven from metal strands with very different specific electrical conductivity. This group includes conductive fibers, yarns, fabrics and finished products thereof [4].

One of the interesting and important things to be monitored is certainly the detection and monitoring of human breath. The frequency and intensity of respiration and the type of respiration are detected as a vertical distribution of breath parameters on the human body [6].

In addition, these new sensor technology materials have unique features such as high sensitivity, an ability to detect large-scale deformations and are flexible, lightweight and easy to manufacture. Moreover, in many cases they can be made multifunctional and have other utility properties. Sensitivity can also be successfully controlled. Their multifunctionality can be represented by the following properties: they can have thermoelectric properties, serve as a passive antenna element or a resistance heating element. Last but not least, they can be used to monitor the sensor fabrication process [7–9].

The first network of carbon nanotubes was made by Walters et al. in an attempt to vaporize graphite using a high energy laser, where the authors dispersed nanotubes into a liquid suspension and then filtered through fine filtration mesh. As a result, pure nanotubes stuck together to form a thin, free-standing, entangled structure that was later named Bucky Paper [9,10].

The aim of this article was a complex study of the electrical conductivity of multiwall carbon nanotube network both in the course of monotonic strain growth and also when loading/unloading cycles were imposed [8]. The multiwalled carbon nanotubes (MWCNTs) network was fixed on a polyurethane dog bond shape test specimen for the creep test. The current proposed solution of polymer composite highly elastic and elongation sensors brought a new generation of these deformation-electrically sensitive converters based on materials from the area of nanocomposite materials. These can be deformed in a wide range of deformations (chest deformation during breathing is in the range of units of percent to about 14%), then with high sensitivity, which can be significantly increased by chemical functionalization [11–15].

In the field of application, cooperation is underway, which leads to the prototype's creation of a breath monitoring system including its own T-shirt integrated sensors, electronics with data transmission and subsequent visualization in a mobile application. In the future, it is possible to consider using this system for sports and fitness clothing, when the sensor will record biometric data, analyze this data for users, which can lead to learning and performing these activities better, it can lead to improved performance of athletes, and the daily activities of the individual. In the present paper the fabrication of flexible and stretchable (carbon nanotubes: CNTs) strain sensor was demonstrated, which in general is applied for human breath monitoring [7].

The article was focused on both the areas, wearable electronics and preparation of highly elastic piezoresistive sensors for the detection of tensile deformation based on an elastic polymer and carbon filler such as carbon nanotubes (CNTs). A highly elastic sensor for a T-shirt was overlaying by the active layer to avoid contact with surroundings and development of the collecting electrode system [9]. The commonly used metal wire strain gauge, which is normally produced in the form of a wire with precise geometry in the elastic film. The ration between resistivity change and deformation is called the gauge factor (GF). The commercial strain sensor has a GF of about 2–5 and an elastic deformation capability in unit %. The sensitivity factor is given by the ability of the wire and foil to react to the applied deformation [11].

2. Materials and Methods

The following raw materials and applied manufacturing procedures were used to achieve the desired functionality.

The materials were also analyzed by X-ray photoelectron spectroscopy (XPS) on TFA XPS Physical Electronics instrument (PHI-TFA, Physical Electronics Inc., Chanhassen, MN, USA) [16,17] at the base pressure in the chamber of about 6×10^{-8} Pa. The samples were excited with X-rays over a 400 µm spot area with a monochromatic Al $K_{\alpha 1,2}$ radiations at 1486.6 eV. Photoelectrons were detected with a hemispherical analyzer positioned at an angle of 45° with respect to the normal to the sample surface. Survey-scan spectra were made at a pass energy of 187.85 eV, the energy step was 0.4 eV. Individual high-resolution spectra for C 1s were taken at a pass energy of 23.5 and 0.1 eV energy step. The concentration of elements was determined from survey spectra by MultiPak v7.3.1 software from Physical Electronics (Physical Electronics Inc., Chanhassen, MN, USA).

Purified multiwall carbon nanotubes (MWCNTs) were received from Sun Nanotech Co. (China). CNTs were synthesized by chemical vapor deposition method of acetylene (CVD). Acetylene was used as a precursor. MWCNTs has the following properties: a diameter of 10–30 nm, length of 1–10 µm, electrical resistivity of 0.12 Ωcm and purity more than 90%. By transmission electron microscopy (TEM, JEOL Ltd., Tokyo, Japan) the diameter of individual nanotubes was found to be between 10 and 60 nm and their lengths were from tens of micrometers up to 3 µm [14]. The maximum aspect ratio of the nanotubes was about 300 [13].

Non-woven polyurethane (PU) porous membranes for MWCNTs dispersion filtration were prepared by electrostatic spinning from a PU solution. Electrostatic spinning from polyurethane dimethyl form amide/methyl isobutyl ketone (DMF/MIBK, 1:3) solution was performed in cooperation with the SPUR a.s. company of the Czech Republic in Zlin. The conditions of electrospinning were as follows: a PU concentration of 16 wt %. Electrical conductivity of the solution was adjusted to 20 µS/cm using sodium chloride, an electric voltage of 75 kV (Matsusada DC power supply, Matsusada, Shiga-ken, Japan) was applied. Temperature of 20–25 °C, and a relative humidity of 25%–35% was controlled during the process (for a detailed schematic of experimental part, Kimmer et al. [12]). Thermal properties were measured by differential scanning calorimetry (DSC, Mettler Toledo, Columbus, OH, USA) on a Mettler Toledo DSC1 STAR System (Mettler Toledo, Columbus, OH, USA). The measurements were performed under nitrogen atmosphere. The first heating was set to erase the thermal history from 25 to 180 °C (heating rate 10 °C·min^{-1}) with a 5 min isotherm, then cooling from 180 to −60 °C (cooling rate 10 °C·min^{-1}), the isotherm at −60 °C for 5 min. The second heating cycle was from −60 to 250 °C (10 °C·min^{-1}) and was intended for analysis of the test material.

A thermoplastic polyurethane (TPU) Desmopan DP 385S (Bayer MaterialScience, Leverkusen, Berlin, Germany). The ultimate strength of TPU of 48.9 MPa, with the strain at break 442.2% and density of 1.20 g/cm^3 were specified by the supplier. Used TPU is high elastics elastomeric polyurethane with an ultimate strength up to 400%. It serves as an elastic base for the MWCNT sensory layer. It is thus an integral part of the sensor, which consists of these two functional layers. By this process a high flexible and elastic sensor is prepared to be capable to measure deformations of a large extent. It is also important that the polymer is thermoplastic. The filter membrane then melts in the preparation process and becomes an adhesive layer between the two components. In conclusion, other polymeric elastomeric matrices can also be used.

$KMnO_4$ oxidation: Oxidized MWCNTs were prepared in a glass reactor with a reflux condenser filled with 250 mL of 0.5 M H_2SO_4, into which 5 g of potassium permanganate ($KMnO_4$) as an oxidizing agent and 2 g of MWCNTs were added. The dispersion was sonicated at 85 °C for 15 h using thermostatic ultrasonic bath (Bandelin Electronic DT 103H, Berlin, Germany). The dispersion was filtered, and then MWCNTs were washed with concentrated HCl to remove MnO_2 and then washed with water until the system attained a pH of 7.

Aqueous dispersion of MWCNTs was prepared by sonication in an apparatus UP 400S from Dr. Hielscher GmbH (ultrasonic horn S7, amplitude 88 μm, power 300 W and frequency 24 kHz, Stuttgard, Germany) for 15 min at room temperature. The nanotube concentration in the suspension was 0.3 wt %. Dispersion also contained surfactants, namely, sodium dodecyl sulphate and 1-pentanol with a concentration of 0.1 and 0.14 M, respectively. Moreover, NaOH aqueous solution was added to adjust the pH to 10. For making an entangled MWCNTs network, a porous polyurethane membrane and a vacuum filtration method was used. About 30 mL of homogenized dispersion was filtered through a funnel of diameter 90 mm. The prepared MWCNTs network was washed several times with deionized 60 °C hot water, afterwards by methanol in situ and dried between two filtration papers for 24 h.

The networks of MWCNTs (pure) and oxidized MWCNTs ($KMnO_4$) were selected for the creep test. The composite stripes 55 mm × 10 mm (L × W) were welded onto the polyurethane bodies and gradually loaded from 0.167 to 1.066 MPa.

A composite stripe of 55 mm × 10 mm with a thickness of 0.08 mm and weight was approximately 2.57 mg was directly welded onto a commercial sport T-shirt to monitor human breathing. The electrical resistance change was measured along the specimen length by the two-point technique using Multiplex datalogger 3498A. The electrodes for a two-point electrical resistance measurement were prepared from very thin Cu wires emailed to sensors with Ag conductive lacquer. The time for hardening of the Ag conductive lacquer after application on the sensor was 2 h. This mentioned technique does not affect the elasticity of the used T-shirt. The electrical resistance measurements as dependence on time were performed during breath and exhale cycles of breathing. The commercial sport T-shirt was chosen from a Czech sports brand (Moira CZ a.s., Strakonice, Czech Republic), Moira, with a composition of 97% polypropylene from the brand Moira and 3% elastane. The T-shirt has to be tight in the entire upper half of the body therefore the size XS–S was chosen. The polyester interlayer strip is necessary for ironing the sensor o the T-shirt.

Two components silicone rubber GMS 2628 from Dawex Chemical s.r.o. (Zlin, Czech Republic) having a hardness score of 26–28 A and with high elasticity was also used to cover the final sensor.

The change in electrical resistance of the MWCNTs network in breath/exhale cycles was measured lengthwise by the two-point method using Multiplex datalogger 34980A (Keysight technologies, Santa Rosa, CA, USA) connected to a PC with a sampling frequency of 10 Hz.

Micrographs using scanning electron microscopy (FEI Nova NanoSEM 450, (FEI company, Hillsboro, OR, USA) were used to observed MWCNTs (pure) and MWCNTs ($KMnO_4$) samples and also a semi product of filtration and final TPU/MWCNTs composites. The samples were fixed with adhesive tape on the aluminum stub.

3. Results

Two types of MWCNTs were used for this article and tested as pure and oxidized, respectively. Based on the results of the creep test, a sample of oxidized MWCNTs ($KMnO_4$) was selected and with values of MWCNTs (pure) was compared. The network of MWCNT (pure) was composed of long entangled tubes (Figure 1A) while the network of MWCNTs (KMnO$_4$) was shortened due to previous oxidation (Figure 1B). Therefore, the number of contacts was maximized, resulting in high sensor sensitivity to strain changes.

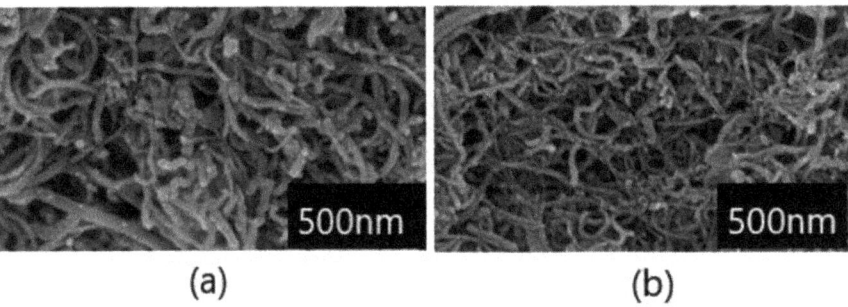

Figure 1. The scanning electron microscope (SEM) micrograph of carbon nanotubes network. The upper surface of entangled carbon nanotube network prepared by the filtering method from multiwalled carbon nanotubes (MWCNTs; pure) and oxidized MWCNTs (KMnO$_4$). (**a**) SEM micrograph of the MWCNTs (pure). (**b**) SEM micrograph of MWCNTs (KMnO$_4$).

The main binding energy peak (284.5 eV) in the XPS spectra of MWCNT was assigned to the C1s-sp^2, while the other ones were assigned to C–O (286.15 eV), C=O (287.1 eV), O–C=O (288.8-289 eV) and C1s-π-π* (291.1–291.5 eV). According to our XPS results of MWCNT the total oxygen content was determined to be 18.8 at % for pure MWCNT and 21.4 for the oxidized one. The sp^3/sp^2 carbon ratios were 2.50 and 1.69 for pure MWCNT and KMnO$_4$ oxidized ones (Figure 2), respectively [7].

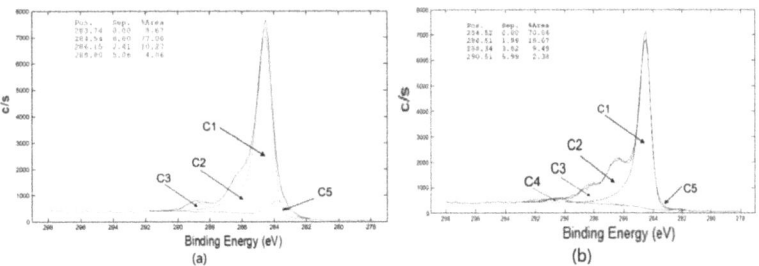

Figure 2. X-ray photoelectron spectroscopy (XPS) on CNT/PU and CNT/PU KMnO$_4$. (**a**) Curve fitting of the C 1s carbon spectrum of the MWCNTs (pure). (**b**) Curve fitting of the C 1s carbon spectrum of MWCNTs (KMnO$_4$).

The resistivities of the network structures were measured to be 0.084 ± 0.003 Ωcm for the network fabricated from pure MWCNTs and 0.156 ± 0.003 Ωcm for the network made of KMnO$_4$ oxidized tubes [18].

Porosities of two principal networks were calculated to be 0.67 and 0.56, for pure and oxidized network respectively [18].

The principle of the sensing is in the reduction of intertube contacts and the creating of cracks. When the composites are elongated, the number of contacts decreases leading to macroscopic resistance of the sensory layer. Oppositely, when the composite relaxed, the number of contacts increases, which leads to a reversible decrease of the composite resistance. As was demonstrated elsewhere [19], oxidation leads to the formation of oxygenated functional groups detached to the CNT surface causing an increase of the contact resistance in CNT junctions. It finally increases sensor sensitivity to the strain when a sharper resistance change is observed when deformed.

Morphology of the prepared composite acting as a strain sensor can be demonstrated by Figure 3. In principle it is the composite created by the three layers as a commercial T-shirt, a CNT/TPU interface layer and two-component silicone rubber [19].

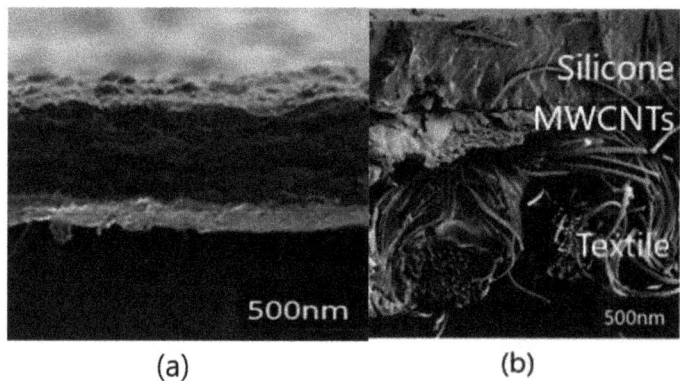

Figure 3. (**a**) The cross-section of a MWCNT film (above) and supporting TPU filtering membrane (underneath). (**b**) MWCNT composite cross-section after the melt welding.

Figure 4 characterizes the DSC analysis of the non-woven thermoplastic PUR membrane. A glass transition (T_g) of the elastomeric component was seen, $T_g = -8.08$ °C. It is also possible to follow two successive areas melting of the material (endothermic process) $T_{m1} = 137.56$ °C and $T_{m2} = 168.38$ °C.

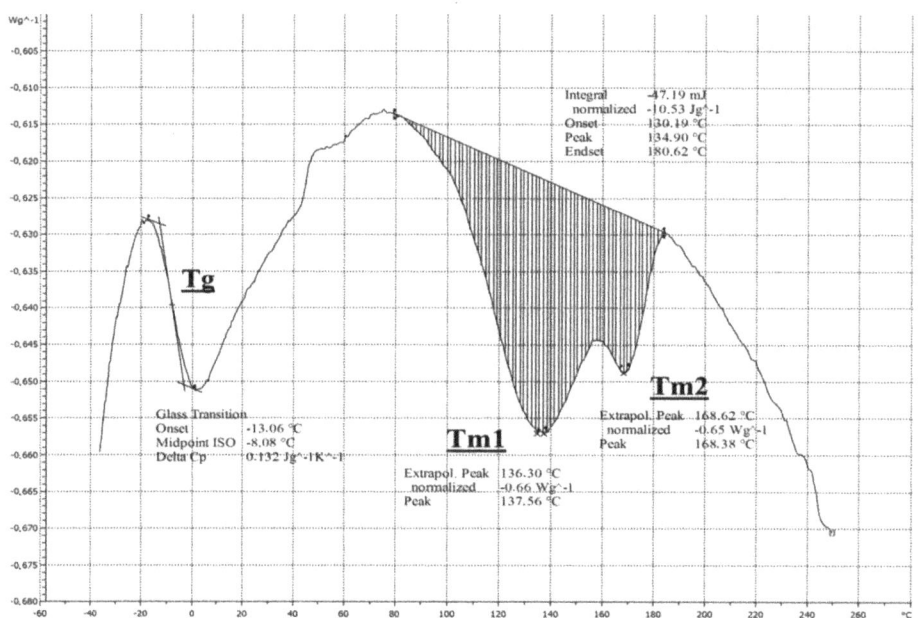

Figure 4. Differential scanning calorimetry (DSC) of a non-woven polyurethane membrane.

In the first case, the sensor of MWCNTs (pure) and oxidized MWCNTs (KMnO$_4$) were deformed by a tensile stress from 0.167 to 1.066 MPa in six extension/relaxation cycles and were compared. The results are shown in Figure 5 as a percentage change in the relative resistance,

$$\Delta R/R_0 = (R - R_0)/R_0 \tag{1}$$

where R_0 is the electrical resistance of the measured sample before the first elongation and R is the resistance during elongation, strain is the relative change in specimen length.

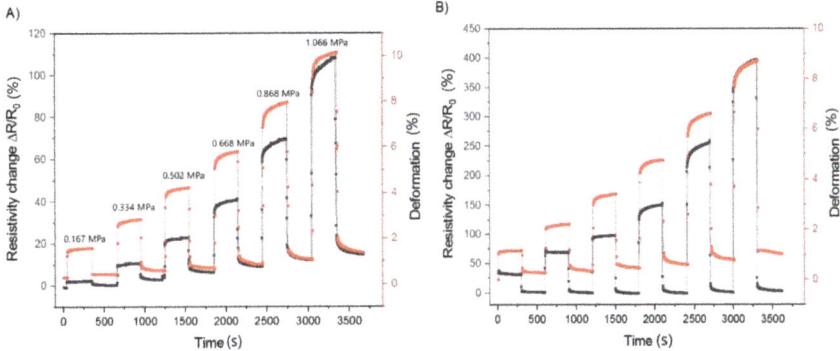

Figure 5. The comparison of the relative resistance change vs. the deformation for the MWCNTs (pure) (**A**) and MWCNTs (KMnO$_4$) (**B**) sensor in six extension/relaxation cycles induced by the tensile stress (from 0.167 to 1.066 MPa). The tensile stresses are the same for both measurements. The deformation is denoted by red circles, the relative resistance changes by black circles.

The elongation periods when the stress was in range from 0.167 to 1.066 MPa, and relaxation when the load was removed was 300 s (Figure 5). The maximum resistance changes of MWCNTs (pure) were 110% and the deformation was 10%. The maximum resistance changes of MWCNTs (KMnO$_4$) were 400% and the deformation was 9%, which implied a fourfold increase in sensitivity.

Further, the sensitivity to applied strain, deformation ε, is defined by the gauge factor (*GF*)

$$GF = (\Delta R/R_0)/\varepsilon \tag{2}$$

Strain represents a relative change in specimen length, as the ratio between the change of specimen length, ΔL, relative to initial length, L_0, $\varepsilon = \Delta L/L_0$. Figure 6 shows the dependence of the gauge factor *GF* on deformation ε.

The issue was the creep test, which means loading by constant extension stress followed by measurement of the specimen deformation and resistance in time. After 5 min of loading, it was followed by relaxation in another 5 min in the off loaded state when the reversible creep occurred in time. The specimen was then loaded with an appropriate higher stress. Finally, there were six consecutive loading/unloading cycles. Elongation of the specimen led to an increase in composite macroscopic resistance and resistance reversible decrease when the specimen was in an off load state and the deformation relaxed. In general, the sensor is then sensitive to deformation and this response is reversible. The measured value of *GF* reached for the example value of 11 for MWCNTs (pure) and of 46 for MWCNTs (KMnO$_4$) at an applied deformation 10% (Figure 6). The principle of sensing is in the formation of a microcrack, which decreases the quantity of intertube contacts, which leads to the macroscopic resistance increase. When the composite relaxes, the microcrack close leading to a reversible resistance decrease. Oxidation incorporates more functional oxygenated groups chemically detached to the CNTs surface leading to a sharper increase of resistance during deformation and a more sensitive sensor to strain. Prestrain leads to the formation of a crack, which is easier opened in the next cycles again leading to higher sensor sensitivity [19].

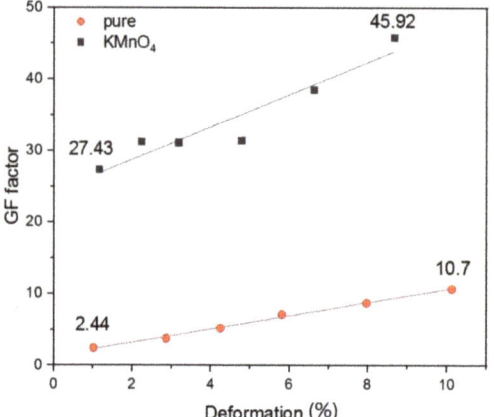

Figure 6. Dependence of gauge factor (GF) with increasing deformation for MWCNTs (pure) and oxidized MWCNTs (KMnO$_4$) sensors.

The next section describes the application of a strain gauge to monitor human breath in two volunteers. Subsequently, different types of breathing were observed, namely deep and normal breathing. As shown in Figure 7, this integrated MWCNTs (KMnO$_4$) sensor was fixed on a T-shirt. Two volunteers A and B were selected to test normal and deep breathing. The maximum deformation of the chest the volunteer (A) was 5% with deep breathing and the change in electrical resistance was 30%. In the volunteer (B), a maximum chest circumference deformation was 7% for deep breathing and a change in electrical resistance was 65% (Figure 8).

The response of the sensor to electrical resistance was therefore very sensitive to deformation. The sensor was reversible and was able to detect breathing in real time.

The stabilizing effect on the resistance extension cycles and the residual normalized resistance change was constant after approximately five cycles. That is, during the first deformations, the network of nanotubes acquired a structure that remained more or less the same regardless of the number of deformation cycles. This mechanical stabilization is advantageous for using the sensor as an elongation sensing element.

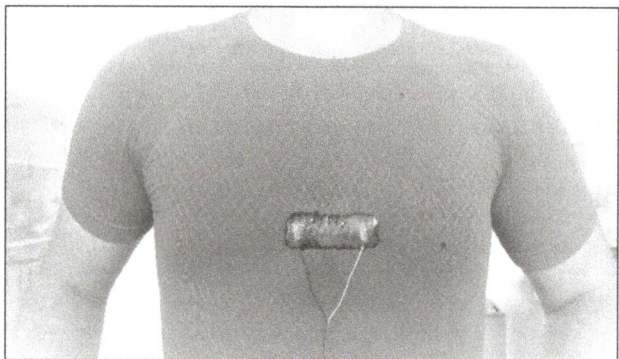

Figure 7. Sensor based of MWCNT (KMnO$_4$) fixed on a sports T-shirt for practical application of a human breath monitoring.

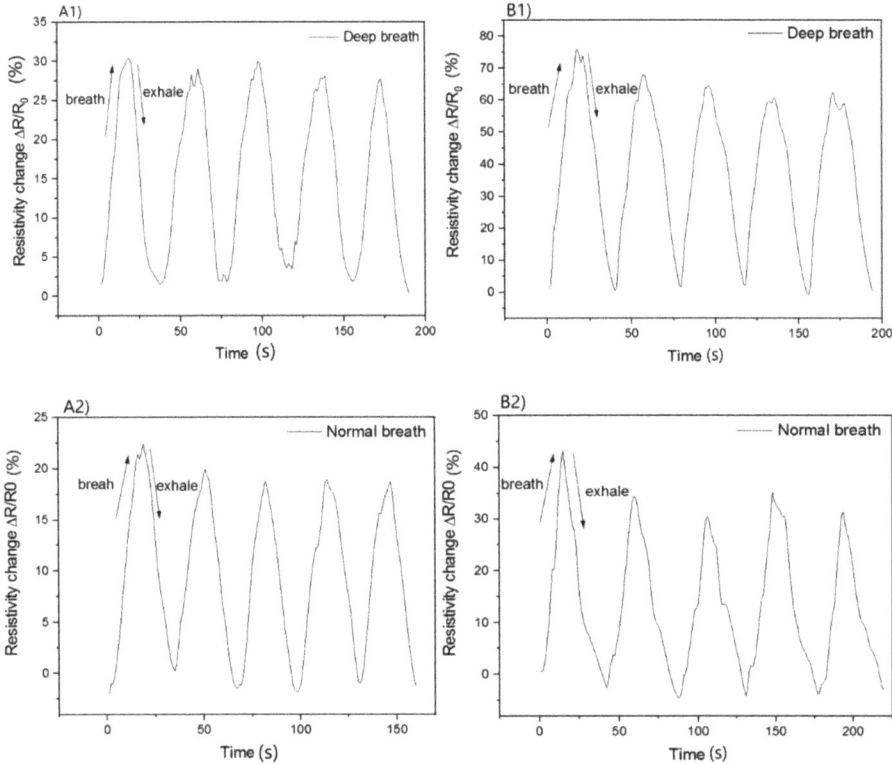

Figure 8. Response to a change in relative resistance, $\Delta R/R_0$, of carbon nanotube sensors (MWCNTs/KMnO$_4$) integrated into the T-shirt to monitor human breath. The comparison of deep breath and normal breath for both volunteers (**A1,A2,B1,B2**).

4. Conclusions

A highly elastic deformable and piezoresistive sensor composed of a network of an electrically conductive entangled carbon nanotubes network built into the elastic commercial T-shirt and encapsulated by elastic silicone was introduced. This material was sensitive enough to be used to monitor human breath. It did not influence any human activity. Respiratory characteristics in two volunteers were compared. The monitoring of breath worked on the principle for the sensors electrical resistance change. Possible application of the sensor in an elastic T-shirt can not only be for athletes, but also for newborns with serious sleep disorders such as sleep apnoea, which can be used to prevent premature death.

Author Contributions: Resources, R.D.; Supervision, P.S., R.O. and J.M. All authors have read and agreed to the published version of the manuscript.

Funding: This research was founded by the Ministry of Education, Youth and Sports of the Czech Republic-Program NPU I (LO1504) and with the support of the Operational Program Research and Development for Innovations co-funded by the European Regional Development Fund (ERDF). The internal grant of TBU in Zlin No. IGA/CPS/2018/005 and IGA/CPS/2019/010 funded from the resources of the Specific University Research. The national budget of the Czech Republic, within the framework of the project CPS-strengthening research capacity (reg. number CZ.1.05/2.1.00/19.0409).

Acknowledgments: This project was supported by the internal grant of TBU in Zlin No. IGA/CPS/2018/005 and IGA/CPS/2019/010 funded from the resources of the Specific University Research. This work was also supported by the Ministry of Education, Youth and Sports of the Czech Republic-Program NPU I (LO1504) and with the support of the Operational Program Research and Development for Innovations co-funded by the European

Regional Development Fund (ERDF) and the national budget of the Czech Republic, within the framework of the project CPS-strengthening research capacity (reg. number CZ.1.05/2.1.00/19.0409).

Conflicts of Interest: The authors declare no conflict of interest.

References

1. Wu, J.X.; Li, L. An Introduction to Wearable Technology and Smart Textiles and Apparel: Terminology, Statistics, Evolution, and Challenges. 2019. Available online: https://www.intechopen.com/books/smart-and-functional-soft-materials/an-introduction-to-wearable-technology-and-smart-textiles-and-apparel-terminology-statistics-evoluti (accessed on 4 February 2020).
2. Yu, L.; Yeo, J.C.; Soon, R.H.; Yeo, T.; Lee, H.H.; Lim, C.T. Highly Stretchable, Weavable, and Washable Piezoresistive Microfiber Sensors. *ACS Appl. Mater. Interfaces* **2018**, *10*, 12773–12780. [CrossRef] [PubMed]
3. Available online: https://www.walmark.cz/magazin/dychani-je-zivot (accessed on 28 February 2018).
4. Grancarić, A.M.; Jerković, I.; Koncar, V.; Cochrane, C.; Kelly, F.M.; Soulat, D.; Legrand, X. Conductive polymers for smart textile applications. *J. Permis. Sagepub* **2017**, *48*, 612–642. [CrossRef]
5. Zhang, X.; Tao, X. Smart textiles: Passive smart. *Text Asia* **2001**, *32*, 45–49.
6. Oliveri, A.; Maselli, M.; Lodi, M.; Storace, M.; Cianchetti, M. Model-based compensation of rate-dependent hysteresis in a piezoresistive strain sensor. *IEEE Trans. Ind. Electron.* **2018**, *66*, 8205–8213. [CrossRef]
7. Benlikaya, R.; Slobodian, P.; Riha, P. Enhanced strain-dependent electrical resistance of polyurethane composites with embedded oxidized multiwalled carbon nanotube networks. *J. Nanomater.* **2013**, *2013*, 327597. [CrossRef]
8. Tadakaluru, S.; Thongsuwan, W.; Singjai, P. Stretchable and Flexible High-Strain Sensors Made Using Carbon Nanotubes and Graphite Films on Natural Rubber. *Sensors* **2014**, *14*, 868–876. [CrossRef] [PubMed]
9. Slobodian, P.; Riha, P.; Saha, P. A highly-deformable composite composed of an entangled network of electrically-conductive carbon-nanotubes embedded in elastic polyurethane. *Carbon* **2012**, *50*, 3446–3453. [CrossRef]
10. Walters, D.A.; Casavant, M.J.; Quin, X.C.; Huffman, C.B.; Boul, P.J.; Ericson, L.M.; Haroz, E.H.; O'Connel, M.J.; Smith, K.; Colbert, D.T.; et al. In-plane-aligned membranes of carbon nanotubes. *Chem. Phys. Lett.* **2001**, *338*, 14–20. [CrossRef]
11. Allaoui, A.; Hoa, S.V.; Evesque, P.; Bai, J. Electronic transport in carbon nanotube tangles under compression: The role of contact resistance. *Scripta Materialia* **2009**, *6*, 628–631. [CrossRef]
12. Kimmer, D.; Slobodian, P.; Petras, D.; Zatloukal, M.; Olejnik, R.; Saha, P. Polyurethane/MWCNT nanowebs prepared by electrospinning process. *J. Appl. Polym. Sci.* **2009**, *111*, 2711–2714. [CrossRef]
13. Slobodian, P.; Daňová, R.; Olejník, R.; Matyáš, J.; Münster, L. Multifunctional flexible and stretchable polyurethane/carbon nanotube strain sensor for human breath monitoring. *Polym. Adv. Technol.* **2019**, *30*, 1891–1898. [CrossRef]
14. Slobodian, P.; Riha, P.; Lengalova, A.; Saha, P. Compressive stress-electrical conductivity characteristics of multiwall carbon nanotube networks. *J. Mater. Sci.* **2011**, *46*, 3186–3190. [CrossRef]
15. Yamada, T.; Hayamizu, Y.; Yamamoto, Y.; Yomogida, Y.; Izadi-Najafabadi, A.; Futaba, D.; Hata, K. A stretchable carbon nanotube strain sensor for human-motion detection. *Nat. Nanotechnol.* **2011**, *6*, 296–301. [CrossRef] [PubMed]
16. Cvelbar, U. Interaction of non-equilibrium oxygen plasma with sintered graphite. *Appl. Surf. Sci.* **2013**, *269*, 33–36. [CrossRef]
17. Cvelbar, U.; Markoli, B.; Poberaj, I.; Zalar, A.; Kosec, L.; Spaić, S. Formation of functional groups on graphite during oxygen plasma treatment. *Appl. Surf. Sci.* **2006**, *253*, 1861–1865. [CrossRef]

18. Slobodian, P.; Riha, P.; Lengálová, A.; Svoboda, P.; Sáha, P. Multi-wall carbon nanotube networks as potential resistive gas sensors for organic vapor detection. *Carbon* **2011**, *49*, 2499–2507. [CrossRef]
19. Slobodian, P.; Riha, P.; Olejnik, R.; Cvelbar, U.; Sáha, P. Enhancing effect of KMnO4 oxidation of carbon nanotubes network embedded in elastic polyurethane on overall electro-mechanical properties of composite. *Compos. Sci. Technol.* **2013**, *81*, 54–60. [CrossRef]

© 2020 by the authors. Licensee MDPI, Basel, Switzerland. This article is an open access article distributed under the terms and conditions of the Creative Commons Attribution (CC BY) license (http://creativecommons.org/licenses/by/4.0/).

Article

Ethylene-Octene-Copolymer with Embedded Carbon and Organic Conductive Nanostructures for Thermoelectric Applications

Petr Slobodian [1,2,*], **Pavel Riha** [3,*], **Robert Olejnik** [1,4] **and Michal Sedlacik** [1]

1. Centre of Polymer Systems, University Institute, Tomas Bata University, Tr. T. Bati 5678, 760 01 Zlin, Czech Republic; olejnik@utb.cz (R.O.); msedlacik@utb.cz (M.S.)
2. Faculty of Technology, Polymer Centre, Tomas Bata University, T.G.M. 275, 760 01 Zlin, Czech Republic
3. The Czech Academy of Sciences, Institute of Hydrodynamics, Pod Patankou 5, 166 12 Prague 6, Czech Republic
4. Department of Production Engineering, Faculty of Technology, Tomas Bata University in Zlin, T. G. Masaryk nam. 275, 762 72 Zlin, Czech Republic
* Correspondence: slobodian@utb.cz (P.S.); riha@ih.cas.cz (P.R.)

Received: 7 May 2020; Accepted: 4 June 2020; Published: 9 June 2020

Abstract: Hybrid thermoelectric composites consisting of organic ethylene-octene-copolymer matrices (EOC) and embedded inorganic pristine and functionalized multiwalled carbon nanotubes, carbon nanofibers or organic polyaniline and polypyrrole particles were used to form conductive nanostructures with thermoelectric properties, which at the same time had sufficient strength, elasticity, and stability. Oxygen doping of carbon nanotubes increased the concentration of carboxyl and C–O functional groups on the nanotube surfaces and enhanced the thermoelectric power of the respective composites by up to 150%. A thermocouple assembled from EOC composites generated electric current by heat supplied with a mere short touch of the finger. A practical application of this thermocouple was provided by a self-powered vapor sensor, for operation of which an electric current in the range of microvolts sufficed, and was readily induced by (waste) heat. The heat-induced energy ensured the functioning of this novel sensor device, which converted chemical signals elicited by the presence of heptane vapors to the electrical domain through the resistance changes of the comprising EOC composites.

Keywords: ethylene-octene-copolymer; carbon nanotubes; carbon fibers; polyaniline; polypyrrole; thermoelectric composites

1. Introduction

Thermoelectric conductive polymer composites convert thermal energy into electricity when there is a difference in temperatures between the hot and cold junctions of such two dissimilar conductive or semiconductive composites. The conversion is based on a phenomenon called the Seebeck effect. The heat supplied at the hot junction (hot side) of the thermoelectrics causes a flow of electric current to the cold side that can be harnessed as useful voltage.

The classical thermoelectrics are made from inorganic materials such as metals (Al, Cu, Ni), metallic alloys (chromel, alumel) and semiconductors (PbT, Bi_2Te_3), which are thermoelectrically efficient, but at the same time expensive, heavy, and some of them materially in short supply. Consequently, alternative organic thermoelectrics are being developed. These organic thermoelectrics do not produce as much energy as the metal ones. The thermoelectric figure of merit is typically only in the range 0.001–0.01 at room temperature. However, the mechanical flexibility, processability, light weight, and low manufacturing costs are attractive features for potential applications for electricity microgeneration e.g., in sensors and electronics.

Petsagkourakis et al. [1] thoroughly reviewed principles and advances in the development of those thermoelectric materials. The electronic properties of the conductive (conjugated) polymers are based on their macromolecular structure and morphology, since they affect the electronic conditions, and therefore the charge transport and the thermoelectric properties. Such conductive polymers have been used mostly for thermoelectric generators and temperature sensors. Hybridizing of such polymers with inorganic thermoelectric materials is another way to enhance their thermoelectric performance. Such hybrids have low thermal conductivity, which is advantageous in energy harvesting. Among others, metals, Bi_2Te_3 or carbon nanoparticles have been used as the inorganic portions of such composites.

Polyaniline (PANI) and polypyrrole (PPy) are two of the most studied conductive polymers owing to their easy and low-cost synthesis, good environmental stability, and simple doping/dedoping processes based on acid–base reactions [2,3]. The PANI conductivity apparently depends on its ability to form polarons, cation radicals [4]. The four double bonds constituting the quinonediimine unit in PANI emeraldine salt convert to three double bonds in a benzene ring and two unpaired electrons, which act as charge carriers. The polarons can eventually spread over the polymer chain to produce a polaron lattice. The polymerization condition has an important effect on the final electrical properties (the conductivity and dielectric loss) of conductive polymers. In the case of PANI, the most used synthesis is the polymerization initiated by an addition of an oxidant, generally ammonium persulfate. This synthetic route leads to the conductive emeraldine salt of PANI, which is formed by the head-to-tail coupling of the monomers. Various fillers such as Ag nanoparticles [5], Bi_2Se_3 nanoplates [6], or thermally reduced graphene [7] can be embedded to improve the properties of PANI.

The thermoelectric properties of PPy may be improved by a controlled synthesis of various nanostructures [8]. Nanotubular-type PPy are synthetized through a chemical polymerization route and treated with various dopants such as hydrochloric acid, p-toluenesulfonic acid monohydrate or tetrabutylammonium hexafluorophosphate, which affect its electrical and thermal properties [9]. Also hybrid PPy thermoelectrics have been prepared such as the PPy nanowire/graphene composite [10], the PPy/multiwalled carbon nanotube composite [11], the PPy/Ag nanocomposite film [12], or the PPy/graphene/PANI nanocomposite with a high thermoelectric power factor [13].

Thanks to the content of oxygen functional groups on their surfaces, the multiwalled carbon nanotubes (MWCNTs) and the carbon nanofibers (CNFs) embedded into ethylene-octene-copolymer (EOC) may affect its thermoelectric power [14]. Besides the oxygen groups, thermoelectric properties of carbon nanotubes can be changed by doping with many other electron donors [15].

Waste heat dissipated from homoiothermic human bodies can be readily used as the source of electrical power for polymer thermoelectrics, which can be used as unobtrusive low cost self-powered sensors and integrated devices for biometric monitoring [16]. Similarly, thermoelectric self-powered temperature sensors based on Te nanowire/poly(3-hexyl thiophene) polymer composite are described in [17]. Alternatively, such thermoelectrics can be used also as a wearable energy harvester turning radiated human body heat into a source of electric energy. The current research progress on flexible thermoelectric devices and conducting polymer thermoelectric materials is reviewed in [18].

In this paper, hybrid thermoelectrics consisting of EOC matrices and embedded pristine or functionalized MWCNTs, carbon nanofibers, and organic PANI and PPy particles were used to form conductive nanostructures with thermoelectric properties, which at the same time had sufficient strength, elasticity, and stability. Their thermoelectric power was measured and discussed in the light of the oxygen content of their inorganic fillers. A practical application of the prepared thermoelectric composites for electricity microgeneration for a self-powered temperature signaling sensor and a self-powered vapor sensor was introduced.

2. Materials and Methods

Purified MWCNTs produced by a chemical vapor deposition of acetylene were supplied by Sun Nanotech Co. Ltd., Jiangxi, China. According to the supplier, the nanotube diameter is 10–30 nm, the length 1–10 µm, the purity ~90% and the volume resistivity 0.12 Ωcm. The diameter of individual nanotubes is between 10 and 60 nm (100 measurements) at the average diameter and the standard deviation is 15 ± 6 nm. The nanotube length is from about 0.2 µm up to 3 µm. They consist of 15 to 35 rolled layers of graphene at the interlayer distance of ca 0.35 nm [19]. The pristine nanotubes are denoted further on as MWCNT(Sun)s.

MWCNTs (BAYTUBES C70 P) were supplied by Bayer MaterialScience AG, Leverkusen, Germany. The nanotube purity is >95 wt %, outer mean diameter ~13 nm, inner mean diameter ~4 nm, length > 1 µm and declared bulk density of MWCNT of agglomerates of micrometric size 45–95 kg/m^3. The pristine nanotubes are denoted further on as MWCNT(Bayer)s.

The carbon nanofibers (CNFs) with trade name VGCF® (Vapor Grown Carbon Fibers) were supplied by Showa Denko K.K., Tokyo, Japan. The fiber diameter was 150 nm, length 10 µm and electrical resistivity 0.012 Ωcm.

The oxygenated MWCNTs were prepared in a glass reactor with a reflux condenser filled with 250 mL of 0.5 M H_2SO_4, into which 5 g $KMnO_4$ and 2 g MWCNTs were added. Then the dispersion was sonicated at 85 °C for 15 h using a thermostatic ultrasonic bath (Bandelin electronic DT 103H, Merck spol. s r. o., Prague, Czech Republic). Thereafter, the product was filtered, washed with concentrated HCl to remove MnO_2, thoroughly rinsed with deionized water and dried [19]. Alternatively, the MWCNTs were oxygenated by HNO_3 as follows: 2 g of the MWCNTs were added to 250 mL of HNO_3 (concentrated) and heated at 140 °C for 2 h. After that, the dispersion was cooled and filtered. The sediment was washed by deionized water and dried at 40 °C for 24 h. The corresponding oxygenated MWCNTs are denoted MWCNT(Sun)H_2SO_4 or MWCNT(Sun)HNO_3 and MWCNT(Bayer)H_2SO_4 or MWCNT(Sun)HNO_3.

For preparing PANI emeraldine salt, 0.2 M aniline hydrochloride was mixed with 0.25 M ammonium persulfate (APS) in water, briefly stirred, and left to polymerize for 24 h at room temperature. Then the PANI precipitate was collected on a filter and washed with 0.2 M HCl and acetone. Polyaniline emeraldine particles were in turn dried in air and then under vacuum at 60 °C for 24 h [20].

Pyrrole monomer (Py, purity ≥ 98%, Sigma–Aldrich Inc., St. Louis, MO, USA) was distilled twice under reduced pressure and stored below 4 °C. Polypyrrole particles were prepared via in situ polymerization of the precooled Py in a system containing surfactant CTAB. The precooled initiator APS was added into the system dropwise and the polymerization was allowed to proceed with stirring for 2 h at 0–5 °C. After being washed with water and ethanol, the PPy powder was dried in air and then under vacuum at 60 °C for 24 h. After drying, both types of polymers were gently ground with mortar and pestle.

The reagents APS (purity = 98%) and HCl (concentration ≈ 35%) were purchased from Sigma–Aldrich Inc., St. Louis, MO, USA. The aniline hydrochloride (purity ≥ 99%) was purchased from Fluka, Buchs, Switzerland, cetyltrimethylammonium bromide (CTAB, purity = 98%) from Lach–Ner Ltd., Neratovice, Czech Republic, acetone and ethanol from Penta Ltd., Chrudim, Czech Republic. The reagents were used without further purification.

The carbon allotrope/EOC composites were prepared by an ultrasonication of dispersions of MWCNTs or CNFs in EOC/toluene solution (5% of EOC in toluene). The chosen concentration of the filler in the composites was 30 wt %, which was well above the percolation threshold. The ultrasonication of the dispersion was done in the thermostatic ultrasonic bath (Bandelin electronic DT 103H, Merck spol. s r. o., Prague, Czech Republic) for 3 h at 80 °C. Then the dispersion was poured into acetone to form a precipitate. The final composite sheets were prepared by compression molding at 100 °C [14]. The organic PANI (PPy)/EOC composites were prepared in the same way as the carbon allotrope composites except for filler concentration, which was 70 wt %.

MWCNTs were analyzed by means of transmission electron microscopy (TEM) using the microscope JEOL JEM 2010 (Jeol Ltd., Freising, Germany) at an accelerating voltage of 160 kV. The sample, the MWCNT dispersion in acetone prepared by ultrasonication, was deposited on a 300 mesh copper grid with a carbon film (SPI, Washington, DC, USA) and dried. The structure of CNFs was observed by means of a scanning electron microscope (SEM) Nova NanoSEM 450, FEI, Lincoln, NE, USA, at operating voltage 10 kV. The sample, CNF dispersion in acetone prepared by ultrasonication, was deposited on carbon targets and covered with a thin Au/Pd layer. For the observations, a regime of secondary electrons was used. The same scanning electron microscope was used to observe the morphology of PANI emeraldine salt and of PPy particles.

The X-ray photoelectron spectroscopy (XPS) signals were measured to obtain information on functional groups attached onto the nanotube surfaces. The XPS signals from MWCNT(Sun), MWCNT(Sun)KMnO$_4$ and MWCNT(Sun)HNO$_3$ network surfaces were recorded using the Thermo Scientific K-Alpha System TFA XPS (Physical Electronics Instrument, Chanhassen, MN, USA) equipped with a micro-focused, monochromatic Al Kα X-ray source (1486.6 eV). An X-ray beam of 400 μm size was used at 6 mA × 12 kV [21]. The spectra were acquired in the constant analyzer energy mode with pass energy of 200 eV for the survey. Narrow regions were collected using the snapshot acquisition mode (150 eV pass energy) enabling rapid collection of data (5 s per region). The narrow region data were as post-processed using the Jansson's algorithm to remove the analyzer point spread function, which resulted in an improved resolution of the spectra for the peak deconvolution [22]. The concentration of elements was determined from survey spectra by MultiPak v7.3.1 software (Physical Electronics Inc., Chanhassen, MN, USA).

Fourier-transform infrared (FTIR) analyses of MWCNTs, MWCNT(KMnO$_4$)s and MWCNT(HNO$_3$)s were performed on the FTIR spectrometer Nicolet 6700 (Thermo Scientific, Waltham, MA, USA). The transmission accessory was used for pristine MWCNT(Sun)s, MWCNT(Sun)KMnO$_4$ and MWCNT(Sun)HNO$_3$ samples in powder form prepared by potassium bromide. FTIR analyses of chemical composition of PANI emeraldine and PPy particles were examined by the above mentioned FTIR spectrometer Nicolet 6700 using the attenuated total reflectance technique with a germanium crystal in the range 600–4000 cm^{-1} at 64 scans per spectrum at 2 cm^{-1} resolution.

The Hall coefficient, the resistivity, and the conductivity of the samples as well as the charge mobility and the charge carrier concentration were measured by means of the HCS 1 apparatus (Linseis Messgeräte GmbH, Selb, Germany) equipped with static 0.7 T field permanent magnets for bipolar measurement. The disc form samples had diameter 20 mm and thickness 2.65 mm. The sample current was set to 4 mA. The thermoelectric power measurement was carried out for all the samples using the set-up illustrated in Figure 1. The schematic diagram shows that the circular composite sample (diameter 20 mm, thickness 2 mm) was placed between two copper electrodes. The ends of each of the Cu electrodes were immersed in thermostatic silicone oil baths set at different temperatures. The temperature at the copper/composite interfaces was measured by a Pt100 temperature sensor. The arising thermoelectric current was measured by the Keithley 2000 Digital Multimeter (Tektronix, Inc. Beaverton, OR, USA).

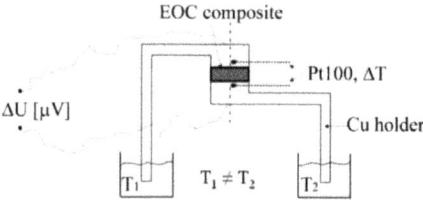

Figure 1. A schematic illustration of the set up for the measurement of electric voltage generated between the hot and cold ends of the Cu/composite/Cu thermoelectric device as a response to a temperature difference.

3. Results

3.1. Characterization of Fillers and Composites

The detailed view of individual pristine MWCNT(Bayer)s, their clusters as well as clusters of the oxidized MWCNT(Bayer)$KMnO_4$ as obtained by the TEM are shown in Figure 2. The wall of the MWCNT(Bayer) consisted of about 15 rolled layers of graphene. The nanotube outer and inner diameter was about 20 nm and 4–10 nm, respectively. There were also defects obstructing nanotube interiors, which were commonly seen in the MWCNT structures. When a cluster of pristine MWCNT(Sun)s was compared with a cluster of wet oxidized nanotubes, a difference in the tube length was visible. The oxidation of MWCNT(Sun)s by $KMnO_4$ caused shortening of the nanotubes, creation of defect sites, and opened ends. A small amount of amorphous carbon after the $KMnO_4$ oxidation process can be also expected [23], although another report showed that oxygenation by $KMnO_4$ in an acidic suspension provides nanotubes free of amorphous carbon [24].

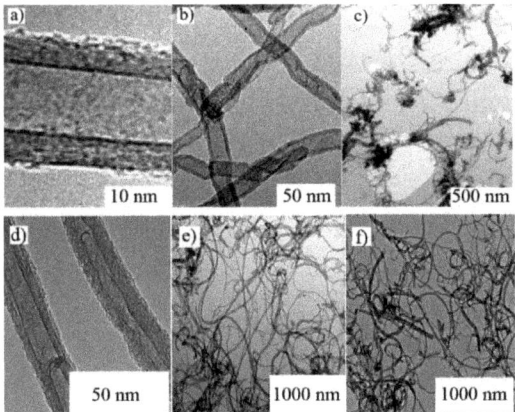

Figure 2. The upper set: Transmission electron microscopy micrographs of MWCNT(Sun)s. (**a**) The structure of an individual pristine nanotube, (**b**) the detail of a nanotube crossing, (**c**) the cluster of pristine nanotubes. The lower set: TEM micrographs of MWCNT(Bayer)s. (**d**) The structure of an individual pristine nanotube, (**e**) the cluster of pristine nanotubes, (**f**) the cluster of oxidized MWCNT(Bayer)$KMnO_4$.

The micrographs of the surfaces of filler layers—PANI and PPy particles as well as MWCNTs and CNFs—are shown in Figure 3. The layers were formed from the respective aqueous dispersion of the fillers on the surface of the interdigitated electrode by the drop method. The particles were globular with a diameter 0.5–1 μm and the PPy particles had narrower size distribution.

The networks of CNFs, MWCNT(Sun)s and MWCNT(Bayer)s resulted from the filtration of their dispersions through non-woven polyurethane membranes. The dispersion, which consisted of CNFs (0.8 mg), or the same amount of nanotubes, dispersed in 530 mL of water with 15.4 g of the surfactant (sodium dodecyl sulphate) and 8.5 mL of the co-surfactant (1-pentanol), was properly sonicated and the pH adjusted to 10 using an aqueous solution of NaOH. The filtrate layer was washed in situ several times with deionized water and finally with methanol.

The cross-section of the composites presented in Figure 4 showed a uniform distribution of the MWCNT(Sun) and CNF filler in the EOC matrix, which together with the filler concentration 30% ensured the electrical and thermal conductivity of the composites.

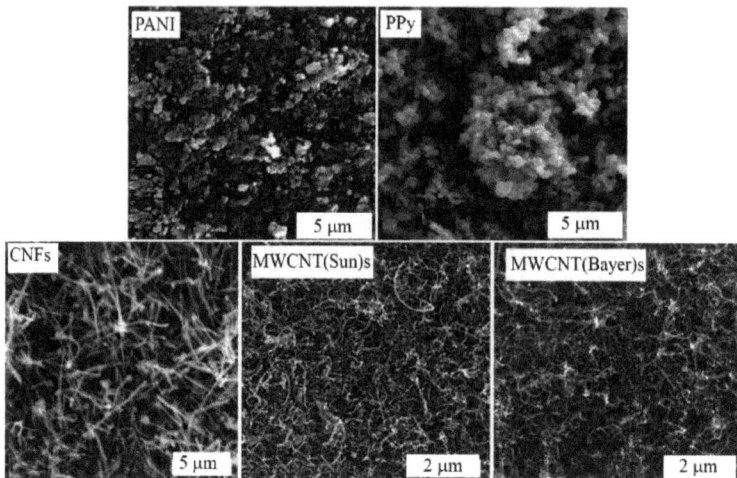

Figure 3. SEM micrographs of a surface of layers made of organic and inorganic fillers. The respective fillers are denoted in the images.

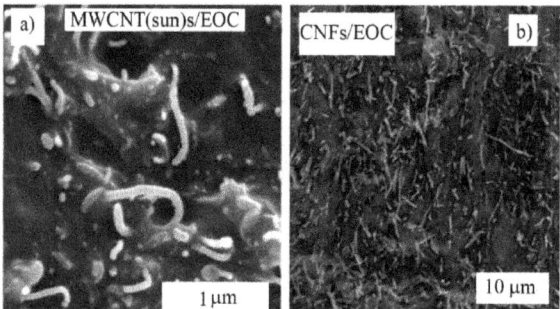

Figure 4. SEM images of cross-sections of carbon allotrope/ethylene-octene-copolymer matrices (EOC) composites. The respective composites are denoted in the images.

3.2. XPS Data

The main binding energy peak (284.5 eV) in the XPS spectra of pristine MWCNT(Sun)s, Figure 5, was assigned to the C1s-sp2, while the other ones were assigned to the C–O (286.2 eV), C=O (287.1 eV), O–C=O (288.6–289 eV) and C1s-π-π* (291.1–291.5 eV). After the oxidation treatment, the intensities of the peaks corresponding to the oxidized carbon bonds increased as seen in Figure 5. FTIR data of the MWCNTs also confirmed the presence of the C–O, C=O and O–C=O functional groups on their surfaces. It is also stated in other studies that MWCNTs treated with $(NH_4)_2S_2O_8$, H_2O_2, or O_3 have higher concentrations of carbonyl and hydroxyl functional groups, while more aggressive oxidants (HNO_3, $KMnO_4$) form higher fractional concentrations of carboxyl groups [25]. Acidic potassium permanganate ($KMnO_4$) is a strong oxidizing agent and produces more surface acidic groups than nitric acid [26]. The increase number of oxygenated functional groups attached on MWCNT(Sun)$KMnO_4$ surfaces significantly increases the contact resistance in the MWCNT junctions of the network structure [27].

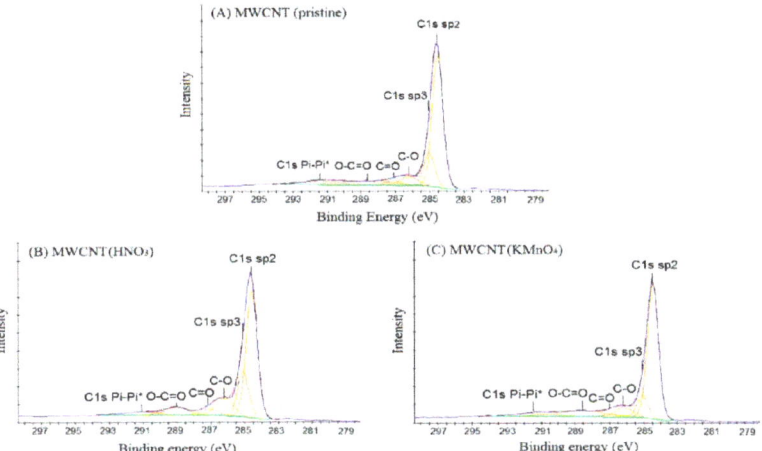

Figure 5. XPS C1s spectra for MWCNT(pristine) (**A**), MWCNT(HNO$_3$) (**B**) and MWCNT (KMnO$_4$) (**C**) samples.

3.3. FTIR Measurements

Table 1 and Figure 6 present the frequencies of some of the functional groups in the FTIR spectra of MWCNT networks. FTIR spectra in Figure 6 from MWCNTs showed a broad peak about 3430 cm^{-1} which was characteristic of the O–H stretch of hydroxyl group. C–H stretching of the pristine MWCNT(Sun) sample was shifted to lower wavelengths for all oxidation treatments. A weak C=O peak at 1705 cm^{-1}, Figure 6, was observed in the pristine MWCNT(Sun) sample, which showed that there was a carbonyl or carboxylic group on its surface. The reason why the pristine MWCNT(Sun) sample had carbonyl and OH groups could be a partial oxidation of the surfaces of MWCNTs during the purification by the manufacturer [28]. A higher shift in the carbonyl stretching mode was seen for MWCNT(HNO$_3$) than for MWCNT(KMnO$_4$). The reason could be a C=O group or other groups that interacted with the C=O group. FTIR spectra in Figure 6 also indicated that there were probably no anhydride/lactone groups on the surfaces of MWCNTs since these groups are usually observed at around 1750 cm^{-1} or higher wavenumber [26,29,30].

Table 1. Summary of FTIR measurements for the pristine and oxidized MWCNT networks.

Possible Assignments	Wavenumber (cm^{-1})		
	MWCNT	MWCNT (HNO$_3$)	MWCNT (KMnO$_4$)
OH stretch	3435	3428	3427
C–H stretch (CH$_2$, CH$_3$)	2908,2840	2980,2880	2978,2890
C=O stretch (carboxyl or ketone)	1705	1726	1710
Intermediate oxidized products—quinone groups	1652	1661,1635	1641
C=C stretch	1559	1580	1569
CH$_2$/CH$_3$ bending	1460	1437	1440
Skeletal C-C tangential motions +C–O stretch	1222	1184	1190
C–O stretch	1082	1084,1049	1087,1046

The peak assigned to the quinone group at 1652 cm^{-1} in MWCNT(pristine) sample was usually shifted to a higher wavelength in oxidized MWCNTs. Coupling effects (i.e., both of inter-molecular and intra-molecular hydrogen bonding with hydroxyl groups) also might be responsible for the downshift in the C=O stretching mode, besides the production of surface-bound quinone groups with extended conjugation [31].

The up-shift in the C=C stretching mode of MWCNTs was observed for both oxidized carbon nanotubes. The highest shift was observed for MWCNT(HNO_3) compared to MWCNT($KMnO_4$). This treatment suggested a change in the structure of the MWCNTs [32].

The C–H (CH_2/CH_3) bending at 1460 cm^{-1} and the peak at 1222 cm^{-1} for the MWCNT(pristine) sample were shifted to a higher wavelength for all oxidized MWCNTs. There was also observed a new band around 1050 cm^{-1} in the FTIR spectra of oxidized MWCNTs, which could be assigned to the alcoholic C–O stretching vibration [33]. Overall, the observed changes in the FTIR spectra of the oxidized MWCNTs confirmed the efficiency of the oxidizing process and the formation of the new oxygen-containing functional groups on the surfaces of the carbon nanotubes.

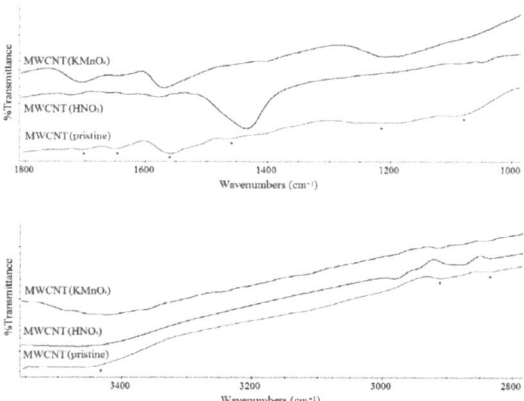

Figure 6. FTIR spectra of the MWCNT samples in the range of 1000–1800 cm^{-1} and 2800–3500 cm^{-1}.

FTIR analysis was also carried out to identify whether the prepared powder was indeed PANI emeraldine or PPy (Figure 7). For the FTIR absorption spectroscopy of the prepared PANI emeraldine, the vibration at 1578 cm^{-1} was attributed to the quinoid ring, while the vibration at 1489 cm^{-1} depicted the presence of a benzoid ring unit [34]. The peak at 1306 cm^{-1} is assigned to the C–N stretching of a secondary aromatic amine. Furthermore, the peak characteristic for the electrically conductive form of PANI emeraldine was observed at 1245 cm^{-1}. This peak was attributed to the stretching of the C–$N^{+\bullet}$ polaron structure [35]. The peak at 1160 cm^{-1} corresponding to the vibrations associated with the C–H of N=Q=N (Q = quinoid ring) also appeared in the spectrum of PANI emeraldine. The band between 913–680 cm^{-1} with a maximum at 823 cm^{-1} was characteristic for an aromatic ring deformation and C–H bond vibrations out of the plane of the ring [36]. These results suggested that the synthesized particles were PANI in the emeraldine state. For the FTIR spectroscopy of the prepared PPy, all the characteristic peaks were in good agreement with the earlier investigations of the same product [37,38]. The vibration peak at 1704 cm^{-1} was assigned to the C=N bond. Peaks at 1550 and 1477 cm^{-1} attributed to the in-plane vibrations of the PPy ring, and the peaks at 1178, 1037 and 783 cm^{-1} attributed to the in-plane bending of the PPy ring were also observed. Finally, the characteristic peak at 900 cm^{-1} was assigned to the out-of-plane vibration of the C_β–H group.

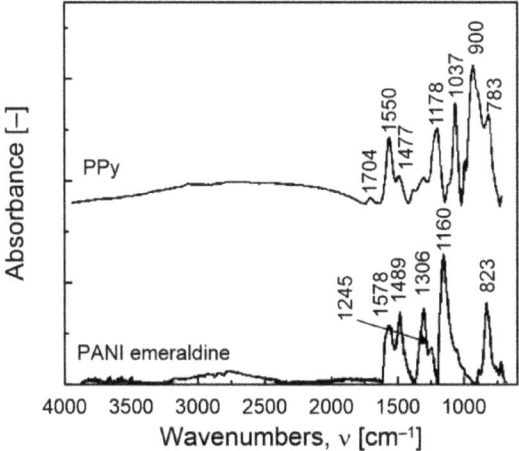

Figure 7. FTIR spectra of the PANI emeraldine and PPy particles.

The experimental test of the electric transport properties of the MWCNT(Sun)pristine/EOC composite at room temperature specified the following values: the sample resistance 0.87 Ω, the resistivity 0.23 Ωcm, the conductivity 4.3 Ω^{-1}cm^{-1} and the magneto-resistance 0.249 mΩ. Further determined properties were the Hall-mobility 7.2 cm^2/Vs, the charge carrier concentration (bulk) 3.7×10^{18} cm^{-3} and the average Hall coefficient +1.67 cm^3/C. The positive sign of the Hall coefficient identified the nature of the composite, i.e., the p-type of drift current when the holes are the dominant current carriers.

3.4. Thermoelectric Power Measurement

Values of induced voltage in response to a temperature difference across the measured samples of the investigated thermoelectric composites are presented in Figure 8. The slope of the linear dependence of the resulting voltage V_{TEV} on the temperature difference ΔT defined the thermoelectric power (Seebeck coefficient):

$$S = (V_{TEV}/\Delta T)_{\text{open circuit}}$$

which evaluated a potential thermoelectric performance. The corresponding thermoelectric power values are summarized in Figure 9. The thermoelectric power of MWCNT/EOC composites was substantially enhanced by the nanotube oxygenation. When compared to the composite with pristine nanotubes (MWCNT(Bayer)pristine/EOC), the respective composite with oxygenated nanotubes achieved a 150% increase of thermoelectric power.

The X-ray photoelectron spectroscopy (XPS) was performed on MWCNTs to ascertain the functional groups attached onto the nanotube surfaces. The oxygen contents (%) of the pristine and oxidized MWCNT samples, as calculated from the XPS spectra, are shown in Figure 9. The comparison of results indicated that all MWCNTs have C–O, C=O and O–C=O groups on their surfaces and that MWCNT(HNO$_3$)s have the maximum percentage of all the oxygen-containing groups. Moreover, the more oxygenated functional groups at the surface of the embedded MWCNTs, the higher the generated electric voltage and the thermoelectric power of the EOC composite as follows from the results in Figure 9. According to the published results on carbon nanotube/thermoplastic polymer composites, positive thermoelectric powers have always been determined [39].

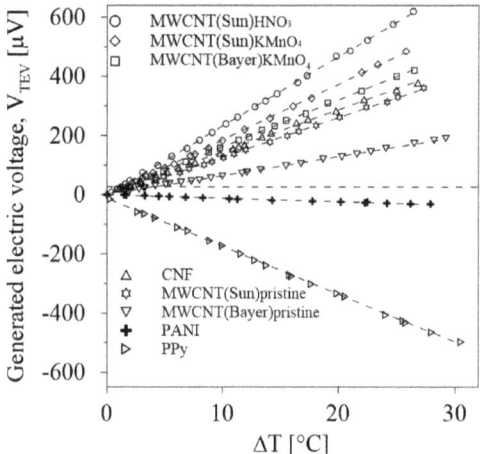

Figure 8. The relations between generated voltage and temperature difference for all investigated EOC composites.

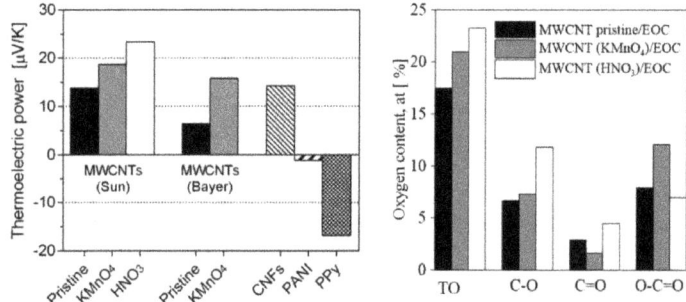

Figure 9. Thermoelectric power for tested EOC composites with indicated embedded fillers (**left panel**). Oxygen content (%) on surfaces of pristine and differently oxygenated MWCNTs (**right panel**). TO denotes the total oxygen amount and the C–O, C=O, and O–C=O the amount of the particular functional groups on the surface of the respective MWCNTs.

3.5. Self-Powered Signaling Sensor of Temperature Change

To demonstrate a possible use of the conductive MWCNT/EOC composites as thermoelectrical materials, a self-powered signaling sensor of temperature change was assembled. The sensor consisted of three conductive strips (thermoelements): one MWCNT(Sun)pristine/EOC and two MWCNT(Sun)KMnO$_4$/EOC composites, Figure 10. The sticky strips were stuck on a PET foil so that the ends overlapped at points A and B. The temperature signal was induced by a finger touch at point A or B, which in turn heated the connection of the strips. The generated electricity was monitored by the Multiplex datalogger 34980A. Even a temperature gradient induced by a mere short touch of a finger sufficed to elicit a detectable signal. The illustrative record of repeated finger touches is presented in Figure 11. Heat transfer from the finger to the hot junction of the sensor modeled a technological situation when such a sensor could monitor for example changes in the generation of waste heat, changes of technological temperature, warming of packaged grocery products, etc.

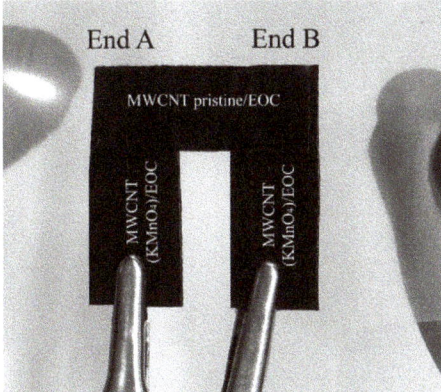

Figure 10. The self-powered signaling sensor of temperature change assembled from MWCNT(Sun) pristine/EOC and MWCNT(Sn)KMnO4/EOC composites.

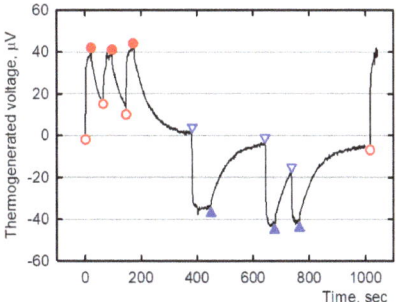

Figure 11. The illustrative time-dependent record of induced voltage in the self-powered signaling sensor by short finger touching at the point A (the finger touch is denoted by open red circles and the finger lift by filled red circles) or at the point B (the finger touch is denoted by open blue triangles and the finger lift by filled blue triangles) of the sensor.

3.6. Self-Powered Vapor Sensor

The induced voltage in the self-powered sensor (Figure 10) was changed not only by heating, but also by ambient organic vapors, Figure 12. As in our paper [14], the self-powered sensor was exposed to saturated organic solvent vapors. In particular, the sensor was placed in a glass bell, which enclosed a layer of liquid organic solvent and its respective saturated vapor. After a chosen time, the sensor was removed from the bell and the effect of vapor desorption on induced voltage was assessed. Such measurements were conducted in saturated vapors at atmospheric pressure, temperature 22 °C, and relative humidity 60%. The temperature gradient in the sensor was induced by means of the resistive heating of the MWCNT network/epoxy composite unit placed under the thermoelement junction. The heating DC voltage was 13 V and the current 0.09 A. The temperature of the hot end was 40.6 °C and of the cold end ambient room temperature.

The variations of the voltage of the self-powered sensor were caused by the changes of the electric resistance of the forming EOC composites through the chemical signals during the respective adsorption/desorption cycles as is illustrated in Figure 9. When the sensor was subjected to heptane, a larger increase of relative resistance resulted in a larger decrease of induced voltage and vice versa. Thus, the sensor output in terms of variation of voltage indicated a good response to the vapor occurrence. Moreover, considering that the sensor did not require a power supply, but it self-produced thermoelectricity from a heat source, which could be waste heat of industrial processes, solar energy, or body heat, etc., then the self-powered vapor sensor appears to be an advantageous alternative to powered vapor sensors. The illustrative thermoelement resistance changes plotted in Figure 10 were quantified by the relative resistance change,

$$S = \frac{R_g - R_a}{R_a} = \frac{\Delta R}{R_a}$$

where R_a represented the initial thermoelement resistance in the air, R_g the resistance when the thermoelement was exposed to a vapor, and ΔR denoted the measured resistance change.

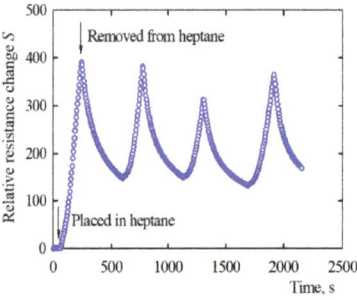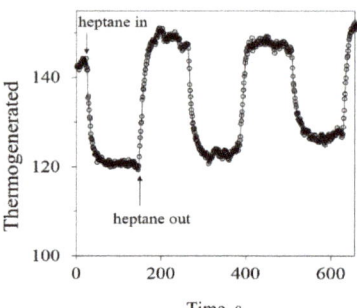

Figure 12. The time-dependent relative resistance change of the MWCNT(Sun)pristine/EOC composite (in %) during adsorption/desorption cycles of exposition to saturated vapors of heptane at 22 °C, the cycle length 600 s. The changes of the thermogenerated voltage induced by the self-powered vapor sensor in the course of three heptane adsorption and desorption cycles at 22 °C.

4. Discussion and Conclusions

This study presents hybrid thermoelectric composites consisting of organic EOC matrices and embedded conductive inorganic pristine and functionalized multiwalled carbon nanotubes, carbon nanofibers or conductive organic PANI and PPy particles. All these composites induced voltage in response to a temperature difference across the respective samples. This response was affected in the case of the composites with inorganic MWCNT nanostructures by the amount of oxygen containing functional groups attached to their surfaces. The more oxygenated were the functional groups at the surfaces of the embedded MWCNTs, the higher was the thermoelectric power of that EOC composite.

According to the published results about carbon nanotube/thermoplastic polymer composites, positive thermoelectric powers have always been determined indicating p-type composites, in which holes are dominant current carriers. The composites with embedded organic PANI a PPy particles induce compensating negative voltage. According to [38], the composites exhibit n-type conductivity specified by dominating electrons as charge carriers.

The chosen combination of the two p-type EOC composites in the experimental thermoelectric microgenerator, which generated sufficient electric current to make the sensors self-powered, was one of many other possible combinations. Other combinations such as p-type together with n-type thermoelements or two n-type EOC composites could have been chosen as well. An illustration of the practical use of the EOC composites as a thermoelectric generator was provided by the self-powered signaling sensor of temperature change and the self-powered vapor sensor. The voltage induced in the sensors, which was in the range of microvolts, did not approach by far the magnitude of induced voltage in the classical metal thermocouples, yet it was sufficient to power the novel sensor device, which converted chemical signals to electric ones. In the self-powered vapor sensor, the resistance varied according to the ambient chemical signals and changed the voltage induced by a source of (waste) heat. The EOC thermoelectric composites thus offer a novel and unique set of properties, which are not readily available in any other material.

Author Contributions: Data curation, R.O.; Investigation, P.S. and M.S.; Writing—original draft, P.R. All authors have read and agreed to the published version of the manuscript.

Funding: This research received no external funding.

Acknowledgments: This work was supported by the Operational Program Research and Development for Innovations co-funded by the European Regional Development Fund (ERDF), the Operational Program Education for Competitiveness co-funded by the European Social Fund (ESF), the National Budget of the Czech Republic within the framework of the Centre of Polymer Systems project (reg. number: CZ.1.05/2.1.00/03.0111), the project Advanced Theoretical and Experimental Studies of Polymer Systems (reg. number: CZ.1.07/2.3.00/20.0104) and by the Fund of Institute of Hydrodynamics AV0Z20600510.

Conflicts of Interest: The authors declare no conflict of interest.

References

1. Petsagkourakis, I.; Tybrant, K.; Crispin, X.; Ohkubo, I.; Satoh, N.; Mori, T. Thermoelectric materials and applications for energy harvesting power generation. *Sci. Technol. Adv. Mater.* **2018**, *19*, 836–862. [CrossRef] [PubMed]
2. MacDiarmid, A.G.; Chiang, J.C.; Richter, A.F.; Epstein, A.J. Polyaniline—A new concept in conducting polymers. *Synth. Met.* **1987**, *18*, 285–290. [CrossRef]
3. Hao, B.; Li, L.C.; Wang, Y.P.; Qian, H.S.; Tong, G.X.; Chen, H.F.; Chen, K.Y. Electrical and microwave absorbing properties of polypyrrole synthesized by optimum strategy. *J. Appl. Polym. Sci.* **2013**, *127*, 4273–4279. [CrossRef]
4. Trivedi, D. Polyanilines. In *Handbook of Organic Conductive Molecules and Polymers*; Nalwa, H.S., Ed.; Wiley: Chichester, UK, 1997; Volume 2, pp. 505–572.
5. Wang, W.J.; Sun, S.P.; Gu, S.J.; Shen, H.W.; Zhang, Q.H.; Zhu, J.J.; Wang, L.J.; Jiang, W. One-pot fabrication and thermoelectric properties of Ag nanoparticles-polyaniline hybrid nanocomposites. *RSC Adv.* **2014**, *51*, 26810–26816. [CrossRef]
6. Mitra, M.; Kulsi, C.; Kargupta, K.; Ganguly, S.; Banerjee, D. Composite of polyaniline-bismuth selenide with enhanced thermoelectric performance. *J. Appl. Polym. Sci.* **2018**, *135*, 46887. [CrossRef]
7. Ube, T.; Koyanagi, J.; Kosaki, T.; Fujimoto, K.; Yokozeki, T.; Ishiguro, T.; Nishio, K. Fabrication of well-isolated graphene and evaluation of thermoelectric performance of polyaniline-graphene composite film. *J. Mat. Sci.* **2019**, *54*, 3904–3913. [CrossRef]
8. Liang, L.R.; Chen, G.M.; Guo, C.Y. Polypyrrole nanostructures and their thermoelectric performance. *Mat. Chem. Front.* **2017**, *1*, 380–386. [CrossRef]
9. Misra, S.; Bharti, M.; Singh, A.; Debnath, A.K.; Aswal, D.K.; Hayakawa, Y. Nanostructured polypyrrole: Enhancement in thermoelectric figure of merit through suppression of thermal conductivity. *Mat. Res. Exp.* **2017**, *4*, 085007. [CrossRef]
10. Du, Y.; Niu, H.; Li, J.; Dou, Y.C.; Shen, S.Z.; Jia, R.P.; Xu, J.Y. Morphologies Tuning of Polypyrrole and Thermoelectric Properties of Polypyrrole Nanowire/Graphene Composites. *Polymers* **2018**, *10*, 1143. [CrossRef]
11. Aghelinejad, M.; Zhang, Y.C.; Leung, S.N. Processing parameters to enhance the electrical conductivity and thermoelectric power factor of polypyrrole/multi-walled carbon nanotubes nanocomposites. *Synt. Met.* **2019**, *247*, 59–66. [CrossRef]
12. Bharti, M.; Singh, A.; Samanta, S.; Debnath, A.K.; Aswal, D.K.; Muthe, K.P.; Gadkari, S.C. Flexo-green Polypyrrole–Silver nanocomposite films for thermoelectric power generation. *Energ. Convers. Manage.* **2017**, *144*, 143–152. [CrossRef]
13. Wang, Y.H.; Yang, J.; Wang, L.Y.; Du, K.; Yin, Q.; Yin, Q.J. Polypyrrole/Graphene/Polyaniline Ternary Nanocomposite with High Thermoelectric Power Factor. *ACS Appl. Mat. Interf.* **2017**, *9*, 20124–20131. [CrossRef] [PubMed]
14. Slobodian, P.; Riha, P.; Olejnik, R.; Benlikaya, R. Analysis of sensing properties of thermoelectric vapor sensor made of carbon nanotubes/ethylene-octene copolymer composites. *Carbon* **2016**, *110*, 257–266. [CrossRef]
15. Nonoguchi, Y.; Ohashi, K.; Kanazawa, R.; Ashiba, K.; Hata, K.; Nakagawa, T.; Adachi, C.; Tanase, T.; Kawai, T. Systematic conversion of single walled carbon nanotubes into n-type thermoelectric materials by molecular dopants. *Sci. Rep.* **2013**, *3*, 3344. [CrossRef]
16. Du, Y.; Cai, K.F.; Chen, S.; Wang, H.; Shen, S.Z.; Donelson, R.; Lin, T. Thermoelectric fabrics: Toward power generating clothing. *Sci. Rep.* **2015**, *5*, 6144. [CrossRef]
17. Yang, Y.; Lin, Z.H.; Hou, T.; Zhang, F.; Wang, Z.L. Nanowure-composite based flexible thermoelectrics nanogenerqators and self-powered temperature sensors. *Nano Res.* **2012**. [CrossRef]

18. Du, Y.; Xu, J.; Paul, B.; Eklund, P. Review: Flexible thermoelectric materials and devices. *Appl. Mater. Today* **2018**, *18*, 366–388. [CrossRef]
19. Slobodian, P.; Riha, P.; Lengalova, A.; Svoboda, P.; Saha, P. Multi-wall carbon nanotube networks as potential resistive gas sensors for organic vapor detection. *Carbon* **2011**, *49*, 2499–2507. [CrossRef]
20. Stejskal, J.; Gilbert, R.G. Polyaniline. Preparation of a conducting polymer (IUPAC technical report). *Pure Appl. Chem.* **2002**, *74*, 857–867. [CrossRef]
21. Benlikaya, R.; Slobodian, P.; Riha, P. Enhanced strain-dependent electrical resistance of polyurethane composites with embedded oxidized multiwalled carbon nanotube networks. *J. Nanomat.* **2013**, *2013*, 327597. [CrossRef]
22. Jansson, P.A. (Ed.) *Deconvolution of Spectra and Images*; Academic Press: San Diego, CA, USA, 1997; pp. 119–134.
23. Hernadi, K.; Siska, A.; Thien-Nga, L.; Forro, L.; Kiricsi, I. Reactivity of different kinds of carbon during oxidative purification of catalytically prepared carbon nanotubes. *Solid State Ionics* **2001**, *141*, 203–209. [CrossRef]
24. Rasheed, A.; Howe, J.Y.; Dadmun, M.D.; Britt, P.F. The efficiency of the oxidation of carbon nanofibers with various oxidizing agents. *Carbon* **2007**, *45*, 1072–1080. [CrossRef]
25. Wepasnick, K.A.; Smith, B.A.; Schrote, K.E.; Wilson, H.K.; Diegelmann, S.R.; Fairbrothe, D.H. Surface and structural characterization of multi-walled carbon nanotubes following different oxidative treatments. *Carbon* **2011**, *49*, 24–36. [CrossRef]
26. Ros, T.G.; Van Dillen, A.J.; Geus, J.W.; Koningsberger, D.C. Surface oxidation of carbon nanofibres. *Chem. Eur. J.* **2002**, *8*, 1151–1162. [CrossRef]
27. Chen, J.; Chen, Q.; Ma, Q. Influence of surface functionalization via chemical oxidation on the properties of carbon nanotubes. *J. Coll. Int. Sci.* **2012**, *370*, 32–38. [CrossRef] [PubMed]
28. Abuilaiwi, F.A.; Laoui, T.; Al-Harthi, M.; Atieh, M.A. Modification and functionalization of multiwalled carbon nanotubes (MWCNT) via Fischer esterification. *Arab. J. Sci. Eng.* **2010**, *35*, 37–48.
29. Fanning, P.E.; Vannice, M.A. A DRIFTS study of the formation of surface groups on carbon by oxidation. *Carbon* **1993**, *31*, 721–730. [CrossRef]
30. Moreno-Castilla, C.; Lopez-Ramon, M.V.; Carrasco-Marın, F. Changes in surface chemistry of activated carbons by wet oxidation. *Carbon* **2000**, *38*, 1995–2001. [CrossRef]
31. Kim, U.J.; Furtado, C.A.; Liu, X.; Chen, G.; Eklund, P.C. Raman and IR spectroscopy of chemically processed single-walled carbon nanotubes. *J. Am. Chem. Soc.* **2005**, *127*, 15437–15445. [CrossRef]
32. Zhang, J.; Zou, H.; Qing, Q.; Yang, Y.; Li, Q.; Liu, Z.; Guo, X.; Du, Z. Effect of chemical oxidation on the structure of sigle-walled nanotubes. *J. Phys. Chem. B* **2003**, *107*, 3712–3718. [CrossRef]
33. Liang, Y.; Zhang, H.; Yi, B.; Zhang, Z.; Tan, Z. Preparation and characterization of multi-walled carbon nanotubes supported Pt Ru catalysts for proton exchange membrane fuel cells. *Carbon* **2005**, *43*, 3144–3152. [CrossRef]
34. Tang, J.S.; Jing, X.B.; Wang, B.C.; Wang, F. Infrared-spectra of soluble polyaniline. *Synth. Met.* **1988**, *24*, 231–238. [CrossRef]
35. MacDiarmid, A.G.; Chiang, J.C.; Huang, W.S.; Humphery, B.D.; Somasiri, N.L.D. Polyaniline: Protonic acid doping to the metallic regime. *Mol. Cryst. Liquid Cryst.* **1985**, *25*, 309–318. [CrossRef]
36. Brožová, L.; Holler, P.; Kovářová, J.; Stejskal, J.; Trchová, M. The stability of polyaniline in strongly alkaline or acidic aqueous media. *Polym. Degrad. Stabil.* **2008**, *93*, 592–600. [CrossRef]
37. Wang, J.G.; Neoh, K.G.; Kang, E.T. Comparative study of chemically synthesized and plasma polymerized pyrrole and thiophene thin films. *Thin Solid Films* **2004**, *446*, 205–217. [CrossRef]
38. Cheah, K.; Forsyth, M.; Truong, V.T. Ordering and stability in conducting polypyrrole. *Synth. Met.* **1998**, *94*, 215–219. [CrossRef]
39. Krause, B.; Barbier, C.; Levente, J.; Klaus, M.; Pötschke, P. Screening of different carbon nanotubes in melt-mixed polymer composites with different polymer matrices for their thermoelectrical properties. *J. Composit. Sci.* **2019**, *3*, 106. [CrossRef]

 © 2020 by the authors. Licensee MDPI, Basel, Switzerland. This article is an open access article distributed under the terms and conditions of the Creative Commons Attribution (CC BY) license (http://creativecommons.org/licenses/by/4.0/).

MDPI
St. Alban-Anlage 66
4052 Basel
Switzerland
Tel. +41 61 683 77 34
Fax +41 61 302 89 18
www.mdpi.com

Polymers Editorial Office
E-mail: polymers@mdpi.com
www.mdpi.com/journal/polymers